GENETIC ANALYSIS OF ANIMAL DEVELOPMENT

GENETIC ANALYSIS OF ANIMAL DEVELOPMENT

ADAM S. WILKINS

Illustrated by Belinda Durrant

A Wiley-Interscience Publication

JOHN WILEY & SONS

New York • Chichester • Brisbane • Toronto • Singapore

Library of Congress Cataloging in Publication Data:

Wilkins, A. S. (Adam S.), 1945–
Genetic analysis of animal development.

"A Wiley-Interscience publication."
Bibliography: p.
Includes index.
1. Developmental genetics. 2. Developmental biology.
3. Embryology. I. Title.

QH453.W55 1985 591.3 85-6460
ISBN 0-471-87662-3
ISBN 0-471-62544-2 (pbk)

10 9 8 7 6 5 4

For Isaac

PREFACE

Understanding the nature of organismal development remains one of the fundamental challenges in the biological sciences. Although numerous questions in other fields of biology, such as neurobiology, are as interesting as those in development, solutions to many of these other problems will first require advances in understanding the mechanisms of biological development.

This book explores the ways one group of investigators, geneticists, approaches the problems of development and, in particular, the special complexities of animal development. Virtually all fields of animal biology are concerned with questions of development in some manner and contribute strategies for investigating it, but genetics occupies a special position. In the first place, it supplies a conceptual framework that is the closest approach we have to a theory of development: the idea that the genes ultimately determine the developmental sequence and that the "logic" of this sequence is to be sought in the genetic "program."

In the second place, the branch of genetics devoted to development provides a uniquely informative set of strategies for investigating the roles of particular macromolecules and the behavior of cells in development. The precision of these strategies in developmental genetics for dissecting and laying bare the components of the developmental process is very great indeed. In fact, recent years have witnessed an accelerating fusion of these genetic approaches with those of molecular biology, immunobiology, and biochemistry. As recently as 10 years ago, developmental genetics could be defined as the study of the effects of genetic changes—mutations—on developmental processes; the methodologies of contemporary developmental genetics have a much broader sweep.

This book evolved from a course I taught in developmental genetics while at Massey University in New Zealand. During this period, I was increasingly struck by the disparity in treatments that eukaryotic molecular biology and developmental genetics have received despite their related subject matter. While there have been several excellent general accounts of developmental problems from the molecular biologists' viewpoint, there have been no comparable discussions for nonspecialists of the approaches that geneticists take to answer these questions. The original idea, in consequence, was to provide a companion book for the molecular treatments, one that would explain the specific methodologies and viewpoints of developmental geneticists.

However, a book dealing with techniques while neglecting the biological content would be very dry. It seemed appropriate, therefore, to introduce both the methods and the concepts of the field through a discussion of the developmental biology of a small number of animal systems, introducing the genetic techniques and ideas as they arose in the biological context.

Specifically, I have chosen to concentrate on the three animal systems that have received the most attention from geneticists: two invertebrates and one vertebrate. They are the nematode *Caenorhabditis elegans,* the fruit fly *Drosophila melanogaster,* and the mouse *Mus musculus*. Although this approach risks loss of generality, these three animal systems span a great range of complexity and phylogenetic distance. To broaden the scope, however, I have included references to comparable phenomena and data from other organisms, where appropriate. In concentrating on a small number of animal systems, there is also the risk that one will draw spurious generalizations. I have tried to avoid this problem, while noting the general conclusions that do seem justified. It goes without saying that even within the self-prescribed limitations, the approach taken is still selective; an encyclopedic account of the developmental biology of the three animals would be much longer than this book.

One further remark on the content is required. As the writing progressed, more and more of the relevant molecular biology had to be brought into the discussion. In some places, the findings complement and buttress the genetics; in others, the two kinds of findings seem to point in opposite directions. Where appropriate, I have ventured my own interpretations; in no case are the latter offered as definitive solutions. They are provided solely as a starting point for thinking further about the issues at hand. In the final chapter, I have tried to draw some of the common themes together and to delineate the conceptual problems posed by some of the findings.

This book attempts to achieve three goals: (1) to serve as an introduction to the techniques used by geneticists to probe animal development; (2) to function as an introduction to the specific biology of and literature pertaining to the three animals currently under most intensive investigation; and (3) to describe the ideas and conundrums of contemporary developmental genetics. I have assumed on the part of the reader only a familiarity with the basic concepts of genetics, of the kind that can be acquired in a first-year university cell biology or genetics course. My hope is that the book will prove helpful to all those seeking either an introduction to or an overview of animal developmental genetics today.

ADAM S. WILKINS

Cambridge, England
September 1985

ACKNOWLEDGMENTS

Many individuals have contributed to the creation of this book, and it is a pleasure to acknowledge their help. The interest and encouragement of Dr. Jonathan Gallant at the beginning of this project was crucial; without his enthusiasm and support, I am not sure that I would ever have made a serious start. There are three other colleagues to whom I owe special thanks: Dr. David Gubb, who was endlessly patient in straightening out my sometimes tortuous prose in the next-to-last draft and with whom I had many informative and lively discussions about *Drosophila* development; Dr. John Sulston, whose help in understanding the intricacies of the biology of *Caenorhabditis* and in providing much-needed figures was tremendously valuable; and Dr. Robin Denell for his comments and suggestions on the completed manuscript. Furthermore, I am very grateful to the following individuals for their careful criticisms of various sections and chapters: Drs. Michael Akam, Michael Ashburner, Sydney Brenner, Martin Chalfie, Jonathan Hodgkin, Martin Johnson, Robert King, Peter Lawrence, John Lucchesi, Anne McLaren, Elizabeth Robertson, Iris Sandler, John White, and Deborah Wilde. Their suggestions were not always followed but were always appreciated; it goes without saying that any faults and errors that remain are entirely my own. In addition, there were numerous other friends and colleagues who contributed information, reprints, or figures. A different kind of appreciation is owed to the Department of Genetics, University of Cambridge, which graciously provided a second home for me during much of the writing.

Finally, I would like to thank my wife Jeanie for her patience, moral support, and sense of humor during the three-year obsessional period in which this book itself developed.

A. S. W.

CONTENTS

PART ONE

PAST AND PRESENT

ONE

INTRODUCTION

. . . it follows that semen will be either blood or the analogous substance, or something formed out of these. . . . And this, too, is why we should expect children to resemble their parents; because there is a resemblance between that which is distributed to the various parts of the body and that which is left over [for reproduction]. Thus, the semen of the hand or of a whole animal *is* hand or face or a whole animal though in an undifferentiated way; in other words, what each of these is *in actuality,* such the semen is *potentially*. . . .

Aristotle, *Generation of Animals,* p. 91

Questions about embryonic development have been intertwined with questions about heredity since the first written speculations on inheritance. To ask why a child resembles its parents involves more than an inquiry about the passage of genes from one generation to the next. It is equally a set of implicit questions about the ways in which genes produce their effects in development. The essential connection between heredity and development was recognized in Aristotle's time; the nature of this connection remains to be fully explained in ours.

This book is about the ways geneticists investigate, and think about, the genetic basis of embryonic development. The major emphasis will be on the work of the last 20 years, but to place the present field of developmental genetics in context, it is necessary to begin with a look at the earlier work, from the beginning of genetic science in 1900 to the seminal discoveries of molecular biology between 1953 and 1961. In the first part of this chapter a brief history will be given of the ways in which geneticists approached—and avoided—the facts of development in the first 60 years of this century. The main currents in the field today and the structure of the book will then be described.

In contemporary terms, the fundamental connection between genes and development is that genes specify the proteins that determine cellular character; the progression of cell types and cell behaviors that comprise development reflect the changes in protein composition that follow from changing gene activities. Ultimately, the kind of organism that a particular fertilized egg gives rise to is determined by its genes, which will be those typical of a given species. The precise details are governed by the particular variants, or alleles, that comprise its individual genetic constitution. However, the relationships between the details of the observable form, or phenotype, of the individual and its underlying genetic constitution, or genotype, are only rarely simple and obvious.

As Mendel discovered, each gene is present in two copies in the mature organism, and certain alleles of each gene can be dominant to, or in other words mask, the expression of other alleles. In other cases, dominance may be partial or the combination of alleles can produce a new trait. In still more complicated genetic situations, the action of one gene can wholly or partially obscure the effect of a different gene, the phenomenon known as "epistasis." Furthermore, many interesting traits are determined not by individual alleles at one genetic locus, but rather by the cumulative or synergistic action of many genes. This is known as "polygenic" or "quantitative inheritance." Finally, the relationships between genes and development are complicated by environmental perturbations. Even very small alterations of environmental conditions can affect the degree of expression of a particular gene. Strong environmental shocks may produce great alterations in gene expression, affecting many gene activities and causing embryos to develop in characteristic but highly abnormal fashions. And beyond such deterministic effects, there is the phenomenon of "developmental noise": small random biochemical perturbations characteristic of any system.

It is this very complex set of interconnections between genes and ultimate developmental outcomes that so delayed the birth of genetics. Genetics is the science of deducing genotypes of parents from the phenotypes of their progeny, but the enormous number and complexity of the processes that intervene between fertilization and the ultimate expression of the genes that create the progeny phenotypes require the inferences to be necessarily indirect and dependent upon many contingent phenomena. It was no small measure of Mendel's genius that he sensed some of these complications in genetic analysis and chose to study hereditary factors whose developmental effects were as simple and direct as could be found. This awareness, in conjunction with his novel statistical approach to inheritance, enabled Mendel to deduce the essential rules of genetic factor transmission. His predecessors failed largely because they had become lost in the wilderness of developmental complexity.

It might have been anticipated that the rediscovery of Mendelian genetics in 1900 would clear the way for a direct use of genetics to probe the role of hereditary factors in biological development. Early twentieth century geneticists were fully aware of the importance of hereditary factors in biological development, and indeed, many of these men had a background in embryology. Foremost among them were William Bateson, the chief proponent of Mendelian genetics, and Thomas Hunt Morgan. It was Morgan who would, in the second decade of the century, synthesize Mendelian genetics and the chromosome theory of inheritance. Given the intense interest in and involvement of these men and their colleagues with questions of development, one might have thought that they would combine their twin interests, genetics and embryology, to explore the ways in which individual genes affect embryological development. Yet this synthesis of interests and approaches did not immediately materialize.

There were two reasons for this failure to address the problem of development. In the first place, the conceptual gap between genes and phenotypes was forbiddingly large. Although the connections between genes (which were not even named as such until 1909) and cellular metabolism were sensed by a few, including Bateson, and although exploration of these connections began long before there was a theoretical framework for understanding these relationships, the concepts were too primitive and the analytical methods were essentially inadequate for approaching the subject. Mendelian genetics was simply the wrong tool for elucidating the facts of embryonic development, and would have been so even if the conceptual framework had been stronger. The province of Mendelian genetics is solely that of the transmission of genes; today it exists outside of and is irrelevant to the ideas of biosynthesis that inform our views of developmental change (Hull, 1974).

The second reason for the neglect of development has to do with the interests and outlook of the early twentieth century geneticists. It is true that many of them had a background in embryology, but nearly all were interested in embryos primarily as mirrors of the evolutionary process, a major preoccupation of post-Darwinian biology. Morgan was an exception; he was greatly interested in the developing embryo as a subject in its own right, but it is evident that he thought that the analysis of how genes achieve their effects was impossible at the time. His response was to leave development for the much more tractable subject of gene inheritance, or transmission genetics. Although he never lost his interest in embryology, and even turned his attention to it again toward the end of his career, both he and most members of his school remained oddly remote from the question of how genes govern development.

Nevertheless, while the early Mendelians focused their attention on the problems of gene transmission and evolution, they were soon tripping over the facts of development. It could hardly have been otherwise, given the genetic foundation of development. The ever wider search for genetic factors that showed typical Mendelian inheritance patterns made it inevitable that developmental phenomena would soon be intruding into studies whose goal was the elucidation of gene inheritance.

One of the earliest findings of this kind involved a test for Mendelian inheritance in a mammal, that of the *yellow* gene in the laboratory mouse.

The initial studies were carried out by Lucien Cuénot, who discovered that the *yellow* allele (symbolized A^y) was dominant to the standard Agouti allele (A), which produces a grayish color. Cuénot observed, to his surprise, that the yellow trait never bred true: intercrossed yellow mice always produced some wild-type progeny. Furthermore, when any yellow mouse was mated to a wild-type, roughly equal numbers of yellow and wild-type progeny were obtained. This finding was explicable if the yellow mice were always heterozygous for the *yellow* and wild-type alleles. However, yellow × yellow crosses never gave the expected 3 : 1 Mendelian ratio for a monohybrid (heterozygote × heterozygote) cross. Cuénot consistently observed a deficit of yellow progeny relative to the expected 75% proportion and smaller than normal litter sizes (Cuénot, 1908). To explain his result, he proposed a developmental hypothesis: that *yellow*-allele carrying eggs could never be fertilized by *yellow*-bearing sperm but only by wild-type bearing sperm. This explanation accounted for the permanently heterozygous condition of *yellow* and the relative deficit of yellow progeny in the monohybrid cross, but not for the reduction in total progeny numbers since there should always be sufficient wild-type sperm to effect fertilization.

In a later and more extensive study, W. E. Castle and C. C. Little (1910) showed that the true ratio in the monohybrid cross was 2 yellow : 1 Agouti. They pointed out that this ratio was exactly the one expected if one quarter of the progeny were homozygous yellow (A^y) and if all of these were lost. Their hypothesis was that *yellow,* which acts as a simple visible dominant mutation when present in one dose ($A^y/+$), acts as a recessive lethal when homozygous (A^y/A^y). This hypothesis of recessive lethality of homozygotes was soon confirmed by others through direct observation of the embryos.

As the testing of factors for Mendelian inheritance patterns progressed, comparable cases in other organisms soon came to light in both plants and animals. In 1912, the Morgan *Drosophila* group reported the first recessive lethal in the fruit fly, this mutant being without a dominant effect when heterozygous. As these kinds of observations accumulated, the nature of the interest they evoked began to change. At first regarded as noteworthy because of the distortion of Mendelian inheritance ratios, the mutants soon came to be seen as interesting in their own right for the developmental effects they produced. When examined closely, these effects were found to be specific for the mutant genes under study, each lethal mutant producing its own characteristic pattern of death. As the specificity of individual lethal actions became ever more apparent, it appeared increasingly likely that each lethal was revealing the specific developmental defects caused by deleting specific gene activities.

It was through such studies of mutant effects that the field of developmental genetics came into being. The growth was slow and piecemeal, and proceeded without a single dramatic discovery or landmark publication of the kind that has so often signaled new departures in genetics. Despite these hazy and undramatic origins, developmental genetics had acquired a recognizable unity by the late 1930s. Most of the early contributions came from Germany, with its strong tradition of experimental embryology, but the landmark work of Sewall Wright in the United States on the basis of coat color in guinea pigs was also of major importance.

The goal of each of the early genetic studies was to reconstruct the developmental function of a particular wild-type gene from the pattern of tissue or organ changes occurring in the mutant. The assumption was that the entire body of work would illuminate the basic biological rules of genetic control of development. Since lethal mutants possess clear, readily observable effects, the primary emphasis in most of this work was on genetic changes that produced death. These included both single gene changes ("point mutants") and those chromosomal rearrangements, such as inversions and translocations, that were found to be associated with a lethal effect. The basic method of analysis consisted of three steps. The first step was the careful delineation of the time of death, the lethal phase, of the mutant under study. The second step was the cataloging of all of the terminal differences between the mutant and the wild-type at the time of death, as revealed by the study of the organs and tissues of the dead embryos. The third phase involved a careful tracing backward in development, using progressively earlier embryos, of these mutant characteristics or "phenes" to the earliest point at which abnormalities could be detected. This initial point of aberrancy was termed the "phenocritical phase." From the initial morphological aberrancy, one then attempted to reconstruct the function of the corresponding wild-type gene during the phenocritical phase.

This method of analysis is a retrospective one—working backward from the known terminal phenotype to the first effects of the mutant condition—and was the core technique of developmental genetics for more than 30 years, from the 1920s through the 1950s. Nevertheless, the method was not successful in illuminating the underlying mechanisms of development, and the period of classical developmental genetics has about it an air of disappointment. Instead of elucidating general principles, the field produced a great number of unrelated observations, each pertaining to the biology of a particular mutant and none definitively identifying the precise role of a wild-type gene in development. If experimental embryologists largely ignored (and even scorned) the contributions of geneticists to embryology during this period, the failure of developmental genetics to achieve its own programmatic goals must be held partly responsible. But in two respects, and belying the appearances, the gains of this period of genetic research were substantial.

The first of these advances consisted of a delineation of the basic difficulties inherent in the mode of mutant analysis. It might appear paradoxical, or even nonsensical, to regard this as progress, but the insights gained provided the foundation for all subsequent analysis. Although there are many potential snares in the interpretation of mutant effects in development, just two kinds will be mentioned here. The first of these is the phenomenon of pleiotropy. As observations accumulated, it became apparent that most mutants do not suffer single developmental changes but rather a whole range of abnormalities; "pleiotropy" means "many ways." In consequence, deducing the primary biochemical defect in development is often a highly complex matter.

Furthermore, there are two different forms of pleiotropy, and understanding any specific pleiotropic response requires that they be distinguished. The first category has been termed direct or "mosaic" pleiotropy. In this form, the defects arise in two or more sites because action of the wild-type gene is required for direct expression these tissues. An instance is provided by the *yellow* allele of the mouse, which in heterozygotes not only affects coat color but also fertility and causes a mild obesity. The second class of pleiotropic response is "relational," in which the mutant defect occurs initially in only one site but triggers a cascade of subsequent developmental defects. An example of relational pleiotropy is that caused by the human blood disease sickle cell anemia, in which the genetic defect is directly expressed only in the red blood cells. This defect reduces the oxygen-carrying capacity of the red blood cells, resulting in a complicated series of metabolic and organ dysfunctions.

A second class of complications in the interpretation of mutant effects can occur if the characteristics of the mutation itself are not properly categorized (Muller, 1932). In general, most deleterious gene mutations involve a deficiency of the relevant gene activity—a deficiency in the amount of product or of active product. Such defects are usually recessive because one dose of wild-type activity is often sufficient for normal development. However, the precise degree of deficiency can make a dramatic difference in the degree or even the kind of developmental syndrome observed. If no detectable wild-type gene activity remains, the mutant is termed an "amorph." In general, amorphs for a vital gene cause sharp, well-defined defects in development, and often an early death of the affected embryos. In contrast, mutant alleles that leave some residual wild-type gene activity, "hypomorphs," will not infrequently show a more normal course of development and may even allow a fraction of the homozygotes to survive.

As for dominant mutants, these too are not of a single type, but rather fall into different categories that must be distinguished. A mutant may be dominant to a wild-type simply because it reduces gene activity and because the single wild-type allele is insufficient to produce normal development; this form of dominance is said to reflect "haplo-insufficiency." More commonly, dominants produce a genuinely novel gene activity. If the activity involves a simple new quality, not present in the wild-type gene, it is termed a "neomorph"; if the activity antagonizes and competes with the wild-type gene activity, it is called an "antimorph." As with recessives, it is critically important to know the kind of dominant one is dealing with in order to interpret the mutant effect correctly. Herman Muller was the first to make these distinctions clearly and to devise the necessary genetic tests to distinguish between the different classes of mutant expression.

Despite the difficulties of interpretation inherent in the retrospective mode of mutant defect analysis, two general conclusions were reached by the late 1940s. These conclusions constituted the second set of advances. The first was that of *site specificity of action* of genes: each gene is expressed and required in one or a few tissues, but generally not in all of them. The second generalization was *phase specificity of gene action:* a particular gene is active only at particular times in development. One can condense these two conclusions into one: biological development is accompanied, and probably driven, by an *ordered sequence of gene expression*. The idea of spatially and temporally patterned sequences of gene expression in development had been sensed by a few in the 1930s but could be expressed with confidence only by the century's midpoint (Hadorn, 1948; Spiegelman, 1948).

Crucial as this idea of spatial and temporal sequences of gene expression is today, it failed to evoke much response by experimental embryologists. The fields of developmental genetics and embryology continued to go their separate ways. The reason is not hard to find. Developmental geneticists talked at length about genes and gene actions, yet they could not visualize or explain what the gene was or how it worked. There had been repeated speculations for decades that particular genes somehow determine particular enzymatic activities, a link that was definitively established by the work of Beadle and Tatum in 1941, but the exact nature of the gene–enzyme connection was obscure. The very idea of changing and changeable gene activities during development, when the gene itself was so remote, seemed more like a figure of speech than a statement of biochemical fact, perhaps even to those who invoked the concept.

The ideas and methods of developmental genetics at midcentury were summarized by Ernst Hadorn in his book *Developmental Genetics and Lethal Factors*. In the 1930s, Hadorn had left the experimentally tractable amphibian embryo for the much more difficult experimental terrain of *Drosophila,* with the explicit goal of applying genetic techniques to developmental questions. By the 1950s, he was one of the major figures in the field. His book was first written in German and published in 1955. Later it was translated, appearing in 1961 in its English language edition. By chance, the second edition appeared just as the whole perspective of the field described in the book was being superseded by a new outlook and a new set of approaches. The dates that span the two editions, 1955 and 1961, virtually bracket the period in which genetics experienced its most fundamental change since its birth as a science. The change, of course, was that brought about by the advent of molecular biology, initiated by the publication of the model of DNA structure by James Watson and Francis Crick in 1953.

The new discipline was concerned with two problems: What is the molecular nature of the gene, and how does the gene work? The Watson–Crick model provided an answer to the first: each gene is a stretch of chemical information, a sequence of nucleotides that encodes and specifies the unique amino acid sequence of a polypeptide. Until the early 1950s, genes had been thought of as entities performing direct activities, much like enzymes. The concept that a gene works solely by specifying the structure of a polypeptide, by encoding its sequence of amino acids, was a major departure.

For those who worked in molecular biology, the post-Watson–Crick era of the 1950s was devoted to figuring out the logic of the genetic code—the number and sequence of nucleotides that specify particular amino acids—and to unraveling the biochemical mechanism of protein synthesis from gene to finished polypeptide. The capstone of this edifice was the realization that there must be an intermediary molecule between the gene and the synthesized protein product. This intermediary carries the information in the gene to the cellular sites of protein synthesis in the cytoplasm. From various considerations, the intermediary molecule was deduced to be a species of RNA, a class of molecules broadly similar to DNA but single-stranded instead of double-stranded. Because of its information-conveying role, this class of RNA was designated "messenger RNA" or simply "mRNA."

How did the revolution in molecular biology affect developmental genetics? The first stages left the field untouched. In Hadorn's book, genes are discussed on every page but DNA is not mentioned once, although the identity of the genetic material with the molecular substance of DNA was well established. But the response to the final step in the problem of protein synthesis was electric. This suddenly heightened interest occurred because of a historical accident: the discovery of mRNA happened to have become bound up with the central problem of developmental genetics—the fact that gene activities are regulated.

Throughout the 1950s, when American and British molecular biologists were grappling with the general problem of protein synthesis, two French scientists at the Institut Pasteur, Jacques Monod and Francois Jacob, were wrestling with an equally mysterious but seemingly smaller problem. The question was this: how does the bacterium *Escherichia coli* turn on the synthesis of three lactose-metabolizing enzymes in response to the addition of lactose "inducer" to the medium? Monod and Jacob and their colleagues had found that the genes encoding these enzymes, which they termed "structural genes," were clustered together on the bacterial chromosome and that their synthesis was controlled coordinately in response to inducer molecules. In recognition of this coordinate control, the three genes were said to comprise an "operon." A second discovery was that mutations in a nearby gene permitted massive synthesis of the three *lac* operon enzymes, even in the absence of inducer. The sole function of this "regulator" gene was to control the switching on and off of the three structural genes. The regulator gene was dubbed the *i* gene, for the inducerless (i^--) phenotype of its mutations. Several observations indicated that this gene specified a "repressor,'' a molecule that physically shut off the expression of the three structural genes, and that the inducer molecules antagonized the function of the repressor, thereby eliciting the synthesis of the three proteins.

By 1960, this first explicit model of gene regulation had been formulated, but several puzzles remained. One was the discovery that enzyme synthesis was apparently regulated by control of the synthesis of an intermediary template molecular species. When inducer was removed, synthesis of this species was shut off immediately but enzyme synthesis continued on the existing template molecules, whose number then declined exponentially through a degradative process.

From the idea of an intermediary template molecule in lactose enzyme synthesis to the idea of a general intermediary molecule in gene-to-protein synthesis is not a large leap—at least in retrospect. The idea of mRNA was born in a conversation one afternoon in April 1960 between Jacob and Monod, on the one hand, and Francis Crick and Sydney Brenner, the chief theoreticians of protein synthesis, on the other (Judson, 1979).

The *lac* operon model (Jacob and Monod, 1961), completed by the idea of mRNA, provided biology with a conception of how genes may be selectively turned on and off. Generalized and stripped to its essentials, the idea is that for each gene or group of clustered genes, specifying enzymes of related function, there is a special regulatory appratus that mediates signals from the environment, in the form of small metabolite co-effectors, to control the synthesis of the proteins. In the specific instance of the *lac* operon, the regulatory apparatus consists of a regulator gene that specifies a repressor molecule, and a specific binding site, the ''operator,'' adjacent to the structural gene cluster, that binds the repressor. The repressor was presumed and later proved to exert its inhibitory action by preventing mRNA synthesis, the copying or ''transcription'' of the structural genes of the operon. Inducer acts to stimulate enzyme synthesis by binding to repressor, releasing the latter from the operator and thereby permitting transcription. The model is diagrammed in Figure 1.1*a* and typifies the process of ''negative control'' or regulation through selective repression and repression-release. A few years after the publication of the model, it was established that some bacterial operons are regulated by ''positive control,'' a system of selective direct *stimulation* of transcription by a bound regulator molecule (reviewed in Englesberg and Wilcox, 1974). A positive control system is illustrated schematically in Figure 1.1*b*.

The operon model was formulated to explain the regulation of genes in some of the simplest organisms, bacteria and their viruses. However, the implications for other organisms, and in particular for development, were apparent. In several papers, Jacob and Monod detailed the principle of regulatory circuitry as the basis of developmental change. The idea is that an initial regulatory step can initiate a second regulated change in gene expression for one or many genes, which can, in turn, produce a third set of changes, and so on, the net effect being a highly ordered sequence of gene

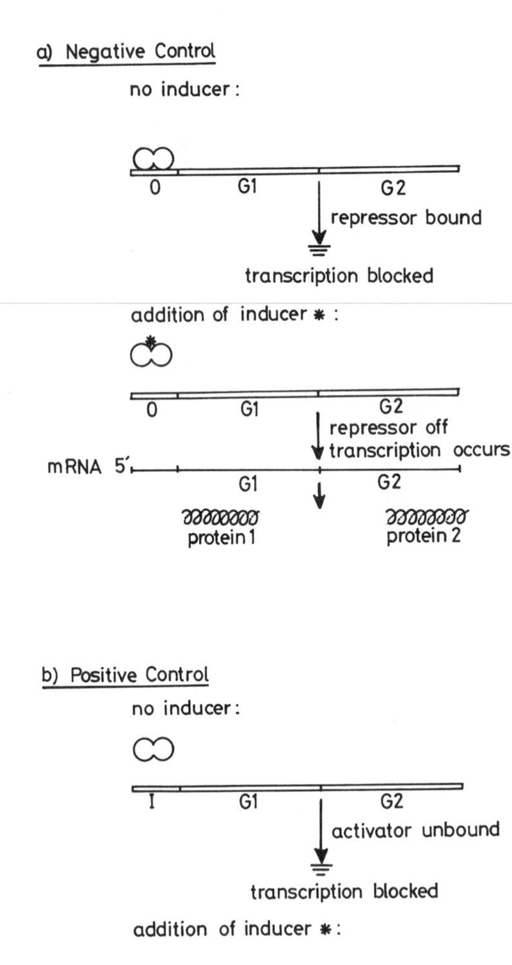

b) Positive Control

no inducer:

I

G1

G2

activator unbound

transcription blocked

addition of inducer *:

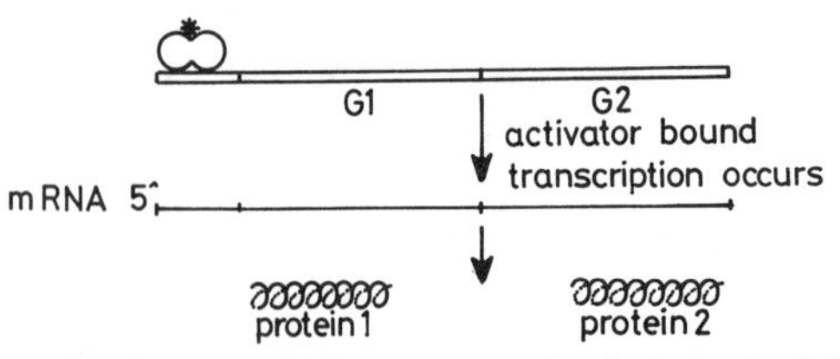

Figure 1.1 Mechanisms for the control of gene expression in bacteria. (*a*) Negative control of transcription: the inducer binds to the repressor, removing it from the operator and allowing transcription. (*b*) Positive control of transcription: the inducer binds to the activator, which binds to the receptor region, facilitating transcription.

expression changes ensuing from a single event (Monod and Jacob, 1961; Jacob and Monod, 1963). Although the general mechanism had been proposed before, Jacob and Monod showed how it could be enacted with only the kinds of regulatory elements demonstrated in bacteria. They also made explicit a notion that had been implicit in the beginnings of molecular biology—that the characteristics of cells are determined by the proteins they synthesize. The corollary is that the source of qualitative differences between cells is found in qualitatively different (and regulated) patterns of protein synthesis. This assumption has underlain much of the subsequent research on differentiation; we shall have occasion to come back to it.

The Jacob–Monod regulatory model marks the boundary between the old and the new developmental genetics. It moved the field from a preoccupation with phenomenology to the beginnings of an analysis of mechanism. Today, just as the concept of the gene as a sequence of nucleotide base pairs has become a part of our reflexive thinking, the Jacob–Monod model is an integral part of the collective imagination of all those working on development. It is virtually impossible to think about the control of gene expression without conjuring up thoughts of regulator genes, small molecule co-effectors, and transcriptional changes. Given the conception's seeming inevitability or even obviousness, it is worth asking what precisely was new in it and why it was perceived to have solved a previously intractable problem.

The principal innovation appears to be the idea of a discrete genetic regulatory apparatus, distinct from the protein coding genes (the structural genes) whose expression creates the measurable phenotypic properties of cells. The idea that there are separate regulatory genes whose sole function is to govern other genes was considered revolutionary. In reality, however, this idea was not entirely novel. The proposition that an independent regulatory system existed to control the expression of other genes had been proposed a decade earlier by Barbara McClintock and was subsequently elaborated by her to explain a set of otherwise inexplicable observations in the genetics of maize (McClintock, 1951, 1956). One cannot know with certainty why McClintock's evidence for regulation failed to excite the same interest as the much later bacterial work, but three factors may have been at work. The primary factor was probably the extreme novelty of the idea of regulation. In 1951, when the gene itself was still hidden in the pre-Watson–Crick mists, and even later when it was less obscure, the idea that genes could literally be switched into different *degrees* of activity was too abstract to be readily grasped. The second element was the fact that McClintock's evidence was presented in the context of maize genetic experiments and nomenclature. To the nonmaize geneticist, this was forbidding territory. Finally, and perhaps fatally, the evidence for gene expression control in maize was inextricably bound up with the then radical notion of moveable or "transposable" genetic elements. In transposing, these maize genetic factors produced genetic changes, which became obvious through their effects on differentiation. By extension, McClintock proposed that the normal sequence of differentiation events involved ordered regular movements of the putative controlling elements.

In the existence of transposable genetic elements, maize appeared freakish, but the suggestion that genetic change produces differentiative change was a common belief. When differentiated cells are cultured, they typically retain many of their recognizable characteristics, even after many generations of growth. The easiest way to account for this stability was to posit that differentiative change entails stable inherited genetic change. The discovery that different somatic cells within the same organism possess equivalent DNA contents (Mirsky and Ris, 1951) eliminated the possibility that differentiation involves large selective deletions of genetic material within different cell types but left open the possibility that some differentiation occurs through selective gene removal or rearrangement. The regulatory logic inherent in the bacterial model rendered all such thinking unnecessary in principle. When the appropriate feedback loops are built in, regulatory circuitry can generate any desired degree of stability.

Nevertheless, the idea of regulation without genetic change was not entirely new, either. One discussion of this concept can be found in a paper by Spiegelman (1948), who, like the Institut Pasteur group, stressed the usefulness of microorganisms for testing ideas about gene control mechanisms. However, the puzzle for Spiegleman was that regulation must take place on the genes in the nucleus, yet the effects are expressed in the cytoplasm; Spiegleman's hypothetical resolution was to export the genes, in the form of "plasmagenes," to the cytoplasm. The existence of mRNA, a feature of the operon model, renders plasmagenes unnecessary and solves the apparent dilemma of action at a distance. In effect, the bacterial models supply what any model of differentiation must supply, an explanation of the two-way interaction between nucleus and cytoplasm that differentiation must entail.

Nearly a quarter of a century has passed since the publication of the Jacob–Monod model, and it is pertinent to ask what its relevance is to development today. The answer depends on whether one is concerned with the molecular mechanics of gene expression in eukaryotes or the general logic of gene expression control. If one is concerned with the details of the molecular mechanism, then the applicability of the model to eukaryotic development may be slight. In the first place, eukaryotic cells are much bigger than bacterial cells, and size alone may make a difference. In small cells like bacteria, diffusion can rapidly carry every regulatory signal to every appropriate site; in eukaryotic cells, which may have thousands of times the internal volume of bacterial cells, changes mediated by diffusion alone may be too slow to be effective for the regulatory task at hand. One alternative for such large cells is to register regulatory changes slowly, but then to "fix" them in some manner, passing such changes on to their (cellular) descendants. An obvious site for such fixation is the eukaryotic chromosome, and a growing body of information suggests that some form of progressive chromosome "modeling" does take place during development.

Furthermore, apart from any novel features inherent in the putative fixation process, eukaryotic cells are in general much more complex than bacterial cells, and the additional layers of complexity may demand regulatory mechanisms not found in prokaryotes. The most obvious difference between the two kinds of cell is that eukaryotic cells possess a true nucleus whereas prokaryotic cells do not. Genetic information in eukaryotes must therefore not only be copied, but the transcripts must be packaged and "processed" to reach the cytoplasm. Regulation of gene expression, in principle, might therefore take place at these additional steps; experimental evidence indicates that it sometimes does (see the review by Brown, 1981). Furthermore, eukaryotic chromosomes are themselves highly complex structures, containing a large complement of basic histone and acidic nonhistone proteins. The former complex with and fold the DNA double helix around ball-like particles called "nucleosomes" to form a so-called nucleosomal fiber. This fiber, in turn, is successively folded and condensed into thicker fibers. The folding and condensation processes are themselves subject to control, which creates possible additional mechanisms for gene regulation.

The final level of complexity in eukaryotic cells is informational—eukaryotic cells contain vastly more DNA than do bacteria. It is clear that much of this DNA does not code for proteins, but whether or not it is functional, or what its functions might be, is unknown. Not only is there seemingly too much DNA for the eukaryotic gene as a whole, but the individual "gene" in eukaryotes is also much larger. The average bacterial gene is only 1000–2000 base pairs (bp) long, with perhaps only 20–40 base pairs serving as binding sites for regulatory proteins of various kinds. The typical eukaryotic gene may be 20,000–40,000 bp, or 20–40 kb, long (1 kb = 1000 bp), with an unknown fraction of the total length devoted to regulation. Within each coding region of a eukaryotic gene, there may be one, two, or many inserted regions or "introns" that have no coding function; the initial or primary transcript of the gene contains the intron sequences, but these are cut or "spliced" out before the transcript is sent into the cytoplasm. Some introns may have functional regulatory roles; others, perhaps the majority, are probably excess baggage. The relatively simple control mechanisms that bacteria employ are not designed to cope with this level of gene complexity.

If one turns from molecular mechanics, however, to the principle of regulatory circuitry, the relevance of the bacterial models is much greater. The fundamental ideas of regulatory switches and sequential, ordered changes in gene expression remain as pertinent today as they were in 1960. Using the DNA–RNA hybridization techniques developed in the 1960s, workers have repeatedly shown that embryonic development is accompanied by an ordered change in transcript populations within cells. Furthermore, while the differentiation of a few cell types, principally the immunoglobulin-producing cells (Tonegawa, 1983), show deletion and rearrangement of their expressed structural genes, most development takes place without irreversible genetic change. One can show this either by molecular tests using DNA hybridization, which show that few if any sequences are permanently lost, or by the biological test of nuclear transplantation. The most extensive tests of the latter kind have been performed in amphibians. In a nuclear transplantation test, the nucleus of a fertilized or unfertilized frog egg is replaced by that of a somatic cell from the same species and the egg is allowed to recommence development. When large numbers of somatic nuclei from different tissue types are tested, the majority are found to be capable of producing descendant nuclei that promote development to the tadpole or adult stages (Gurdon et al., 1975). (There may be some loss of totipotency of nuclei by the adult stage [DiBerardino, 1980], or in earlier stages in certain frog species [King and Briggs, 1956], but these losses of capacity may reflect either problems of chromosome replication or a degree of irreversibility in chromosome "modeling" processes rather than genetic loss.) If most somatic nuclei contain a complete set of genetic instructions, it follows that development and differentiation involve regulated changes in gene expression rather than changes in genes themselves.

The net effect of the Jacob–Monod model on developmental genetics was to shift attention from the observable developmental effects of mutations to the underlying effects on gene expression. Within the terms of the model, mutations could be classified, in principle, into either those in structural genes or those in regulatory genes; the immediate challenge was to devise ways of making these assignments. In addition, the whole battery of new molecular techniques for assessing changes in protein and RNA populations could be and were applied to the analysis of wild-type and mutant development. These activities continue to be a major element in developmental genetics. A secondary, subtler change concerned the way that biologists began to view the "logic" of developmental control. Development began to be seen as a "progam" encoded in the genome. The consequence was a new emphasis on the genome as a source of information not just for the organism but for the investigator as well.

But the story of developmental genetics since the early 1960s has been more than one of the fertilizing effect of ideas from molecular biology. The field has simultaneously been experiencing its own internal revolution, one owing little to molecular biology. Some of the techniques and approaches have been borrowed from microbial and molecular genetics, but the principal ones derive from the earlier classical period. The focus of this new work is the individual cell and the developmental constraints that govern the behavior of cells during development. The genetic methods that comprise the heart of this program of investigation have been collectively termed "clonal analysis" (Nothiger, 1976).

In essence, clonal analysis involves tracking the mitotic descendents of a cell—its *lineage*—during development by means of a cellular characteristic created by a genetic "marker." All that is required for such tracking is that the marked cellular characteristic be distinguishable from the background cells and that it be expressed independently, or "autonomously," of the characteristics of the surrounding cells. The shape and location of the marked clone then provide information about the initial cell's developmental role or "fate." If the original marked cell has been placed by some means in a new location within the embryo or within a different embryo, the development of the clone can then provide a hint about that cell's repertoire of developmental capacities, its "potencies."

With this focus on the individual cell as the unit of development, clonal analysis bridges the gap between the molecular analysis of developmental change, with its primary emphasis on subcellular events at the level of the genome, and that of traditional experimental embryology, with its emphasis on the aggregate behavior of cell groups, called "embryonic fields" or "blastemas." Indeed, clonal analysis can remove much of the ambiguity associated with the older methods of experimental embryology. The typical experiment in embryology consists of transplanting a macroscopic piece of tissue from a defined site in an embryo at a defined stage to a different site in either the same or a different embryo, and scoring the developmental response of the transplant or of the recipient region.

Whatever the results, there will always be uncertainties in the interpretation, because it cannot be known how much of the response is caused by the intrinsic capacities of the individual cells expressed autonomously or by the collective or "system" interactive effects arising from the transplanted cells acting in concert. Furthermore, it generally cannot be determined whether the cells in the transplant have responded homogeneously or heterogeneously, or what the sources of the difference, if they exist, might be—cell lineage history, physical position per se, or other factors. (For a complete discussion of the problems of interpretation, the reader should consult Conway et al., 1980.) By tracking individual mitotic lineages in developing primordia, much of this ambiguity can be eliminated. By examining large numbers of marked clones, it becomes possible to establish whether given constituent clones all have the same intrinsic developmental capabilities and to evaluate separately the roles of position and interactive effects.

To a large extent, the findings of clonal analysis define the phenomena that ultimately must be explained by the molecular studies of gene expression control. There are two broadly different forms of clonal analysis. The first involves physically introducing genetically marked cells from one embryo into a second embryo. Such genetically composite animals are termed "chimeras" after the legendary Greek monster that had the head of a lion, the body of a goat, and the tail of a dragon. The second method involves inducing a genetic difference in one or more cells in a single individual. Such individuals, in which all cells derive from the same zygote, are said to be "genetic mosaics." Each approach has its strengths and limitations, as will become apparent in the following chapters. However, the versatility and usefulness of clonal analysis for probing developmental questions can be illustrated with its application to the first developmental question addressed in studying *Drosophila*. This problem was the mechanism of sex determination.

The earliest work with *Drosophila* showed that males and females consistently differ in their chromosomal composition (or "karyotype"). Both sexes are identical for three pairs of chromosomes, the autosomes, but differ in a fourth pair, the sex chromosomes, with females possessing two X chromosomes and males one X and a morphologically distinguishable chromosome, the Y. Because the Y is not found in females, the simplest explanation of the sex difference is that possession of the Y confers maleness, its absence producing femaleness. This first hypothesis was refuted, however, when Calvin Bridges, working in Morgan's laboratory, discovered the existence of rare female flies possessing two Xs and a Y (XXY flies) and occasional males possessing a single X and no Y (XO) flies). Further studies of flies with altered autosomal and sex chromosome compositions led to the formulation of the "chromosome balance" theory of sex determination, the idea that sex is determined by the ratio of X chromosomes to autosomes (Bridges, 1925). (This mechanism is not universal; in mammals, sex *is* determined by the presence or absence of the Y rather than by the number of X chromosomes.)

Sexual phenotype is a complex character, however, involving several external and internal structures of the fly. Does the sex-determining chromosomal ratio act in each and every one of these sites, or does the correct development of sexual phenotype involve only a sex-setting "decision" in one or two of these sites? The answer was provided by the study of certain naturally occurring genetic mosaics that were part male and part female. These "gynandromorphs" consist of large sectors of female and male tissue, composed respectively of XX and XO tissue. Morgan and Bridges (1919) found that cellular and regional phenotype always followed cellular chromosome composition. Sexual phenotype is therefore always determined autonomously within cells as a function of sex chromosome balance; in the fruit fly, there are no critical sex-determining sites or pervasive systemic influences. The origin of the gynandromorphs was postulated to be in the loss of one of the X chromosomes from one of the cells of the early cleavage embryo.

In 1929, A.H. Sturtevant extended the study of gynandromorphs to the first thorough clonal analysis of early development in an embryonic system. In examining an eye color mutant of *Drosophila simulans,* Sturtevant found that the mutation caused the frequent loss of chromosomes in one of the early cleavage divisions of the embryo. Embryos that lost one of the major autosomes—either the second or third chromosome, by convention—were inviable, but the female embryos that lost either an X chromosome or one of the tiny autosomes (the fourth chromosome) survived. In embryos with two X chromosomes, it was the maternally derived X that was preferentially lost. The key finding was that up to 3% of the viable embryos were gynandromorphs, produced by loss of the maternal X in the first, second, or third cleavage division of XX embryos (Sturtevant, 1929). By mating his mutant females to males with sex-linked markers—recessive bristle and pigmentation mutations on the X—he was able to delineate XO patches of tissue unambiguously, since loss of the maternal X permitted expression of the recessives on the remaining X chromosome. With large numbers of gynandromorphs generated in the mutant and clear marking of XO areas, Sturtevant showed that any area of the gynandromorph individuals could be XO and that whenever an area included sex-specific structures in the fly, the area was male.

These results amply confirmed the earlier conclusion that sex determination within each cell is autonomous. Moreover, Sturtevant was able to take the clonal analysis two steps further. Firstly, because the gynandromorphs showed that the dividing line between male and female tissue could fall anywhere, it was reasonable to assume that the line was much less likely to fall between two structures whose progenitor cells were close together at the time of cell fate specification than if they were far apart. The frequencies of separation by the male–female boundary in the adult gynandromorphs were thus inverse measures of the distances between progenitor cells and could be used as measures of these distances in the early embryo. Secondly, the findings permit an assessment of the developmental states of the first cleavage nuclei in *Drosophila*. In insect embryos, the cleavage divisions take place in a syncytium. As divisions occur, nuclei move outward toward the periphery, becoming cellularized only at the end of cleavage. Since any early nucleus serendipitously distinguished by the loss of an X could give rise to any set of structures on the surface of the adult fly (see Fig. 6.11), it followed that (1) the initial orientations of the nuclei must not be fixed and (2) the developmental fate of the first nuclei must not be specified. Rather, the first fate specifications must occur either as the nuclei approach the surface or during the cellularization process. This work, which we shall return to, illustrates the tremendous potential power of clonal analysis. By examining clones that express their cellular markers only at the *end* of development, it is possible to make some fundamental deductions about the *earliest* events in embryogenesis.

If tracking of cell lineages were the only function of clonal analysis, it would be an essential technique in developmental genetics. However, the technique has an even broader application: it can be used with mutations that affect cellular development behavior. To probe the effect of a particular mutation on a clone's development, a marker mutation is used so that cells expressing the developmental mutation are simultaneously visibly marked. By comparing clones with and without this second marker, effects of the developmental mutation on clone behavior can be distinguished. Furthermore, by isolating the cells of the mutant genotype against a background of normal cells, the complications of relational pleiotropy can be eliminated.

If developmental genetics today is very different from that of a quarter of a century ago, it is largely clonal analysis that has made it so. While the early influx of ideas and techniques from molecular biology reshaped the ways in which developmental biologists thought about genes and gene actions in development, the findings of clonal analysis have created a body of information on cell growth and cell lineages in development that previously did not exist. In consequence, the focus of attention in the field has shifted primarily from the embryo as a whole to the cell and the cell group in development. In addition, a third kind of activity is now beginning to make itself felt—that of recombinant DNA technology, the ability to clone genes and to use these cloned sequences to define the structure and activity of specific genes. The techniques of recombinant DNA analysis make it possible to answer many of the questions about genes that the older science could not touch.

This book is about developmental genetics today, its methods and its ideas as applied to animal systems. I have chosen to present the material by centering the discussion on the developmental biology of the three animals that have received the most attention from developmental geneticists. The intention behind this approach is to provide both a firm biological context for assessing the issues that arise and to give the reader the necessary background for delving more deeply into the literature of these systems, which are likely to remain the principal model systems for some time to come.

The three animals to be examined are the small nematode, *Caenorhabditis elegans;* the fruit fly, *Drosophila melanogaster;* and the laboratory mouse *Mus musculus*. Each has special advantages and disadvantages as an experimental organism, and each has had a distinctive history in developmental genetics. The mouse and *Drosophila* have had long and nearly coextensive histories as experimental organisms in genetics but differ in their histories as subjects of developmental interest. The mouse may be regarded as the doyen of animal genetics. Mouse experimentation began with the work of Cuénot in France and William Castle in the United States in the early 1900s. Fruit fly research began in several laboratories, including Castle's, before Morgan took it up in 1908, but it was the work of Morgan's group that truly launched it. With the mouse, developmental studies and speculations have almost always accompanied the genetic work, from the first experiments on. In *Drosophila,* apart from the pioneering work of Sturtevant, serious investigation of its developmental biology did not commence until the late 1930s with the important studies of Hadorn (1937) and Poulson (1940). The case of the nematode, *Caenorhabditis,* is different. Relative to the fruit fly and mouse, it is an organism almost without a past (from a geneticist's point of view). It was chosen by Sydney Brenner for its potential as a simple animal model system suitable for the genetic probing of developmental processes. Work on *C. elegans* began in Brenner's laboratory in England in 1965.

The main body of this book, Chapters 2 through 7, is devoted to the developmental biology and essential developmental genetics of these animals. Because the life history of each of the three organisms differs in numerous respects from those of the other two, and because early development poses special questions of its own, the material has been divided into individual chapters according to organism and stage of development. Chapters 2 through 4 describe early embryonic development and its genetics. Each of these chapters discusses the special role of oocyte structure and composition, specified by the maternal genome, in the process of early cellular fate restrictions in the embryo. Chapters 5 through 7 describe in a parallel fashion the later events of embryonic development, beginning at the approximate point at which development passes from maternal control to that of the embryonic genome. The final section of the book deals with several general issues and problems: pattern formation (Chapter 8), the applications of recombinant DNA analysis to problems in developmental genetics (Chapter 9), and the prospects for achieving a general understanding of the genetic foundations of development (Chapter 10).

A word of caution concerning terminology is in order. Developmental biology is replete with terms, many of them deliberately borrowed from psychology (Oppenheimer, 1967), that describe the general capabilities or developmental responses of embryonic cells under various experimental manipulations. Some of the more familiar ones are "determination," "induction," "commitment," "regulation," "competence," "organization," and "individuation." These terms have become irreplaceable through frequent usage. However, indispensability is not necessarily the same thing as adequacy, and categorization under such broad behavioral headings often leads to ambiguity regarding what is being described. Perhaps the tradition of adopting terms from psychology was bound to cause trouble; to categorize one set of mysterious processes by analogy to an even more mysterious set of phenomena is hardly guaranteed to produce clarity. The fundamental difficulty, however, is that each term represents a shorthand operational definition of a complex cellular and developmental property displayed under a particular set of experimental conditions. The temptation is to equate, unconsciously, the label with some kind of fixed or even universal set of molecular properties, forgetting the contingent, operational nature of the classification. Thus, cells classified as committed to a pathway in one test might well exhibit greater developmental flexibility (lack of commitment) under another. Furthermore, if one is comparing the same general property such as commitment in two different cellular populations, the possibilities for error are particularly great if one assumes that the two states are equivalent.

The concept of "determination," which figures importantly in Chapters 2, 3, and 4, deserves special note. It is a general fact of embryonic development that as embryogenesis proceeds, cells lose some measure of an initially large set of developmental capabilities ("potencies") and become intrinsically restricted to following a particular line of development toward a specific outcome ("fate"). To put the matter more precisely, cells lose the capacity to produce descendant cells with a large developmental repertoire. The restriction shows up as a reduced capacity of the descendant cellular progeny to diverge widely in their differentiative capacities. Determination refers to this restriction of developmental potency and consists of some fixed, somatically heritable limitation of ability, scored when the cells' progeny are tested for their differentiative abilities in a new environment. It can be shown in several organisms by clonal analysis that cells in situ within the developing organism experience restrictions in terms of the structures or regions to which their descendants can contribute. It is tempting, therefore, to say that such cells have become determined. However, unless the cells have been removed from their normal environment and tested, either in isolation or in a new location, it cannot be judged whether the restriction is inherent in the descendants of the particular cell or is a highly labile restriction that happens to be enforced by the local cellular environment. To avoid this confusion, I have endeavored to use the term "determination" only when there is some independent test showing that the restriction is inherent to the cells. In instances where this is not clear, I have used the more neutral term of "assignment" of the developmental pathway.

There is, of course, no general solution to the problem of terminology at present. Developmental geneticists are interested in the genetic basis of the phenomena discovered and described by experimental embryologists; they must therefore use the operational terms of embryology. The only safeguard is to remain aware of the contingent and operational nature of the descriptive terms and to be cautious in ascribing either general or particular molecular properties to cells on the basis of these categorizations.

PART TWO

OOCYTE AND EMBRYO

OOGENESIS AND THE PROGRAMMING OF EARLY DEVELOPMENT

. . . we must hold fast to the fact that the specific formative energy of the germ [i.e. the early embryo] is not impressed upon it from without, but is somehow determined by an internal organization, inherent in the egg and handed on intact from one generation to another by cell division. Precisely what this organization is we do not know.

E.B. Wilson, *The Cell in Development and Heredity* (*1925*), p. 1035

The fascination of the growing embryo lies in its progression toward ever greater complexity. Silently, continuously, the embryo shapes itself into an increasingly intricate entity.

Nowhere is this progression toward higher organization more rapid, or more significant with respect to later development, than in the first stages of embryogenesis, after the sperm has entered the egg. As soon as fertilization occurs, the intracellular contents of the egg begin a series of movements and transformations that culminate in the fusion of male and female haploid pronuclei. Nuclear fusion is immediately followed by a series of cell divisions, termed "cleavage divisions," that create successively smaller cells, the blastomeres, within the original zygotic mass. At the end of cleavage, the multicellular blastula stage embryo immediately begins to transform itself by a series of cellular movements, characteristic for each embryo type, into the first multilayered embryonic stage, the gastrula. It is usually within the blastula stage that the cells of the embryo are first assigned different fates, and it is during the gastrula stage that they acquire different states of determination.

A century of experimental embryology has demonstrated that for most, perhaps all, animal embryos, these first events of development are controlled by the components of the egg external to the nucleus (Wilson, 1925; Davidson, 1976). These components set the pattern of cleavage divisions and establish those first differences in prospective cell fate that set the stage for all subsequent embryonic changes. The sperm, of course, contributes half of the genetic material for the new embryo, and may participate in other ways, but it is the female gamete that sets the pattern of the first stages of development.

The fundamental questions about early embryonic development therefore become questions about the biochemical organization of the precursor cell of the embryo, the oocyte, and about its formation, the process of oogenesis. Because the oocyte is the most complex cell in the adult animal, analysis of the roots of early development is a correspondingly challenging problem. The structural complexity of the egg reflects the fact that this cell acts as the precursor of the embryo in a dual capacity. On the one hand, the oocyte provides the physical framework of the early embryo and must serve as a storage vessel for all the components required for cleavage and the biosynthetic activities of these first stages. These functions may be briefly termed those of "storage." On the other hand, the oocyte provides the biochemical blueprint for the establishment of those first critical cellular differences that translate into differences of fate and state of determination; its structural heterogeneity prefigures the cellular heterogeneity of the early embryo. This role of biochemical prepattern may be termed that of "morphogenetic instruction."

These complexities in composition and structure do not develop autonomously. Oogenesis requires specific activities of associated somatic or sister germ line cells, and often necessitates long periods of completion, up to months or years in some animal species. This developmental complexity, in turn, probably reflects an underlying intricacy of genetic expression, involving numerous gene activities in several cell types that occur in a highly ordered sequence.

There are essentially two different kinds of questions that geneticists ask about oogenesis and early development, and it is important to distinguish between them. The first concern the *genetic basis* of oogenesis and, in particular, the distinctiveness of the gene set expressed during oogenesis relative to other cell differentiations. The basic question is this: to what extent is the biological distinctiveness of the oocyte a reflection of a uniquely expressed gene set? In principle, the oocyte might possess *no* qualitatively unique constituents but derive all of its special properties from a unique combination and/or arrangement of gene products, each of which is shared with one or more cell types. In reality, there are very likely to be at least some unique oocyte constituents. One has already been described: the small (5 S) ribosomal RNA in the oocytes of the frog *Xenopus laevis* is different from the comparable small RNA in the non-gamete-producing cells, the somatic cells, of this animal (Ford and Southern, 1973).

The second set of questions deals explicitly with the *action* of the oocyte gene products in generating the unique characteristics of the oocyte, and in particular with the "morphogenetic instructions" sequestered within the egg. What are the molecules that comprise this system, how are they arranged in the cleavage stage embryo, and how do they exert their effect(s)?

OOCYTE STORAGE FUNCTIONS

Before examining the principal genetic approaches to early development, it is helpful first to take a closer look at the properties of oocytes in terms of their principal functions. The most striking physical characteristic of the oocyte is its size. Oocytes are invariably larger than the typical animal cell of the animal from which they derive. Large size is demanded for fulfillment of the storage function, and a major fraction of the biosynthetic activities of oogenesis is directed to stocking the oocyte with materials that will be required. These requirements are often met by the activities of numerous associated somatic cells and/or sister germ line (gamete-forming) cells that act as feeders for each developing oocyte.

Several kinds of constituents are stored, each associated with a particular requirement of the early embryo. The first group includes all the nutritional or metabolic reserves that will be required by the embryo during its initial phase of growth. These reserves are generally in the form of "yolk," a complex mixture of proteins, phospholipids, and neutral fats. In the yolk-rich (megalecithal) eggs of the fishes, reptiles, and birds, the yolk comprises the bulk of the mature oocyte. The oocyte and cytoplasm are restricted to a small disc on top of the comparatively huge yolk mass itself. The second category of stored component is the set of molecules required directly for the cell cycles of cleavage. These may include reserves of tubulin for the assembly of the mitotic spindles during cleavage and the various enzymes, factors, and nucleotides required for the rounds of nuclear DNA replication that precede each cleavage division. The third group of stored constituents are those required for the extensive program of protein synthesis that takes place in early embryos. These components include a very large store of ribosomes and a diverse set of informational RNAs (Davidson, 1976).

The informational RNAs are defined as those RNAs that are not universal constituents of the protein synthesis apparatus (namely, the ribosomal RNAs and the transfer RNAs that bring amino acids to the ribosomes). However, the nature of the information they carry is not clear in all cases. Nevertheless, analysis of the informational RNAs can provide a preliminary answer to the question about the distinctiveness of the expressed gene set in oocytes. In particular, the transcripts of the so-called single copy portion of the genome should be the most informative. The eukaryotic genome consists of two major classes of DNA sequence: so-called single copy or unique sequences, ranging from hundreds to thousands of base pairs in length but each present only once in each haploid genetic complement, and the repetitive or midrepetitive sequences, often 100 or a few hundred base pairs in length, that are present in numerous families of closely related copies in each haploid genome (Britten and Kohne, 1968). The majority of protein coding sequences are expected to be in the single copy fraction, but both kinds of sequence are transcribed and oocytes contain large stores of transcripts of these two classes of sequence.

Because the single copy transcripts include the great majority of the expressed structural genes, measurements of the total cytoplasmic content of the stored single copy transcripts can provide minimal estimates of the total number of expressed genes. A comparison of this kind between the oocyte and other cell stages and cell types, for the sea urchin, is shown in Figure II.1. The results are given in terms of sequence "complexity," defined as the total length of the diverse single copy sequences represented. The striking result is the relatively and absolutely large sequence content of the oocyte in comparison to later stages and cell types. If one takes the length of an average protein coding sequence as 1500 bp, then the sea urchin oocyte would seem to have the capacity to code for about 24,000 different proteins

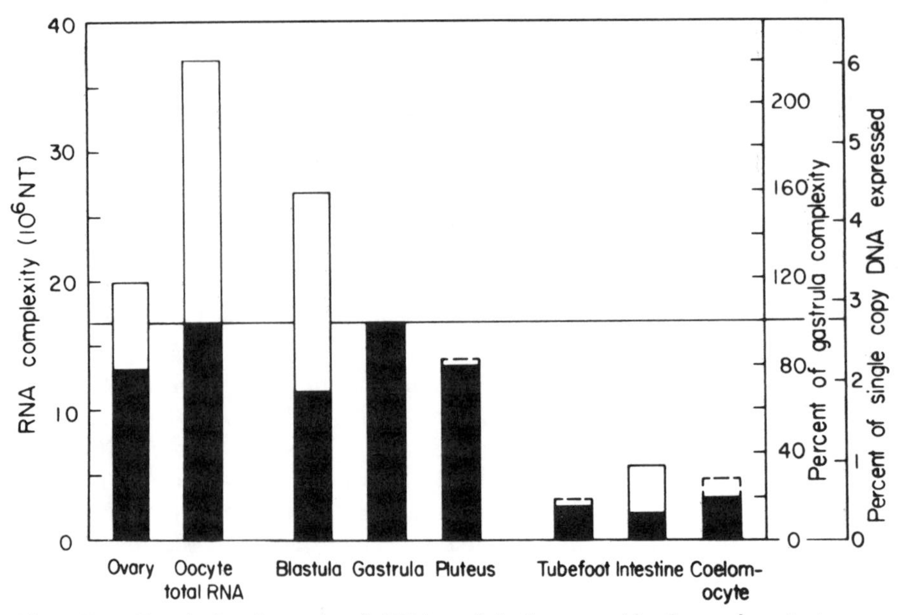

Figure II.1 Complexity of cytoplasmic RNA pools in the sea urchin, *Strongylocentrotus purpuratus*. The height of each bar gives the total complexity of cytoplasmic RNA hybridizing to single-copy DNA; the black portion gives that part found in gastrular stage embryos. (Reproduced with permission from Galau et al., 1976; copyright MIT Press.)

(Galau et al., 1976). Furthermore, many of these sequences (shown in black) are shared with the gastrula stage and with subsequent stages and cell types; these overlaps in transcript set are produced by recurrent synthesis rather than persistence from the oocyte (Hough-Evans et al., 1977). A certain minority fraction of the set of gastrular stage sequences, equivalent to about 1000–1500 genes' worth, are shared by all stages and cell types examined. These ubiquitous sequences may represent genes coding for the proteins that carry out the universal, essential steps of metabolism in cells; such genes are said to perform "housekeeping" functions.

The results seem to suggest that the specialness of the oocyte results from the extensiveness of its expressed gene set rather than from a uniquely expressed subset of genes. However, two findings put this interpretation in a different light. The first is that all the oocyte sequences stored in the cytoplasm are found to be transcribed at much later stages, perhaps in all cell types, but the majority of these sequences are retained in the nucleus. The set of cytoplasmic transcripts therefore reflects a complex set of transcriptional and posttranscriptional processing events instead of directly mirroring transcriptional patterns (Davidson and Britten, 1979). The second point bears on the assumption that the cytoplasmic single copy transcripts are used directly for translation.

It is certain that some portion of this RNA pool is truly informational, providing a direct capacity for the synthesis of proteins needed by the embryo. However, much of the maternal transcript pool in sea urchin and amphibian eggs is in the form of large RNA molecules, containing interspersed repetitive sequences and resembling unprocessed nuclear transcripts (see the review by Davidson et al., 1983b). These large transcripts cannot be translated by the protein synthetic apparatus and may be translatable only after further processing. The function of much of the maternal RNA pool is therefore obscure; these transcripts are a selected portion of the total nuclear RNA pool, in terms of their sequence content, but they may function primarily as a metabolic reserve of nucleotides for future nucleic acid synthesis (L. Hereford, personal communication). The only certain test of a gene sequence's function is a genetic one: if a mutation in that gene produces autonomous cell dysfunction, then that gene is needed in development.

MORPHOGENETIC INSTRUCTIONS

Despite such ambiguities in interpreting the biological roles of stored components in the oocyte, the procedures for detecting or measuring these components are at least usually straightforward. The oocyte is treated as if it were homogeneous and the component of interest is purified. However, investigating the organizational properties that give rise to the morphogenetic instruction system in the egg is a more difficult matter. Central to the analysis is the recognition that oocytes are not homogeneous, but possess or give rise to significant regional intracellular differentiation. It is the particular features of this cytoplasmic (or membrane) diversification that create the crucial differences in the fates of blastomeres.

In common with many other developmental events, these first fate-setting events involve interactions between nuclei and cytoplasm, but these early interactions are unusual in several respects. Firstly, there is the very fact of spatial differentiation within an initial single cytoplasm, which triggers different responses in different nuclei. Furthermore, the responsible cytoplasmic (or membrane) constituents are synthesized under the direction of one set of genetic instructions, that of the maternal parent, yet act on nuclei that carry a potentially different set of alleles. This second genome is that of the zygote, which is composed of equal contributions from both parents. In consequence, the embryonic nuclei of the blastomeres are initially under the control or influence of maternal constituents and maternally encoded processes, and only later in development do these nuclei begin to exert complete control over the activities of the cells that house them. The timing of this transition from maternal to zygotic genome dependence will, of course, depend on the particular animal system and its developmental program. When both the maternal and zygotic genomes are wild-type, the transition is imperceptible, but when either the maternal or the zygotic genome is homozygous mutant for some vital gene, the deviation from normal development can be quite marked. The alteration can take place at the beginning of embryonic development, if the maternal genome is mutant; or during the period of transition, if the zygotic genome is defective for the gene function; or well after the transition, depending upon when the gene product is needed and when its supplies are exhausted. The genetic complexities are discussed further below.

In thinking about the fate-specifying system built into the egg, it is essential to note that what have been termed the "morphogenetic instructions" within the ovum usually consist of two independent molecular subsystems (Wilson, 1925). It is the combined operation of these two subsystems that creates blastomeres with differing fates. The first is composed of the regulatory molecules themselves, the "morphogenetic determinants," as they were designated by August Weissman late in the nineteenth century. The second subsystem is the one that directs the placement and orientation of mitotic spindles during cleavage: it consists of all those intracellular and intercellular factors that govern the pattern of cleavage. This built-in program for cleavage has been named "topogenesis" (Nigon, 1965), to distinguish it from the set of gene regulatory molecules of the other subsystem. Topogenesis can include such gross physical factors as yolk mass (which exerts a viscous, retarding effect on cleavage when present in bulk), but for many egg types, the topogenetic program is based on a subtler biochemical organization within the egg (Wilson, 1925). Topogenesis determines the size and placement of the blastomeres and thereby the partitioning of the morphogenetic substances.

The proof that topogenesis and morphogenesis are organized by different molecular elements is that these processes can be uncoupled in certain oocytes by particular experimental treatments. An example of such uncoupling is provided by the embryo of the insect *Smittia,* in which the normal specification of fate can be dramatically altered in the absence of any detectable effect on cleavage. In *Smittia,* it has been found that any one of several treatments of the pre- or early cleavage embryo abolishes the normal pattern of anteroposterior (a-p) segment specification, yielding mirror-image, double abdomen larvae (Kalthoff, 1979) (Fig. II.2). These treatments include exposure to a focused ultraviolet (UV) beam at the anterior end of the cortex (without exposure of the cleavage nuclei), puncture of the anterior end, or centrifugation along the a-p axis of the egg. The effect is always an ordered switch in the fate of the anterior-half cells and does not involve a change in the cleavage pattern, which in *Smittia,* as in *Drosophila,* is a sequence of syncytial nuclear divisions accompanied by an outward migration to the cell periphery.

The converse uncoupling, in which a normal expression of developmental potential occurs despite the suppression of cleavage, can also take place. In normal ascidian embryos, the enzyme activity acetylcholinesterase (AcE) is expressed solely in the muscle cells of the neurula-stage embryo, as detected by histochemical staining. This activity usually appears approximately eight hours after fertilization, when the embryo contains several hundred cells. If cleavage is suppressed by means of treatment with the drug cytochalasin B, at any point in cleavage from the 2-cell to the 64-cell stage embryo, and the normal developmental time is allowed to elapse in the presence of the drug, the enzyme not only appears on schedule but is expressed *only in those progenitor blastomeres that would have given rise to the muscle cells* had cleavage been allowed to continue (Whittaker, 1973, 1979) (Fig. II.3).

Although both the topogenetic and morphogenetic systems of the egg are little understood, it is the latter that pose the most important questions. One of these questions concerns the stage in development when the "morphogenetic determinants" are laid down: are they prelocalized in the oocyte, or do they come to occupy their final positions after fertilization? The answer is that there are a variety of behaviors. In some species, the determinants are prelocalized; in others, they assume their final positions just after fertilization; while in still others, localization occurs during cleavage (see the discussion by Freeman, 1979). Nor do all the different kinds of determinant within an oocyte necessarily become positioned at the same time. In *Drosophila,* the germ line determinants appear at the posterior end of the egg during the final stages of oogenesis (Illmensee et al., 1976), while the cytoplasmic instructions for somatic structures become localized only late in cleavage (Schubiger and Newman, 1981).

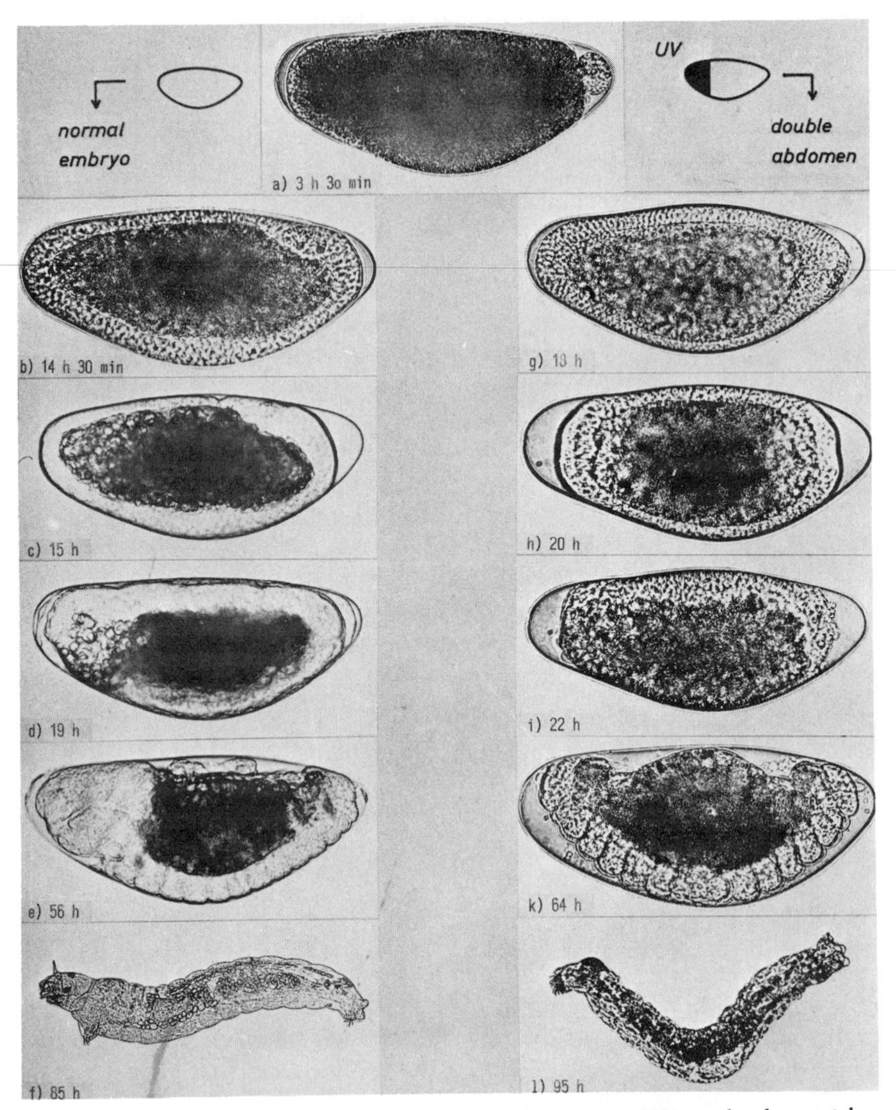

Figure II.2 Induction of double-abdomen larvae in *Smittia*. The wild-type developmental sequence is shown on the left, that of UV-induced reversed-abdomen embryos and larvae on the right. (Kindly provided by Dr. K. Kalthoff; reproduced with permission from Kalthoff, 1979.)

The most significant questions about morphogenetic determinants, however, concern their molecular nature and mode(s) of action. The principal difficulty in this analysis involves the way these substances are assayed. The properties and positions of these molecules are always deduced indirectly and retrospectively, on the basis of the ways the mitotic descendants of the original cells behave, long after the primary actions of the determinants themselves have been completed. Whether the existence and location of a

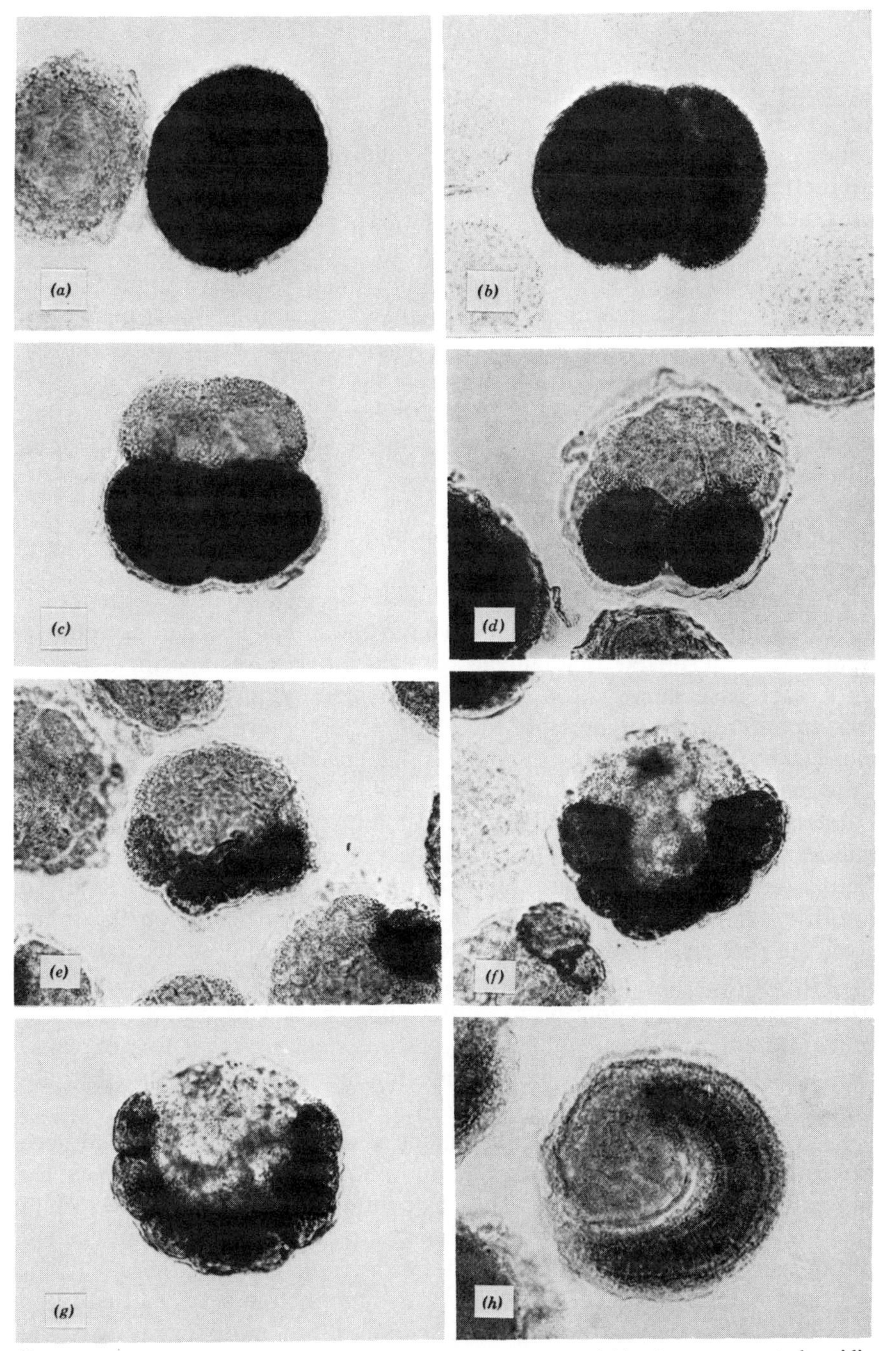

Figure II.3 Segregation of acetylcholinesterase-forming potential in cleavage-arrested ascidian embryos. AcE activity (dark staining), assayed at 15–16 hr of development at 18°C, in *Ciona* embryos arrested in cleavage at the following stages: (*a*) 1-cell; (*b*) 2-cell; (*c*) 4-cell; (*d*) 8-cell; (*e*) 16-cell; (*f*) 32-cell; (*g*) 64-cell. (*h*) A 12-hr control larva. (Kindly provided by Dr. J.R. Whittaker; reproduced with permission from Whittaker, 1979.)

particular determinant is inferred from the capacity for "self-differentiation" of isolated blastomeres, or conversely, from the absence of specific tissues or structures following selective removal of blastomeres, the measured effect is always considerably removed from the time of action of the substance of interest. Inevitably, the interpretations are colored by the particular analytical procedures used and by the underlying assumptions made.

The ascidian egg illustrates the problem. It is a textbook example of the "mosaic" egg. The mosaic egg is one in which specific blastomeres are assigned specific roles early in cleavage, with little capacity for interchangeability of fates between them when cells are removed or injured. The localization of acetylcholinesterase-forming potential, described earlier, is a biochemical aspect of one such highly localized capability. The contrasting form of egg organization recognized in classical embryology was the "regulative" type, in which early blastomeres can, under certain circumstances, substitute for one another; echinoderm and amphibian eggs possess this property.

The classical experiments of E. G. Conklin on ascidian development (see the review by Davidson, 1976, chap. 7) have traditionally been interpreted to mean that the fertilized ascidian egg segregates five tissue-specific determinants, each associated with a specific pigmented cytoplasm or cortical region, to different blastomeres of the embryo. However, the same observations can be explained by the hypothesis that specification of the five tissues is produced by the action of three sequentially formed molecular gradients (Catalano et al., 1979). The difference in interpretation is significant. If the traditional explanation is correct and the pattern of cell fates is brought about by five qualitatively different determinants, the matter can be viewed from the perspective of conventional regulatory models. If, on the other hand, the pattern of cell fates is produced by a system of molecular gradients, the crux of the matter becomes the manner in which different cells "read" different concentrations of the substances and become directed along different developmental pathways. Contemporary regulatory explanations, based on prokaryotic models, are for the most part ill-suited to account for gradient actions.

As in the case of ascidians, the conflict of interpretations sketched here between localized determinants and morphogenetic gradients arises for many different oocyte systems. The basic difference is schematized in Figure II.4. Although the localized determinant model is generally associated with mosaic eggs and the gradient model with regulative eggs, there may not be a fundamental difference in the mechanism of action between these two kinds of eggs but only in the degree of spread of the "morphogens" and the rapidity with which cell assignments become fixed. In mosaic eggs, cleavage may sequester determinants early, while in regulative eggs the definitive segregation may take place only later in cleavage (Wilson, 1925; Davidson, 1976).

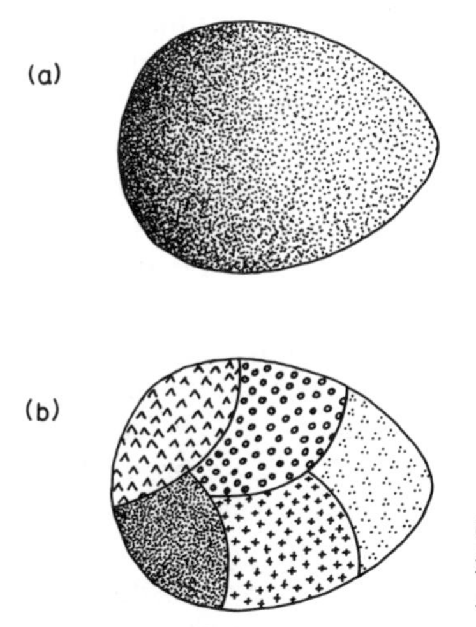

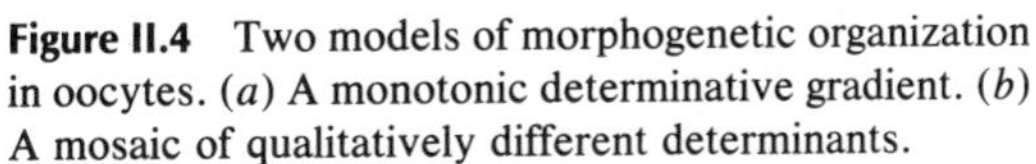

Figure II.4 Two models of morphogenetic organization in oocytes. (*a*) A monotonic determinative gradient. (*b*) A mosaic of qualitatively different determinants.

This suggestion may dissolve the seeming dichotomy between regulative and mosaic eggs, but it leaves a fundamental question unanswered: whether there are determinants that comprise *qualitatively distinct* cell instructions, or whether it is their different *amounts* that create qualitatively different fates. Neither experimental embryology nor biochemical analysis has been capable of answering this question.

GENETIC ANALYSIS

Genetic analysis, in contrast, has the potential to resolve this issue, at least in those animal systems in which wholesale mutant isolation is feasible. It has this potential because the alternative morphogenetic models lead to very different genetic predictions. If the localized determinant model is correct, then it should be possible to find mutants that produce individually aberrant determinants. Such mutants should produce eggs that always give rise to *specific* or localized abnormalities of structure or of germ layer derivatives, while producing initially normal embryonic development throughout the rest of the system. If, however, the morphogenetic system consists of one or more molecular gradients, mutants of the fate-specifying system should always show *generalized* effects on morphogenesis, either over large regions or throughout the early embryo. The analysis can be complicated by the existence of mutants defective in genes for general metabolic functions, which can create generalized disturbances. However, such mutants can be eliminated from consideration by various secondary tests. The genetic analysis of morphogenetic systems is feasible in both the nematode and the fruit fly, whose embryos seem to show a mosaic form of determination, because of the ease of mutant isolation in these animals.

The point of departure for such analysis is the collection of mutants that are affected in oogenesis. Such mutants will, in general, exhibit female sterility, and those whose effects are limited to oogenesis will show no other outward mutant phenes. Not all female-sterile mutants are defective in oogenesis, of course, some being behavioral or egg-laying mutants, but inspection of each mutant can usually quickly identify the putative oogenesis defectives. Among the true oogenesis defectives, some will be defective directly in the germ line cells (the oocytes or oocyte precursor cells) and other will be defective in one or more of the accessory helper cell types.

Oogenesis defectives can also be classified with respect to the number of eggs that are laid. Those with little or no egg production, low fecundity mutants, are generally defective in some early step of oocyte production. Fecund mutants produce normal numbers of eggs, but these eggs yield aberrant development. This latter group is the most informative for probing the intricacies of the relationships between oocyte structure or composition and subsequent embryonic development. The defining genetic characteristic of these mutants is that defective embryonic development ensues only when the maternal parents are homozygous for the (recessive) mutations, irrespective of the genotype of the paternal parent. Such mutants are termed "maternal effect mutants."

The combinations of crosses that identify maternal effect mutants are as follows (where *m* stands for the mutation and + for the wild-type allele):

1 $+/+$ females $\times$ m/m males $\rightarrow$ $+/m$ progeny, normal development.

2 $m/+$ females $\times$ $+/+$ males $\rightarrow$ $+/m$ and $+/+$ progeny, normal development.

3 m/m females $\times$ $+/+$ $\rightarrow$ $+/m$ progeny, defective or altered development.

Thus, crosses 1 and 3 both yield heterozygous progeny but give dramatically different results. Development is unaffected when the fathers are homozygous for the mutation but is defective when the mothers are.

Why should the critical genotype be that of the (diploid) mother and not that of the (haploid) egg itself? In the first place, the helper cells in oogenesis are diploid, and any mutational lesions that create their effects through these cells must do so via the diploid maternal genotype. In the second place, the oocyte itself is diploid throughout most of oogenesis. For the great majority of oocyte types, it is only after fertilization that the oocyte completes meiosis to produce a haploid pronucleus. (Technically, an oocyte becomes an egg cell only when meiosis is completed.) In mothers that are heterozygous for an oogenesis mutation (cross 2), the wild-type allele usually supplies enough of the normal product to permit full embryonic development. It is because heterozygous mothers can produce viable embryos that it is possible to isolate m/m mothers in the first place; they are produced in crosses between heterozygous ($m/+$) mothers and either homozygous mutant (m/m) or heterozygous ($m/+$) fathers and can develop normally because of the initial supply of wild-type gene product in their maternal parent.

One of the first maternal effect mutations to be discovered illustrates the basic genetic properties of such mutants. The mutation affects the direction of shell coiling in snails. When one views snail shells from the top, one can see that the shell spirals in either a clockwise pattern, "dextral coiling," or in a counterclockwise pattern, "sinistral coiling." The internal organs follow these respective patterns of coiling. Thus, the two forms, when found in the same species, are mirror images. In the species *Limnaea peregra,* the most common form is dextral but a variant form is sinistral. The sinistral mutation behaves as a simple Mendelian recessive, but its inheritance pattern reveals that the direction of shell coiling is not a function of the individual snail's own genotype but of that of its maternal parent (Boycott and Diver, 1923; Sturtevant, 1923). The two reciprocal crosses that reveal this maternal-dominated inheritance pattern are shown in Figure II.5. When the eggs of dextral ($+/+$) individuals are fertilized by sinistral allele (s)-bearing sperm, all the progeny are dextral. When the reciprocal cross takes place, the fertilization of eggs from s/s individuals by $+$-bearing sperm, all the progeny are sinistral in phenotype, although dextral in genotype ($+/s$), as in the previous cross. When the dextral genotype F_1 progeny from both crosses self-fertilize themselves—snails being hermaphroditic and capable of self- or cross-fertilization—both F_2 broods are dextral in phenotype (in accordance with the maternal genotype). However, self-fertilization of these dextral phenotype F_2 individuals reveals a classic 3 : 1 segregation of dextral : sinistral broods in the F_3, each brood showing the maternal genotype of the parent from which it derived.

What is the developmental basis of this maternal effect on shell coiling? It stems from a maternally based difference in cleavage. In gastropods, cleavage is spiral: the first two divisions produce a quartet of nearly equal-sized blastomeres within one plane but the third cleavage divides each blastomere into a large "macromere" and a small "micromere," with the micromeres situated obliquely and in a plane above the macromeres. The micromeres are therefore offset with respect to their larger sister cells, thus giving a spiral pattern. The dextral and sinistral alleles create mirror-image cleavage spindle placements with respect to one another, beginning with the first cleavage. In third cleavage embryos from dextral genotype mothers, the result of this mirror-image symmetry is that the micromeres are placed clockwise with respect to the macromeres, while in embryos from sinistral mothers, the micromeres are placed counterclockwise. Since the internal organs are derived from the large "D" macromere, the mirror-image symmetrical placement of this macromere with respect to the other blastomeres determines the mirror-image shell coiling seen in the two forms (Fig. II.6). The genetic function that is ultimately responsible for these shell coiling patterns is clearly involved in topogenesis; it somehow determines the positioning of the mitotic spindles during cleavage.

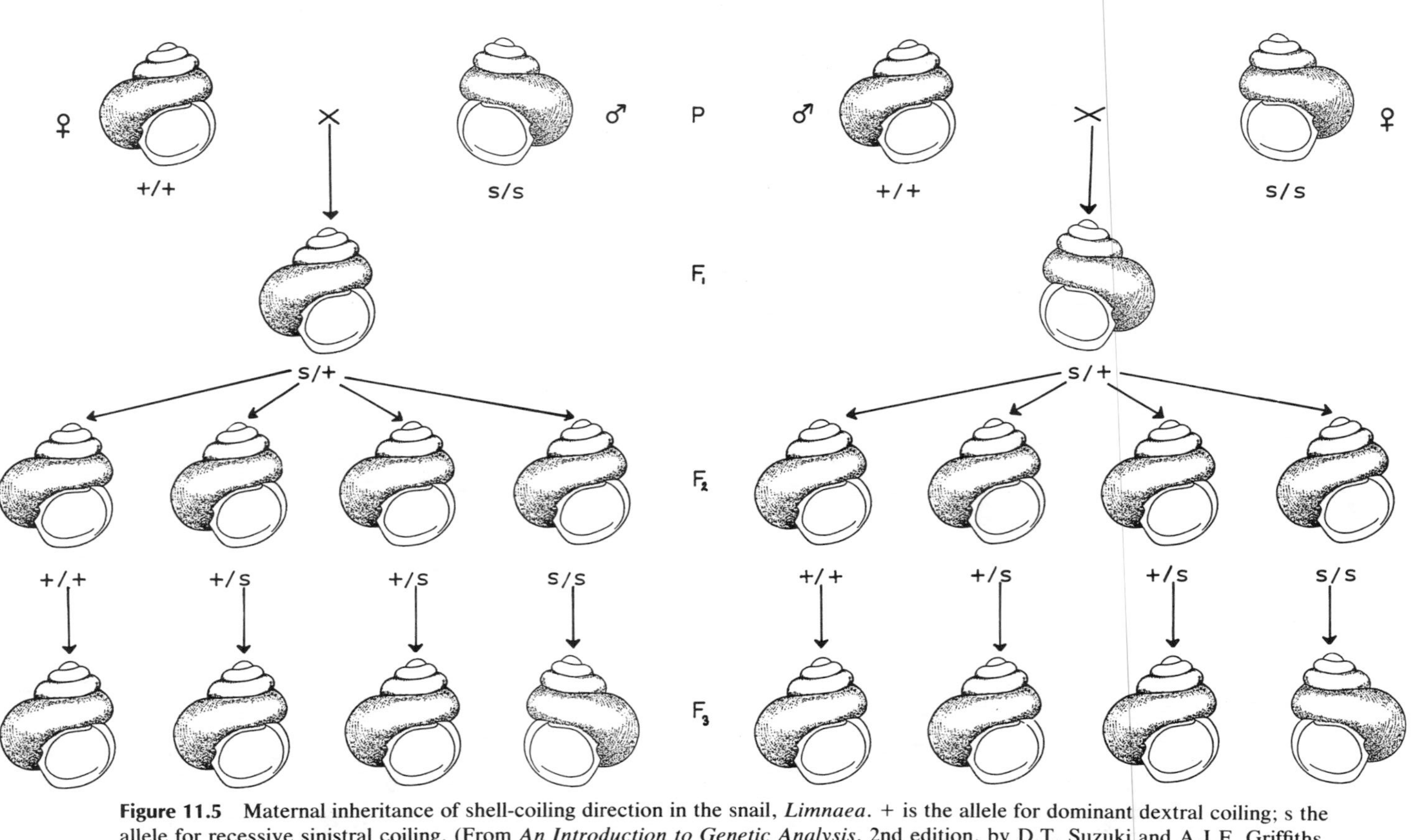

Figure 11.5 Maternal inheritance of shell-coiling direction in the snail, *Limnaea*. + is the allele for dominant dextral coiling; s the allele for recessive sinistral coiling. (From *An Introduction to Genetic Analysis*, 2nd edition, by D.T. Suzuki and A.J F. Griffiths. W.H. Freeman and Company. Copyright © 1981).

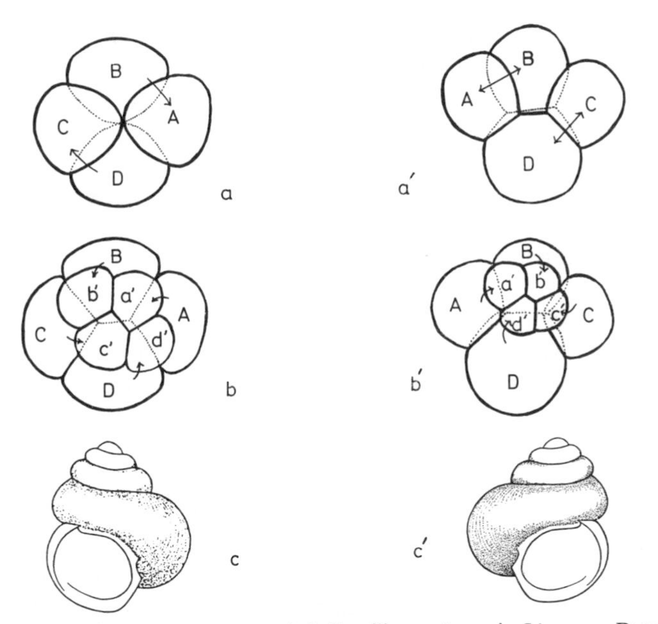

Figure II.6 Early embryonic cleavage and shell-coiling patterns in *Limnaea*. Details provided in text. (Adapted from Morgan, 1934. © 1934, Columbia University Press. Reprinted by permission.)

This topogenesis gene in *Limnaea* is an example of a *strict* maternal effect gene. The mutant has no other effect on the biology of the animal, and as far as can be determined, is not expressed at any stage other than oogenesis. It is also one of the few examples of a maternal effect mutation that, while altering development, does not cause it to be defective.

There is another large category of oogenesis gene functions, those that are expressed maternally during oogenesis and during later stages of development as well. Completely deficient, or amorphic, mutants of such genes would be standard lethal mutants, but if the gene product for any such gene is required in much larger amounts in early development rather than later, a hypomorphic mutant of such a gene might appear to be a maternal effect mutation (Bischoff and Lucchesi, 1971). Such "leaky" mutations would permit the survival and development of homozygous female progeny, but these females would subsequently produce defective progeny because of their inability to supply sufficient gene activity during oogenesis. Such maternal effect mutations do indeed exist, and can be denoted as "partial" maternal effect mutations to distinguish them from the previous class.

An example of a partial maternal effect gene is *deep orange* (*dor*) in *Drosophila*. The original mutant was observed to have two effects: it altered the eye color of the fruit fly from dark red to orange and it had a maternal effect on development (Counce, 1956). *dor* is similar to other maternal effect mutations in showing its effect on embryonic development only when the female parent is homozygous, but it differs from most of the strict maternal effect mutants in that a small proportion of the embryos from *dor/dor* mothers × wild-type fathers complete development and these survivors are invariably found to be heterozygous (+/*dor*) daughters. Mapping reveals the gene to be sex-linked, that is, to be located on the X chromosome. In consequence, matings of *dor/dor* females with +/Y males yield two kinds of progeny, heterozygous (+/*dor*) daughters and *dor*/Y males. (In X-Y sex-determining systems, sons always get their X chromosome from their mothers, and daughters get one X chromosome from their mother and the other from their father.)

The fact that some daughters survive while the *dor* sons never survive must mean that the wild-type (+) allele in the female progeny has some activity that *can make up for the oocyte deficiency*. [Rescue can also be provided by injection of wild-type cytoplasm into homozygous *dor* embryos from homozygous mutant mothers (Garen and Gehring, 1972; Marsh et al., 1977).] The expression of the wild-type allele by the incoming sperm shows that expression of the gene is not limited to oogenesis. In fact, expression of *dor* probably takes place throughout development. More severely defective mutants of the gene show not only the eye color and female sterility phenes but also effects on larval viability, adult female longevity, and a lethal interaction with a second eye color mutation (*rosy, ry*) (Bischoff and Lucchesi, 1971).

The case of *dor* is not unusual; many genes in *Drosophila* can produce a spectrum of mutant alleles with differing phenotypic effects. Perhaps the best understood is the *rudimentary* (*r*) gene, which was first identified on the basis of its mutant's effects on wing shape. The *r* mutants also have maternal effects, and both the wing and maternal effects are explicable in terms of this locus' role in pyrimidine metabolism: the gene encodes a large single polypeptide that specifies three essential enzyme activities in pyrimidine metabolism. Not surprisingly, the maternal effect can be alleviated directly by pyrimidine injection into the early embryo (Okada et al., 1974a).

The examples of *dor* and *r* illustrate the difficulty in trying to analyze morphogenetic systems in eggs by the study of maternal effect mutants. In order to find the mutants of interest, one may initially have to sort through a large collection of partial maternal effect mutants, many of whose defects are in general cell functions. It is the remaining mutants that provide the essential source material for a genetic analysis of morphogenetic systems.

TWO

CAENORHABDITIS ELEGANS

FROM OOCYTE TO EARLY EMBRYO

The ease of handling of the nematode coupled with its small genome size suggests that it is feasible to look for mutants in all of the genes to try to discover how they participate in the development and functioning of a simple multicellular organism.

S. Brenner (1974)

Caenorhabditis elegans, a small, free-living nematode, occupies a special place in the bestiary of developmental genetics. Among animals complex enough to have a true neuromuscular system and a corresponding set of behaviors, it is one of the very simplest. The adult is just over 1 mm in length and consists of only a few thousand cells. Furthermore, its generation time is extremely brief (just over 52 hours at 25°C), which is a great advantage for experimental work.

Furthermore, this organism is readily accessible to genetic and molecular analysis. Developmental and behavioral mutants can be obtained with ease, and, once isolated, such mutants can be readily mapped and classified by a variety of genetic techniques. More than 200 genes have already been mapped and subjected to at least preliminary genetic analysis; this number corresponds to about 5% of the functions of the genome. Further, the genome is sufficiently simple to lend itself to a thorough molecular analysis by recombinant DNA techniques. As methods for isolating desired genes become more refined, more and more of the genome will be open to combined genetic and molecular characterization.

Finally, it is possible to trace the origins of every somatic cell during the animal's development by a variety of microscopic and computer reconstruction techniques. These methods permit a complete cataloging of cellular behaviors during the development of both the wild-type and mutant strains.

It is this combination of inherent simplicity and accessibility to genetic, molecular, and developmental analysis that makes *C. elegans* a particularly favorable subject for the genetic dissection of development. In this chapter, the first, maternally directed, stages of development in this animal will be examined and the genetic strategems that have been employed to investigate them discussed. Chapter 5 will consider the later stages of development, which are governed by the zygotic genome.

THE LIFE CYCLE IN OUTLINE

In a wild-type population of *C. elegans,* most adults are self-fertilizing hermaphrodites. However, there are rare males that can mate with the hermaphrodites. The life cycle of the hermaphrodite is summarized in Figure 2.1. The complete sequence, from fertilized egg to emergence of the adult, includes five distinct phases of approximately equal duration. These periods are the initial stage of embryogenesis, which takes place inside the egg shell, and four successive larval stages. Each larval stage transition is preceded by a 2-hr period of reduced activity, termed the "lethargus," which is followed by a "molt," the shedding of the larval cuticle.

Oocyte formation takes place in the ovary of the hermaphrodite, and fertilization takes place as the oocytes pass through a special sperm receptacle, the spermatheca. The first cleavage divisions of the embryo take place

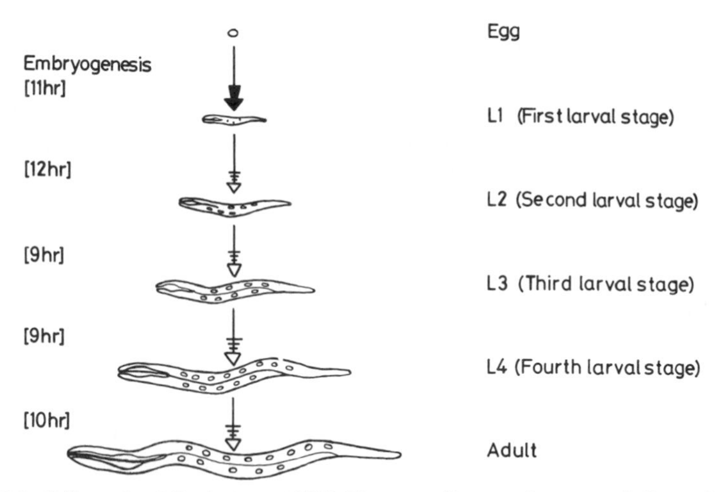

Figure 2.1 Life cycle of *C. elegans* at 25°C. The cross-lines on the arrows indicate the periods of lethargus.

as the zygote first moves out of the spermatheca and then down the oviduct toward the external opening, the vulva. Egg laying usually occurs at the 30-blastomere stage, about three hours after fertilization. Cleavage continues, and is accompanied by a progressive diminishment of cell size, with little obvious cellular differentiation until the embryo has reached the 540-cell stage, approximately five and a half hours after fertilization (Schierenberg et al., 1980). The end of cleavage is followed by the second phase of embryogenesis, lasting for five to six hours, which consists of a sequence of cellular differentiations and morphogenetic movements that transform the embryo from a round ball of cells into an elongated, twisted early larval form (the "pretzel" stage). Some of the intermediary morphogenetic stages during this transformation have been descriptively labeled the "comma," "tadpole," "plum," and "loop" (Krieg et al., 1978); a few of the stages of embryogenesis are shown in Figure 2.2.

The newly hatched first stage larva (L1) is very similar in external morphology to the adult, but only one-quarter as long. The four stages of larval growth involve two principal kinds of change: a substantial increase in size (through cellular growth and division) and the attainment of sexual maturation (involving the development of the gonads and the secondary sexual characteristics).

In addition to the sequence of events in the main life cycle, there is an alternative pathway that the organism can take, namely, dauer larva formation. Development of the dauer larva, a nearly dormant larval form, takes place at the second molt if the animals have been subjected to starvation or overcrowding. The dauer larva does not feed and is inactive except for a touch stimulus response. Dauer larvae are slightly longer but only half as wide as normal second stage larvae (Fig. 2.3).

GENETICS AND GENOME STRUCTURE

The diploid chromosomal constitution of the *C. elegans* hermaphrodite consists of 12 chromosomes: 5 pairs of autosomes and 2 X chromosomes. The genotype is symbolized as 5AA + XX. The rare males have 11 chromosomes, the set consisting of 5 pairs of autosomes and a single X. The male genotype can be symbolized as 5AA + XO. Male progeny result from occasional errors in pairing and separation that occur during meiosis in gamete formation. Such errors of meiotic nondisjunction produce gametes that lack X chromosomes. When such nullo-X gametes are fertilized by normal single X-bearing gametes, the result is an XO individual, a male. The frequency with which males are obtained can be considerably enhanced, however, by the use of certain mutations that cause an elevated rate of nondisjunction (Hodgkin et al., 1979). In a normal population, it occurs with a frequency of about 0.2%.

Self-fertilizing hermaphroditic organisms possess some advantages for

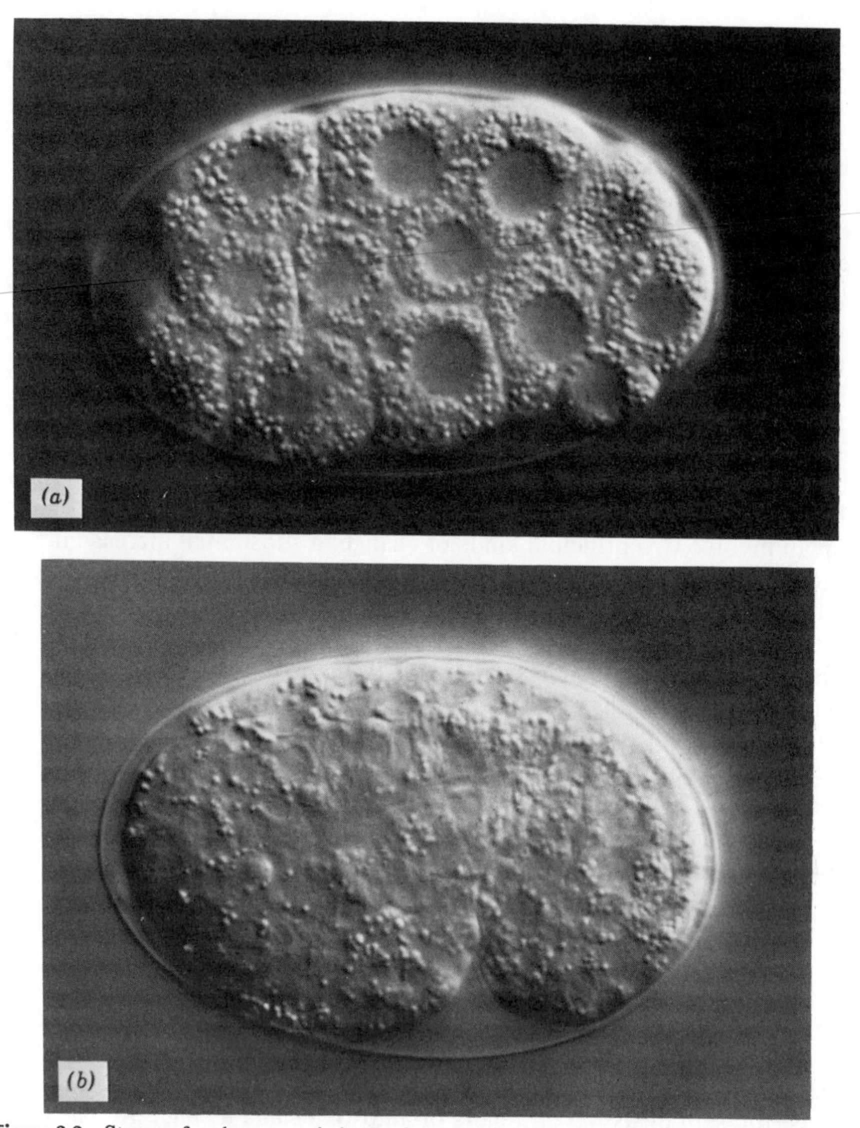

Figure 2.2 Stages of embryogenesis in *C. elegans*. (*a*) A 28-cell stage, beginning of gastrulation; (*b*) comma; (*c*) plum; (*d*) late pretzel. (Kindly provided by Dr. J.E. Sulston; reproduced with permission from Sulston et al., 1983).

genetic analysis but pose a potential problem. The advantage is that mutants can be readily isolated as homozygotes, without extensive backcrossing. For any hermaphrodite that is heterozygous for a new mutation, one-fourth of its progeny will be homozygous for the mutation. In contrast, organisms with a conventional sexual system must be put through two crosses—an initial outcross and then a cross between siblings—to isolate homozygotes of new

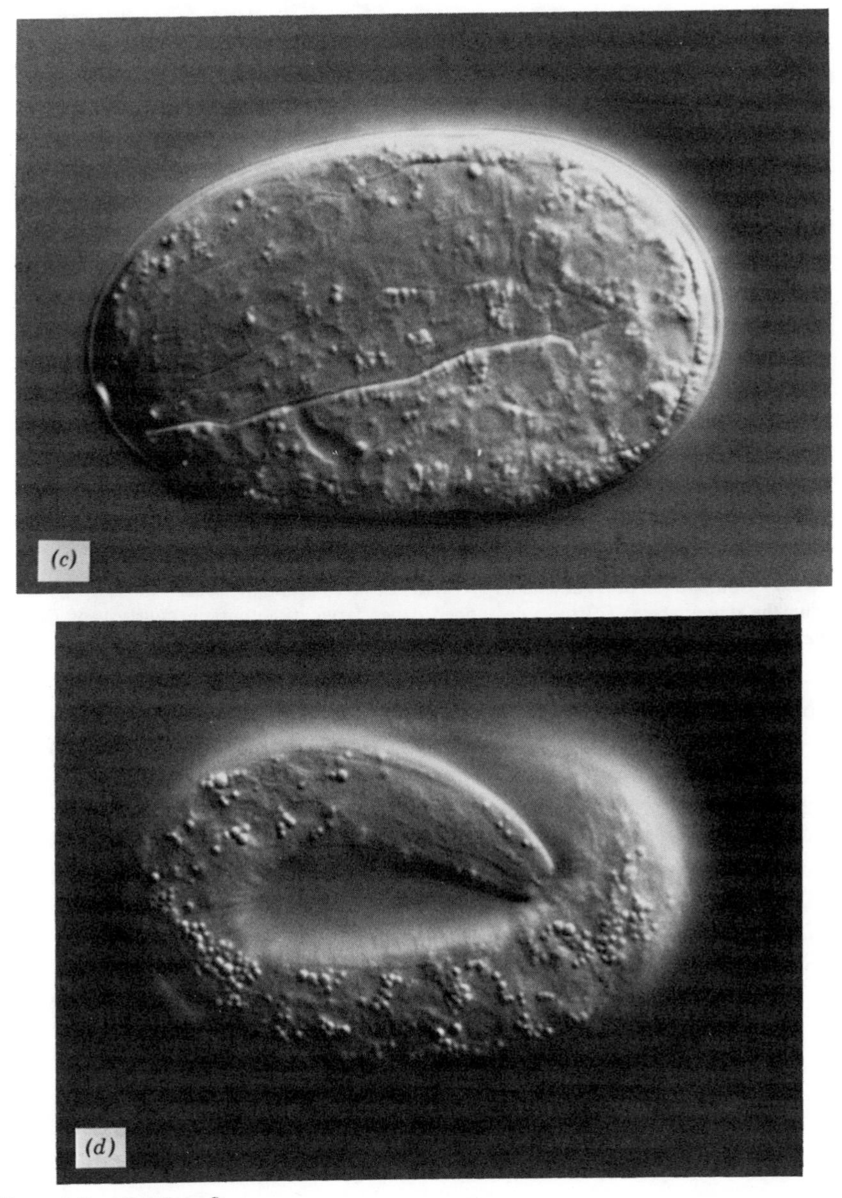

Figure 2.2 (*Continued*)

recessive mutations. This difference between the two sexual systems is illustrated in Figure 2.4. A second advantage of the hermaphroditic system in *C. elegans* is that many of the mutants obtained are sufficiently abnormal in their morphology to make mating difficult. Because fertilization in the hermaphrodite is internal, stocks of such mutants can be readily perpetuated.

The potential disadvantage of the self-fertilizing hermaphrodite system is

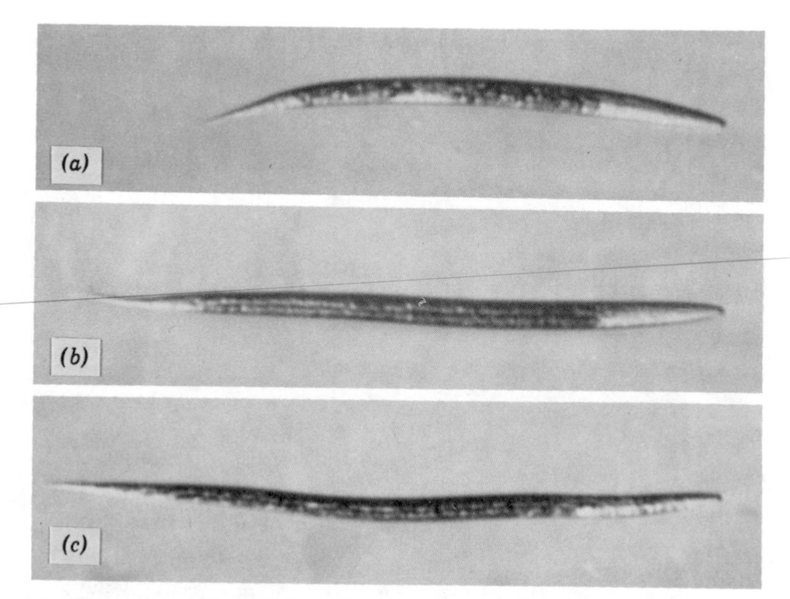

Figure 2.3 Comparison of L2 and dauer larva. (*a*) L2 larva beginning second molt; (*b*) L2 larva entering the dauer stage; (*c*) fully developed dauer larva. (Photograph courtesy of Dr. D. Riddle; from Golden and Riddle, 1984.)

the obverse of its strength: if the mutant hermaphroditic individuals are altogether incapable of outcrossing, one cannot perform any of the standard genetic tests on them, since all of these tests require crosses of some sort. Fortunately, the existence of males that can mate with hermaphrodites circumvents this problem in *Caenorhabditis*. When males are mated with homozygous mutant hermaphrodites, half of the sperm of the male are nullo-

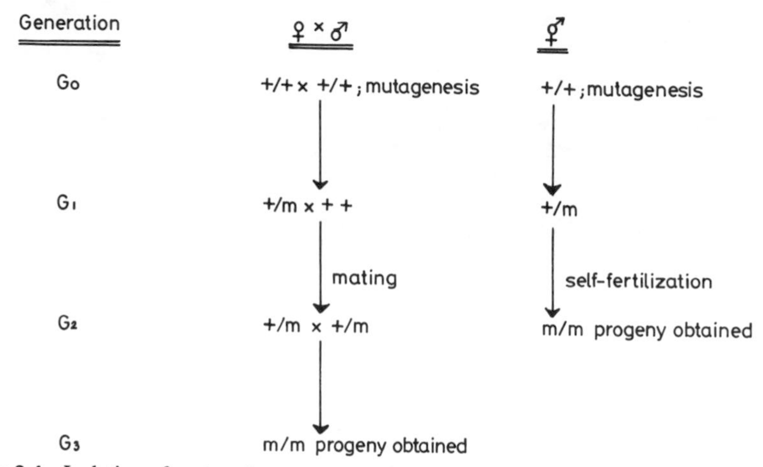

Figure 2.4 Isolation of mutant homozygotes compared in (*a*) hermaphroditic (right) and (*b*) standard sexual systems (left).

X. Thus, approximately half of the progeny will be males, and these will be heterozygous for the mutation originally carried in the mutant hermaphrodites. In such crosses, however, care must be taken to ensure correct discrimination of the outcrossed progeny from any progeny produced by self-fertilization. This can be accomplished by use of a second distinguishing marker in the original hermaphrodite stock. In outcrossings from this stock, none of the outcross progeny will display this second recessive marker. In *C. elegans,* this complication is further minimized because mating greatly suppresses self-fertilization in the hermaphrodite through preferential utilization of the male's sperm (Ward and Carrel, 1979).

The standard procedures for genetic analysis of *C. elegans* were developed by Brenner (1974). An expanded and revised treatment has been given by Herman and Horvitz (1980). Although the organism lacks the impressive array of visible traits that can be altered by mutation, such as those found in *Drosophila,* it can give rise to several obvious variants, which can be observed with a dissecting microscope. A few of the more useful morphological variations are "dumpy," "small," and "long" phenotypes, symbolized respectively as Dpy, Sma, and Lon. Another class is "blister" (Bli), which is distinguished by fluid-filled blisters in the cuticle. A few mutant types are shown in Figure 2.5.

Equally useful are certain behavioral variants that can be spotted on the basis of their abnormal movements. The wild-type adult moves with a regular, sinusoidal movement in the dorsoventral plane. The movements can be either forward or reverse. Following mutagenesis and segregation of the homozygotes, a relatively common class of mutants is one that have less regular movements. These have been categorized broadly as "uncoordinated," or Unc, mutants, and are affected in either the neural or muscular systems but possess normal pharyngeal muscle movements and hence can feed and therefore survive. A second class with altered movement are the "roller" (Rol) mutants. These mutants rotate around the long axis of the body, producing a circular motion. When cultured on a lawn of bacteria, the standard food, Rol mutants can be distinguished as those that make craters in the lawn.

In addition to phenotypic characterization, a detailed genetic characterization of any new mutant is required. The characteristics of the mutation that must be determined include its recessivity/dominance, the chromosome it is located on, and its precise position ("map location") on that chromosome. In addition, the identity of the affected gene with respect to other mutations that produce similar phenotypes and that map in nearly the same position must be established. All of these genetic tests are essentially similar to those in other animals, but involve slight modifications because of the hermaphroditic nature of the sexual system in this organism.

To determine whether a mutation, *a*, is recessive to the wild-type, homozygous mutant hermaphrodites, *a*/*a*, are crossed with wild-type males, symbolized +/+, and the cross progeny, +/*a*, are scored for the mutant pheno-

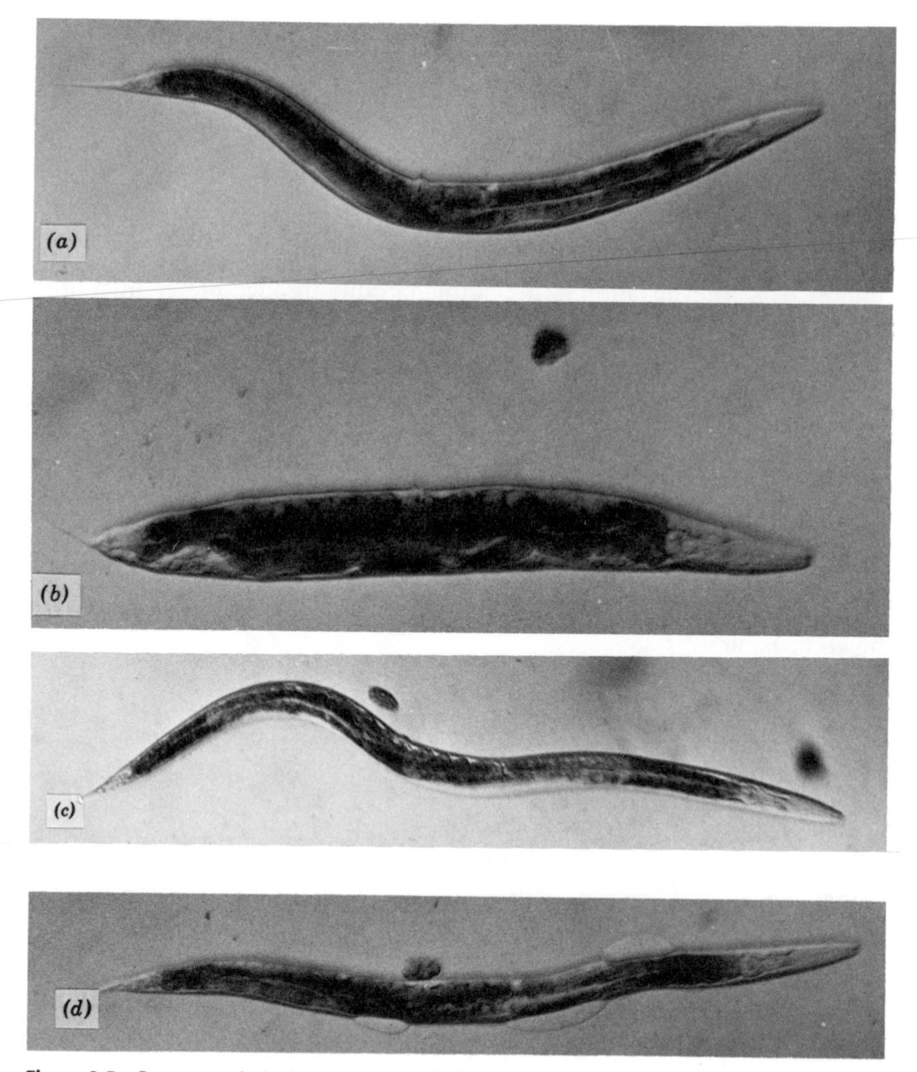

Figure 2.5 Some morphological mutants of *C. elegans*. (*a*) An *sma* mutant; (*b*) a *dpy* mutant; (*c*) wild-type; (*d*) a *bli* mutant. Magnificantion: X133. (Photographs kindly supplied by J.A. Hodgkin.)

type. Most mutations are found to be recessive to the wild-type or weakly semidominant. If only mutant sons and wild-type daughters are obtained in such crosses, then the mutation is identified as a sex-linked one, that is, on the X chromosome.

If the mutation is not on the X, then it must be on one of the five autosomes. Each chromosome is sufficiently small to comprise a single linkage group; therefore, if a new mutation is shown to be linked to a marker on a known linkage group, its linkage group is thereby identified. To perform this test for linkage to a marker, the *trans* double heterozygote, $a+/+b$, is constructed in which b is a mutant gene producing a different phenotype on that linkage group. If segregation from the hermaphrodite produces the classical dihybrid 9 : 3 : 3 : 1 segregation ratio, the genes cannot be linked and the new mutation must be on one of the other chromosomes. If an excess of wild-type progeny and a deficit of the double mutant phenotype are found, then linkage of the two genes is indicated. Once a gene is identified as being on a particular chromosome, its genetic distance with respect to other genes on that chromosome is established by genetic mapping procedures. However, genetic mapping in a self-fertilizing hermaphrodite involves some complexities not seen in standard sexual systems (see the Appendix).

Mutations producing similar phenotypes that are found to map at or near the same position on a linkage group may be either in the same gene or in neighboring genes with a similar function. To determine whether two mutations are in the same gene or not, a standard "complementation test" is performed. The rationale behind this test is that if the mutations are in different, although neighboring, genes, the double heterozygote will have one functional dose of each gene and will be wild-type, whereas if the mutations are both in the same genetic unit, the double heterozygote will have no wild-type copies of the gene and will be phenotypically mutant (Fig. 2.6).

The rules of nomenclature for *C. elegans* have been summarized by Horvitz et al. (1979). Briefly, phenotype classes are given by three-letter designations, beginning with a capital letter (e.g., Dpy, Sma) and gene names are given by three-letter, italicized designations, all in lowercase (e.g., *dpy-1*, *lon-2*). For ease of reference, the linkage group number (I to V and X) can be

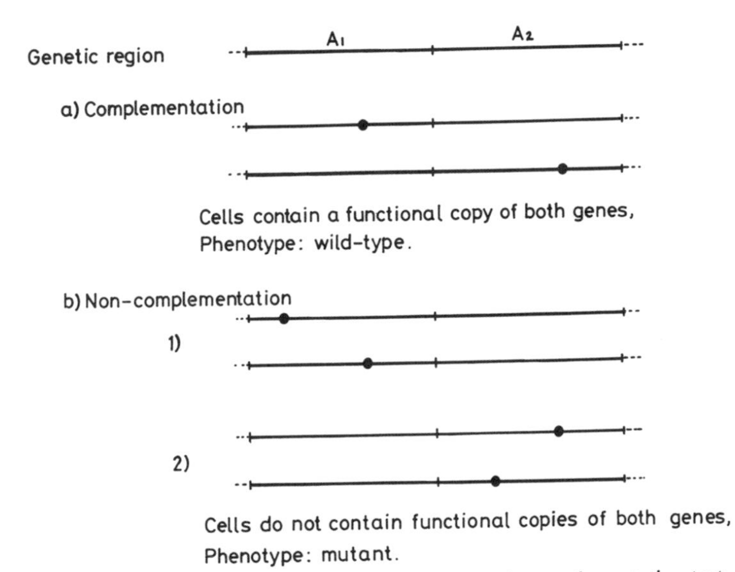

Figure 2.6 Diagrammatic representation of a genetic complementation test.

appended (e.g., *dpy-1 III*). A partial map of the *C. elegans* genome is given in the Appendix.

Any genetic map, of course, is purely a formal representation of the relative distances between genes, based on the relative frequencies with which recombination events during meiosis separate genes. A genetic map provides no precise information about the physical locations of particular genes on chromosomes or of the physical details of gene organization. For such information, one must turn to molecular studies of genome structure. With respect to the genome of *C. elegans,* little can be stated with certainty.

The few observations that are solid concern the overall size of the genome and the general pattern of interspersion of single-copy and repetitive sequences. The total genome size is 8×10^7 base pairs per haploid genome complement, the smallest animal genome yet measured (Sulston and Brenner, 1974). The number of genes contained within the genetic complement is unknown but is probably between 2000 and 4000. The lower estimate was given by Brenner (1974) on the basis of the average mutation rate per gene and the total chromosomal mutation rate to lethal mutations for the X chromosome; the calculation was based on the simplifying assumption that every gene has an essential function, one required for viability of the animal. This assumption is not literally true, since there are now documented cases of genes whose complete mutational inactivation does not produce lethality for the individual organism, but rather other phenotypes (Meneely and Herman, 1979; Greenwald and Horvitz, 1980). If one assumes that there might be as many as three times the number of genes as there are indispensable vital genetic functions (see, for instance, Chapter 3) than there may be a total of 5000–6000 genes. A number of this order suggests, in turn, an average gene size of 10,000–15,000 bp if the haploid genome of 8×10^7 is completely "filled" with genes. A gene of this size seems perfectly typical for eukaryotes. Much of a genetic unit of this size might be taken up by introns or other sequences of unknown function; for most genes, the portion devoted to protein coding would be relatively small.

A question that recurs in considering gene structure concerns the role of the midrepetitive sequences. In *C. elegans,* these sequences average about 660 bp in length and are interspersed, on average, every several thousand base pairs of single copy DNA (Emmons et al., 1980). Studies of cloned DNA sequences suggest that each family of related interspersed sequences may be present only about 10 times per haploid genome (Emmons et al., 1979); renaturation studies of DNA suggest a somewhat higher figure. For years, one line of speculation has centered on the possibility that each distinct midrepetitive sequence family acts as a set of *cis*-regulatory sites for the coordinate control of a group of protein coding genes (Britten and Davidson, 1969). Recently, evidence has come to light suggesting that there may be common regulatory *cis*-control sites for certain gene groups, but the sites that have been identified are considerably smaller than the several hundred base pair long midrepetitive units identified by renaturation studies (Davidson et al., 1983a). The existence of small sites for binding regulatory molecules is consistent with what is known of such interactions in bacteria; it is difficult to see how individual proteins could recognize the base sequence information in much longer regions. The function, if any, of the large midrepetitive sequences in eukaryotes remains an open question.

OOGENESIS AND THE EARLY EMBRYO

The classical work of Theodor Boveri on the nematode embryo around the turn of the century established this embryo as a mosaic system. Its first four cleavage divisions create a small set of blastomeres, none capable of substituting in development for any of the others and each destined to give rise to one of the major cell lineages of the animal. Boveri also obtained the first evidence for a discrete cytoplasmic determinant, one that is responsible for the establishment of the germ line. It is a tribute to Boveri's ability and to the difficulties of analyzing this embryonic system that comparatively little has been learned subsequently about the mechanisms underlying early cell fate specification in the nematode egg. Fortunately, a few clues to the operation of the topogenetic and morphogenetic systems of the egg have recently come to light from various forms of embryo manipulation and analysis. These will be discussed after we have first examined the developmental biology of egg formation in *C. elegans*.

Oogenesis

Caenorhabditis has one of the smallest eggs that have been studied, measuring 40×60 μm. Its formation has been described in detail by Hirsh et al. (1976) and is summarized below.

Oogenesis takes place in the ovaries of the hermaphrodite, each ovary occupying the dorsal and distal arm of each U-shaped gonad (see Fig. 5.9). The ovary forms a tapered cylinder of densely packed peripheral nuclei surrounding a central core of cytoplasm. The 1300 nuclei coating the inner surface of the cylinder are partially separated from each other by cell membranes but are open to the syncytial core. At the loop region of the gonad, the nuclei begin to move into the proximal arm, and each nucleus, surrounded by cytoplasm, becomes enclosed within a cell membrane. As the 10–14 oocytes move down the proximal arm in single file toward the spermatheca, they increase in size. Figure 2.7 shows a cross section of the distal arm and the loop region. The first stages of meiosis occur as the nuclei progress through the gonad. In the distal quarter of the ovary, the nuclei are predominantly in interphase; in the remainder of the distal arm, they are in pachytene (the stage of closest apposition of homologues) and are entering the loop. There, the nuclei progress to diakinesis, remaining in that phase until fertilization.

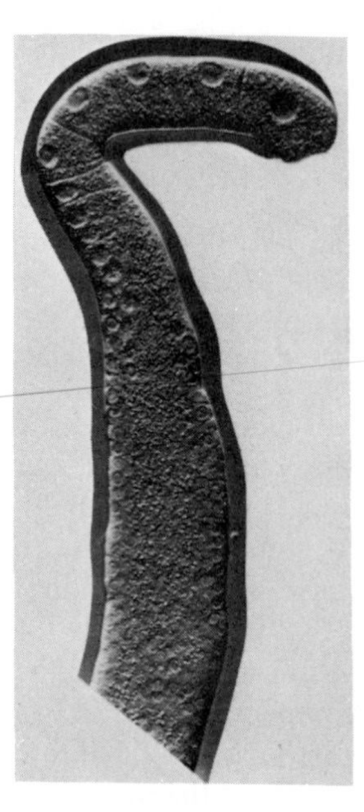

Figure 2.7 Ovary and loop region in an adult hermaphrodite of *C. elegans*. (Photograph kindly provided by Dr. D. Hirsh; reproduced with permission from Hirsh et al., 1976.)

Comparable syncytial arrangements of primordial germ cells are extremely common in the animal kingdom. They have been observed in polychaete worms, crustaceans, insects, birds, and mammals. During oogenesis in *Caenorhabditis,* and probably in other animals, this form of organization harnesses the synthetic activities of many nuclei for the rapid assembly of oocytes. The mature oocyte is approximately 250 times larger than a distal arm cell, and at 20°C, one oocyte is developed on average from its precursor cell every 40 min. This accumulation of oocyte material shows a distinct regional specialization. At the distal tip of the ovary the cytoplasm is filled with ribosomes and mitochondria, but toward the loop region there is an increasing concentration of yolk bodies and lipid droplets, the kinds of cytoplasmic inclusions that are typical of oocytes. The yolk proteins themselves, a small group of glycoproteins (Sharrock, 1983), are synthesized not in the ovary itself but in the intestine, from which they are secreted into the body cavity and taken up by the ovary (Sharrock and Kimble, 1983). The thin epithelial sheath cells that enclose the ovary may be active in the synthesis of nonyolk materials or as conduits of nutrients for biosynthesis.

The sequence in which the oocyte acquires its critical biochemical asymmetries must be related in some manner to the distinctive regional composition of the gonad. Somewhere within the gonad, the oocyte must acquire those properties that generate the topogenetic and morphogenetic systems of the early embryo. Unfortunately, the cytological observations merely indicate possible sites within the gonad where this action may take place. The proximal arm of the gonad is one such site, as oocytes move in a single stream toward the spermatheca. This linear arrangement may facilitate the development of some polarity or future developmental axis within the oocyte. The other possibility is that the significant biochemical heterogeneities are established in the loop region, as each nucleus plus its associated core cytoplasm is packaged into an individual cellular unit.

Fertilization takes place in the spermatheca, as oocytes are moved into it one at a time, and triggers numerous cytoplasmic and nuclear changes. The fertilized egg rapidly develops a vitelline membrane and a hard, transparent, chitinous shell, while the nucleus is activated to complete meiosis. The female pronucleus is located eccentrically and distally from the male pronucleus, and the two nuclei migrate toward one another, the former traversing the greater distance before fusion. Several forms of evidence indicate that the movements of the female and male pronuclei are dependent on the action of the cytoskeleton of the egg (Strome and Wood, 1983). The cytoskeleton is an elaborate set of protein fibers and filaments that crisscross the cytoplasm; all eukaryotic cells have such a cytoskeleton, although the details of its composition and structure vary greatly between different cell types. As the egg moves out of the spermatheca and down the oviduct, the cleavage divisions begin.

Early Cleavage: The Topogenetic System

The events of early cleavage reveal the remarkably precise organization of the nematode egg. The first four to six divisions normally occur within the oviduct, but eggs can be released at the one-cell stage and observed in vitro (Hirsh et al., 1976; Deppe et al., 1978).

In nematodes, the first four cleavage divisions of the zygote suffice to generate the basic organization of the developing animal. These divisions sequentially create or reveal the basic axes and polarities of the embryo and generate a special set of six blastomeres, each of which gives rise to one of the six major cellular lineages of the larva. These key lineage precursor cells have been termed "embryonic blast cells" (Laufer et al., 1980) and arise through an invariant pattern of early cleavage divisions.

The sequence of the first four divisions is shown in Figure 2.8. The designations of the embryonic blast cells—AB, MS, E, C, D, and P_4—are those employed by Boveri (1899) to denote the comparable blastomeres in *Ascaris*. The details of the six lineages founded by these cells will be described in Chapter 5, but one important feature deserves mention here. For nearly a century, it had been believed that the embryonic blast cells had a strict germ layer identity, each giving rise solely to cells of ectodermal, mesodermal, or endodermal provenance. The recent complete analysis of embryogenesis in *C. elegans* (Sulston et al., 1983) has shown this to be untrue; three of the lineages, as depicted in the figure, produce both ectodermal and mesodermal cells.

The geometry of the first cleavage divisions, constituting the pattern of topogenesis, is depicted in Figure 2.9. The very first division sets apart the

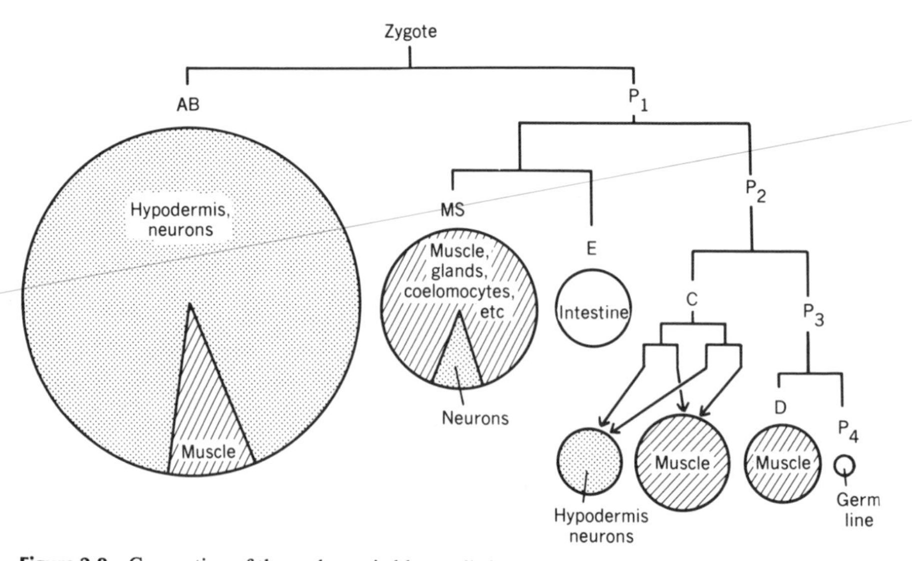

Figure 2.8 Generation of the embryonic blast cells in the first cleavage divisions of the embryo. Ectodermal derivatives indicated by stippling, mesodermal tissues by striping. (Kindly provided by Dr. J.E. Sulston; reproduced with permission from Sulston et al., 1983.)

first lineage, that of the AB line, and establishes the anteroposterior (a-p) axis and polarity of the embryo and of the future adult. The sister cell from this first division, P_1, is always smaller than the AB cell. Its division, in turn, is also asymmetric and generates a large embryonic blast cell, EMS (sometimes symbolized as EMSt), and a smaller cell, P_2. This second division, which is also a-p but with a dorsoventral (d-v) component, establishes the dorsoventral polarity of the embryo. The EMS cell divides to give two blast cells, the E cell and the MS cell, while P_2 divides asymmetrically to give the C blast cell and the smaller P_3 cell. The fourth cleavage involves the division of P_3 into the D blast cell and P_4, the latter being the direct precursor blast cell of the germ line. The beginnings of left-right (l-r) asymmetry take place in the third division by oblique divisions within the AB lineage. In each P-cell division, the P daughter is always smaller than its sibling cell, and by the 16-cell stage, all major lineages have been established.

The orientation and timing of the ensuing divisions are strictly controlled (Deppe et al., 1978; Sulston et al., 1983). Each cell in every lineage shows a fixed orientation of division, either a-p, d-v, l-r, or oblique, but showing a major directional component with respect to the major axes. The division program is thus strictly constant until the end of cleavage, when the embryo consists of about 540 viable cells. (A certain number of additional cells are formed but are eliminated by programmed cell death.) A sequence of cellular migrations begins at the 28-cell stage, when the two cells of the E lineage (the precursor cells of the intestine) migrate internally. This movement is the first

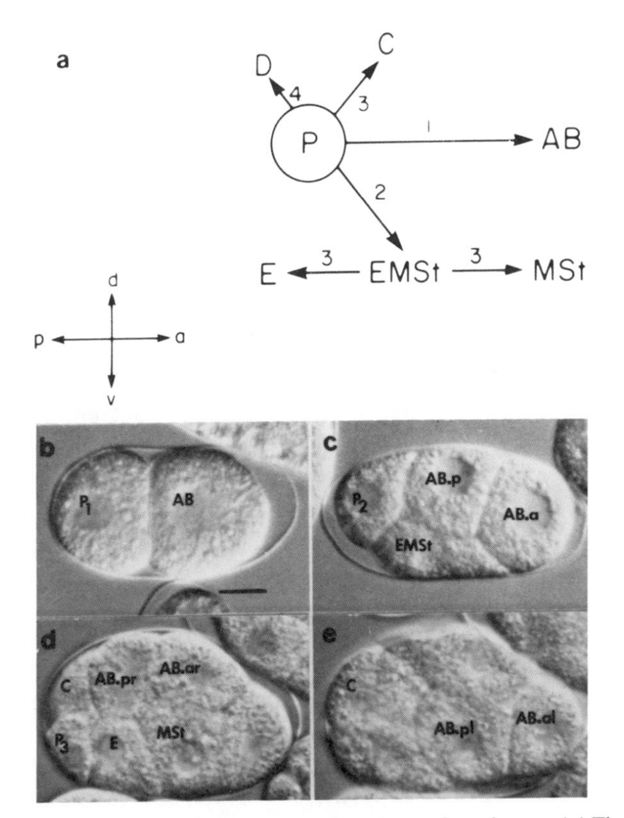

Figure 2.9 Pattern of topogenesis in first cleavage divisions of *C. elegans*. (*a*) The directions of cleavage and the positions of the daughter cells following division; (*b*) two-cell embryo; (*c*) four-cell embryo; (*d*) eight-cell stage focused in the upper plane; (*e*) eight-cell embryo focused in the lower plane. All embryonic cells are designated by their blast cell precursor (e.g., AB, E), followed by the sequence of division positions of the cells from which they derive; for example, AB.al is the left-hand daughter of the anterior daughter of blast cell AB. (EMSt = EMS cell.) (Figure kindly provided by Dr. J. Laufer; reproduced with permission from Laufer et al., 1980; copyright MIT Press.)

step in nematode gastrulation, a sequence of cellular invaginations and external cell migrations. Throughout cleavage, there is little marked cellular differentiation, with the notable exception of the E lineage, whose eight cells at the 150- to 200-cell stage can be recognized by the presence of fluorescent rhabditin granules (particles composed of tryptophan breakdown products) (Babu, 1974).

This sequence of precise cleavage divisions indicates the existence of an underlying program of topogenesis. To what extent is the observed division pattern the result of autonomous cellular properties of individual blastomeres, and to what extent is the pattern determined by cellular interactions between blastomeres or between cells and the enclosing egg shell? The process has been analyzed by bursting eggs and allowing the early embryos to develop free of the shell (Laufer et al., 1980). Eggs are burst by applying pressure through a cover slip to the underlying embryo. In this process, one or more of the blastomeres will lyse, producing partial embryos. These embryos continue to grow and divide if placed in a medium containing nematode coelomic fluid.

Observation of the partial embryos reveals two characteristic division patterns. Partial embryos derived from AB or E embryonic blast cells show successive divisions at a constant oblique angle; the result is initially a helical cell array and ultimately a ball of cells. Partial embryos derived from the P lineage, however, show a predominantly linear division pattern. The data suggest that the cleavage program of the intact embryo reflects the combined action of these two intrinsic division programs, in conjunction with interactions between cells of different lineages and possibly the egg shell. Some of the cellular interactions inferred from the study of partial embryos appear to result from purely spatial constraints—the C blast cell, for instance, seemingly pushed to one side by the cells of the AB lineage—while others may involve selective migrations of one cell type over or around cells of other lineages. One example of the latter is the migration of AB cells over and around E and MS cells, since the partial embryos consisting wholly of AB cells display no comparable cell movements. Such cell-guided cell movements presumably involve and require specific cell surface differences between lineages (Laufer et al., 1980).

The basis of the difference in division pattern between the P and other lineages is unknown but may be related to size, given that P cells in the first divisions are always smaller than their sister embryonic blast cells. This hypothesis is supported by the finding of one maternal effect mutant in which the first division is symmetrical, yielding two equal-sized blastomeres. In this mutant, both cells exhibit the helical AB-type pattern seen in partial embryos (cited in Laufer et al., 1980). The observations, taken as a whole, show that the topogenetic sequence is complex, involving several factors and mechanisms.

The Morphogenetic System of the Egg

In classical embryology, the mosaic behavior of an embryo—the inability of blastomeres to substitute developmental roles—was traditionally equated with a mosaic organization of morphogenetic determinants in the egg. This inference is unwarranted because the presence or absence of regulative ability in blastomeres need reflect nothing of the initial process by which fates were originally imposed. In the case of the nematode egg, however, at least the germ line appears to be founded by a qualitatively discrete cytoplasm. The evidence is connected with Boveri's discovery of "chromatin diminution" in the nematode *Ascaris*. In the embryo of this animal, determination of the germ line is always coupled to the inheritance of a particular distinguishable cytoplasm: at each step in cleavage, this cytoplasm is passed to one blastomere at a time, with all of the other blastomeres undergoing the process of partial chromosome fragmentation named by Boveri, chromatin diminution. Redistributing this cytoplasmic material to additional blastomeres by experimental manipulation spares all the cells that inherit it from diminution (Boveri, 1910). In normal development, only one blastomere at the 16-cell stage receives this cytoplasm, and this single cell, possessing a full chromosome complement, gives rise to the germ line. The behavior of the germ line specific cytoplasm fits the traditional picture of a qualitatively unique morphogenetic determinant.

C. elegans shows neither visible signs of chromatin diminution in the somatic cell lineages nor evidence of selective DNA loss in these cell lines when tested by DNA analytical methods (Sulston and Brenner, 1974; Emmons et al., 1979). However, the early germ line cells do show characteristic cytoplasmic granules that are not evident in the other somatic lineages (Krieg et al., 1978). These granules possess a unique antigenic determinant, detectable by immunofluorescence, using a fluorescent compound-tagged antibody. Initially, this unique germ line antigen is present throughout the egg but is rapidly segregated to the posterior pole after fertilization and is subsequently concentrated in the cells of the P lineage (Strome and Wood, 1982); the few granules that remain outside P_4 are apparently degraded. The initial segregation of the P granules to the posterior pole is mediated by some process dependent upon the functioning of the actin filaments in the cell, one major contractile component of the cytoskeleton of the egg, rather than by the microtubules, a second major component of the cytoskeleton (Strome and Wood, 1983). Whether the P granules cause germ line determination or are merely associated with its segregation is unknown.

It is also unknown whether cell fate specification in the other lineages is associated with qualitatively distinct cytoplasmic components. One line of evidence suggests that it is not. When small holes are punched in blastomeres with a laser microbeam, cytoplasm is extruded, but a large percentage of such blastomeres can heal, permitting subsequent development (Laufer and von Ehrenstein, 1981). Even when large amounts of cytoplasm and membrane are removed from given blastomeres, many embryos survive and develop into normal fertile adults. This shows that despite the inability of whole blastomeres to substitute for one another, individual blastomeres apparently have a high tolerance to damage. Even more significantly, when cytoplasms are mixed by laser-mediated fusion, there is no transfer of developmental capacity between cells (Wood et al., 1983).

Perhaps the most striking clues to the operation of the morphogenetic system lie in the relationship between blastomere size and division rate to fate assignments. The possible importance of size is indicated by the asymmetry of daughter cell sizes in every P cell division, in which the P daughter is always smaller than its blast cell sib. In contrast, initial divisions within every blast cell lineage are symmetric with respect to daughter cell size. It is difficult to escape the impression that the geometry of division or the size of the daughter cells is connected in some way to the establishment of differences in developmental potential, at least between the P lineage and the other five.

Another connection between the early division process and the morphogenetic instructions of the egg involves the rate of cell divisions. In every P cell division that generates a somatic embryonic blast cell, the P cell daughter exhibits a different cell division rate from that of the sibling blast cell, whose progeny, in turn, show synchronous and characteristic division rates. Apparently a direct relationship exists between these lineage-specific division rates and the point of origin of the precursor blast cell within the egg: the more anterior the position of origin, the faster the rate of division of that blast cell's progeny (Deppe et al., 1978). Furthermore, the relative cell division rates are maintained within the respective lineages even after cells have migrated to new positions with respect to one another. This implies that cells of each lineage possess an internal clock that autonomously regulates the rate of cell division.

These observations have led to the suggestion that there is some kind of graded property within the egg that sets cell division rates within the blast cell lineages. This gradient evidently runs along the a-p axis, with the developmental potential of each lineage related to the rate of division imposed by the gradient (Deppe et al., 1978; von Ehrenstein et al., 1979). This proposal constitutes a gradient hypothesis for the segregation of developmental potential in a mosaic egg. The specific prediction of the model is that maternal effect mutants altered in developmental potential will show generalized alterations in the rates of cell division within the different lineages. The classical mosaic model of egg organization, in contrast, makes no predictions about cell division rate and developmental potential. Its primary genetic prediction is that a class of maternal effect mutants exists whose initial primary defects are solely within specific embryonic blast cell lineages.

GENETIC ANALYSIS OF EARLY DEVELOPMENT

To analyze oogenesis and early development requires a collection of maternal effect mutants. The search for such mutants in *C. elegans* must involve a screening for mutants that are viable but sterile as hermaphrodites. However, the maternal effect mutants are only a subset of the class of hermaphrodite steriles. Because the hermaphrodite supplies sperm to its own eggs, infertility can also result from a sperm defect. The mutants of interest are among those that lay fertilized eggs that experience aberrant development. However, even among these embryogenesis defectives, only a fraction are maternal effect mutants; the rest are mutant in zygotic genome functions.

The identification of embryogenesis defectives must therefore be followed by screening to identify the maternally defective mutants.

In most screens, a substantial proportion of the isolated mutants are hypomorphs. Because hypomorphic alleles frequently give variable or weak effects, there is an inclination to discard these mutants and save only those with little or no residual activity, the amorphs. If applied to the isolation of maternal effect mutants, this procedure would involve the following problem: any individual homozygous for an amorphic maternal effect mutation would be incapable of producing progeny. The mutation would therefore be lost. Only if heterozygous sibs had been preserved could the mutation be preserved in a stock. However, the need to save sibs of all tested individuals would more than double the work. In fact, the propagation of any extremely defective mutations involves constant maintenance of the mutations in heterozygous carriers. Furthermore, if all hypomorphs were discarded, all maternal effect mutations in functions required at stages later than oogenesis would be excluded, because their isolation as maternal effect mutants requires some residual gene activity of the mutant gene at postoogenetic stages. Therefore, to identify as many as possible of the genes required for a given developmental process, every mutant obtained should be saved.

The goal is to maximize the initial recovery of mutants without having to maintain large numbers of heterozygous siblings of suspected mutants. One strategy for doing so is to search specifically for *conditional lethal* mutants. These are mutants that produce lethality under one standard environmental condition, the *restrictive* state, but permit survival under an alternative condition, the *permissive* state. Stocks of such mutants can be maintained as homozygotes under the permissive condition but analyzed for their developmental effects under the restrictive one. Many mutant defects are conditional under some set of environmental conditions or with certain genetic backgrounds (Hadorn, 1961). However, identification of mutations in all or most of the genes that affect a particular developmental stage cannot depend on finding just the right set of conditions for each isolated mutant. The restrictive/permissive conditions must be chosen so that mutants displaying disjoint behavior will be recovered. Analysis is then performed by comparing their behavior under permissive and restrictive conditions, with all other variables of environment and genotype held constant.

The most generally useful form of conditional lethality involves growth temperature. Mutants are isolated that survive at one temperature but die or manifest an abnormality at a different temperature. It is customary to employ the opposite extremes of the temperature range for growth as the restrictive and permissive temperatures. Thus, a heat-sensitive lethal allele homozygote can be maintained indefinitely as a stock by growing the organism at the lower temperature but analyzed for its lethal effect at the higher temperature. The use of such temperature-sensitive or ts alleles has a second advantage: through limited pulse exposures of the mutant to the restrictive temperature at different stages of development, it is often possible to identify the approximate times of action (or synthesis) of the wild-type protein during development.

The third reason for choosing temperature sensitivity as the form of conditional lethality is that the great majority of protein-coding genes can mutate to give ts alleles. In consequence, the selection of temperature-sensitive mutants can, in principle, allow the identification of mutations in most or all genes affecting a particular developmental process. The reason is that every wild-type protein has an optimal thermal stability within the normal temperature growth range. Alterations in the amino acid sequence of the protein, produced by point mutations, can lead to an altered stability of the polypeptide chain in its active conformation; at the restrictive temperature, the mutant protein becomes denatured. This susceptibility to denaturation may occur preferentially during synthesis of the protein chain, in which case the mutant is said to be "temperature-sensitive in synthesis" or *tss* (Sadler and Novick, 1965). Alternatively, the temperature sensitivity may result from susceptibility of the correctly folded protein chain, in which case the protein is said to be "thermolabile" or *tl*. Unfortunately, it is not always possible to distinguish between these alternatives, which can lead to some ambiguities in interpretation (see below).

At present, the genetics of embryonic development and of maternal effects in *C. elegans* is almost entirely based on temperature-sensitive mutants; the remainder of this chapter is primarily a discussion of these mutants and what has been learned from them. The analyses have not proven as informative as was hoped; the molecular basis of the nematode egg's morphogenetic system still remains a mystery. However, in one other respect, the mutants have been immensely valuable, namely, in elucidating the relationship between the gene set expressed in oogenesis and the gene sets used in later stages of development. The analysis of the mutants will be presented here in some detail because it illustrates the versatility of ts mutants in probing gene expression and gene product function in development.

Mutant Isolation

A procedure for obtaining large numbers of ts mutants has been described by Vanderslice and Hirsh (1976) and is outlined in Figure 2.10. Late larvae are exposed to the mutagen ethyl methyl sulfonate (EMS), which induces a high frequency of point mutations, and their progeny are allowed to self-fertilize to produce homozygotes. These progeny include a fraction of induced ts homozygotes. Individual larvae are then allowed to form isolated larval broods, which are tested at the permissive and restrictive temperatures. Putative ts mutants are identified as those that accumulate one predominant developmental stage at the restrictive temperature but that show all stages at the permissive temperature.

In the particular scheme shown, 16°C was chosen as the permissive temperature and 25°C as the restrictive temperature. A representative sample of

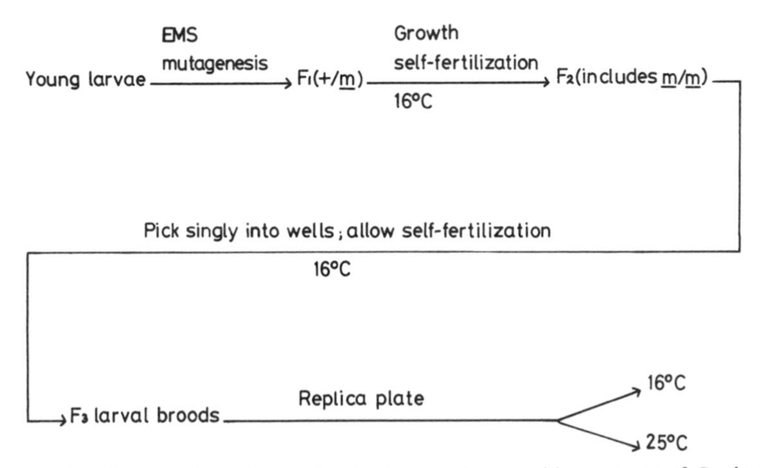

Figure 2.10 Mutagenesis and screening for temperature-sensitive mutants of *C. elegans*.

mutant phenotypes is shown in Table 2.1. The classes of particular interest for oogenesis are those that produce no eggs and those that produce fertilized eggs that develop poorly or not at all. Those that lay no eggs are presumably defective in formation of the gonad and are termed *gon* mutants. Those that lay fertilized eggs incapable of further development are defective in embryogenesis and are designated either *zyg* (for zygote defective) or *emb*.

To date, three extensive searches for ts embryogenesis-defective mutants have been made. Following Cassada et al. (1981), these three sets will be

Table 2.1 Classes of *C. elegans* ts Mutants from One Screening[a]

Phenotype at Restrictive Temperature	Frequency	Designation	Presumptive Defect
ts mutants			
No offspring or eggs	112 (50%)	*gon*	Gonad formation or function
Unfertilized eggs only	24 (11%)	*sp*	Spermatogenesis
Fertilized eggs only	21 (9%)	*zyg* = *emb*	Embryogenesis
Intermediate growth stage	48 (22%)	*acc*	Larval-essential function
Gross morphological defects	8 (4%)	*mor*	Larval or general function
Sterile but morphologically normal progeny	10 (4%)	*abF₁*	Early germ line function
	Total: 223 (100%)		

Source. Data from Hirsh and Vanderslice (1976).
[a] Wild-type, all stages present.

denoted respectively as B (Hirsh and Vanderslice, 1976; Wood et al., 1980), HC (Miwa et al., 1980), and G (Cassada et al., 1981). Most of the mutants have been mapped, and extensive complementation tests between the three sets have been carried out. Altogether, these mutants identify 54 genes.

The map positions of some of these genes are shown in the Appendix. It is evident that the genes are scattered around the genome. With the exception of the X chromosome, which has only two (*emb-15* and *let-2*), there is a fair sprinkling of these genes on all linkage groups. However, the distribution within linkage groups, particularly on chromosome 3, is markedly skewed, showing distinct clustering. The basis of this clustering is unknown at present. It may either represent a recombinational "cold spot" and hence poor recombinational separation of genes that are physically distant, or the map positions may reflect actual physical distances between the genes; in the latter case, it would reflect some sort of genetic specialization within this chromosome. Only the physical isolation and molecular characterization of these regions of the genome by recombinant DNA techniques will permit discrimination between these two alternatives.

The mutant isolations have also revealed an abnormally high mutability of at least three of the genes. These three genes are *let-2* on the X chromosome and *emb-9* (= *zyg-6*) and *emb-5* on chromosome 3. (The first of these was originally identified by an absolute lethal allele by Meneely and Herman, hence its designation.) The genetic basis of this high mutability is as obscure as that of the clustering phenomenon. The genes may be disproportionately large; the mutation rate of these genes per nucleotide by EMS may be 10-fold higher than that of the other genes; mutants of these genes may be 10-fold more likely to be ts; or growth at 25°C may be particularly sensitive to any diminution of activity of these genes. If the last explanation is the correct one, one would predict that such critical gene dosage effects would be reflected in semidominance of the mutants; this latter characteristic is in fact shared by all the *emb-9* mutants.

Analyzing the Program of Gene Expression

The set of *emb-zyg* mutants provides the source material for approaching the first question about oogenesis posed at the beginning of this chapter: how distinctive is the gene set expressed in oogenesis compared to gene sets expressed in later stages? To answer this question, each mutant must first be individually analyzed for its pattern of gene expression during development. This categorization is provisionally provided by means of three simple genetic tests (Wood et al., 1980). These tests are comparable to the standard crosses that define maternal effect mutants in conventional sexual systems but are adapted to the genetic peculiarities of the self-fertilizing hermaphrodite (Table 2.2). The results of the crosses establish whether the gene performs a maternal or zygotic genome function and, if maternal, whether strict or partial. Since a maternal function in a self-fertilizing hermaphrodite might

Table 2.2 Genetic Tests of *Emb* Mutants to Establish the Time of Action[a]

Selfing or S-test To determine whether the function must be expressed by the zygotic genome.

$$+/m \quad 25°\text{, self-fertilization} \longrightarrow \begin{matrix} +/+ & +/m & m/m \\ 1 : & 2 : & 1 \end{matrix}$$

Results and interpretation: If all genotypic classes survive, then expression of the wild-type gene function by the maternal genome is sufficient. If the m/m class dies during embryogenesis, then expression of the wild-type function by the zygotic genome is necessary.

Rescue or R-test To determine which maternal effect functions are also zygotically active and capable of rescuing progeny from maternally defective hermaphrodites.

$$m/m \times +/+ \xrightarrow[\text{25° cross}]{} m/+$$

Results and interpretation: If all $m/+$ progeny are inviable, then the function is a strict maternal. If some cross progeny are viable, then the maternal effect function is provisionally classified as one that is expressed by the zygotic genome as well. Definitive classification is provided by the H-test.

Heterozygous rescue or H-test To determine whether rescue of a maternal effect mutation is by the wild-type gene itself or via the function of this gene during sperm production.

$$m/m \times +/m \xrightarrow[\text{25° cross}]{} \begin{matrix} m/m & +/m \\ 1 : & 1 \end{matrix}$$

Results and interpretation: If only the $+/m$ class is rescued, then the wild-type allele is required in the zygote itself for rescue. If both classes survive, then wild-type function is necessary for sperm production to supply a needed sperm component for the zygote.

[a] For details, see Wood et al. (1980).

involve the animal's sperm, the crosses are designed to determine whether a given gene is required in oogenesis or in the sperm's contribution to the zygote.

First, the selfing or S-test establishes whether the mutant is viable as a homozygote when the wild-type allele has been expressed during oocyte formation. If 25% of the eggs of the heterozygous hermaphrodite ($+/m$) fail to hatch, then the m/m class is presumed to be lost, implying that the expression of the wild-type allele is required in the zygotic genome. If all eggs hatch, then maternal expression of the wild-type allele is sufficient to promote normal development and the gene is identified as a maternal effect function. Second, the rescue or R-test is used to distinguish the partial from the strict maternal mutants. Rescue of zygotes from m/m mothers by a +-bearing sperm signifies that there is expression of the paternal + allele during embryogenesis. Hence, the gene's expression is not restricted to oogenesis, that is, it is a partial maternal. Failure to rescue indicates that the mutant is a presumptive strict maternal.

Finally, the heterozygous rescue or H-test distinguishes among the rescuable mutants—between those that are spared because the + allele was expressed in the zygote and those that are rescued because the + allele was expressed during sperm formation. The H-test serves as a check on the possibility that defective embryogenesis might reflect the loss of some essential postfertilization sperm component. An example of the latter is something brought in by the sperm that is required for sperm pronucleus movement. A mutant defect in any such component should not occur when sperm develop from heterozygous ($+/m$) spermatogonial cells. Therefore, rescue of zygotes from m/m mothers by m-bearing sperm from heterozygous fathers indicates that the gene is essential for a *paternal* function.

The three crosses generate five possible categories of embryonic-essential function, defined in terms of the gametic and zygotic patterns and the requirements of gene expression. Each class can be designated by a two-letter notation (Miwa et al., 1980), as summarized in Table 2.3. The S-test generates the first letter of the classification scheme and denotes whether parental expression, designated *G* (for the Greek word *goneis*), is sufficient for survival or whether expression by the zygotic genome, designated *Z,* is essential. The second letter of the classification is given by the combined results of the R- and H-tests. The three categories of parentally expressed functions are as follows: GM, maternal expression of the + allele is essential (a strict maternal); GZ, maternal or zygotic gene expression of the + allele suffices to rescue embryogenesis (a partial maternal); GP, paternal expression of the + allele is required. The two categories of zygotic gene expression are as follows: ZZ, zygotic genome expression essential for embryogenesis; ZM, zygotic *and* maternal expression required to promote development.

What patterns of gene expression are actually found when the tests are applied to the *emb-zyg* mutants? The complete classification has been re-

Table 2.3 Mutant Expression Classes in *C. elegans*

Class Symbol	Description
GM	Strict maternal; maternal expression necessary and sufficient
GZ	Partial maternal; zygotic expression rescues the early embryo
GP	Paternal expression in sperm production required
ZZ	Zygotic genome expression essential
ZM	Both maternal and zygotic genome expression required

ported only for the B mutants (Wood et al., 1980) and the HC mutant set (Miwa et al., 1980), but the results form a consistent picture. Both sets of mutants reveal a high proportion of maternally expressed genes. For mutants of the B set, 21 of 24 genes give surviving progeny when there has been prior maternal expression. For the HC set, 8 out of 9 genes give surviving progeny. The frequency of strict maternals, GM, is also high in both sets, 11 of 21 mutants in the B set and 8 of 9 genes in the HC set, giving strict maternal mutants. (The higher percentage of GMs in the HC set may reflect a difference in the selection procedure. The HC mutants were classified as *emb* only if the same number of eggs were produced at permissive and restrictive temperatures.) Conversely, the number of strict zygotic-essential functions, ZZ, was low in both sets, being 2 out of 24 in the B set and 1 out of 11 in the HC set. In the combined group, only one mutant was ZM, requiring both zygotic and maternal expression, and only one was GP, requiring a paternal sperm function. The classification in the largest set (G) is less complete but also shows a very high frequency of maternal expression, and only a few, 5 out of 40 mutants, exhibit a strict requirement for zygotic genome expression (with 2 of these in the same gene, the highly mutable *emb-9*) (Cassada et al., 1981).

The classification of the mutants by the S-, R-, and H-tests categorizes the *emb-zyg* genes with respect to their patterns of expression during oogenesis and early embryogenesis. Most of the identified genes are expressed during oogenesis, and in many cases this expression is both necessary and sufficient to allow mutant embryos to survive. Are these genes, however, restricted in their expression to oogenesis/early embryogenesis, or are they expressed at other times as well? Neither the initial classification of the mutants nor the genetic crosses establishes whether these genes are required for later developmental stages or processes. To determine such requirements, it is necessary to shift homozygotes to the restrictive temperature at successive stages after embryogenesis and to examine the animals for any additional mutant phenotypes that appear.

When this test was carried out, most of the mutants exhibited secondary, postembryonic defects. These secondary phenes consisted of gonadogenesis (Gon) defects, cessation of growth and accumulation at the L2, L3, or L4 stages (the Acc phenotype), and various morphological defects arising during larval development (the Mor phenotype). Of the 21 ts mutants of the B set, only 2 displayed the Emb phenotype alone. Of the remaining 19, 8 showed both Gon and Acc phenotypes, 9 showed the Gon phenotype without other defects, and 2 had Mor defects. Among the 10 HC mutants, which defined 8 *emb* genes, mutants of 2 of the genes also showed the Gon secondary phenotype. Finally, in the G set, 30 of the 35 maternal effect mutants showed secondary mutant phenes.

In most cases the presence of secondary mutant phenes is unlikely to be caused by secondary mutations in the stocks. All stocks were purified of potential second site mutations by several rounds of backcrossing and reisolation. In the few instances in which reversions of the Emb phenotype were selected, the secondary phenes were found to be simultaneously reverted. The various developmental defects produced by temperature shift in most mutants must therefore reflect the requirement for expression of these maternal effect genes at later stages of development. Such postoogenetic activity is to be expected in the partial maternal (GZ) class but, surprisingly, appears to be as common in the strict maternal (GM) class. For instance, 9 of the 11 strict maternals of the B mutant set show secondary phenes, and the 2 *emb* mutants of the HC set that produce Gon defects are in the strict maternal class.

The genetic classification scheme, based on the S-, R-, and H-tests, must therefore be only an approximate and incomplete guide to the expression of the maternal effect genes, providing no clues to requirements at later periods of development. At best, it gives only a relative quantitative measure of gene expression during early development. It is easy to imagine that a gene classified as GM, a strict maternal, is also expressed by the zygotic genome, but in too slight a degree to rescue the maternal deficiency. Furthermore, the assignment of a maternal effect mutant to either the strict or partial maternal class is in some cases a function of the particular mutant *allele* under examination.

The mutant alleles *B117* and *B189* provide an example. These mutations map at the same chromosomal position and do not complement. By definition, they must be mutant alleles of the same gene (*zyg-6*). Yet, by the genetic tests, *B189* is a GZ function (a partial maternal) while *B117* is in the ZZ class (an essential zygotic function) (Wood et al., 1980). The genetic tests are designed to provide information about the expression of the *wild-type* gene, but in this case different mutant alleles give different classifications for this gene. The simplest explanation of the disparity is that it stems from a difference in the degree of leakiness of the alleles, with *B189* possessing more residual activity than *B117*. The survival of the *B189* homozygotes, when segregated from +/*B189* mothers, is attributable to the + allele activity in the mother combined with the residual allelic activity in the homozygotes themselves. In contrast, the *B117* homozygotes lack this residual activity and die.

Evidently, the genetic classification of patterns of gene activity is both approximate and subject to error. A more precise means of delineating either the time of gene expression or the activity of the gene product is required. Fortunately, the temperature sensitivity of the mutants provides a direct method for assessing the times of gene product action. Phased exposures of a mutant to the restrictive temperature can establish the specific period(s) during which temperature inactivation of the gene product produces the mutant phenotype. For all *tl* mutants, this temperature-sensitive period (TSP) is the approximate period in which the gene product is active and required in the cell (Suzuki, 1970)—unless the product is synthesized long before its use, in which case the TSP measures the period from the beginning of its synthesis to the end of its required use. For *tss* mutants, the TSP measures the period of synthesis of the gene product. The TSP is therefore not an unambiguous property. However, its delineation is often surprisingly informative.

How does one actually measure a TSP? One might determine a TSP by first growing the ts mutants at the permissive temperature and then placing different batches of each at the restrictive temperature for sequential periods during development. These "pulses" of high temperature would serve to localize approximately the period or periods of temperature sensitivity. Depending on the initial results, one could then appropriately shorten or lengthen the pulse duration to determine the precise limits of the TSP. If there are multiple TSPs, they could be clearly delineated by this method. The difficulties are that the experiment would have to be done several times in a row to obtain an accurate placement of the TSP and that the combined upshifts and downshifts might introduce some developmental artifacts.

An alternative method for determining the TSP is the *reciprocal temperature-shift experiment*. Two sets of synchronously developing animals are placed at the restrictive and permissive temperatures, respectively, at the beginning of development. At fixed intervals, one batch of each is shifted quickly to the other temperature. The fraction of normal or surviving individuals following each timed shift is scored at the end of the experiment. The reciprocal temperature shifts generate reciprocal survival curves that together delimit the TSP.

The logic of the experiment can be illustrated with the simplest case, that of a mutant that synthesizes a TL protein during the same period in which the protein is required to act. The beginning of the TSP is defined by a series of downshifts from restrictive to permissive temperatures. Imagine that a group of embryos or eggs begins to develop at the restrictive temperature and that equal numbers are shifted downward to the permissive temperature at successive periods. Those that are downshifted before the critical developmental period will all survive, while those maintained at the restrictive temperature during this period and then downshifted will fail to execute the necessary cellular function and die. (In the latter case, death may not follow instantaneously but may occur at a much later stage.) In the downshift series, therefore, the developmental period in which the survival rate decreases from 100% to 0% defines the period of irreversible developmental defects, an interval called the "defective execution stage" (Hartwell et al., 1970). The 50% survival point, the midpoint of the curve, can be taken as the average time of defective execution and the point of inflection of the curve as the *beginning of the TSP*.

The temperature-upshift curve gives the converse result and defines the end of the TSP. In this series, mutant embryos shifted to the restrictive temperature before the protein is synthesized will all have defective protein during the critical period and suffer a lethal defect. In contrast, all embryos shifted after this period will survive. Again, upshift during the critical period gives fractional survival, the proportion depending upon the time of upshift, the synchrony of the animals, and the fraction of expressed activity necessary to ensure survival. In upshift, then, the phase in which survival begins to increase and goes to 100% is the *normal execution stage* (Hartwell et al., 1970), during which the gene activity is required. The midpoint of the upshift survival curve marks the midpoint of the normal execution stage, and the beginning of the rise in survival represents the *end of the TSP*. The two survival curves from the reciproal shift experiments together delimit the TSP as the region of overlap, where survival is changing in both curves.

A hypothetical example, for the situation described, is shown in Figure 2.11*a*. For both curves, survival is plotted as a function of developmental stage rather than time. This standardization is necessary because the rates of normal development at the permissive and restrictive temperatures differ. These rate differences are independent of the ts effect of the mutations being tested. In wild-type *Caenorhabditis,* development at 25°C is twice as rapid as at 16°C (Byerly et al., 1976). The most reliable method for determining the stage of development is direct observation of the wild-type and of mutants, since the mutants' growth rates even at the permissive temperature are often somewhat slower than that of the wild-type.

Besides the simplest situation described above, other possibilities exist; these are shown in Figure 2.11*b* and *c*. In Figure 2.11*b,* the defective execution stage comes well after the normal execution stage. In terms of the formal logic of the experiment, the result seems paradoxical, since the *beginning* of the TSP, defined as the beginning of the downshift curve, seems to come *after* the presumed *endpoint* of the TSP, defined as the end of the rise in the upshift curve. However, assuming that the staging at the two temperatures has been performed correctly, there is a simple biological explanation: the wild-type function is normally carried out at a comparatively early point in development, but the protein is synthesized over a much longer period and can carry out the correct function later than the normal time. In this situation, defective execution—the point of irreversible damage, as measured by the downshift curve—would indeed follow the completion of gene product action at the normal early time. The diagram in Figure 2.11*c* shows the converse result, with defective execution substantially preceding the normal execution point. This result also presents a seeming puzzle. How can the failure to perform a cellular function produce a deleterious effect before that process is normally completed? One possibility is that the wild-type protein serves as a check on some process, extended in time, which if completed prematurely causes aberrant development.

Figure 2.12 illustrates the measurement of TSPs for three *emb* mutants by the reciprocal temperature-shift method. The three mutations are two alleles of a GM, strict maternal, function (*emb-5*) and a single mutant of a ZZ, zygotic-essential, function (*emb-9*). For both alleles of *emb-5* the TSP occurs early, roughly between the 8- and 30-cell stages of development, while the TSP for the ZZ function, *emb-9*, occurs after cleavage is completed (between

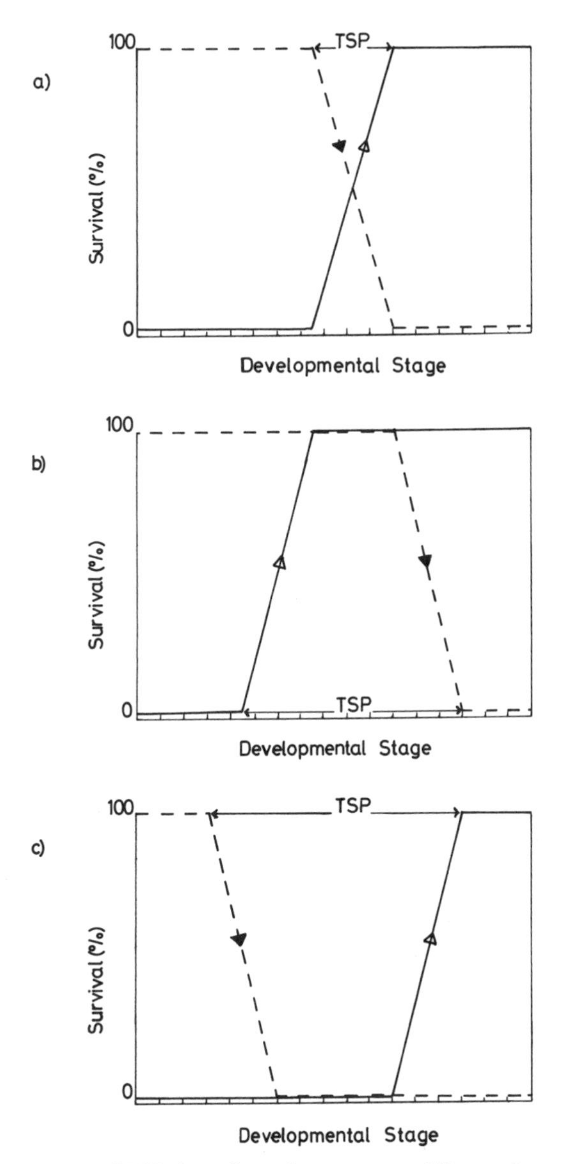

Figure 2.11 Measurement of TSPs in reciprocal temperature-shift experiment: some patterns. Details described in the text. Dashed line, downshift survival curve; black arrowhead, defective execution point; solid line, upshift survival curve; open arrowhead, normal execution point. (After Miwa et al., 1980.)

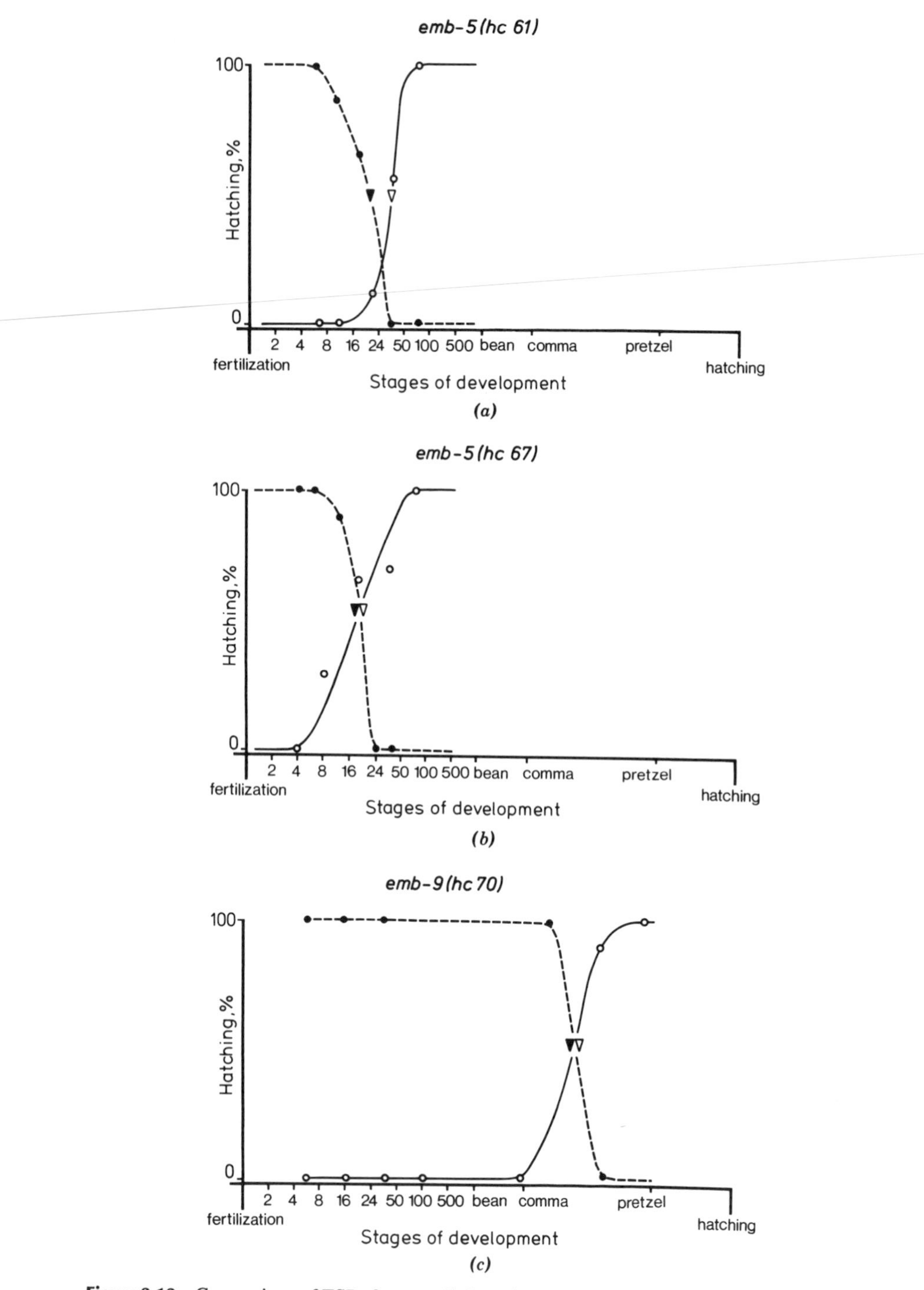

Figure 2.12 Comparison of TSPs for two allelic strict maternal mutants (of *emb-5*) and for that of a zygotic-essential function (*emb-9*). Symbols as in Figure 2.11. Each point represents at least 10 eggs. Results are discussed in the text. (Figure kindly provided by Dr. J. Miwa; from Miwa et al., 1980).

the comma and pretzel stages of development). This difference in TSP between the mutants of the two genes agrees with the simple expectation that GM functions, expressed by the maternal genome, and whose products are stored in the egg, are most likely to be needed in early development, the phase dominated by the activities of maternally produced components. Zygotic genome-expressed functions, in contrast, are more likely to act later in development, as the genome of the embryo takes over direction. In general, the TSPs of the various mutants bear out this difference in timing: strict maternals generally have their first TSPs during oogenesis or early cleavage, while partial maternals show later and often prolonged TSPs, and strict zygotic functions have TSPs located distinctly within the period of embryogenesis and often late (Wood et al., 1980; Schierenberg et al., 1980).

Apart from such general differences between classes of mutants, the detailed comparison of allelic mutants of the same gene can sometimes be informative with respect to the relative severity of the mutant defects or to the mode of action of the wild-type gene. The comparison of the two *emb-5* mutants is a case in point. The respective TSPs for the two mutants (Fig. 2.12) are essentially similar, with the normal and defective execution stages occurring at nearly the same time, around the 24 to 28-cell stage, but *hc61* shows defective execution (the downshift curve) slightly preceding normal execution.

Microscopic examination of the living embryos of these two mutants at the restrictive temperature confirms these results and pinpoints the defect as occurring in the endodermal (E) cell lineage. In normal embryos, two E cells, consisting of the anterior division product of the E cell, E.a, and the posterior division product, E.p, migrate internally from their initial posteroventral location at about the 28-cell stage. Following migration, the two cells divide to yield four E lineage cells, an anterodorsal and anteroventral pair (E.ad and E.av) and a parallel posterodorsal and posteroventral pair (E.pd and E.pv). These cells give rise to the corresponding four sections of the adult intestine (e.g., E.ad gives rise to the anterior, dorsal row of intestinal cells). In both *emb-5* mutants, *hc61* and *hc67,* in contrast, the E cell divisions occur precociously relative to the wild-type (Schierenberg et al., 1980). This premature division is followed by a prolonged migration phase, at the end of which the four E cells find themselves in an abnormal anatomical environment, apposed to an additional D cell and four extra MS cells. Furthermore, the E.ad and E.av cells are now in reversed positions (Fig. 2.13). Inevitably, the end result is aberrant intestinal development.

This sequence of events occurs in both mutants, but a detailed comparison reveals a subtle difference that bears on the accelerated defective execution in *hc61*. In the latter, E.a and E.p divide before migration begins, while in *hc67,* division occurs during migration. The relative difference in the time of defective execution therefore signals a genuine difference in the behavior of the two mutants. In several respects, *hc61,* seems to be the more severely defective allele, showing more extreme secondary phenes; the accelerated

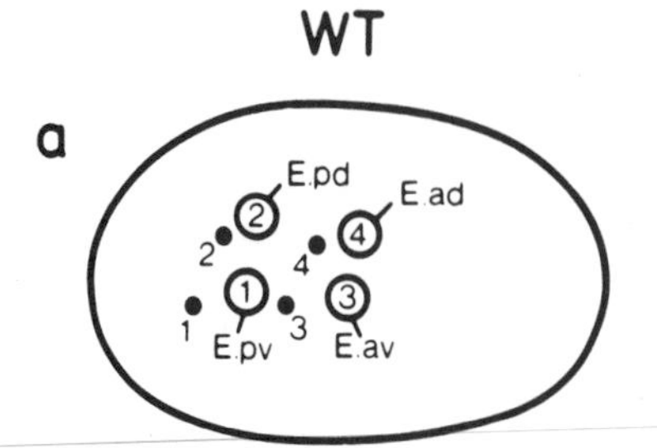

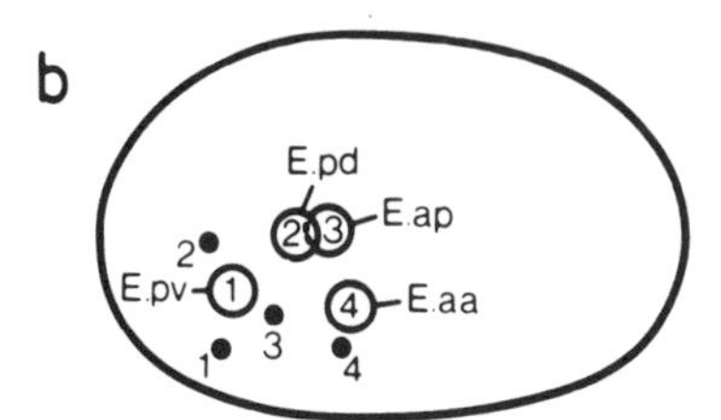

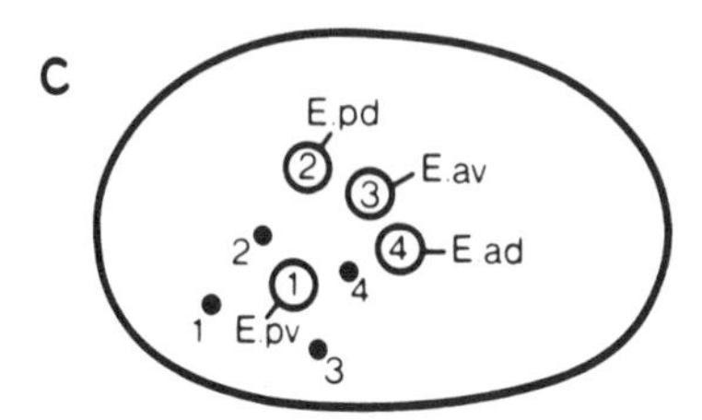

Figure 2.13 Alteration of E lineage cell behavior in *emb-5* mutants. The black dots indicate the positions of the four E lineage cells just after the division of E.a and E.p; the open circles, their positions just before the next division. Orientation: anterior to right and dorsal to top. (Cell designations are as noted in the legend of Fig. 2.9.) In *hc61,* the division of E.a is a-p rather than d-v; E.ap therefore corresponds to E.av and E.aa to E.ad. (Reproduced with permission from Schierenberg et al., 1980.)

inward movement of its cells is one example. The wild-type *emb-5* gene product appears to influence the gastrulation process directly and may act to delay the second division in the E cells relative to the migration process.

The interpretation of TSPs is usually somewhat ambiguous. The principal uncertainty involves the nature of the ts mutant defect, whether it is of the *tl* or *tss* kind. When there is a large temporal difference between the synthesis of a protein and its utilization, the basis of a ts defect in that protein becomes an important factor in interpreting the nature of the TSP. A thorough discussion of the technical and analytical problems in studying ts mutants is given in Hirsh (1979) and Wood et al. (1980). However, the analysis of the *emb-5* mutants shows how informative this approach, in combination with detailed developmental analysis, can be. The central point emerging from the study of TSPs of *emb* mutants of *C. elegans* is that the measured initial TSPs correlate reasonably well with the predictions from the genetic classification. In general, maternal effect mutants show initial TSPs in oogenesis and early cleavage, while strict zygotic-essential expression genes are frequently expressed later.

The properties of the *emb-zyg* mutants as a whole provide a preliminary answer to the first general question about oogenesis, the distinctiveness of the gene set expressed in oogenesis versus the sets of later stages and cells. Although the 54 *emb-zyg* genes are a small subset of the entire genome, they are probably representative of the genes expressed in oogenesis. The great majority are maternal effect genes, and even among those that are identified as nonmaternal by the biological tests, some are probably expressed maternally but to an extent insufficient to rescue the embryo.

The striking fact about the identified maternal effect genes is that nearly all are also expressed in cell types and stages following oocyte formation; there is little evidence for a highly discrete set of oogenesis genes.

This genetic characterization of oogenesis gene functions in *C. elegans* is reminiscent of the sea urchin cytoplasmic informational RNA data (Fig. II.1), which also show a broad overlap between the gene set expressed in oogenesis and that expressed in other cell types. Nevertheless, for definitive comparisons in the nematode, one needs to know the precise numbers of these various gene sets and their degrees of overlap. The genetic data do not indicate the precise number of oogenesis genes, but they can be used to estimate a minimum number. The method is based on the simplifying assumption that if all of the genes required in a given developmental pathway are equally mutable, then the total number of genes required is given by the Poisson distribution. This method provides an estimate of the total gene number by comparing the number of loci with multiple mutations to those represented by single mutations. The simplifying assumption of equal mutability of *emb* genes is necessary to make this estimate. If the three highly mutable genes are excluded, therefore, the calculation can be performed (Cassada et al., 1981). The figure obtained is 211 genes, or about 5% of the total genome, assuming this comprises about 4000 genes (Table 2.4).

There is a second way to estimate the number of oogenesis genes. If it is assumed that all genes give the same proportion of ts mutants, and that all *emb* genes are oogenesis genes, then the proportion of ts mutants that are *emb* mutants will give the proportion of the genome expressed in oogenesis. This fraction is between 10% (Wood et al., 1980) and 28% (Cassada et al., 1981), the latter figure corresponding to about 1100 *emb* genes.

However, even 1100 is likely to be an underestimate of the number of genes expressed in oocyte formation. In the first place, the *emb* mutants are defective in genes essential for embryogenesis; any inactivating mutation in any of these genes constitutes a lethal event for the line carrying it. However, in addition to such absolutely essential genes, there may be a number of "fine-tuning" genes (Williams and Newell, 1976). Mutational inactivation

Table 2.4 Calculation of the Number of Embryogenesis-Essential Genetic Functions

Poisson expression: $P_n = \frac{m^n e^{-m}}{n!}$ where m = average number of mutations per gene

P_n = proportion of genes with n mutations

$$P_0 = e^{-m} \qquad P_1 = me^{-m} \qquad P_2 = \frac{m^2 e^{-m}}{2}$$

$$A = \text{number of loci with two mutations} = N \times \frac{m^2 e^{-m}}{2} = 6^a$$

$$B = \text{number of loci with one mutation} = N \times me^{-m} = 44^a$$

$$\text{Ratio: } A/B = \frac{Nm^2 e^{-m}}{2Nme^{-m}} = m/2$$

$$6/44 = m/2;\ m = 12/44 = 0.27$$

To find X, the number of genes with no mutations (the undetected *emb* loci), compute from the P_0/P_1 ratio, since $P_0 = P_1/m$. Hence,

$$X = B/m = 44/0.27 = 162$$

Total number of *emb* functions = 162 + 56 identified (of normal mutability) + 3 hypermutable = 211.

[a] Figures from Cassada et al. (1981).

of any of these genes would not entail lethality for the organism, and such genes would be undetectable in standard *emb* mutant hunts. In the second place, there might be a large class of genes that are transcribed during oogenesis but whose gene products function only *later* in development. This strategy of expression may be employed in the sea urchin, in which many genes may be continuously or repeatedly transcribed from early development on to permit the requisite accumulation of their protein products for later stages of development (Hough-Evans et al., 1977). The fact that much of the maternal cytoplasmic transcripts of the sea urchin may require additional processing is consistent with the idea of such delayed utilization.

If early transcription with delayed utilization takes place in *C. elegans*, then there should be larval stage-essential genes, recognized by their *Acc* or *Mor* phenotypes, that are expressed maternally. There is some evidence for the existence of such genes. In a test of 12 different acc_2 mutants (those that accumulate in the L2 stage), the mutant homozygotes of 4 were found to be rescued by + allele expression in heterozygous mothers (Wood et al., 1980). The results do not identify the kinds of components that are being stored in the eggs—whether mRNAs, proteins, or final metabolic products—but they show that there is maternal expression (transcription) of genes whose action is apparently not required until the first or second larval stage. As in the genetic classification of maternal versus zygotic expression in the *emb* mutants, these biological tests for maternal expression score only full developmental rescue and might not detect low levels of maternal expression. The proportion of maternally rescuable larval stage genes in any such test is therefore likely to be an underestimate of the true level of expression.

The various studies provide a self-consistent picture of widespread maternal gene expression during oogenesis in *C. elegans* and little support for the idea that the unique properties of the oocyte derive from a uniquely expressed gene set in the developing oocyte or ovary. Ultimately, the genetic findings in *C. elegans* must be supplemented by biochemical or molecular studies of gene expression patterns. Such studies have been hampered by the impermeability of the nematode egg shell to radioactive precursors and by the difficulties in obtaining large numbers of synchronously developing eggs and embryos for development.

One approach that partially circumvents these problems is a squash technique (Gossett and Hecht, 1980). The method is similar to the burst embryo method of Laufer et al. (1980): the embryo is squashed under a coverslip and the embryo bursts through the egg shell, leaving a cellular monolayer in good preparations. This monolayer can then be subjected to any biochemical technique applicable to single cells. One such method is that of in situ hybridization, which permits measurement of mRNA transcript contents in cells. A version of the in situ technique that has been applied to the nematode makes use of the fact that many mRNAs have a "tail" of polyadenylic acid (poly A) at their 5′ terminus. These poly A^+ mRNAs can be detected by binding the complementary molecule, polyuridylic acid (poly U). When squashed embryos are exposed to tritium-labeled poly U under appropriate conditions and then subjected to autoradiography, the grain counts in the autoradiograms reveal the relative cellular concentrations of poly A tails and hence of poly A^+ mRNA molecules in the cells of the embryo.

The results of one such experiment are shown in Figure 2.14. The key finding is that the early embryo contains rich cytoplasmic stores of poly U binding sites, presumptive stored mRNA molecules. From the 90- to the 125-cell stage, these binding sites first appear in the nucleus and then greatly increase in number, a change that indicates the increasing activity of the zygotic genome in the cells of the embryo. The latest period for the onset of nuclear gene expression is at the 90- to 125-cell stage, since low levels of mRNA are not detected by this method. This finding confirms that the first part of embryogenesis is exclusively or predominantly under the control of maternal components stored in the embryo.

With tritiated poly U as the molecular detection device, only total transcript contents are measured. However, the method can be adapted to examine transcript contents for individual genes by employing highly radioactive cloned gene sequences. The application of in situ hybridization techniques to

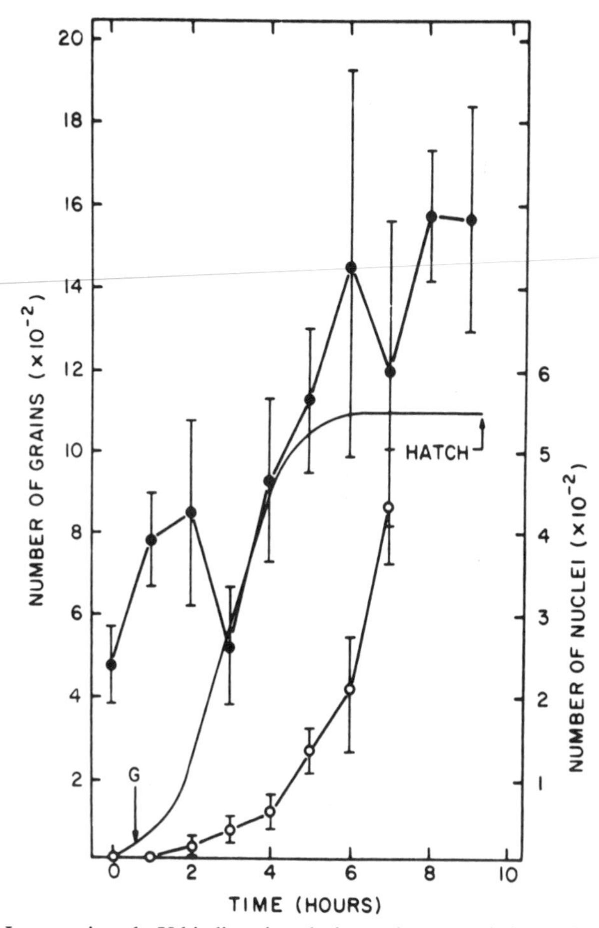

Figure 2.14 Increase in poly U binding sites during embryogenesis in *C. elegans*. The closed circles represent the grain counts over the entire squash; the open circles, the grain counts over the nuclei; the solid line indicates the increase in nuclei per embryo; G marks the time of gastrulation. The slight decrease in total grain count at 3 hr may be an autoradiographic artifact reflecting the deposition of the cuticle at this time. (Reproduced with permission from Hecht et al., 1981.)

the measurement of individual gene transcripts in *Drosophila* cells is described in Chapter 9. Another approach to the biochemical analysis of embryos is to track individual protein products by histochemical or immunochemical methods (Gossett et al., 1982).

The Morphogenetic System: Analysis and Mechanism

In principle, ts maternal effect *emb-zyg* mutants of *C. elegans* should be as useful for probing the morphogenetic system built into the oocyte as they are for ascertaining patterns of gene expression. However, the analysis of determinative mechanisms involves special problems. While every ts *emb* mutant reveals something about gene expression during development, only a small fraction of any mutant set would be expected to be directly affected in the positioning or functioning of the fate-setting system. Because nearly all *emb* mutants exhibit some specificity in their developmental defectiveness, the challenge is to discriminate between those altered in the fate *instructional* system and those that are defective in their *implementation* of these instructions or even of the metabolic or cell division requirements of the developmental program.

Furthermore, it is not even clear what criteria should be used to make this discrimination or even which class of mutants to concentrate on. If the sole function of the molecules of the instructional system is to establish different developmental capacities in the cells of the cleavage stage embryo—and this is the usual assumption—then the maternal effect mutants without secondary mutant phenes require the sharpest scrutiny. If, on the other hand, the determinant molecules play other roles in the development or growth of the organism, then mutations that affect their production or function would be general lethals or partial maternals, depending on the stringency of the post-oogenesis requirements and the severity of the mutant defects. Because it is not known whether the instructional molecules that make up the morphogenetic system are unique to the oocyte or not, either group of maternal effect mutants might harbor the mutants of interest. The only clue is a completely external one, from *Drosophila,* in which most of the known maternal mutants of the morphogenetic system are strict maternals without secondary phenes (Chapter 3).

The difficulty in analyzing the morphogenetic system of *Caenorhabditis* by genetic means is in reality twofold. The first problem is the matter mentioned above of discriminating primary instructional defects from secondary developmental defects, and the second is the number of mutants. Consider the question of mutant numbers first. The number of genes involved in specifying components of the morphogenetic system is, of course, unknown, but it could be a small fraction of the number of genes expressed during oocyte formation. If the number of oogenesis genes is on the order of a few thousand, then the present sample of 54 identified *emb* genes (of which a few are not expressed maternally) is almost certain to be inadequate.

The problem of discrimination is best understood in terms of the predictions of the two major models, the mosaic determinant model and the gradient model. The former posits that there are distinct, qualitatively different instructional molecules and predicts that mutational inactivation or alteration of one determinant should leave the initial steps of development of the other lineages essentially unaffected. The identification of mutants affected in single embryonic lineages would provide strong support for the mosaic determinant model. None of the mutants reported to date fits this prediction. However, the failure to find such mutants, when the entire set of mutants is small, is uninformative; a much larger set of mutants would have to be examined before this result could be considered significant. In contrast to the mosaic model, the gradient model predicts the existence of maternal effect mutants producing alterations throughout the entire set of six fundamental embryonic lineages or, at least, in several lineages at a time. The great majority of maternal effect mutants whose progeny defects have been described (Wood et al., 1980; Schierenberg et al., 1980), have just this property. Unfortunately, mutational lesions in genes for general metabolic or cell division functions would also be expected to have this effect; indeed, it seems probable that most of the *emb* mutants are defective for such general cell functions. Devising criteria for distinguishing such mutants from mutants of a gradient system is a challenging problem.

However, the specific cell division rate-gradient hypothesis of Deppe et al. (1978) leads to a specific prediction. The hypothesis is that the gradient of lineage cell division rates running from anterior to posterior as a function of the site of origin of the lineages is connected to the settings of developmental potential in these lineages. The model predicts a class of maternal effect mutants with altered gradients such that recognizable lineages arise in *uncharacteristic* locations in the embryo (e.g., germ line cells at a nonposterior location) and that these "new" lineages would show the characteristic cell division rates associated with these cell types in the wild-type embryo. Such mutants would be in the general category of "homeotic mutants," those that replace particular cells or structures with cells or structures typical of another location. Homeotics are common throughout the world of insects, including *Drosophila* (and are discussed in Chapters 3 and 6) and in *Caenorhabditis* postembryonic development (Chapter 5). Finding maternal effect homeotics in the nematode may not be an easy matter, but if they were identified, they would provide a definitive test of the cell division gradient hypothesis.

The gradient of division rates within the early embryo is still the only clue to the relationship(s) between the six cellular lineages. In closing this discussion of the morphogenetic system, it may be helpful to consider briefly how the cell division rate might be related to the establishment of developmental potential in the early embryo. If there is a relationship between these two cell properties, it almost certainly does not involve cell division itself, but rather something correlated with the cell division rate. The basis for this conclusion is that some developmental potentials can be segregated correctly *in the absence of cell division*. We have seen an instance of this in one mosaic system, the ascidian (Fig. II.3). *Caenorhabditis* is a second. The experiment involved tracking a marker of the E lineage—the rhabditin granules, which can be detected by their fluorescence in the cells of the intestinal lineage—in burst embryos that were exposed to mitotic spindle and cleavage inhibitors. A result fully comparable to the ascidian results is obtained. Rhabditin granules appear in the cleavage blocked embryos in normal amounts and approximately on schedule, and only in those blastomeres that would have given rise to the E lineage had cleavage taken place (Laufer et al., 1980). For instance, when cleavage is blocked at the two-cell stage, only the P_1 cell develops rhabditin granules, while if division is blocked at the four-cell stage, only the EMS blastomere shows the characteristic fluorescent pattern. Measurement of the DNA content by staining with the DNA reagent Hoechst 33258 showed that the nuclei of the division-inhibited cells are polyploid, indicating that the treatment blocked cell division specifically and not nuclear DNA replication. The molecular nature of the segregated potential remains to be determined, but if the ascidian results are a guide, the property being segregated is a capacity for gene expression and not preformed, masked gene products.

If determination of lineages is linked in some way to a cell cycle process other than division itself, the obvious candidate is nuclear DNA replication. Replication may be important in either of two distinct fashions. It might simply be *permissive,* allowing gene expression to occur. For the ascidian embryo, Satoh and Ikegami (1981) have presented evidence that expression of acetylcholinesterase (AcE) activity in presumptive muscle cells, whether cleavage blocked or normal, requires a critical number of nuclear DNA replication cycles (eight) from the time of fertilization, and have argued that a "quantal" eighth division creates a state of competence for interaction with the determinants. For the amphibian embryo of *Xenopus,* a certain number of rounds of replication (12) is also required for transcriptional activity. The elegant experiments of Newport and Kirschner (1982a,b) showed that replication is essential to titrate out a given amount of general transcriptional inhibitor. Such permissive modes of DNA replication involvement in the control of gene expression may play some part in *C. elegans* biology but do not readily lend themselves to an explanation of differerential settings of gene potential by differential rates of replication.

Von Ehrenstein et al. (1979) have argued that chromosomal DNA replication plays an *instructive* role in setting these potentials. They have suggested that differential rates of replication accompany differential cell division rates in the early embryo and that these differential rates are a function of the placement of specific histone variants on or near replication origins, the chromosomal sites where replication is initiated. The rate of replication is known to be primarily a function of the number of available replication origins (Callan, 1974; Blumenthal et al., 1974), and the availability of origin sites may be determined by the placement of certain histone species (Wilkins, 1976; Hand, 1978). The relationship between replication rate and gene expression potential might then be a consequence of which chromosomal origins are active, either through direct or secondary effects on transcription.

The model is speculative but leads to some specific predictions about replication rates in different cell lineages and the role(s) of histones in controlling these rates. The question of the involvement of DNA replication and the setting of gene expression recurs with respect to determination in the early *Drosophila* embryo and postembryonic development in *Caenorhabditis;* we will return to this question later in examining these subjects.

THREE

DROSOPHILA MELANOGASTER

FROM OOCYTE TO BLASTODERM

When Thomas Hunt Morgan left embryology for genetics he deserted *Hydra*, planaria, sea urchins and amphibia for *Drosophila*. The fruit fly was recognized, and still is recognized, as an ideal organism for the study of eukaryote genetics, but it now seems that it will turn out to be equally suitable for answering many questions concerned with development.

H. Schneiderman and P. Bryant (1971)

There are several factors that have contributed to the *Drosophila* renaissance of recent years. First, as noted by Schneiderman and Bryant, the genome of the fruit fly is intermediate in size and complexity between that of prokaryotes and mammals. The combination of a relatively simple genome and a complex developmental program is favorable for the analysis of the molecular biology of development. A second contributing factor has been the development of methods of cell culture (Hadorn, 1965), nuclear transplantation (Illmensee, 1972), and microsurgery (Bryant, 1971), which permit the application of the essential techniques of experimental embryology to the study of this small animal. However, the principal reason for the choice of *Drosophila* for developmental studies is the genetic legacy of the earlier studies, bequeathed by Morgan and his colleagues. The exquisite precision of genetic analytical techniques designed for this organism provides an opportunity for the study of genetic control in development that is unmatched by any other animal system.

THE LIFE CYCLE IN OUTLINE

The life cycle of *Drosophila,* like that of *Caenorhabditis,* involves a period of embryogenesis, completed within the egg shell, and a sequence of larval growth stages or "instars." (There are three such larval stages in the fruit fly.) The final development of the adult or "imago" from the last larval stage is, however, considerably more complex than the larval–adult transition of the nematode and entails a dramatic pupal metamorphosis. The difference in complexity between the nematode and the fruit fly is further reflected in their constituent cell numbers. At hatching, the first larval stage of *Drosophila* consists of approximately 10,000 cells (Madhavan and Schneiderman, 1977), nearly 20 times that of the *C. elegans* L1 stage. The adult fruit fly consists of over a million cells, while the adult *Caenorhabditis* has only 2500–3500 cells and nuclei in all. The life cycle in *Drosophila* wild-type strains takes 10–12 days at 25°C and is diagrammed in Figure 3.1.

The first phase of embryogenesis in *Drosophila* consists of a sequence of syncytial nuclear cleavage divisions, which convert the fertilized egg into an ellipsoid monolayer syncytium of 6000 nuclei surrounding a yolk mass. The first nine cycles of nuclear division take place within the interior of the egg, and the final four divisions take place in the outer cytoplasm or "periplasm" (Turner and Mahowald, 1976; Zalokar and Erk, 1976). At the tenth division, a few nuclei bud off into cells at the posterior end of the embryo. These "pole cells" later give rise to the germ line of the animal. At the end of cleavage, the syncytial monolayer is converted into a cellular monolayer as cell membranes grow down and around the nuclei. The embryo is now at the cellular blastoderm stage. Blastoderm is reached approximately 3.5 hr after

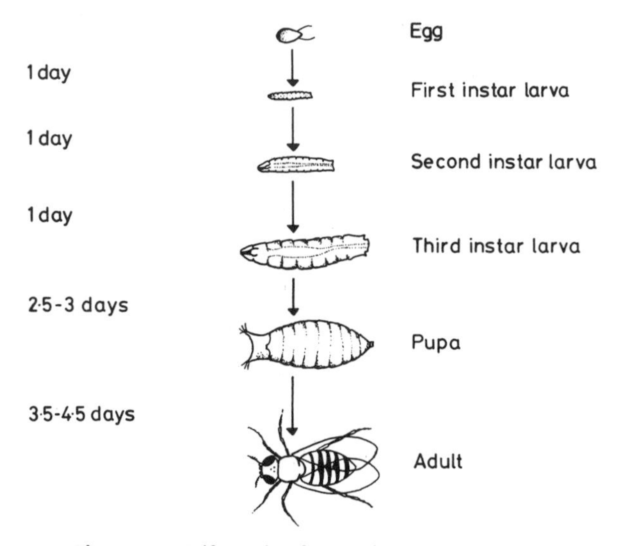

Figure 3.1 Life cycle of *D. melanogaster* at 25°C.

fertilization and marks the point of zygotic genome activation. The blastoderm stage is rapidly succeeded by a series of cellular invaginations and movements of cell sheets that constitute gastrulation. Further events of morphogenesis and differentiation, over a period of more than 17 hr, create the first instar larval form, which hatches from the egg case about 24 hr after fertilization.

The three stages of larval development are outwardly devoted to larval growth. However, the larva contains cells that later give rise to the structures of the adult fly; the larval growth period is also one of rapid multiplication of most of these cells. The presumptive imaginal cells are arranged in clusters in specific sites on the surface of the larval body in two kinds of cell groups, "imaginal discs" and "abdominal histoblast nests." The imaginal discs give rise to all the external head and thorax structures, the internal and external genitalia, and much of the musculature of the imago. The abdominal histoblast nests generate the outer surface of the abdominal segments of the adult (except for the eighth segment, which is formed in part by the terminalia). The positions of the various imaginal cell groups and the structures they give rise to are shown in Figure 3.2.

The imaginal cell groups are set apart from the surrounding larval cells during the first half of embryogenesis but become cytologically identifiable only in the first larval instar. The disc and histoblast cells are visibly different from the surrounding larval cells in having smaller nuclei and an undifferentiated, epithelial appearance. A fundamental cytological difference between imaginal and larval cells is that the former are diploid cells capable of repeated rounds of cell division, while the majority of larval cells grow without cell division. Many larval cells are polyploid, while in others the chromosomes continue to undergo rounds of replication, but these chromosomes remain undivided after each round of DNA replication, becoming multistranded or "polytene." Within the group of imaginal cells, the disc cells differ from the abdominal histoblast cells in that they divide during the larval stages; the histoblast cells commence their cycles of cell division only during the pupal stage.

During the pupal period, most of the larval cells undergo dissolution (histolysis). The form of the imago gradually takes shape as each imaginal disc progressively unfolds and differentiates. During the final hours of the pupal period, the various disc structures (eyes, legs, wings, etc.) knit together to produce the adult. The sequence of differentiative and morphogenetic changes is triggered at the end of the larval period by the secretion of the hormone 20-hydroxyecdysone from the larval brain.

GENETICS AND GENOME STRUCTURE

The genetic system of *Drosophila melanogaster* is conventional for an outbreeding animal. Its genome is organized into three pairs of autosomes and

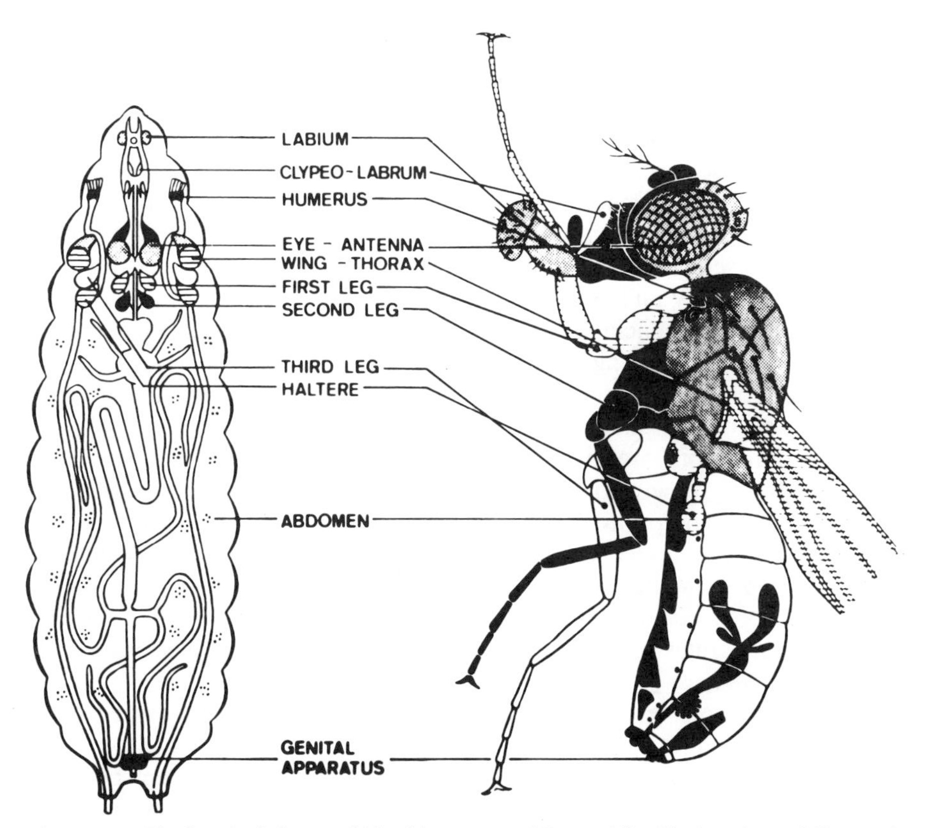

Figure 3.2 The imaginal discs and histoblast nests of *Drosophila*. The locations of discs and histoblast nests (the latter shown as clusters of dots) shown in the third instar larva on the left and their derivatives in the adult on the right. (Reproduced with permission from Nothiger, 1972.)

two sex chromosomes, females having the constitution 3AA + XX and males 3AA + XY. The X chromosome is designated as the first chromosome, the two large metacentric chromosomes as the second and third chromosomes, and the comparatively tiny autosome as the fourth chromosome. The sex chromosomes have certain structural peculiarities. The X chromosome material to the right of its centromere (as the chromosome is conventionally drawn) is largely densely staining or heterochromatic and devoid of essential genetic functions; the left arm contains the X-chromosome genes and even more heterochromatin. The Y chromosome appears wholly heterochromatic in all cells except spermatogonia but contains six loci that are essential for male fertility (Gatti and Pimpinelli, 1983).

Apart from their chromosomal difference, male and female *D. melanogaster* differ in their capacity for recombination. In standard laboratory strains, males completely lack recombination, which takes place only in females.

Three-quarters of a century of *Drosophila* work has produced a wealth of mutations affecting visible traits. Many of these mutant lines are of direct interest for developmental studies. These mutants range from those producing small differences in eye or body pigmentation, or in bristle form or pattern, to those producing profound developmental changes. Particularly interesting are the homeotics, which substitute one imaginal structure for another (e.g., a leg for an antenna). Thousands of mutants of *Drosophila* have been discovered and characterized; many are enumerated and described in Lindsley and Grell (1968).

The genetic nomenclature of *Drosophila* is more complex than that of *Caenorhabditis,* reflecting its longer history. Recessive visible mutants are usually given a one-, two-, or three-lowercase letter notation that is an abbreviation of the phenotype description (e.g., *r* for *rudimentary*), and dominant mutations are given designations with an initial uppercase letter (e.g., *Cy* for *Curly wings*). The wild-type allele is given by a + superscript (e.g., r^+, Cy^+), and particular mutant alleles are also given by a superscript (e.g., r^{49}). Lethals are generally denoted by an "l", followed by the number of the chromosome bearing the lethal, and an individual allele designation (e.g., $l\,(3)\,c^{43ts}$ is a particular ts lethal allele on the third chromosome). The nomenclatural rules for the various kinds of chromosome aberrations and rearrangements are given in the Appendix.

There are two reference systems for describing gene locations in *D. melanogaster.* The first system is that of genetic map distance, calculated from meiotic recombination frequencies, with respect to the ends of each chromosome. For example, the location of $l\,(3)\,c^{43ts}$ is 3-49.1 ± 1.0, or about 49 map units from the left end of chromosome 3. A genetic map of *D. melanogaster,* showing commonly used visible marker genes and some of the genes that will be mentioned in this book, is given in the Appendix.

The second system used to designate gene locations is cytogenetic. It is based on the distinctive appearance of the polytene chromosomes of the larval salivary gland cells. When these chromosomes are spread and stained, they are readily distinguishable by their large size and characteristic banded pattern. The chromosomes of a salivary gland cell are shown in Figure 9.5. The width of these chromosomes reflects the numerous, side-by-side chromatids of which they are composed—up to 1024 chromatids by late third instar—and their length results from their relative absence of condensation. The source of the banding pattern is unknown, but each chromosome region has a *distinctive banding pattern.* Furthermore, by using a variety of genetic stratagems, it has been shown that each genetic function has a fixed location in a particular band or (less darkly stained) interband region. Each gene can be described by its site on the salivary chromosomes, and each salivary chromosome represents a magnified version of the euchromatic portions of a standard chromosome. (The heterochromatic portions of chromosomes are substantially underreplicated in third instar polytene chromosomes.) In the nomenclatural system of Calvin Bridges (1935), the genome is divided into

102 sections. Each section consists of six subsections, labeled A to F, and each band within a subsection is given a number; gene locations are given in terms of band number. Thus, the gene for the enzyme alcohol dehydrogenase, *Adh,* maps to 35B2.3. The designation 2.3 indicates the resolution of the cytogenetic gene localization, in this case to within band 2 or 3 in section 35B. The relationships between the genetic and cytogenetic maps are summarized in Figure 3.3.

When one turns from the characterization of the genome by genetic or cytogenetic means to the exploration of its molecular organization, many questions arise. The fundamental problems are those encountered in the exploration of any eukaryotic genome; we have touched on some of them in considering that of *C. elegans*. They concern the number of genetic functional units in the genome, the nature of regional organization of gene groups within chromosomes, and the role of the various repetitive sequences that are scattered around the genome. Although these questions remain unanswered, as they do for other eukaryotes, there is a substantially larger data base with which to approach them in *Drosophila*. The most recent review of the *Drosophila* genome structure is that of Spradling and Rubin (1981), and much of what follows is summarized from their paper.

Total DNA content and gross sequence composition have been determined by DNA renaturation studies. The haploid genome content is 1.65×10^8 bp; the genome is thus about 2.0 times the size of that of *C. elegans*. Of this, approximately 74% renatures as single-copy sequences, 10% as midrepetitive sequences, and 16% as highly repeated or "satellite" DNA. The last group of sequences is composed of four major species, each consisting of a small oligonucleotide repeat present thousands of times. They are located predominantly or exclusively in the centromeric heterochromatin. The midrepetitive sequences, in contrast, are scattered around the genome, al-

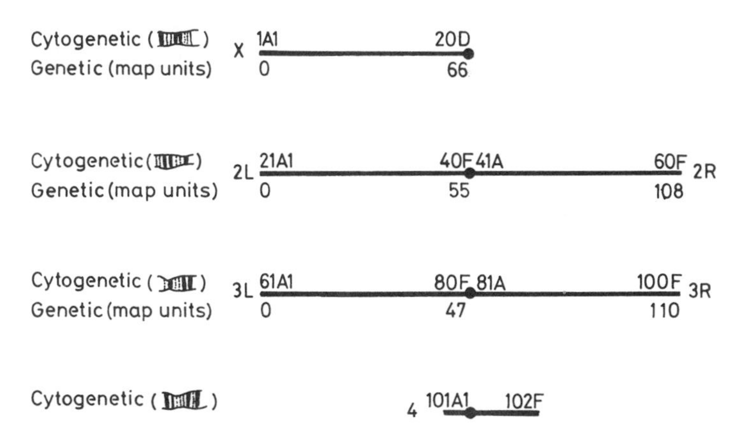

Figure 3.3 The cytogenetic (top) and genetic (bottom) maps of the *Drosophila* chromosomes. Details in text.

though about 10% appear to be localized in the centromeric heterochromatin.

Much debate has centered on the question of how many genetic functions exist in the *Drosophila* genome. The controversy stems from a disparity in the estimates produced by genetic means and those produced by molecular analyses. The genetic tests suggest that there is one genetic function, defined as a genetic complementation group, per (polytene chromosome) band (see Judd et al., 1972; Lefevre, 1974; Woodruff and Ashburner, 1979). Since there are approximately 5000 bands in a polytene chromosome complement, this yields an estimate of about 5000 genes for the total genome. In fact, several bands contain more than one complementation group (Young and Judd, 1978), and a more realistic estimate of gene number, defined as number of complementation groups is probably 6000 to 10,000 (Judd, 1977).

However, the genetic tests detect only those genes whose products are required either for viability or fertility, or whose absence causes a visibly mutant phenotype. There are several reasons for suspecting the presence of genes whose presence would not be detected in these ways. In the first place, null alleles of a large proportion of enzyme coding loci produce no phenotypic change (O'Brien, 1973). Flies that are completely deficient in these gene activities may be perfectly viable and fertile under standard laboratory conditions and show no visible effects. (Some or all of these genes might have important effects on fitness in the wild.)

In the second place, the regional density of genetic functions that have been cloned is always greater than that predicted on the basis of the one band–one gene relationship. These cloned regions have for the most part been of genes that specify abundant cellular products, and so may be a nonrepresentative sample. On the other hand, cloning of large chromosomal regions that may be substantially more representative of the genome as a whole also reveals the presence of more genetic functions than there are bands (Hall et al., 1983). Finally, saturation hybridization studies, designed to detect all mRNAs complementary to single-copy DNA, detect considerably more than 5000 distinct mRNA sequences; the numbers are closer to 15,000–17,000. These analyses will be discussed in more detail in Chapter 6. The molecular evidence thus supports a total gene number considerably in excess of 5000, perhaps on the order of 15,000–20,000. A gene number of 20,000 or so would imply that the average gene size in *Drosophila* is approximately 7000–10,000 bp, a fairly typical size for a eukaryotic gene.

A second question about genome organization concerns the organization within the genome of genes with a related cellular function. A variety of different patterns have been observed. Some gene groups, such as the genes for the histone proteins and the ribosomal RNAs, exist in single clusters. Other gene groups are widely dispersed, and still others, such as the tRNAs, are partially dispersed and partially clustered. For the groups of genes that specify enzymes of single metabolic pathways, there are no reports of operon-like organization. On the other hand, two groups of genes that seem particularly important in development for the control of segmental phenotype form clusters within a relatively small portion of chromosome 3; these genes will be described in detail in Chapter 6.

The third general question about genome organization concerns the biological function of repetitive sequences. The highly repeated satellite sequences can be deleted in large blocks without producing major effects on viability; they may have a role in promoting chromosomal pairing during meiosis, but they appear to be inessential for the individual organism.

Whether the midrepetitive sequences have an important role in the expression of the genome is an open question. The present evidence suggests that they have no major function in the organization or expression of the essential genes of the organism. There are two principal classes of midrepetitive sequence that can be distinguished: large, approximately 5000 bp long, "mobile" elements and shorter, about 300 bp long, sequences. The former may be derived from initial DNA copies of RNA retroviruses that have become integrated into the genome. The source of the smaller, midrepetitive elements is unknown. Significantly, however, individual elements of both classes appear to be able to shift location in the genome at relatively high rates, as shown by the techniques of molecular mapping described in Chapter 9 (Strobel et al., 1979; Young, 1979). Yet, these shifts seem to be without obvious phenotypic effect. Furthermore, related lines or species often show markedly different numbers or locations of midrepetitive elements without exhibiting any corresponding developmental alteration (see, for instance, Dowsett, 1983). This "fluidity" of midrepetitive elements suggests that none have an obligate role in the control of genetic expression. It also casts doubt on the idea that the midrepetitive sequences function as carriers of shorter sequences that act as obligate *cis*-control elements (Davidson et al., 1983a).

There is abundant evidence that the insertion of some repetitive elements within genetic units, which occurs occasionally, can disrupt or alter their activity (Bender et al., 1983b). In this respect, they may be highly similar to McClintock's maize controlling elements. However, the great majority seem to exist in the genome without affecting the normal regulatory machinery of the cell. The same conclusion may pertain to most of the midrepetitive elements in other organisms. It can be stated with higher certainty regarding *Drosophila* because the individual midrepetitive elements can be located in the chromosomes with precision.

The structure and function of individual genes are as much of a puzzle as these properties of the genome as a whole. From genetic analysis alone, there appear to be two categories of gene: simple and complex. Both kinds have introns and similar sequences for initiating transcription and processing of transcripts. They differ in size and in the complementation behavior of their mutants.

For simple genes, most mutations give a similar phenotype and do not complement in diploids. In contrast, different mutations within a complex locus may differ in phenotype yet not show complementation. For instance, heterozygotes of mutations *a* and *b* within a complex locus (genotype: *a*/*b*) may have either the *a* phenotype or the *b* phenotype or both. If, for example, the heterozygote has the *a* phenotype, this implies that the *a*-bearing chromosome has b^+ activity, while the *b*-bearing chromosome has neither *a* nor *b* activity. The behavior of complex loci suggests that they consist of multiple bits of genetic information and that inactivation of some may inactivate expression of some or all of the remainder. A possible molecular explanation for the behavior of complex loci will be discussed in Chapter 9. A comparison of the genetic characteristics of simple versus complex loci can be found in Finnerty (1976) and Judd (1976), respectively.

OOGENESIS IN DROSOPHILA

Oocyte formation is a considerably more complex process in *Drosophila* than it is in the nematode, involving many more cell and tissue types that interact in a complex and involved sequence. The developmental biology of the female gonad is discussed in detail in King (1970), and the biology of oogenesis in the fruit fly has been most recently reviewed by Mahowald and Kambysellis (1980) and by Mahowald (1983).

The structure of the ovary and of the accessory organs of the female reproductive system in *Drosophila* is shown in Figure 3.4. The bulk of the

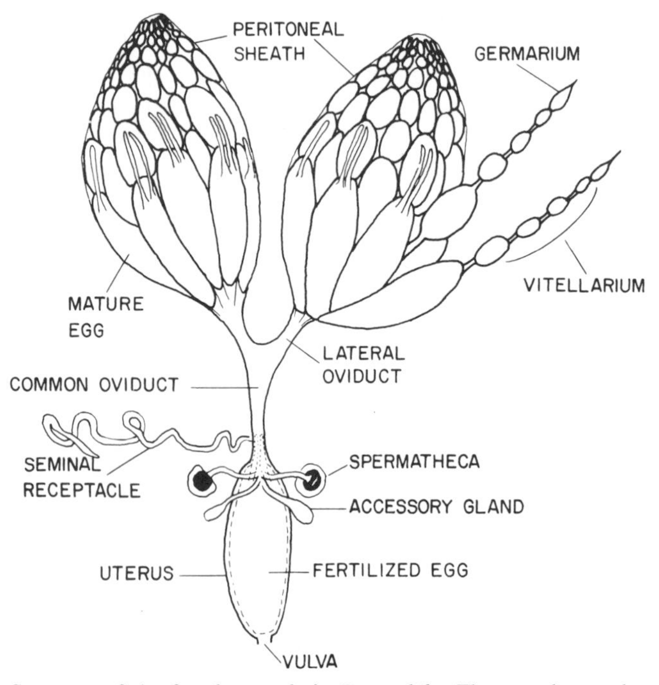

Figure 3.4 Structure of the female gonads in *Drosophila*. The gonads are shown from the dorsal side; two ovarioles are shown on the right, plucked loose from the mass, and the uterus is shown in its distended, egg-carrying form. (Kindly supplied by Dr. A. Mahowald; from Mahowald and Kambysellis, 1980.)

gonad consists of the two ovaries. These are connected via separate oviducts to the common oviduct, which empties into the uterus. The ovaries have a composite cellular origin. The germ line cells, which include the oocytes, are formed from the pole cells set aside during late cleavage. The cells that surround each developing egg chamber and make up the walls of the egg chambers are derived from somatic mesoderm. The tapered structures that encase the egg chambers are termed "ovarioles," and each ovary consists of a cluster of 15–17 ovarioles. Mature eggs are released from the posterior ends of the ovarioles into the oviducts, and fertilization occurs in the uterus. Because sperm can be stored in the spermatheca and seminal receptacles, fertilization of the egg may take place sometime after mating.

The tapered anterior tip of the ovariole is called the "germarium" and is the site of origin of the pro-oocyte. The pro-oocyte arises in a characteristic and complex set of cell divisions that begins with an oogonial stem cell. Each stem cell first divides to produce two offspring, a daughter stem cell, and a so-called cystoblast cell. Each cystoblast, in turn, gives rise to a cluster of 16 interconnected cystocytes by four sequential mitotic divisions, the 16 cystocytes comprising a "cyst." Two of these 16 cells become pro-oocytes and one of them eventually the oocyte. The 14 sister germ line cells become "nurse cells," whose function is to synthesize cytoplasmic materials and transport them to the growing oocyte. The 16-cell cyst, surrounded by a monolayer of follicle cells, is termed the "egg chamber." The sequence of divisions, from stem cell to early cyst, is summarized in Figure 3.5.

The process by which two of the 16 cystocytes are singled out to become pro-oocytes provides an illustration in *Drosophila* of a direct relationship between cell lineage and cell fate. The key to this relationship is found in the pattern of intercellular canals that connect the 16 cystocytes. These intercellular canals are the result of incomplete cell division, and each marks the site of an earlier cleavage furrow. Examination of the pattern and number of these canals shows that eight cells within each cyst are connected to only one other cell, four cells are connected to two cells each, two cells are connected to three other cells, and finally, two cells have four canals and are connected to each other and to three other cells as well. It is invariably one of the two four-canal cells that becomes the pro-oocyte. Since each canal is formed by incomplete cytokinesis division, it follows that the two cystocytes with four canals contain the canal that was the first to be generated by the cystoblast. Both of these cells enter the oocyte developmental pathway, since both synthesize synaptinemal complexes (an essential structure for chromosome pairing) and commence meiosis. However, by the time the 16-cell cyst has passed the middle of the germarium, one pro-oocyte loses its synaptinemal complex and switches to the nurse cell developmental pathway, leaving a single pro-oocyte (Carpenter, 1975).

It is not known what factors determine the selection of the definitive pro-oocyte. However, the chosen cell is usually the most posterior. This fact might suggest that an oocyte determinant, initially attached to the posterior cortex of the cystoblast parent cell, determines the selection (Brown and

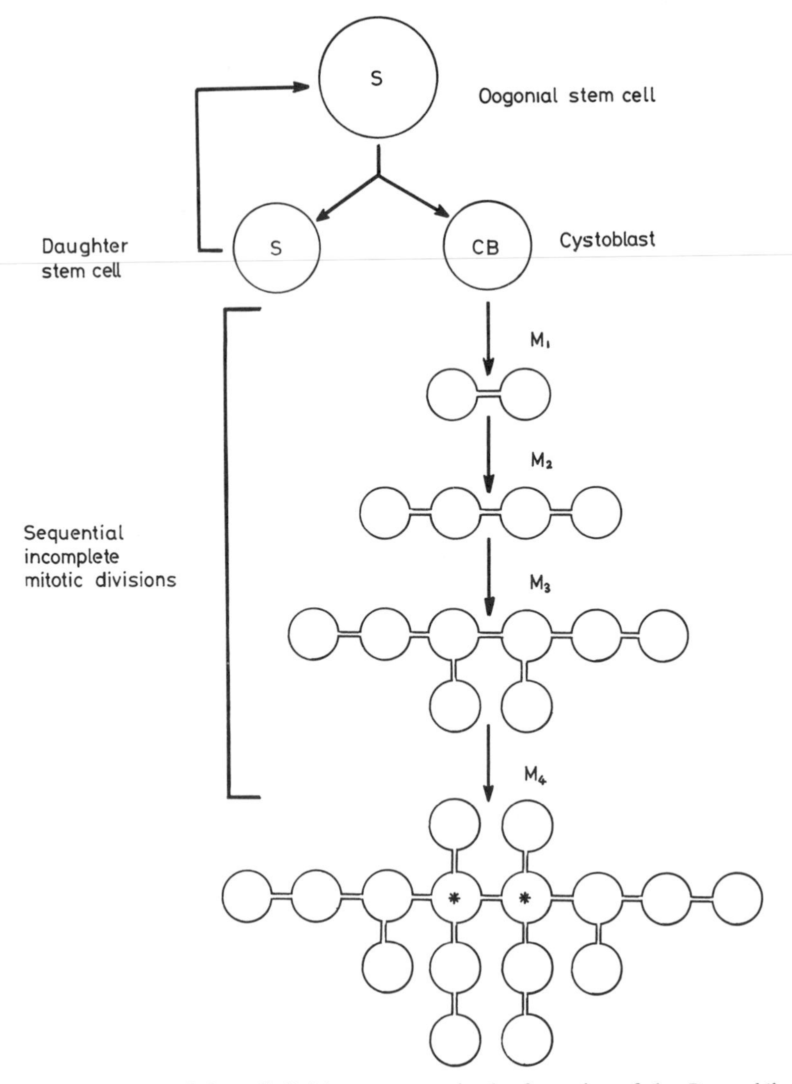

Figure 3.5 Summary of the cell division sequence in the formation of the *Drosophila* egg chamber. The two four-canal cells are starred. Details in text. (Adapted with permission from King et al., 1982.)

King, 1972). Alternatively, a determinative influence of the posterior position itself might somehow select one of the two four-bridge cells. In either event, the posterior positioning of the oocyte with respect to the nurse cells defines the a-p polarity of the egg chamber and later of the egg itself; these first signs of a-p polarity are probably significant for later development.

Each ovariole consists of two distinct regions: the germarium, in which the egg chambers are first formed, and the vitellarium, in which the major phase of oocyte growth and development takes place. The final stages of egg chamber development in the germarium involve the covering of the cyst by a monolayer of somatic "prefollicle" cells. Within the first 30 hr of entering the vitellarium, these 80 somatic cells divide another four times to give 1200 follicle cells, which constitute the definitive somatic cell covering of each cyst.

The development of the *Drosophila* oocyte is conventionally divided into 14 stages (King, 1970) (Fig. 3.6). During the first six stages, oocytes and nurse cells remain roughly the same size as each other, while increasing in volume by approximately 40-fold. Each volume doubling takes about 9 hr. During stage 7, the egg chamber elongates and growth temporarily slows. Beginning in stage 8, the process of vitellogenin (yolk protein) accumulation begins. Some vitellogenin synthesis takes place in the follicle cells of the ovary (Brennan et al., 1982), but, as in *C. elegans* and vertebrates, most of it occurs outside the ovary. For *Drosophila,* the principal site of yolk protein synthesis is the fat body; the yolk polypeptides are released into the hemolymph and taken up by the developing oocyte.

From stage 8 on, growth of the oocyte is rapid with respect to the nurse cells. Between stages 8 and 12, the oocyte increases in volume 1500-fold, with a doubling time of only 2 hr. Part of this increase is at the expense of the nurse cells, which, beginning in stage 11, empty their contents into the oocyte. As a result, by the end of oogenesis, the oocyte occupies the entire interior of the egg chamber. Stages 13 and 14 involve the secretion of the outer protective coverings of the egg—the vitelline membrane and the chorion—and the progression of the oocyte into metaphase of meiosis I. The oocyte remains in this stage until fertilization. The total increase in oocyte volume during development is approximately 90,000-fold. The mature oocyte, shown in stage 14, is considerably larger than the *Caenorhabditis* egg and measures 140×500 μm.

The development of the oocyte involves two kinds of cellular interaction. The first set of interactions are those between the oocyte and the nurse cells. In more primitive insects, the oocyte synthesizes a large part of its final mass. However, the fruit fly oocyte is largely the passive recipient of the biosynthetic activities of the nurse cells. The nurse cells themselves are polyploid, reaching 512 and 1024 times the haploid DNA content depending on their position in the chamber. This degree of polyploidy gives the nurse cells several thousand genomes for synthesizing needed gene products (proteins, ribosomes, mRNAs, etc.) for the oocyte. The oocyte nucleus is not completely inactive during the developmental sequence; it shows a brief period of transcriptional activity during stages 9 and 10, but the nurse cells provide by far the greater part of the store of gene products.

Transfer of materials from nurse cells to oocyte is effected through intercellular channels, which are large enough to transmit mitochondria. However, despite this size, there may be some selectivity in the molecular species that are transported. When intact stage 10 egg chambers were incubated in the presence of [^{35}S]methionine and then dissected into halves, the ante-

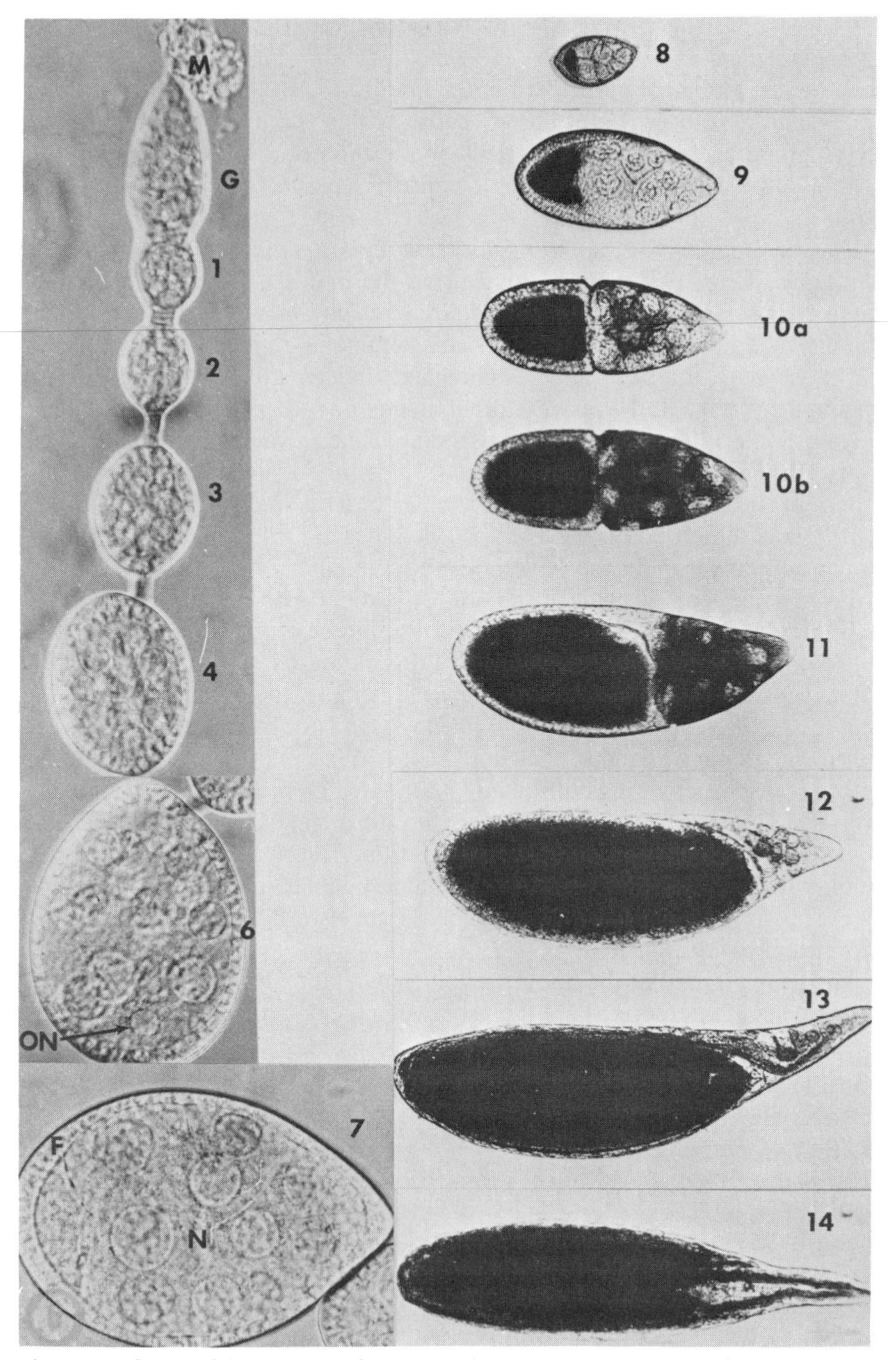

Figure 3.6 Stages of development of the *Drosophila* egg chamber; the classificatory scheme is by King (1970). Stages shown on the left with Nomarski optics (X380); stages shown on the right are the vitellogenic stages, shown by bright field microscopy (X100). The oocyte is on the left in the vitellogenic stages depicted. M, muscle sheath (stripped away); F, follicle cell layer; N, nurse cells; ON, oocyte nucleus. (From Mahowald and Kambysellis, 1980; kindly provided by Dr. A. Mahowald.)

rior halves consisting of nurse cells and the posterior halves containing the ooplasm, certain prominently labeled protein species specific to the anterior (nurse cell) halves were found (Gutzheit and Gehring, 1979). These protein species must be selectively retained within the nurse cells. Because the space occupied by the nurse cells later becomes the anterior half of the egg, it is possible that this molecular difference has some significance for the a-p polarity of the egg or embryo.

The second class of cellular interactions within the egg chamber are those that take place between the germ line cells—the oocyte and nurse cells—and the somatic follicle cells. These follicle cells also become polyploid between stages 6 and 10, although to a smaller degree than the nurse cells. The requirement for polyploidy in these somatic cells is presumably also related to biosynthetic needs: the secretion of the vitelline membrane and then the layers of the chorion by an enveloping monolayer of cells.

The follicle cells undergo a characteristic set of migrations between stages 7 and 13 that are significant for the final morphogenetic stages of egg development. Between stages 7 and 10, most of the follicle cells migrate posteriorly, coming to surround the oocyte within the egg chamber and leaving only 80 or so thin epithelial cells covering the nurse cells. During stages 9 and 10, a small group of follicle cells, the "border cells," move from the anterior end of the egg chamber between the nurse cells and through the center of the cyst and come to lie at the anterior border of the oocyte. These cells secrete the micropylar cone and canal through which the fertilizing sperm later enters (Bishop and King, 1984). The covering layer of follicle cells first secretes the vitelline membrane during stages 9–11. It then switches to secreting the successive layers of the chorion, the "shell" of the *Drosophila* egg. Polarized clusters of follicle cells are also responsible for forming certain distinctive chorionic structures, the two long tapering appendages of the anterior dorsal surface (Fig. 3.15*a*) and a chorionic plaque at the posterior end of the egg (Turner and Mahowald, 1976). The proteins that comprise the chorion have been analyzed by Waring and Mahowald (1979).

The net effect of all the activities of the accessory cells (both nurse and follicle cells) is to provide the egg with a distinctive surface and well-defined a-p and d-v axes or polarities. As with the nematode egg, these features foreshadow the corresponding axes of the future embryo. For the investigator, these external signs are often a useful guide to the internal organization of the egg during later embryonic development. Internally, there are few signs of such organization. The only distinctive internal cytological marker is a set of particles at the posterior end of the ooplasm, termed the "polar granules." These structures first form as small bodies (less than 0.2 μm in diameter) in stage 9 oocytes and gradually increase in size (up to 0.5 μm in diameter) in the later stages of oogenesis (Mahowald, 1962). Initially, the granules contain both RNA and protein, but lose the RNA component during cleavage in the early embryo. Because of their distinctive appearance and location, and their inclusion in the pole cells during late cleavage, they are possible determinants of the germ line. This subject will be discussed at the end of the chapter.

THE *DROSOPHILA* EGG: STOREHOUSE OF GENETIC MESSAGES

The egg of *Drosophila,* like all mature oocytes, is rich in stored RNA molecules. The predominant RNA component is RNA, the bulk of which is transferred in the form of completed ribosomes to the oocyte during the period of nurse cell breakdown. However, a small but significant portion of the egg RNA is mRNA. Poly A^+ mRNA alone comprises 1.6% of the total egg RNA (Lovett and Goldstein, 1977), and nonpolyadenylated RNA may make a further contribution. The bulk of the maternally produced, stored mRNA is presumably derived from transcription of nurse cell nuclei during the stages of egg chamber growth. However, some fraction may be derived from the oocyte nucleus during the comparatively brief period of its activity in stages 9 and 10. The total informational complexity of the oocyte mRNA pool is equivalent to 1.2×10^7 bp of DNA, approximately 10% of the total single-copy DNA (Hough-Evans et al., 1980).

Presumably, part of this mRNA pool is used to code for proteins that are required in early embryogenesis. For example, the stored tubulin mRNA molecules presumably serve this purpose. Although tubulin protein amounts to 3% of the total protein of the mature egg (Loyd et al., 1981), these protein stores may not be sufficient to meet the demands for tubulin during early embryogenesis. The egg also possesses a large store of tubulin mRNA, amounting to approximately 2% of the total poly A^+ mRNA population (Loyd et al., 1981). This mRNA is almost certainly required for the further de novo synthesis of tubulin during cleavage. Inhibitors of protein synthesis rapidly bring nuclear division to a halt in zygotes (Zalokar and Erk, 1976).

The demand for translatable maternal mRNA may be particularly high during the cleavage divisions because zygotic genome expression is practically nonexistent until blastoderm formation. Nevertheless, much of the maternal mRNA remains untranslated during this first phase of embryogenesis and may be employed only after activation of the zygotic genome in postblastoderm stages. Indeed, a considerable portion of the maternal poly A^+ mRNA remains unutilized during the first part of embryogenesis. By as late as 3.5 hr, the time of blastoderm formation, 33% of this mRNA pool is still unattached to ribosomes and is found packaged in special storage ribonucleoprotein (RNP) particles with a sedimentation value of 30–50S.

The existence of a large sequestered population of mRNAs raises some interesting questions. Do they comprise a unique class of sequences that are programmed to be utilized at particular times after blastoderm formation? Even if they are a representative sample of the set of maternal sequences, how late do they persist in development and can they make a significant contribution to development?

The nature of the informational content of the sequestered messages has been investigated in two different fashions and has received two somewhat different answers. Goldstein (1978) compared the translated poly A^+ mRNAs, those that are part of polyribosome structures, and the sequestered (RNP) fraction by means of RNA-DNA hybridization and concluded that there are no sequences uniquely present in the sequestered fraction at either 0.5 or 2.5 hr of development. This finding might imply that the unused RNP sequences simply provide a surplus storage capacity. Mermod et al. (1980) approached the question by comparing the in vitro translation products obtained from poly A^+ mRNAs attached to polyribosomes and those from the RNPs, using a two-dimensional gel separation of the labeled, synthesized polypeptides. They found general similarity in the translation products from the two mRNA populations but detected a small number of species found preferentially in the RNP messenger set of mature oocytes. This result suggests that there is some selective control of maternal mRNA translation, that is, some degree of *posttranscriptional control* of gene expression.

The discrepancy between the two studies probably reflects the different techniques used. The saturation hybridization method provides a rough measure of the number of different sequences being expressed, while the in vitro translation method registers only those mRNA species present in high concentration. With the latter method, only large changes in the *quantity* of particular mRNA species are detected. If there were qualitatively similar sets of messages in the free and RNP mRNA populations, but certain abundant mRNAs were preferentially sequestered in RNP particles, the two sets of results would be compatible. The possible significance of purely quantitative shifts in gene expression will be considered below.

After zygotic genome expression is activated at blastoderm formation, the mRNA pools in the developing embryo become increasingly dominated by zygotic genome transcripts. It therefore becomes increasingly difficult to trace maternal mRNAs by biochemical means from blastoderm formation. To determine whether these mRNAs make a *functional* contribution to the developing organism at postblastodermal stages, it is necessary to resort to genetic stratagems.

One approach is to manipulate the maternal dosage of genes and determine the effects on viability, either in wild-type embryos or in those that are zygotically deficient for the maternally expressed genes. (The latter kind of experiment is the maternal R-test described in Chapter 2.) Several studies of these kinds have been performed, and the results agree in two respects. Firstly, a very large part of the maternal genome is expressed in oogenesis, and the gene products are required until late embryogenesis. Secondly, a normal (diploid) maternal gene dosage for many genes is required or beneficial for ensuring normal development. When the maternal gene dosage is insufficient, an increased zygotic genome dosage can often supplement and repair the maternal deficiency (Garcia-Bellido and Moscoso del Prado, 1979; Robbins, 1980; Garcia-Bellido and Robbins, 1983).

Garcia-Bellido and Robbins (1983) also showed that the expression of much of the maternal genome during oogenesis is required to ensure both the viability of progeny and the production of eggs. Homozygous deficiencies in the maternal genome for many genes reduce both the number of eggs produced and the viability of the embryos that result from the few eggs that are produced. The experiments involved the induction of homozygous deficient clones in the germ line, a procedure that will be explained later in this chapter.

However, these genetic tests do not establish the nature of the products that are stored in eggs. They might be maternal mRNAs, or they might be the proteins translated from these mRNAs or even the small metabolites that the proteins manufacture. The same uncertainty in interpretation of late larval rescue by maternal gene expression in *Caenorhabditis* has been previously noted.

One kind of observation that could resolve the question of whether maternal mRNAs are being passed on would be the finding of an increase in enzyme activity encoded by a gene present in the maternal genotype but not in that of the embryo itself. Gerasimova and Smirnova (1979) followed this strategy to track the accumulation during development of maternal forms of two enzymes specified by genes on the X chromosome. The genes were 6-phosphogluconate dehydrogenase (*Pgd,* 1–0.65) and glucose-6-phosphate (6GPD) (the *Zw* gene, 1–63). To distinguish maternally specified from zygotically specified forms of the two enzymes, they employed alleles that produce enzymes of slightly different electrophoretic mobility. Such "electromorphs" can be readily distinguished following starch gel electrophoresis and staining of the enzyme bands for activity. (An illustration of such electromorph separation can be seen in Fig. 4.8.) The maternally encoded forms of the two enzymes were designated the A forms, and the paternal forms were either null alleles or the electromorphic B varieties. To ensure that the progeny would inherit the A-type products without inheriting the A genes themselves, attached X mothers were used and only male progeny were followed. Because the two Xs of the mothers are tied together in attached X females, half of their gametes contain no maternal X chromosome; when such nullo-X gametes receive the paternal X, they develop into sons with no maternal X chromosome in their genotype.

The time course of enzyme activity in the male progeny of two crosses was followed:

1 $\hat{XX}(ywfZw^{A}Pgd^{A})/Y \times y^{+}w^{+}f^{+}Pgd^{-}Zw^{-}/Y$
2 $\hat{XX}(ywfZw^{A}Pgd^{A})/Y \times y^{+}w^{+}f^{+}Pgd^{B}Zw^{B}/Y$

In both crosses, male progeny were distinguishable by their darkened mouth hooks, produced by the y^{+} allele of their paternal X chromosome. In cross 1, male progeny received only the $Pgd^{-}Zw^{-}$ chromosome of their fathers; their sole capacity for synthesis of either of the two enzymes must have lain in maternal mRNA stores. The time course of appearance of the two enzymes in these mutant male progeny shows a 25- to 30-fold increase in both enzymes. This increase took place between early embryogenesis and the late larval period. All of the activity produced was of the maternal A type. In cross 2, in which zygotic genome expression of the paternal X could contribute enzyme activity, the observed increases were much larger, being on the order of 150- to 200-fold greater than the values at the start of development and consisted primarily of the paternal, electromorph B forms. To determine whether the increased activity in cross 1 might be the result of some form of progressive enzyme activation, Gerasimova and Smirnova measured the absolute amount of 6GPD protein by immunoprecipitation; the increase in protein was found to match the increase in enzymatic activity. These results provide the strongest evidence to date that maternal mRNAs can persist and function until late in *Drosophila* development.

These results concerning maternal expression in oogenesis in *Drosophila* are reminiscent of the findings regarding gene expression in the nematode and the transcription data in the sea urchin. Oogenesis in the fruit fly seems to be characterized by a very high level of gene expression in terms of both the number of genes expressed and the requirement for full maternal dosage of many of the expressed genes. The finding of comparable patterns in evolutionarily distant organisms suggests that a particular biological stratagem may be widely used. In this instance, the results of the different oocyte systems suggest that oogenesis is often accompanied by a massive, general expression of genes.

CLEAVAGE, BLASTODERM FORMATION, AND ACTIVATION OF THE ZYGOTIC GENOME

Fertilization in *Drosophila* occurs as the mature eggs are released into the oviducts. A single sperm enters the egg cytoplasm, and fertilization releases the oocyte from metaphase of meiosis I, triggering progression of the oocyte through the remainder of meiosis.

The haploid male and female pronuclei do not unite after completion of meiosis II, but become closely apposed and undergo the first cleavage in parallel (Sonnenblick, 1950). Immediately following this first cleavage, whose axis is oriented at random within the egg (Parks, 1936), nuclear fusion (syngamy) takes place to produce two diploid nuclei. As successive syncytial nuclear cleavage divisions occur, the daughter nuclei move outward through the yolk to the periphery, each nucleus enclosed in a jacket of cytoplasm. The propulsive force for this movement may be provided by the rhythmic contractions of the ooplasm that accompany cleavage (Fullilove et al., 1978; Foe and Alberts, 1983).

The first nine divisions occur with great rapidity, approximately every 9.6

min at 24°C (Rabinowitz, 1941). By the end of the ninth division, the nuclei have reached the surface and pooled their associated cytoplasmic coats with the outer, periplasmic layer. At this stage, 10–20 prominent bulges at the posterior mark the position of as many cleavage nuclei. These bulges pinch off to form the pole cells, which undergo two more divisions to produce about 50 per embryo (Underwood et al., 1980b). The pole cells later migrate dorsally and then invaginate to enter the embryo.

The remaining surface syncytial nuclei undergo four more cleavage divisions in the periplasmic layer (Foe and Alberts, 1983). Altogether, there are 13 cleavage divisions, which populate the surface layer with about 6000 nuclei. The final number of nuclei shows a surprisingly wide range, from

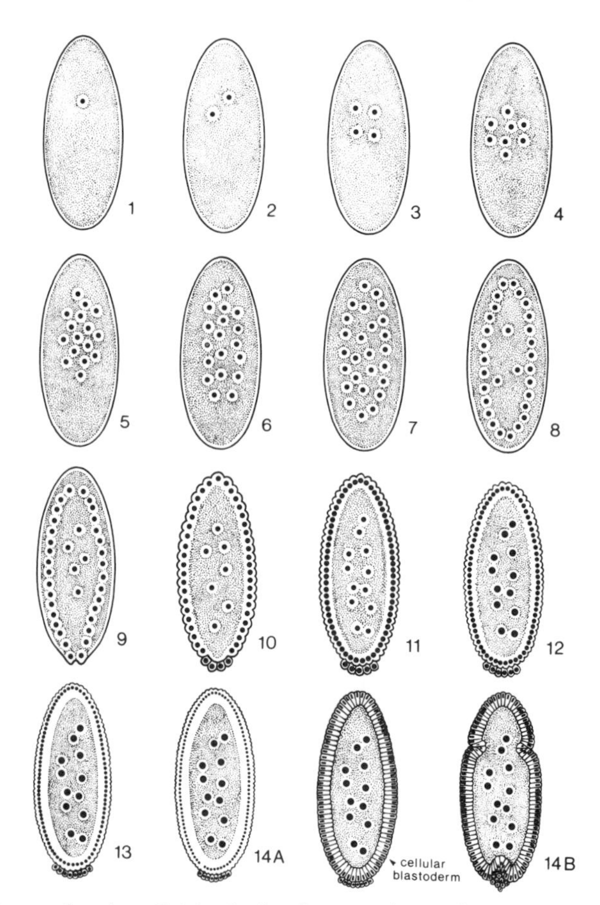

Figure 3.7 Stages of nuclear division in the cleavage of *D. melanogaster*. The number to the left of each embryo represents its developmental stage and the total number of nuclear division cycles experienced by the dividing nuclei. Anterior at top, posterior at bottom. Solid black circles, nuclei; stippled areas, yolky cytoplasm; clear areas, yolk-free cytoplasm. See text for details. (Figure courtesy of Dr. V.E. Foe; reproduced from Foe and Alberts, 1983, with permission.)

4600 to 6300 per embryo (Turner and Mahowald, 1976). Cleavage lasts for about 2.5 hr; its stages are shown in Figure 3.7.

The stage following the 13th division is termed "syncytial blastoderm" and lasts for about 50 min. During this period, membranes grow downward from the cell surface to enclose each nucleus within a long columnar cell 6 μm wide and 30 μm deep (Fullilove and Jacobson, 1971). The completion of the process leads to the "cellular blastoderm" stage. Shortly after formation of the cellular blastoderm, the embryo begins a series of cell movements and invaginations that give rise to the three germ layers and eventually to the segmented form of the larva.

The blastoderm marks the point of transition, not only between the syncytial and cellular phases of development but also between the maternally dominated part of embryogenesis and the beginning of the zygotic genome direction of development. This critical transition takes the form of a rapid activation of transcription in the previous quiescent embryonic nuclei. Activation of zygotic genome transcription has been detected in a variety of ways: autoradiographically, as a rapid increase in the labeling of nuclei in permeabilized embryos (Zalokar, 1976); as an increase in the nuclear binding of labeled poly U under nucleic acid hybridization conditions, signifying a rise in the nuclear titer of poly A^+ mRNAs (Lamb and Laird, 1976); and by a greatly increased frequency of transcriptional complexes in spread chromatin (McKnight and Miller, 1976). Before the syncytial blastoderm stage, transcriptional activity in nuclei is undetectable by standard radioisotope labeling techniques (Zalokar, 1976). The activation of zygotic genome expression is necessary for the completion of blastoderm formation: eggs injected with the RNA synthesis inhibitor α-amanitin at the end of the ninth cleavage division fail to cellularize (Gutzheit, 1979).

In contrast to transcription, protein synthesis is active throughout early development, even in unfertilized eggs (Zalokar, 1976), and interference with protein synthesis by standard inhibitors of translation abruptly stops cleavage (Zalokar and Erk, 1976). Given the inactivity of the embryo's nuclei, this high level of protein synthesis must be taking place on maternal mRNA templates. The first cells formed, the pole cells, are particularly active in protein synthesis.

The onset of substantial transcription in the embryonic nuclei is detectable within the first 20–30 min of the syncytial blastoderm stage. It includes the activation of rRNA synthesis from the clustered RNA cistrons in the nucleolus organizer, and of scattered nonribosomal transcription units (TUs) throughout the genome. Each TU is signified by a gradient of RNP fibrils of increasing size along a stretch of chromatin. The two classes of TU are distinguishable by their sizes and by the clustering and morphology of the fibrils (McKnight and Miller, 1976). The newly activated nonribosomal TUs average 2.93 $\pm$ 1.86 μm in chromatin length and range up to 8 μm in length. Assuming a typical packing ratio (μm of B-DNA per μm chromatin) of 1.6 and given that there are 3000 bp/μm DNA, these chromatin lengths correspond to about 14,000 ± 8,900 and 38,000 bp, respectively. Values in this range correspond to typical average lengths of polytene chromosome bands and may represent TUs of band-length genes. The mechanisms of zygotic genome activation at syncytial blastoderm are unknown, but simple migration away from cytoplasmic repressors is unlikely to be the explanation because blastodermal nuclei are surrounded by cytoplasm carried along from the interior (Foe et al., 1982). Interestingly, the pole cell nuclei remain untouched by this wave of transcriptional activation; they persist in a state of transcriptional quiescence (Zalokar, 1976).

Despite the impression conveyed by these data of a massive and complete activation of transcription in embryonic nuclei following cleavage, there is in reality a low level of zygotic genome expression during cleavage. Examination of spread chromatin from early embryos shows a low frequency of RNP fibril gradients in early embryos, signifying the existence of early TUs (McKnight and Miller, 1976). These first TUs can be detected as early as the 64-nuclei stage and can be identified as nonribosomal TUs by their fibril morphologies; they are distinguished from the later nonribosomal TUs by their size. Measuring only 0.91 ± 0.4 μm of chromatin in length, their sequence lengths are only about 4300 ± 1900 bp. Recently, molecular techniques have permitted the identification of gene sequences that are transcribed in preblastoderm embryos (Sina and Pellegrini, 1982). Transcription rates of these sequences are low and, for the most part, the transcripts are short relative to later transcripts. Several of these sequences appear to be restricted in their expression to preblastoderm stages; others are also expressed at later times. However, given the very low rates of transcription prior to blastoderm, it is probable that the developmental program of the preblastoderm embryo is predominantly under the control of maternal genome products. The findings described below indicate that by blastoderm the maternal program has impressed firm commitments on the cells or nuclei of the embryo.

DETERMINATION IN THE BLASTODERM EMBRYO

The evidence for distinctive commitments and restrictions of blastodermal cells comes from several sources. One class of experiments involves selective cell ablations at the blastoderm stage and scoring of the developmental consequences. Selective cell killing or removal of cells from the blastoderm produces specific, localized defects both in the larva (Lohs-Schardin et al., 1979b; Underwood et al., 1980a; Schubiger and Newman, 1981) and in the adult (Bownes and Sang, 1974a,b; Lohs-Schardin et al., 1979a). By correlating the patterns of observed defects with the positions of the inflicted cell damage or removal, it is possible to construct detailed "fate maps" of progenitor cells for the different larval and imaginal disc precursors; fate mapping procedures and their results will be discussed in some detail in Chapter 6. The pertinent point here is that the cells of the blastoderm have characteristic fates and relatively little capacity to substitute for their neighbors in development.

However, the cell ablation experiments do not prove that particular cells are "determined" to form particular structures. To assay states of determination, cells must first be removed from their normal locations and cultured for several cell generations; only if the progeny cells form the structures that they would have given rise to in situ can the tested cells be regarded as determined.

The first experiment to test for states of determination in blastodermal cells was carried out by Chan and Gehring (1971). They assayed anterior and posterior halves of blastoderm embryos to establish which structures could be formed by each half-embryo region. The procedure involved three steps: blastoderm embryos from a genetically marked strain were first bisected into anterior and posterior halves; the cells of the two sets of half-embryos were cultured separately and only then exposed to conditions that provoke differentiation of imaginal structures; and finally, the tested implants were scored for their inventory of imaginal structures.

To culture blastoderm cells, the embryos were dissociated by gentle homogenization in a balanced saline solution and then injected into the abdominal cavity of wild-type female adult flies. This in vivo culture method allows prospective imaginal cells to grow without differentiating. To obtain differentiation of the cultured pieces, Chan and Gehring removed the implants from the adult hosts, after a growth period of 10–14 days, and transplanted them into the body cavities of early third instar larvae. Exposure to ecdysone during pupation of these host larvae stimulates differentiation of the imaginal implants, and the tested fragments are then directly examined for the disc structures to which they give rise. To ensure that absence of a particular disc type in the cultured implants was not an artifact produced by the procedure, Chan and Gehring first combined the anterior and posterior halves with a comparable mass of whole blastoderms of a second distinctive genotype before homogenization (Fig. 3.8). With the difference in cuticular markers of the test and control cells, any structure could be assigned unambiguously either to the tested half-blastoderms or to the control embryos. (The host animals, both larvae and adults, were of the wild-type genotype, permitting detection of any contribution from this source.)

Anterior half-blastoderms, with the exception of one doubtful case, yielded only head and thoracic imaginal structures (44 out of 45 implants), and posterior halves produced only abdominal and thoracic derivatives (26 classifiable implants). The experiment demonstrates that some form of determination to "anteriorness" or "posteriorness" has occurred by the blastoderm stage. The results also provide a crude fate mapping of the cells that give rise to imaginal primordia: cells of the blastoderm that give rise to head structures are located solely at the anterior end of the embryo; those that develop into imaginal thoracic structures are in the middle; and the abdominal structure precursors are located in the posterior half of the embryo.

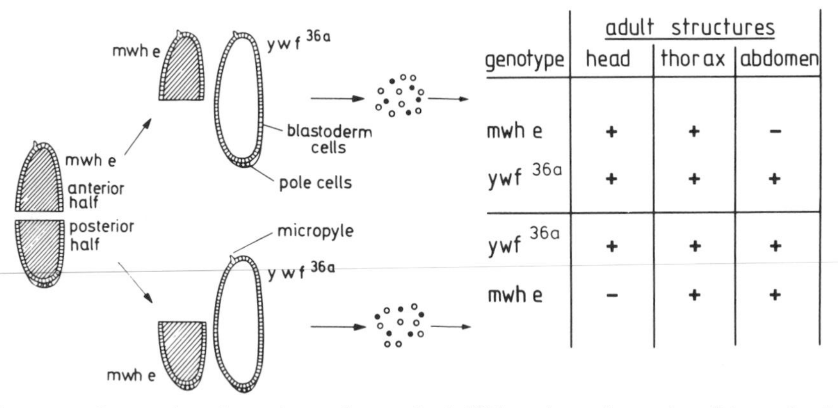

Figure 3.8 Preparation of anterior- and posterior-half blastoderms for testing of determinative states. Genetically marked embryos (*mwh e*) at the blastoderm stage are bisected, and the anterior and posterior halves are separately mixed with whole carrier embryos (genotype: $y\ w\ f^{36a}$). The mixtures are dissociated and filtered to make cell suspensions, which are concentrated by centrifugation and injected into adult females for in vivo culture. The developmental contributions of experimental and control cells are shown on the left. (Figure kindly provided by Dr. W. J. Gehring.)

The experiment does not indicate the degree of restriction in blastodermal cells beyond this rough regional characterization, but other results suggest that the narrowness of assignment may be quite sharp. These experiments involve transplantations of *single* blastodermal cells from genetically marked embryos into host blastodermal embryos; the cells were allowed to grow inside the host embryos, and were then scored after eclosion of the hosts. Transplantation was made either into the comparable location in the host ("homotopic" transplantation) or into a different site ("heterotopic" transplantation) (Illmensee et al., 1976; Illmensee, 1978). The genetic marker employed in these experiments was the autonomous cell marker gene *mal,* whose wild-type function is to specify an essential activator of the enzyme aldehyde oxidase (Finnerty, 1976). mal^+ donor cells in a mal^- donor background differentially express aldehyde oxidase and can therefore be readily detected by histochemical staining.

In homotopic transplants, surviving implants were found integrated into the corresponding host imaginal structures (Fig. 3.9*a*,*e*). The heterotopic transplants, however, provide the critical test of the stringency and specificity of determined cellular states. In the majority of cases, where the implant was recovered, the donor tissue was found to be in the form of unintegrated implants within the body cavity of the host characteristic. The structures formed were those characteristic of the *donor site of origin* (Fig. 3.9*b*,*d*). Most of these structures were those of one disc type. In a few instances, derivatives of two neighboring discs, leg and wing, were found (Fig. 3.9*b*). However, some single imaginal precursor cells of the blastoderm normally give rise to discs of two types if these discs are in the same segment. In one

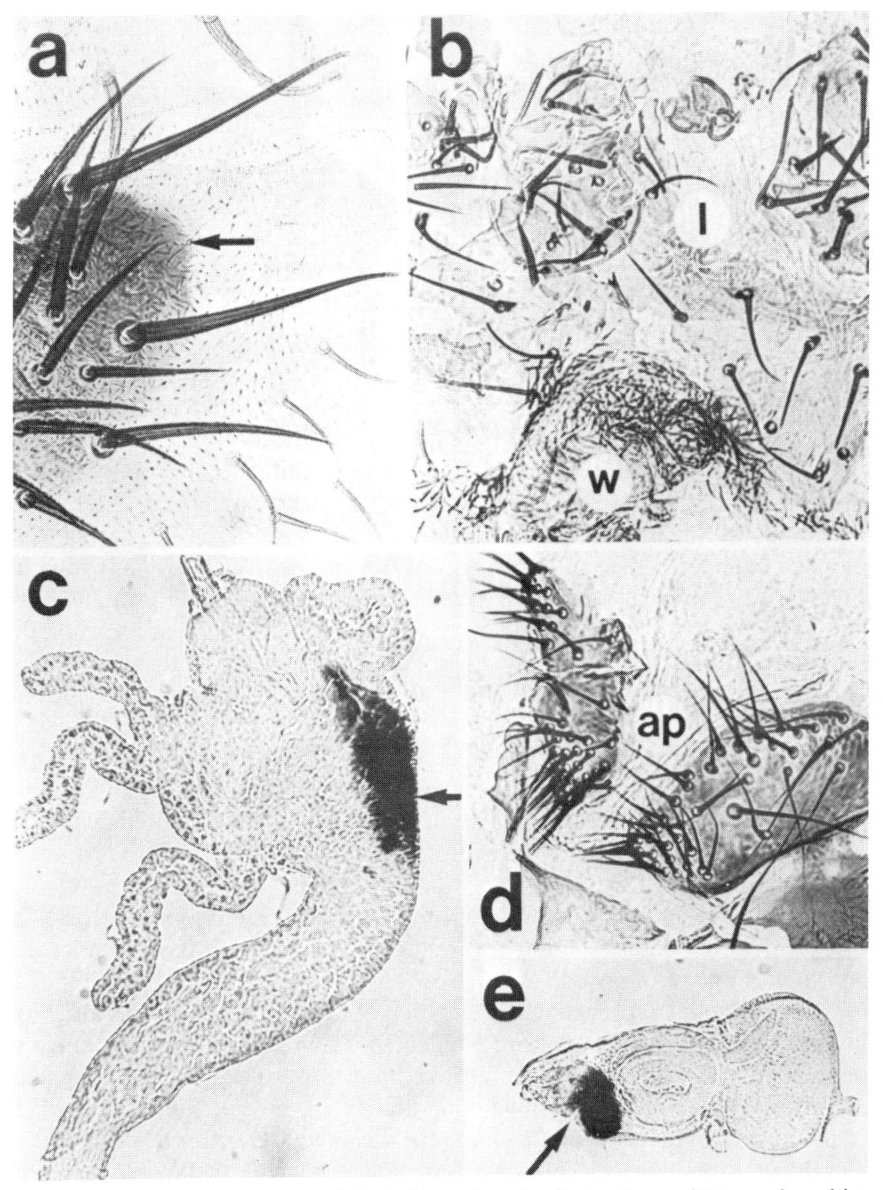

Figure 3.9 Determinative states of single blastodermal cells (*mwh e mal*$^+$) transplanted into host (*y sn mal*) blastoderm embryos, assayed by their developmental outcomes. The donor cell contributions from homotopic (*a* and *e*) and heterotopic (*b*, *c*, and *d*) transplantations, following metamorphosis, are shown. (*a*) Abdominal region of a female chimera, with cells derived from a donor (*mal*$^+$) shown as dark staining (X250); (*b*) donor-derived, nonintegrated transplant of wing (w) and leg (l) tissue from a midlateral blastoderm cell transplanted to the posterior pole of a recipient embryo (X150); (*c*) a larval anterior midgut clone (*mal*$^+$) from a midventral blastoderm cell placed anteriorly (X75); (*d*) nonintegrated male genital structures (ap, anal plate) from a posterior cell placed anteriorly (X150); (*e*) integrated donor (*mal*$^+$) clone in eye antennal disc. (Reproduced with permission from Illmensee, 1978.)

heterotopic transplant, a single midventral cell gave rise to structures not characteristic of its normal fate (Fig. 3.9*c*); this may be an instance of developmental lability. For the most part, the single cell transplants indicate that single blastodermal cells have a narrow range of intrinsic developmental capabilities.

To explore the molecular basis of these differences in determinative state, it would be extremely helpful to have molecular markers for these differences. Unfortunately, there has been little success to date in identifying such differences. Some evidence for pole cell specific protein(s) has been obtained (Graziosi and Roberts, 1975; Gutzheit and Gehring, 1979), as well as evidence for the synthesis of a protein specific to all blastodermal cells (Gutzheit, 1979), but molecular markers of anterior-posterior blastodermal cell differences (or of finer differences within the blastodermal cell layer) have not been detected in *Drosophila*. In another dipteran, *Smittia,* such searches have been successful. Distinctive abundant marker proteins of both the anterior and posterior halves of the blastoderm have been found. Furthermore, the synthesis of these proteins can be changed in parallel with induced changes of fate in either half of the embryo, such as that shown in Figure II.2 (Jackle and Kalthoff, 1980, 1981). Although such proteins have not been reported in *Drosophila,* it would be surprising, in view of the relatedness of all diptera, if they did not exist. Such early molecular markers of determination, in combination with the ability to manipulate the determinative system by genetic means (see below), could prove extremely useful.

Determination, as defined by the techniques of experimental embryology, is a cellular property: cells are restricted to follow certain paths and prohibited from following others. However, if one wants to understand the molecular basis of the phenomenon, it is crucial to know whether there is a set of nuclear properties that correlate with the cellular behavior. Presumably, determination involves some restriction in the capacity for gene expression. Questions about determination may reduce to the nature of these restrictions and how they are imposed.

The traditional approach to the question of nuclear states in determination involves nuclear transplantation. Nuclei from somatic cells at some stage of development are placed in an unfertilized or fertilized egg and their capacity to participate in subsequent developmental events is tested, usually in place of the host oocyte or zygote nucleus. (In the syncytial embryos of insects, a variant of this approach is to add the donor nucleus to the host embryo without removing the host nucleus; successfully developing embryos are nuclear transplant chimeras.) If the transplanted nucleus can give rise to any adult structure, it is said to be "totipotent" and inferred to have experienced no irreversible genetic changes.

The principle of nuclear transplantation experiments is easy to describe but the interpretation of the results is not always straightforward, particularly for the majority of test nuclei that do *not* successfully supplant the zygote nucleus in development. Most of these failures are attributable to the failure of the donor nuclei to adjust to the cytoplasmic environment of the egg, in particular to the imposed rapid rate of replication that takes place in the egg. Damage during manipulation might also be a contributory cause. However, some fraction may be produced by irreversible genetic change or by some rigid form of chromosome modeling that cannot be erased by the egg cytoplasm.

In *Drosophila,* numerous nuclear transplantation experiments have been performed (reviewed by Illmensee, 1976, 1978). Although few transplanted nuclei have carried development directly to the imago stage, a significant finding is that nuclei from blastodermal or postblastodermal cells are as capable of supporting development to successive stages as nuclei from cleavage stage embryos. Since some degree of determination takes place at blastoderm, this seems to suggest that the process of determination involves no fundamental alteration of nuclear state but is "carried" solely in the cytoplasm. In this view, the nuclear state that accompanies cellular determination is a highly labile one, perhaps fully comparable to the transient regulatory states that exist in prokaryotes.

This inference does not necessarily follow. Determination might well involve the tight binding of regulatory molecules and/or certain kinds of chromosomal folding patterns, affecting gene expression. In this circumstance, self-perpetuation of the determined state would involve recurrent synthesis and binding of the regulatory molecules and/or self-templating of the chromosomal folding pattern. (For one view of how such states may be perpetuated, see Weintraub et al., 1977). If a nucleus whose chromosomes possess such properties is placed in a cytoplasm that lacks the requisite molecules for perpetuating this state, and then asked to replicate, its chromosomes will lose their distinctive properties through simple dissociation of the bound molecules or titration in successive rounds of replication. This may well be the situation experienced by a somatic nucleus placed in the very different cytoplasmic environment of the fertilized egg.

If nuclear states are fairly rigidly set or require several rounds of replication to be erased, then a determined nucleus placed in a cytoplasm where it could not rapidly replicate would retain its state. Kauffman (1980) performed such a test with heterotopic nuclear transplantations at the early syncytial blastoderm stage. Using genetically marked nuclei to distinguish cells derived from donor nuclei, he found that cells arising from the transplanted nuclei almost always display the developmental fates typical of the donor sites from which the nuclei were derived and not those of the recipient blastodermal location in which they were placed.

The culture and assay procedure in this experiment was similar to that of the Chan-Gehring (1971) experiment: following nuclear transplantation and cellularization, the recipient embryos were cut in half and each section that contained transplant material was cultured in the abdomen of a host adult fly (of a third distinguishable genotype) for 8–10 days, then injected into a larva and allowed to metamorphose along with the larval hosts. Of 73 scorable implants, 15 were found to be of the donor genotype, and of these, 14 were of structures typical of the donor site from which the nuclei were taken. Since ample time was allowed for exposure of nuclei to counterdetermining influences before cellularization, the results suggest two conclusions. The first is that by syncytial blastoderm, nuclear commitments have been imposed. The second is that these nuclear commitments are stable in the absence of exposure to or replication in early egg cytoplasm.

To sum up: the first determinative events have occurred by the cellular blastoderm stage; the specificity of these determinative assignments is unknown but may be fairly high; finally, these restrictions on cellular capacity may have moderately stable nuclear state correlates. One central task for a genetic analysis of development in *Drosophila* is to identify the maternal mechanism that creates these states.

GENETIC ANALYSIS OF OOGENESIS IN *DROSOPHILA:* AN OVERVIEW

Analysis of the development and properties of the mature oocyte of the fruit fly is a formidable task, judging from the number of cell transformations and cellular interactions required for oocyte formation. This developmental sequence is schematized in Figure 3.10, which summarizes the major events in the pathway. The underlying genetic complexity of oogenesis is considerably greater than indicated by the diagram; for the completion of each step, symbolized by a single line, there are many genes whose activity is required.

The genetic dissection of oogenesis in *Drosophila* can be approached from either of two standpoints. It can be treated primarily as an exercise in applying genetic stratagems to unearthing the critical transition points in a complex and interesting developmental pathway. Alternatively, the focus can be narrowed to understanding those aspects of the egg that give rise to the morphogenetic system of the embryo. These two approaches are complementary: understanding the morphogenetic system in depth requires an understanding of how the oocyte develops. In the section that immediately follows, we will first look at oogenesis as a developmental pathway, and the various genetic methods used to probe it will be described. In the next section, we will examine the morphogenetic system of the egg as revealed by genetic analysis.

The source material for a genetic dissection of oogenesis in *Drosophila* is a large set of female sterility (*fs*) mutants. The great majority of these mutants are defective in some aspect of oocyte formation—some obviously so, others in more subtle fashion. These oocyte defectives can initially be classified into non-egg layers or sterile egg layers. The first category are, for the most part, defective in some early or intermediate stage of egg formation; the second are predominantly maternal effect mutants, whose defects prevent normal embryogenesis from occurring.

As in *C. elegans,* the mutants obtained must be screened and categorized

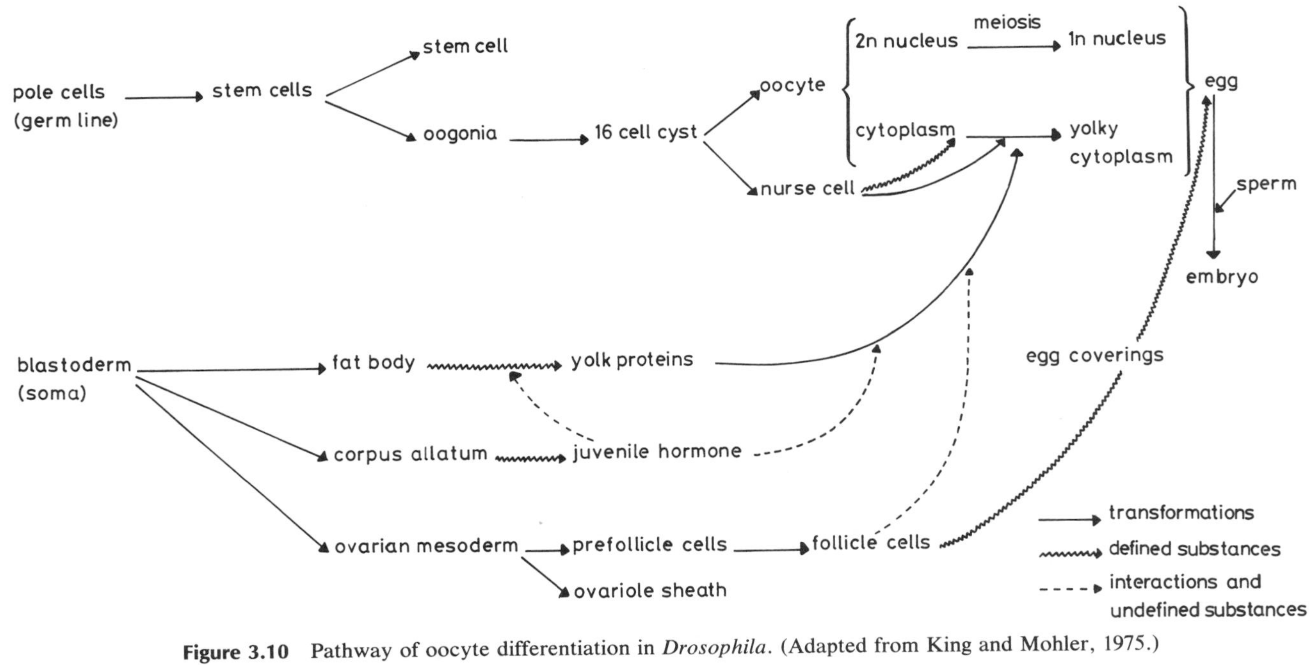

Figure 3.10 Pathway of oocyte differentiation in *Drosophila*. (Adapted from King and Mohler, 1975.)

to select the ones of interest. Some *fs* mutants make normal eggs but are behavioral defectives, incapable of mating. Others make normal eggs and can mate but do not lay the eggs; these are oviposition defectives. The majority have defects in oogenesis or early embryogenesis. However, a substantial proportion of the latter are pleiotropic maternal effect mutants, such as *dor* and *r*, whose mutant defects are hypomorphic mutations in general cellular functions. As in *Caenorhabditis,* many mutants must be examined in order to identify those that are specifically informative about oogenesis.

The method used to obtain oogenesis mutants in *Drosphila* is rather different from that employed in *Caenorhabditis*. Since *Drosophila* is not hermaphroditic, the major effort is invested in making stocks homozygous for newly induced mutations. In contrast to the nematode system, the final screening step for presumptive oogenesis defectivity is comparatively easy. In particular, the well-defined developmental progression in oocyte development (Fig. 3.6) provides a series of developmental landmarks with which any mutant can be compared.

The particular series of crosses used to obtain the mutants is determined by whether one is looking for autosomal or sex-linked mutants. Representative screening methods for the two different classes are shown in Figure 3.11. Both rely on the use of specially constructed chromosomes termed "balancer chromosomes" for the isolation of perpetuation of the obtained *fs* mutants. Balancer chromosomes are distinguished by dominant mutations (e.g., *Curly wing, Cy,* or *Bar eye, B*) and carry multiple inversions. The inversions function to suppress recombination. The function of balancers is to permit the conservation of linkage groups (chromosomes) bearing mutations without recombinational loss of the mutations. Thus, any heterozygote for a balancer and a mutation-bearing chromosome (Bal/*m*) will produce only two kinds of gametes, balancer-bearing gametes and mutation-bearing gametes. If the dominant mutations are themselves lethal when homozygous or if the balancers bear recessive lethals in addition to their dominant markers, and if the homologous chromosome also bears a lethal, then such a balancer stock is a self-perpetuating heterozygote. Any recessive lethal can be maintained indefinitely in *trans* to a lethal-bearing balancer, and this is the method by which lethals, including female steriles (which are lethal for a line, although not for the carrier female), are perpetuated as stocks.

The essence of the schemes shown in the figure is to transmit single mutagenized chromosomes into individual flies; the male and female progeny of these flies are then crossed to produce female mutant homozygotes. New chromosomal lines that produce female sterility specifically are then further characterized by dissection and microscopic examination to determine the degree of ovary development and the presence or absence of eggs in the ovarioles. For those *fs* mutants that produce eggs, the stage of arrested development of the eggs is determined. From the observed characteristics, it is sometimes possible to make hypotheses about the nature of the

a) Isolating autosomal fs mutants

G1 Bal-A / Bal-B ♀♀ × +/+ ; EMS-treated ♂♂

G2 3 Bal-A / Bal-B ♀♀ × Bal-A / "+" single ♂♂

G3 3 Bal-A / Bal-B ♀♀ × Bal-A / "+" single ♂♂

G4 Select Bal-A / "+" ♀♀ × Bal-A / "+" ♂♂ siblings

G5 Select "+" / "+" ♀♀ ; × to wild-type ♂♂

fertile stocks-discard; sterile stocks-characterize

b) Isolating sex-linked fs mutants

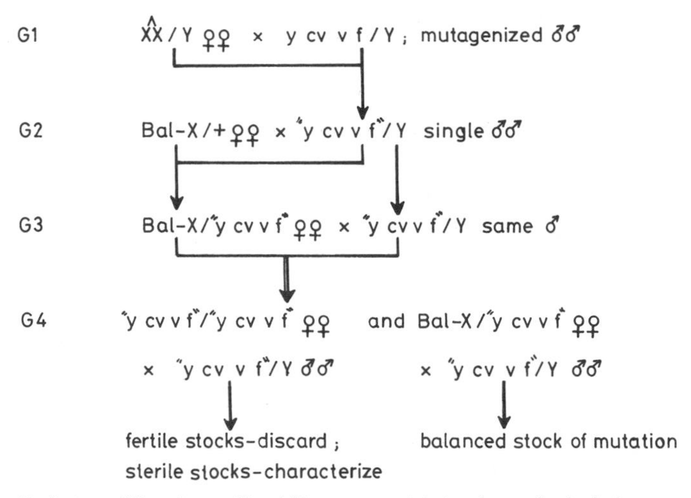

Figure 3.11 Isolation of female sterility (*fs*) mutants. (*a*) A scheme for isolating autosomal *fs* mutants (based on that of Bakken, 1973); (*b*) a procedure for isolating sex-linked *fs* mutants (from Mohler, 1977). Bal-A and Bal-B, autosomal balancer chromosomes with different dominant marker mutations; +, mutagenized wild-type chromosome; X̂X, attached X chromosome; Bal-X, an X balancer chromosome carrying a dominant *Bar* mutation; *y cv v f*, an X-chromosome with four recessive marker mutations.

mutant blocks. Thus, in mutants that are found to accumulate eggs at stages 7 and 8, the defect is most probably within the program of vitellogenesis, or yolk production, since stage 8 marks the beginning of yolk accumulation in the egg. Further examination by electron microscopy can sometimes suggest the probable cellular site of the initial defect. However, the visual evidence always requires corroboration by other means.

An intensive effort in several laboratories (Bakken, 1973; Gans et al., 1975; Rice and Garen, 1975; Mohler, 1977) has yielded a large number of new *fs* mutations, many in previously undetected genes. However, for the reasons enumerated earlier, many are not informative about the distinctive features of oocyte differentiation, or are defective in vitellogenesis, which involves a distinctive but readily understandable feature of the egg. Nevertheless, when these relatively uninformative mutant categories are subtracted, a large number of mutants remain that are potentially instructive about the distinctive control events in oogenesis. One characteristic that most of these mutants show is autonomous expression in ovaries: when ovaries are transplanted into the abdomens of wild-type females, they continue to produce defective egg chambers. This characteristic indicates that the defect is a consequence of some localized cellular deficiency within the ovary.

A particularly interesting class of mutants are a group affected in the first events in oocyte differentiation, the so-called ovarian tumor mutants. These mutants do not produce true invasive tumors, but rather egg chambers filled with hundreds to thousands of young cystocytes (in place of the normal egg chamber containing 15 nurse cells and 1 oocyte). Five genes that can mutate to produce the ovarian tumor phenotype have been identified. Two are on the second chromosome (*fes* and *nw*), and the other three are sex-linked (*fu, otu,* and *fs(1)1621*) (Table 3.1). All of the mutants show ovary auton-

Table 3.1 Ovarian Tumor Mutants of *Drosophila melanogaster*

Gene	Map Position	Other Characteristics	Reference
female sterility (*fes*)	2–5	—	King (1970)
narrow (*nw*)	2–83	Wing defects	King (1970)
fused (*fu*)	1–59.5	Wing defects	King (1970)
otu (formerly *fs(1)231*)	1–22.7	—[a]	Gans et al. (1975); King and Riley (1982); Bishop and King (1984)
fs(1)1621	1–11.7	—	Gans et al. (1975); Gollin and King (1981)

[a] A bristle defect was reported in the initial allele, otu[1], but is not seen in other mutants of this locus.

omy (King and Bodenstein, 1965; Wieschaus et al., 1981; King and Riley, 1982).

The developmental defects in these strains appear to reflect failures of cell division control within the cysts. As noted earlier, development of the wild-type cyst is produced by a set of four mitoses each followed by incomplete cytokinesis, beginning with the cystoblast progenitor cell that results in a set of 16 cystocytes connected by 15 intercellular canals, of which one four-canal cell develops into the oocyte. The tumorous phenotype seems to be a consequence either of the failure to arrest cytokinesis or to generate four-canal cells. The mutants provide a means of examining the study of developmental mutants: whenever a mutant uncouples two processes or features that are normally found together, one may infer that these processes or features are not obligately linked in wild-type development.

The observations also support the idea that cytokinesis must be incomplete for differentiation of an oocyte to occur. Johnson and King (1972) examined egg chambers under conditions of maximal expressivity of *fes* and found that 35–40% of all divisions completely separate the daughter cells, in contrast to the situation in the wild-type, where cystocyte division is always incomplete. One result of fes^+ gene action, at least, must be to suppress the completion of cytokinesis within cysts. The mutants *fs(1)1621* and *otu* also show elevated frequencies of completed cell division in cysts (King, 1979; Gollin and King, 1981; King and Riley, 1982). Like *fes*, *fs(1)1621* shows a reduction in mutant expressivity when flies are grown at 18°C. At this temperature, egg chambers of mixed phenotype are produced, in which cells are typically tumorous and others are nurse cell-like. This result suggests that the cellular commitment to differentiate is made independently by each cystocyte rather than by the egg chamber as a whole.

In wild-type cyst development, there are invariant relationships between the capacity to form oocytes and the presence of four-canal cells and the possession of 15 nurse cells. The tumorous mutants can be used to test which of these relationships are crucial for oocyte development. Such tests are possible because the severity of the mutant phenotype can be adjusted by manipulation of environmental conditions and background genotype. By varying expressivity, it is possible to determine whether normal oocyte development is correlated with the frequency of the normal division or intercellular canal pattern.

For several of the mutants, conditions that shift egg chamber development away from the "tumorous" phenotype toward the wild-type pattern include an increase in the heterozygosity of the background genotype and a decrease in the growth temperature from 25°C to 18°C. When egg chamber development in the *fes* mutant is examined under these conditions, it is found that approximately 2% of the chambers contain yolky oocyte-like cells, an 80-fold increase over the standard *fes* frequency at 25°C (King et al., 1961). It was invariably the case that the yolky, oocyte-like cells had four canals and that none of the cells with three or two canals accumulated yolk

(King, 1970). However, oocyte-like cells were found in chambers containing as few as 10 or as many as 30 nurse cells. Thus, while commitment of a cell to differentiate into an oocyte seems to require four canals, a particular number or arrangement of nurse cells within the egg chamber is not critical. The results suggest one possible explanation of the mutant defects in terms of cystocyte growth. In wild-type development, mitotic divisions proceed without substantial growth until each of the 16 cyst cells has a volume of about 50 μm^3. This might result in a nucleocytoplasmic ratio that is too high to support further division; by the time additional growth has occurred, commitment to nurse and oocyte cell formation could inhibit further division (King, 1970). If growth in the tumorous mutants is accelerated relative to division, the minimal inhibitory cytoplasmic volume may never be attained, with a consequent continuation of cell division. The alternative possibility is that the wild-type alleles of these genes specify products that directly or indirectly inhibit the physical process of cell separation, cytokinesis.

The formation of the 16-cell cyst and the selection of one of the initial four-canal cells as the definitive pro-oocyte take place in the germarium, the germarial phase lasting for 3 days (at 25°C). However, the major phase of oocyte growth and development occurs in the vitellarium. During the 5.5 days that the oocyte is in the vitellarium, it experiences its most complex set of intercellular reactions, first with the nurse cells and then with the premature and mature follicle cells. Potentially, mutations that affect these interactions can help to define those processes by which the oocyte acquires its definitive embryogenesis-directing properties. However, the very complexity of these interactions guarantees that interpretations of the mutant effects will be hedged about with uncertainties.

For instance, microscopic observation or biochemical tests might indicate that nurse cells were aberrant in a particular mutant and were the source of the developmental effect. However, this conclusion could be false; the initial defect might be in the follicle cells, with only a secondary but observable consequence in the nurse cells. Similarly, apparent follicle cell defects might be secondary consequences of developmental aberrations originating elsewhere. In fact, the analysis of ts mutant defects shows that there is a large degree of overlap in the expression of essential genes between follicle cells and nurse cells (King and Mohler, 1975). For a hypomorphic mutant of such a jointly expressed gene, an apparent mutant specificity in one or the other cell type may therefore reflect only the residual expression of the mutant gene and the relative requirements for that gene activity in the two cell types.

Unambiguous assignment of cellular sites of gene action is clearly a complex matter. Indeed, this problem of distinguishing *initial* sites of gene action is encountered in the developmental analysis of mutant effects whenever cellular interactions between two or more tissue types are involved. The solution to the problem involves a genetic technique: the experimenter transplants cells or pieces of tissue between the mutant and the wild-type to create reciprocal genetic chimeras in which one tissue is mutant and the other is wild-type. If the mutant gene is acting in one cell type but not in the other, the reciprocal chimeras should reveal the site of action. One chimera will show the full mutant genotype—the one in which the expressing tissue is mutant—while the other chimera, possessing wild-type expressing tissue, will be normal. The principle of this experiment is illustrated in Figure 3.12. We have already seen the utility of chimeras for answering questions about

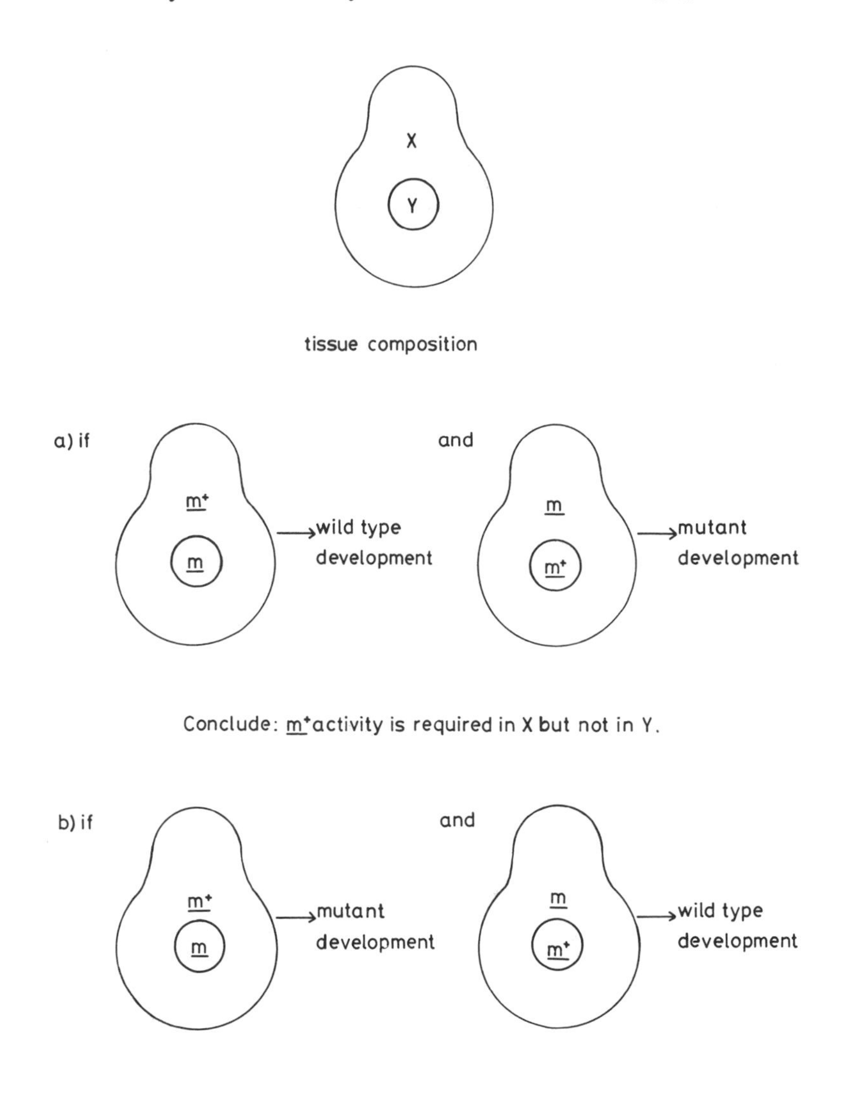

Figure 3.12 The reciprocal chimera experiment: outcomes and interpretations. X and Y represent the two component tissues of an organ; m^+ and m, the wild-type and mutant alleles, respectively, of the gene whose site of action is under test.

the properties of cells and nuclei, as in the single blastodermal cell transplants of Illmensee and the nuclear transplantation experiments of Kauffman. The reciprocal chimera method illustrates the usefulness of chimeras for answering a second question, that of sites of gene action. The site of mutant gene effect is inferred to be the site of the wild-type allele's activity in normal development.

For the analysis of sites of gene action in oogenesis, the ability to create reciprocal chimeras is made comparatively easy by the fact that all the germ line cells of the ovary derive from a very small initial group of cells, the pole cells. To create chimeric egg chambers, one transplants pole cells of one genotype into the posterior pole of an embryo of a second genotype; the transplanted donor cells give rise to donor germ line cells that are enveloped by somatic follicle cells of the host genotype. To create reciprocal chimeras between the mutant and the wild-type, one performs the transplantation in both directions: mutant pole cells into wild-type hosts and wild-type pole cells into mutant hosts.

This type of analysis will be illustrated with two different sex-linked *fs* mutants: *ocelliless* (*oc*) (1–23) (Johnson and King, 1974) and *fs* (*1*) *K10* (1-0.5) (Wieschaus et al., 1978). In contrast to the ovarian tumor mutants, *oc* and *K10* may be considered mutants of functions required late in oogenesis. Both mutants produce eggs, but these eggs exhibit aberrant chorions characteristic of each mutant. In addition, both mutants are ovary autonomous; their cellular sites of defect, therefore, must be in either the germ line or follicle cells or both.

The formation of the chorionic layers is one of the final events in oogenesis, taking place between stages 11 and 14. These layers are secreted in sequence by the underlying follicle cells. First, the inner layer or endochorion is laid down in stage 12, and then the outer layer or exochorion is deposited by the follicle cells on top of the endochorion in stages 13 and 14. A section through the chorion of a mature oocyte is shown in Figure 3.13*a*. The endochorion consists of a thin inner layer over the vitelline membrane and a much thicker outer compartmented layer containing many air spaces. These endochorionic spaces probably ensure provision of an air supply for submersed eggs. The exochorion is less electron dense than either endochorionic layer. In stage 14 oocytes it is covered by the squamous follicle cells, which are sloughed off during ovulation, leaving behind a characteristic array of hexagonal imprints.

In *oc,* the whole chorionic architecture is highly disrupted (Fig. 3.13*b*). The inner endochorionic layer is completely missing; the outer endochorionic layer, with its struts, is greatly diminished and lacks air spaces; and the exochorionic layer is less homogeneous in appearance and is separated from the follicle cells by a layer of debris. *oc* is pleiotropic but unlike many pleiotropic *fs* mutants, the adult phenotype (which includes the absence of the eyespots from which the mutant takes its name) is probably unrelated to

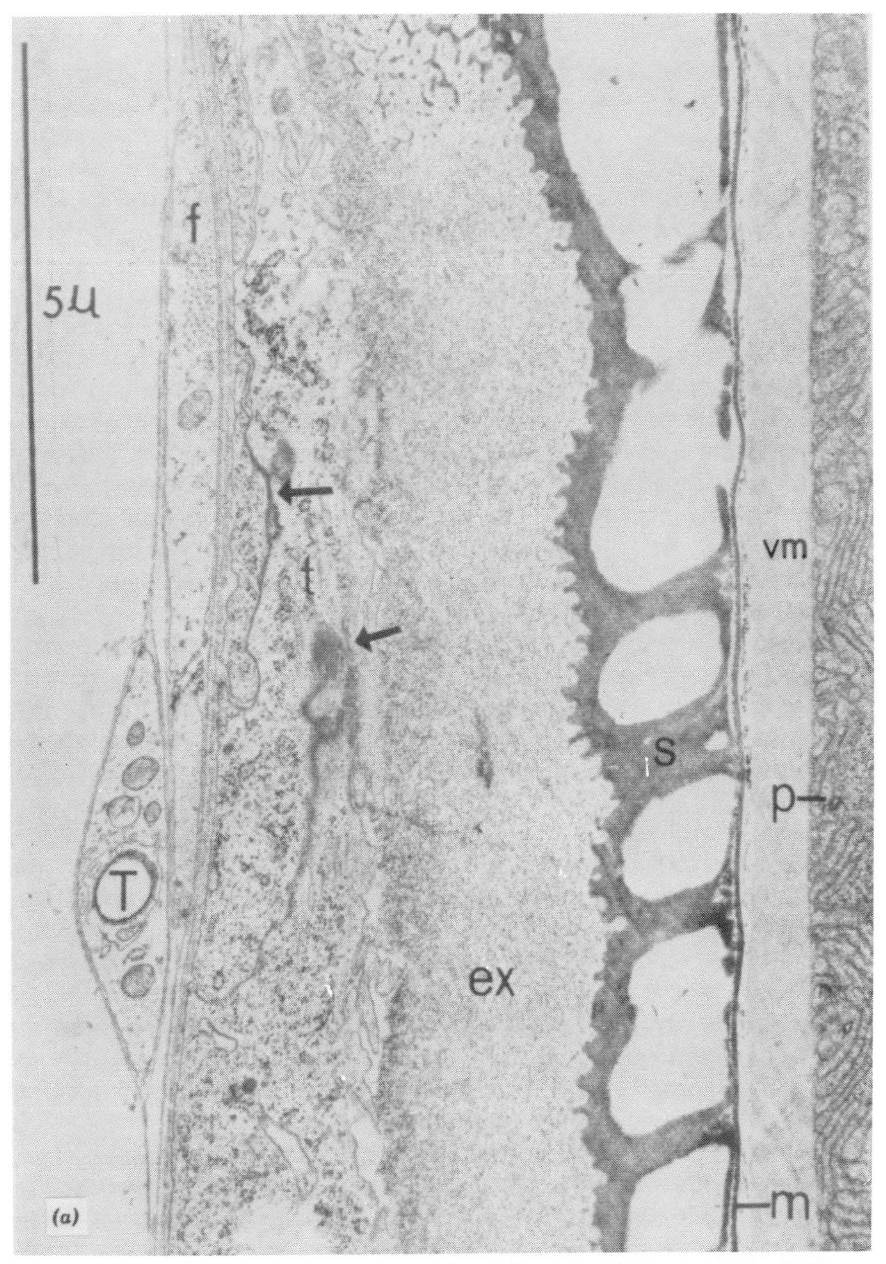

Figure 3.13 Chorion ultrastructure in eggs of (*a*) wild-type and (*b*) *oc*. ex, exochorion; m, chorionic membrane; s, strut connecting two layers of endochorion; vm, vitelline membrane; f, follicle cell. (Reproduced with permission from *Int. J. Insect Morphol. Embryol.* vol. 3, Johnson and King, Oogenesis in the ocelliless mutant of *Drosophila melanogaster* Meigen,'' © 1974 Pergamon Press.)

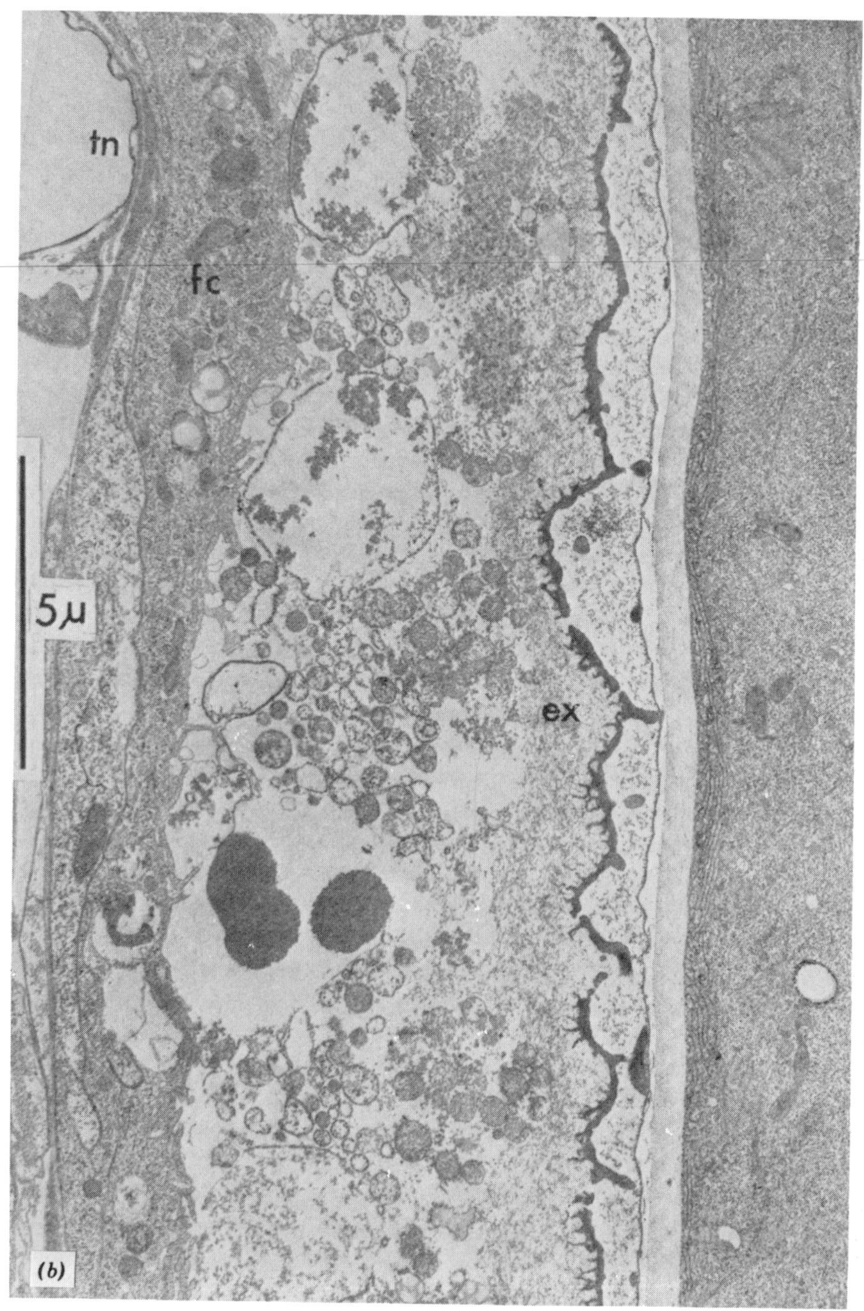

Figure 3.13 (*Continued*)

the oogenesis defect; the nature of the *oc* mutation will be described in Chapter 9.

The experiment to determine whether *oc* acts directly in the chorion-secreting follicle cells or indirectly through the germ line cells was performed by Underwood and Mahowald (1980) and is diagrammed in Figure 3.14. The top part of the figure shows the genotype of the *oc*-bearing strain. Like all recessive lethal or sterility mutations, *oc* is maintained as a hetero-

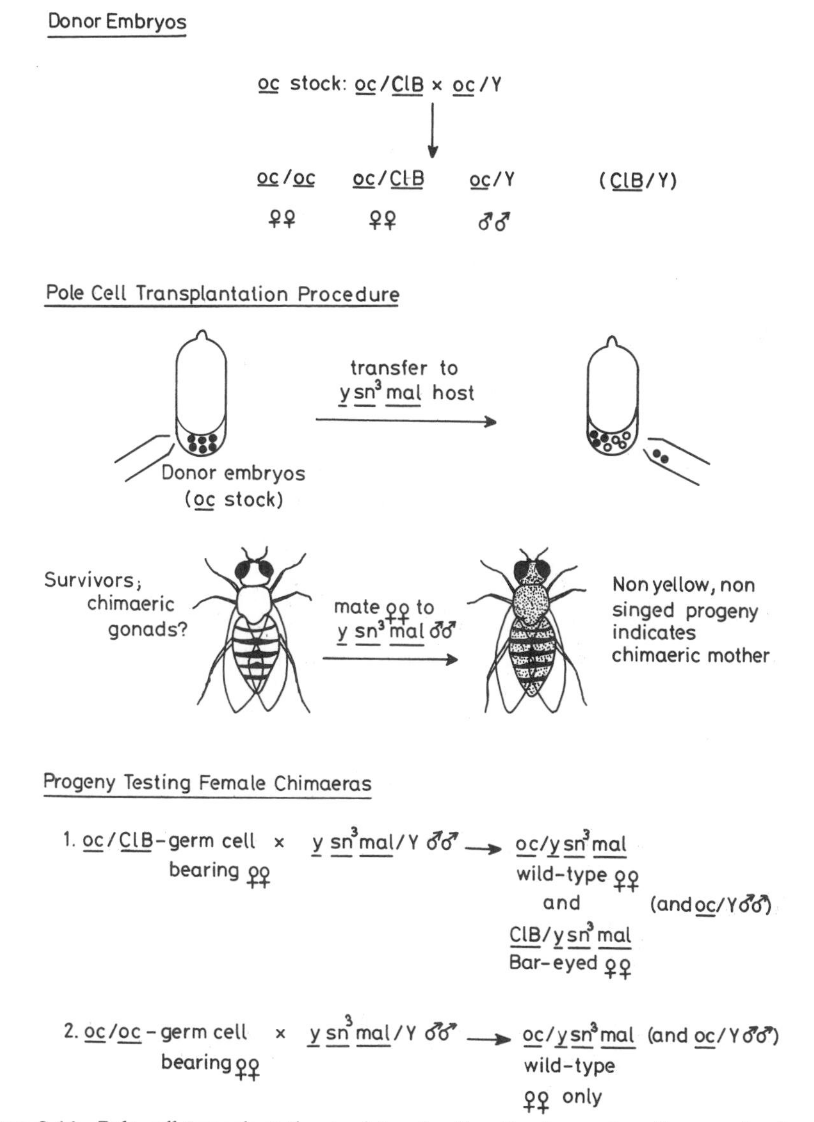

Figure 3.14 Pole cell transplantation to determine the site of *oc* action. See text for details. (*ClB* is a standard X chromosome balancer, carrying a recessive lethal and a *Bar* mutation.)

zygote over a marked balancer chromosome. Mating within this balanced heterozygous stock routinely produces three kinds of viable individuals in a 1 : 1 : 1 ratio. (The males with the balancer chromosome, *ClB,* die because of the lethal, *l,* on this X chromosome.) The three classes of viable progeny are *oc/oc* sterile females, *oc/ClB* fertile females (the balancer chromosome being wild-type with respect to the *oc* defect), and *oc*/Y fertile males (the *oc* mutation having no effect on male fertility).

The critical experiment involves transplanting pole cells from *oc/oc* female embryos into +/+ female embryos and determining whether or not the donor cells can give rise to viable progeny. If *oc* oocytes enveloped by + follicle cells produce progeny and in normal numbers, then the mutant defect must be in the follicle cells; if the *oc* germ line cells cannot produce progeny when surrounded by normal somatic cells, then the mutation must involve a germ line defect, presumably in the nurse cells.

This experiment is complicated by two factors. In the first place, half of the pole cell transplantations inevitably fail. The reason is that all "heterosexual" chimeras, in which female (XX) pole cells are transferred to male (XY) blastoderm hosts or vice versa, never produce functional donor germ line cells. Only transplantations between embryos of the same sex allow donated pole cells to contribute to the germ line (van Deusen, 1976). Because embryos cannot be sexed rapidly before transplantation, the number of transplantations must be proportionately scaled up to obtain an adequate sample. Furthermore, at the blastoderm stage (when the pole cell transplantations are performed) the embryos of all three diploid genotypes in the *oc*-containing stock are indistinguishable phenotypically. Therefore, the pole cell transfers must be done "blind," not just with respect to sex but to genotype as well. The donor embryos of interest, the *oc/oc* female embryos, comprise only a third of the total. The procedure that is adopted in such situations is to transplant pole cells from a large number of embryos and then to determine, by progeny analysis of the surviving hosts, if donor cells of the genotype of interest made any contribution to their germ line. To distinguish donor cells from host germ line cells, the two lines are differentially marked with respect to some obvious cuticular character(s).

Underwood and Mahowald (1980) performed the experiment by transferring donor pole cells from embryos of the *oc* stock into blastodermal host embryos of the *y* sn^3 *mal* genotype and then mating the survivors to flies of the latter genotype. Successful transmission of donor cells is shown by the appearance of nonyellow, nonsinged progeny (singed is a mutant bristle phenotype). The discrimination of female hosts bearing *oc/ClB* donor cells from those bearing *oc/oc* germ line cells (the ones of interest) is done by comparing the ratio of *oc* sons (without ocelli) to Bar-eyed daughters; the former should produce approximately equal numbers of both, the latter only *oc* sons.

The results are given in part A of Table 3.2. The critical finding is given in the first line: of 43 surviving fertile female chimeras, a total of 11 had *oc/oc*

Table 3.2 Reciprocal Pole Cell Transplantations to Test Site of the *oc* Defect

Test Group	Number of Embryos	Genotype		Number		
		Host	Donor[a]	Fertile	Sterile	Chimera (%)
A	205	*y sn*^{3} *mal/y sn*^{3} *mal*	*oc/oc*	43	5 (11)	11 (26)
			oc/ClB			5 (11)
		y sn^3 *mal*/Y	*oc*/Y			11 (31)
				36	5 (12)	
			ClB/Y			4 (11)
B	191	*oc/oc*	*y sn*^{3} *mal/y sn*^{3} *mal*	0	14 (100)	2 (29)[b]
		oc/ClB		28	1 (3)	9 (31)[b]
		oc/Y	*y sn*^{3} *mal*/Y	8 (100)	0 (0)	3 (38)

Source. Data from Underwood and Mahowald (1980).
[a] Donor genotype determined by progeny types (see text).
[b] Determined by staining for *mal*$^+$ in ovaries.

functional germ line cells. Among these chimeras, the progeny counts showed a substantial (10–67%) proportion of donor cells, with none of the progeny being Bar-eyed females. In contrast, five of the chimeric females gave equal numbers of Bar-eyed daughters and *oc* sons and thus had received *oc/ClB* pole cells (Table 3.3).

oc is a recessive mutation and a presumptive amorph or hypomorph. (The evidence for hypomorphism will be described in Chapter 9.) The pole cell transplantation experiment shows that the wild-type function reduced or abolished by *oc* is not required in the germ line. Part B of Table 3.2 shows that mutant expression in the soma, however, prevents progeny production. In this chimera experiment, in which *oc*$^+$ pole cells were transplanted to *oc/oc* hosts, there is no problem of genotype identification; all the surviving *oc* female embryos can be directly identified by their adult phenotype (the ocellar and bristle defects). This kind of transplantation produced no fertile chimeras (B, line 1), while transfer of *oc* pole cells into phenotypically wild-

Table 3.3 Progeny Tests of Chimeric Female Flies Derived from *oc* Stock Pole Cell Transplantations

Genotype of Donor Cells	Number of ♀ Chimeras	Total Number of Progeny	+ ♀	*B* ♀	*oc* ♂
oc/oc	11	2231	176	0	144
oc/ClB	5	1019	65	55	62

Source. Data from Underwood and Mahowald (1980).

type hosts gave 31% chimerism among the surviving fertile female hosts. Evidently, the wild-type function is required in the follicle cells for normal development.

The *fs* mutant *K10* also affects chorionic development, but in a very different fashion. *K10,* unlike *oc,* does not affect the basic ultrastructure of the chorionic layers, but rather the external patterning of the chorion and, in particular, the chorionic markers of d-v polarity.

The principal landmark of the dorsal surface in normal *Drosophila* eggs is a pair of long dorsal appendages that are secreted during the final stages of oogenesis. In *K10* eggs, these appendages are greatly extended around the embryo both laterally and ventrally, being distally fused in the ventral midline; together they form an incomplete ring, with only the topmost part of the

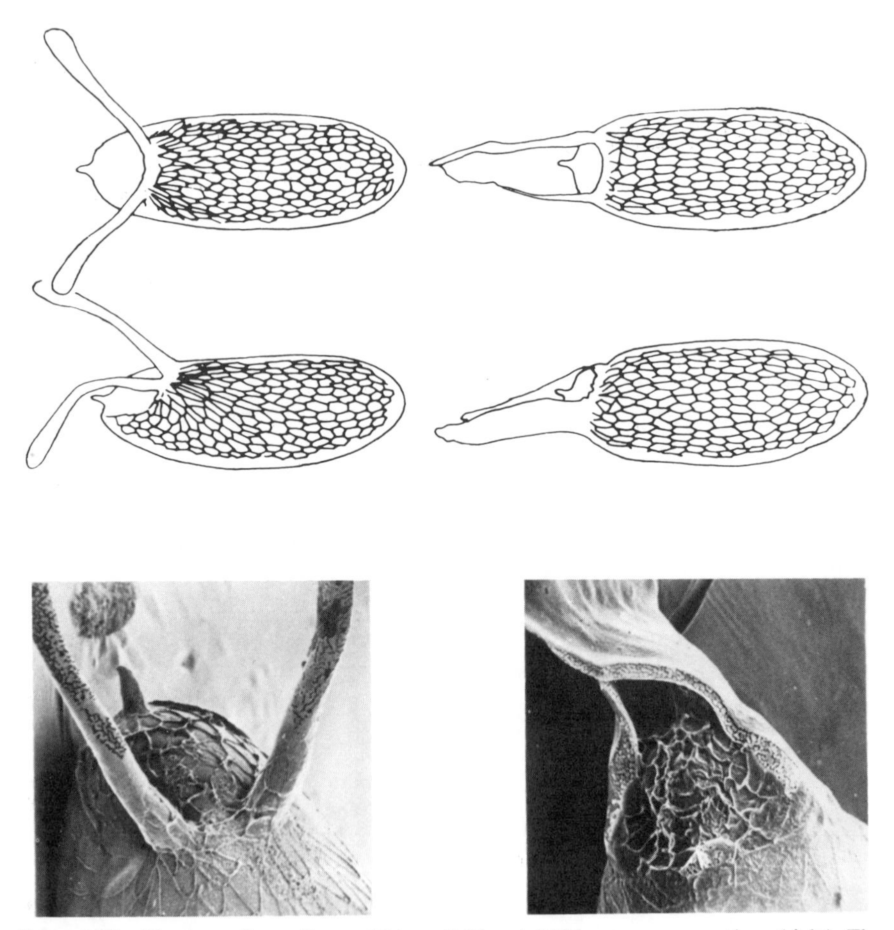

Figure 3.15 Chorions of eggs from wild-type (left) and *K10* homozygous mothers (right). The top part shows camera lucida drawings of follicle cell imprints from the dorsal and lateral sides. The bottom part shows scanning electron micrographs of the dorsal appendages. (Reproduced with permission from Wieschaus, 1979; figure kindly provided by Dr. E. Wieschaus.)

dorsal surface left unencircled (Fig. 3.15). The extension of these structures effectively produces a "dorsalization" of the chorionic pattern, at least at the anterior end of the oocyte. This interpretation is borne out by an examination of the follicle cell "footprints," the traces left by the follicle cells when they are sloughed off at the end of oogenesis. In normal eggs, these imprints are considerably elongated on the dorsal side relative to the ventral and lateral sides. In the anterior portion of *K10* eggs, the ventral and lateral footprints display the elongated character typical of wild-type dorsal footprints. (Whether there is comparable dorsalization in the posterior end is difficult to judge because the differences in the wild-type pattern at the posterior end are less pronounced.) All eggs of *K10* mothers show the mutant phenotype but differ in the degree (expressivity) of this phenotype, with 5% of the eggs having dorsal appendages that extend only partway around the circumference. Such eggs may represent only partial dorsalization (Wieschaus, 1979).

Because the chorion is secreted by the follicle cells, it might be expected that *K10,* like *oc,* is expressed by these cells. This hypothesis can be tested by the reciprocal pole cell transplantation procedure. Since the *K10* phenotype is obvious in eggs, the scoring procedure is more direct than for *oc*. Surprisingly, *K10* is expressed in the germ line (Table 3.4). When wild-type pole cells were transplanted into *K10* hosts, 18 surviving homozygous female *K10* recipients were obtained; of these, 9 laid wild-type eggs in addition to *K10* eggs (part A). Thus, normal eggs can develop from wild-type pole cells even when the follicle cell layer is homozygous *K10*. In the reciprocal

Table 3.4 Reciprocal Pole Cell Transplantations between the *K10*-Carrying Strain and the Fertile ($y\ w\ f^{36a}$) Strain

Injected Host Embryos (Number)	Donor Genotype	Host Genotypes	Surviving Hosts	Hosts Giving Donor Progeny
A. Transplantation of control pole cells into embryos of the *K10* strain				
161	$y\ w\ f^{36a}/$ $y\ w\ f^{36a}$	*K10/ClB*	16	8
		K10/K10	18	9 (*K10* eggs + *y w f* progeny)
	or			
	$y\ w\ f^{36a}$/Y	*K10*/Y	14	9

Injected Host Embryos (Number)	Surviving Embryos	Chimera Classes	Inferred Donor Cell Genotype
B. Transplantation of *K10* germ line cells into embryos of the fertile strain			
262	35	4 laid *K10* eggs	*K10/K10*
		4 yielded *Bar*-eyed progeny	*K10/ClB*

Source. Data from Wieschaus et al. (1978).

transplantation—of *K10* pole cells into wild-type hosts—4 of the 36 surviving hosts laid typical *K10* eggs in addition to wild-type eggs (part B) (Wieschaus et al., 1978). The results show that the *K10* egg phenotype follows the genotype of the germ line and not that of the follicle cells. Presumably the gene acts through the nurse cells in a manner that affects follicle cell disposition and chorion synthesis; in the absence of the wild-type function, the follicle cells in the anterior half of the egg chamber assume a dorsal-characteristic disposition and chorion secretion pattern. Some evidence, to be discussed in the next section, suggests that the chorion pattern produced by the *K10* defect is symptomatic of a more profound dorsalization within the egg that affects the course of embryogenesis. It appears that the wild-type allele is essential for the normal establishment of d-v polarity within the egg.

Because of the distinctive phenotype of the *K10* egg, this mutant can be used as a genetic marker to probe the normal development of the female germ line and the genes that affect it. These second generation studies of *K10* have employed a different form of clonal analysis, that of induced mitotic recombination. This technique involves the induction of homozygous mutant cells within a background of heterozygous (wild-type phenotype) cells by means of x-ray irradiation of heterozygous developing animals. Using *K10* as a cellular marker of the germ line, induced mitotic recombination has been employed in two ways. In the first, it has been used to probe the early and intermediate steps of ovary development during which the germ line is proliferating. In the second, the mutant has been employed to determine the cellular sites of action of other *fs* mutants affected during oogenesis. These studies are of direct interest for the biology of oogenesis. At the same time, they can serve as an introduction to the use of mitotic recombination, the single most useful form of clonal analysis in *Drosophila*.

As its name implies, mitotic recombination is the exchange of chromosome segments between paired homologous chromosomes in mitotically dividing cells. In contrast to standard meiotic recombination, which accompanies the production of haploid gametes, mitotic recombination produces diploid recombinant cells; it can take place in either somatic cells or premeiotic diploid germ line cells. Its occurrence in eukaryotes is not universal but takes place in the cells of those that, like *Drosophila,* have extensive pairing of homologous chromosomes. X-rays stimulate mitotic recombination by creating chromosome breaks, facilitating exchange of segments between the paired chromosomes. Like meiotic recombination, mitotic exchange takes place between replicated chromosomes when each chromosome consists of two chromatids tied together at the centromere.

In homozygous wild-type cells, the occurrence of mitotic exchange should be without genetic consequence. However, in cells that are heterozygous for recessive mutations, mitotic recombination creates homozygous diploid cells that develop into marked clones. When the recombining cell is heterozygous for just one recessive genetic marker over a wild-type homologue, recombination gives rise to a clone homozygous for the mutation and a homozygous wild-type clone. However, if the cells are heterozygous for genetic markers in *trans* on the two homologues, recombination can produce two marked daughter cells that grow into two marked clones; such double clones are termed "twin spots." The phenomenon of mitotic recombination is diagrammed in Figure 3.16 and explained in more detail in the legend.

One principal use of mitotic recombination is to measure the growth kinetics of a tissue or a structure. The procedure involves inducing marked cells at successive stages of development and measuring the relative sizes of the marked clones at the end of development. From the rate of change in clone size, one can ascertain the cell division rate and, with a few assumptions, estimate the absolute number of cells present in the tissue or structure at each stage. The basic premise is that the proportion of marked cells in a single clone at the end of development is an inverse measure of the number of precursor target cells present at the time of recombinant cell induction (Fig. 3.17).

This principle has been used to measure the growth kinetics of the female germ line, using *K10* as a clonal marker. The reasoning behind the experiment is that the fewer the number of germ line precursor cells at the time of

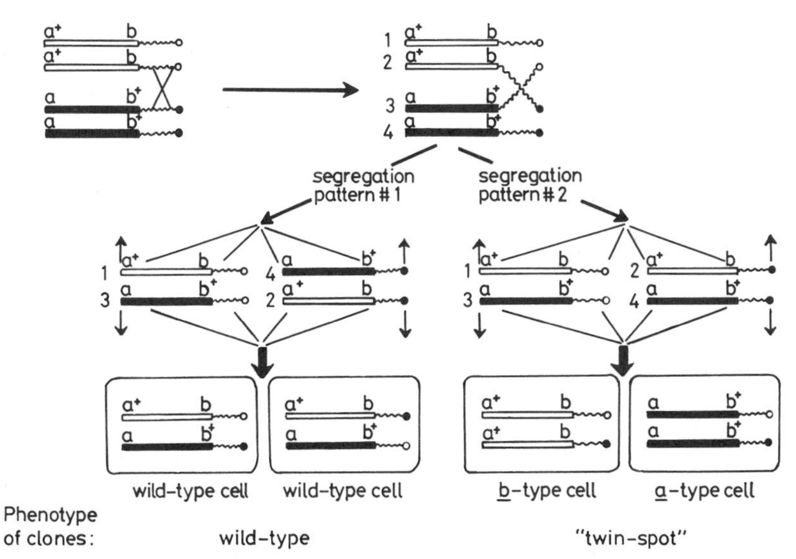

Figure 3.16 Mitotic recombination in a *trans* double heterozygote. Exchange always occurs in the postreplicative (four-stranded) stage, and preferentially in the centromeric heterochromatin (as shown). Following exchange, the chromosomes can segregate in either of two equally likely patterns. In segregation pattern 1, both nonrecombinant chromosomes are segregated to one daughter cell and both recombinants to the other; the consequence is the production of two genetically wild-type daughter cells. In segregation pattern 2, each daughter cell receives one recombinant and one nonrecombinant chromosome; the two daughters therefore become homozygous for the two opposing genetic markers, generating a twin spot. For a recombination between *a* and *b*, all the daughters from segregation 2 would be heterozygous (and wild-type) for *b*, and half would be homozygous for *a;* the latter would generate single *a* clones.

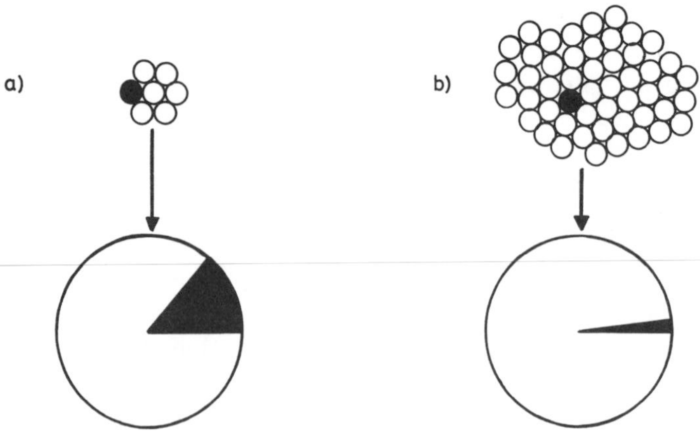

Figure 3.17 Clone size as a function of cell number. In (*a*), one out of seven progenitor cells is homozygous for the marker; in the fully developed structure, approximately 1/7th of the total area shows the marker. In (*b*), 1 out of 50 cells, in a later stage of the same primordium, is homozygous for the marker; only 1/50th of the surface of the developed structure displays the marker phenotype.

clonal induction, the larger will be the clutch of *K10* eggs issuing from any female with a mosaic germ line. Conversely, the greater the number of precursor cells, the smaller will be the marked clutch. Since only one-half of the daugher cells of mitotically recombinant cells can be homozygous for a given marker (see Fig. 3.16), the number of target cells in that germ line mosaic individual should be:

$$1/(2 \times \text{fraction of the clutch that has the K10 phenotype}).$$

The first step in the experiment is to create heterozygous female embryos by crossing *K10* males (which are unaffected by the mutation) with +/+ females. The heterozygous embryos are then irradiated in batches at successive developmental stages, and the progeny are allowed to develop to adulthood. Finally, the fraction of *K10* eggs is measured for each mosaic female and the clone size is calculated. (Females that lay single *K10* eggs are not counted, since such single eggs probably arise spontaneously late in development within single egg chambers.)

The validity of this genetic method for ascertaining cell numbers can be checked by applying the technique to the blastoderm stage, where the number of germ line precursor cells—the pole cells—can be directly counted. Their number is approximately 40–50. Of these, approximately half migrate successfully to the gonad, without dividing, the remainder being lost en route (Underwood et al., 1980b). The expected number of germ line precursor cells at blastoderm is therefore about 20–25. The estimate from mitotic recombination studies is about seven germ line precursor cells (Wieschaus and Szabad, 1979). Considering the number of simplifying assumptions in the genetic method, the agreement between the two estimates is surprisingly good. (In *Drosophila,* the estimates from mitotic recombination are often two- to threefold lower than the probable actual number of cells whose population is being measured; some of the reasons for these discrepancies will be discussed in Chapter 6.)

If the assumption of inverse proportionality between clone size and number of target cells is correct, then as the number of germ line cells in the gonad begins to increase from the end of embryogenesis (Sonnenblick, 1950), the estimated clone sizes within mosaic females should correspondingly decrease. This is exactly what is observed (Wieschaus, 1978b; Wieschaus and Szabad, 1979). The method also permits an estimate of the number of oogonial stem cells in mature ovarioles. The final estimated number of germ line precursor cells, oogonial stem cells, is about 100 per female, a value that is reached in the early pupal period. With 2 ovaries per female and 17 ovarioles per ovary, the number of oogonial stem cells per ovariole in the adult must be about 3 (100 stem cells/ 2 ovaries per female × 17 ovarioles per ovary = 3); this number agrees with estimates derived by direct electron microscopy (Carpenter, 1975).

Because egg chambers are produced in linear sequence within ovarioles, the genetic marking method can further be used to estimate the pattern of activity of these stem cells. One constructs doubly heterozygous cells, for *K10* and *mal* in *trans,* and then induces mitotic recombination to produce a twin spot within the oogonial stem cell pool; the genotype is shown in Figure 3.18. (*mal* eggs are detected either by their absence of aldehyde oxidase, the histochemical test described earlier, or by their giving rise to brown-eyed progeny, from which the name *maroon-like* is derived, because of a *mal* maternal effect on eye color.) When the sequence of *K10* and *mal* eggs within twin-spot ovarioles is examined, it is found that there is no regular pattern of appearance of either type. Evidently, stem cells are active in irregular bursts, rather than working in strict sequence during the lifetime of the adult.

The second application of *K10* as a marker to probe oogenesis has been in determining the cellular locus of action of a number of other sex-linked *fs* mutations. By constructing double heterozygotes for *K10* and the *fs* mutation under test, with the two mutations located in *cis* (i.e., linked on the same chromosome), and then irradiating with x-rays to induce mitotic recombination, doubly homozygous germ line clones can be obtained (Fig. 3.19).

There are three possible outcomes with respect to the production of *K10* eggs in individuals that are germ line mosaic for such clones: (1) If the linked *fs* mutation is an early germ line block, then the number of *K10* eggs should be greatly decreased when compared to the simple *K10* heterozygote. (2) If the second *fs* mutation is active in a late germ line function, then the great majority of mitotically induced *K10* eggs will also show this defect. (Most mitotic recombination events occur as single exchanges in or near the cen-

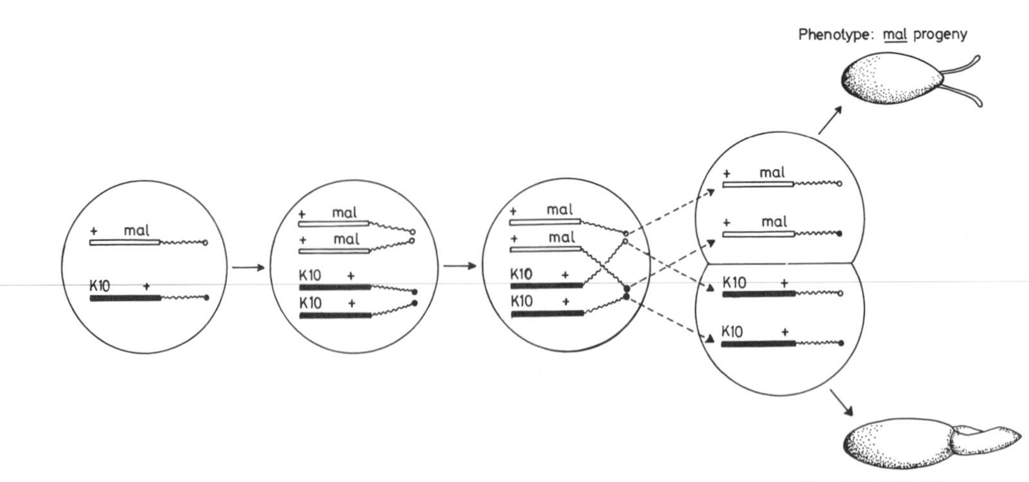

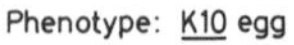

Figure 3.18 Mitotic recombination in the female germ line: production of egg chambers homozygous for *mal* or *K10*. The *trans* heterozygous genotype is shown on the left. In the absence of recombination, all egg chambers of this genotype will be phenotypically wild-type. If mitotic exchange between *mal* and the centromere takes place during the period of stem cell proliferation, stem cells homozygous for one or the other of the two markers will be produced. A homozygous *mal* stem cell will generate *mal* egg chambers, detectable by the absence of staining for aldehyde oxidase or when allowed to develop by the production of a brown-eyed fly (from the *mal* maternal effect). A *K10* stem cell will generate *K10* egg chambers, distinguished by their chorionic dorsalization. (Adapted from Wieschaus, 1978a.)

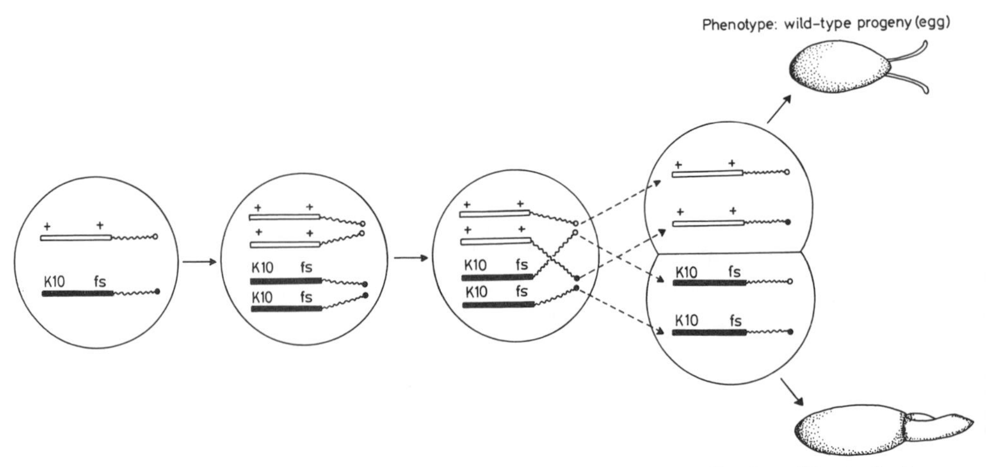

Figure 3.19 Using mitotic recombination to determine the cellular site of action of X chromosomes *fs* mutations. The experimental strategy is described in the text. (Adapted from Wieschaus, 1978a.)

tromeric heterochromatin; hence, most recombinant clones homozygous for the centromere-distal *K10* gene will also be homozygous for the second *fs* mutation.) (3) If the linked *fs* mutation is not expressed in the germ line but rather in the follicle cells, then the *K10* egg frequency will be about that observed in the absence of the additional *fs* mutant; the *K10* germ line clones homozygous for the somatic cell defect will produce eggs that are unaffected by their homozygosity for this second mutation. (However, the relatively small fraction of recombination events that occur between *K10* and the second mutation, producing *K10* homozygotes that are still heterozygous for the second mutation, must be allowed for in interpreting the results.)

Of 12 sex-linked *fs* mutations tested in this fashion, 7 were found to have no effect on *K10* egg frequency or appearance, presumably being follicle cell defectives; 3 were classified as mutant in early germ line functions; and 2 could not be clearly categorized (Wieschaus et al., 1981). (Genes required in both nurse cells and follicle cells could give an ambiguous result in the test.) Although this method provides the same kind of information as the reciprocal pole cell transplantation technique, it is simpler to carry out; oogenesis functions can therefore be much more quickly classified as to cellular site than by the transplantation method. Because K10 is on the X chromosome, the method is presently limited to X-linked *fs* mutants. However, if autosomal markers that produce eggs as distinctive as those made by *K10* homozygotes become available, it should be possible to extend the method to classifying autosomal *fs* mutants.

MATERNAL EFFECT MUTANTS: GENERAL

Among the numerous *Drosophila fs* strains that have been isolated are many maternal effect mutants (Bakken, 1973; Gans et al., 1975; Rice and Garen, 1975; Mohler, 1977). Altogether, these mutants identify approximately 100 maternal effect loci, but the complete set of genes that can mutate to give a maternal effect phenotype must be considerably larger. From the Poisson distribution (see illustration of method in Table 2.4), King and Mohler (1975) have estimated that there are approximately 645 maternal effect loci in the genome. Using a similar approach but a different data base, Gans et al. (1975) have estimated this figure as 1000 ± 500. Although much greater than the number of identified maternal effect genes, both figures are probably substantial underestimates of the number of genes that function in oogenesis, for the reasons mentioned earlier (p. 71). Of the maternal effect functions that have been isolated, there is little general clustering of their loci; the genes are found on all chromosomes on which they have been sought (the X, second, and third), and with an apparently random distribution. However, one interesting exception to this generalization is a cluster of five *fs* mutants on the second chromosome, all of which seem to display differential interactions with centromeric heterochromatin in creating their maternal effects (Sandler, 1977).

Many maternal effect mutants are partial maternals, being mutant in genes expressed both in oogenesis and in other developmental stages. These mutants show the expected wide variety of phenotypes and will not be discussed further here. The mutants of potentially greatest interest for analyzing the morphogenetic system of the embryo are the strict maternals, those classified as having no expression outside oogenesis. The various strict maternals that have been identified in the *fs* mutant hunts are grouped and categorized phenotypically by the scheme of Zalokar et al. (1975) in Table 3.5.

Two qualifications concerning this classification should be noted. First, many of the mutants produce eggs that vary in the severity of the embryonic defects and the point of arrest. Second, where different mutant alleles for the same gene (defined by their lack of complementation) are known, the mutant syndromes sometimes differ with respect to the strength or apparent nature of the developmental defect. For instance, the sex-linked mutants 1162 and 383 do not complement and are therefore allelic. They are classified as blastoderm defectives because 1162 forms embryos with abnormal blastoderms and 383 is allelic to it. However, 383 forms apparently normal em-

Table 3.5 Collections of *Drosophila* Strict Maternal Effect Mutants: Developmental Phenotypes[a]

	Number of Complementation Groups[b]					
Chromosome (Reference)	Fertilization Block	Syngamy/ Haploid Development	Early Cleavage	Late Cleavage/ Blastoderm	Gastrula	Variable or Late Arrest
X chromosome (Gans et al., 1975)	4	5	4	6	3	3
Second chromosome (Bakken, 1973)		1	4	2		
Third chromosome (Bakken, 1973)	3		1			1
Third chromosome (Rice and Garen, 1975)				3		

[a] As classified with respect to the primary arrest point by the criteria of Zalokar et al. (1975).
[b] As defined in standard complementation tests, except in the study by Rice and Garen (1975), where mutations were identified by mapping as belonging to different regions and hence genes.

bryos that arrest in late embryogenesis (Zalokar et al., 1975). As in *C. elegans,* different allelic pairs sometimes resemble different kinds of maternal effect functions, the more severely defective mutations appearing to be strict maternals and the less severely defective partial maternals.

For most of the strict maternals, the defects are seemingly generalized ones of nuclear or cellular behavior during early embryogenesis. Many show early defects in syngamy or in nuclear division during cleavage, while others that show major blastodermal or postblastodermal defects also show abnormalities in late cleavage. None of the mutants from *fs* screens has the characteristics expected of a mutant defective in a determinant for a specific imaginal disc, segment, or germ layer.

For the most part, the primary defects seem to be in the metabolic requirements for cleavage. Indeed, several of the mutant syndromes are similar to those observed when permeabilized *Drosophila* embryos are exposed to various drugs that interfere with macromolecular synthesis or cell division (Zalokar and Erk, 1976). For example, one X chromosome mutant, *fs(1)1459,* forms a normal-appearing syncytial blastoderm but then proceeds to lose nuclei from the surface to the interior. This effect is mimicked by treatment with the actin microfilament antagonist cytochalasin B. This comparison suggests that the mutation might affect microfilament function, with consequent inhibition of cellularization. Other mutants show clumped chromatin in early cleavage, an effect mimicked by treatment with the transcription inhibitor actinomycin D; these genetic defects may therefore involve direct or indirect interference with RNA synthesis.

Interestingly, some of the drug-induced and mutant defects are not spatially homogeneous, but exhibit some form of a-p polarity. Two examples are provided by third chromosome temperature-sensitive maternal effect mutants described by Rice and Garen (1975). Both of these mutants show a polarized defective cellularization at blastoderm. Thus, embryos produced from *mat(3)3* mothers remain acellular in the posterodorsal region, while *mat(3)6* is acellular at blastoderm in the central region. A second form of a-p polarity is revealed in the occurrence of polarized patterns of mitosis or "mitotic waves" in one, and perhaps two, of the mutants (Zalokar et al., 1975). This effect is also produced by DNA synthesis inhibitors (Zalokar and Erk, 1976). With the chemically induced mitotic waves, the mitotic pattern is generally that of a more advanced mitotic stage at one or both poles, grading to earlier stages toward the middle. The occurrence of such polarized effects, whether of cellularization or mitosis, may signal the existence of some underlying a-p property within the egg. This possibility and its potential significance will be discussed later.

The mutant set presents little evidence for a mosaic of determinants. The defects all appear to be general ones of metabolism or cell division, a picture very similar to that for the nematode strict maternal *emb* mutants. Such negative evidence neither disproves the mosaic determinant model nor proves the involvement of gradients in determination. However, the nature of the morphogenetic system in *Drosophila* has recently been resolved by the discovery of a special class of "coordinate" maternal effect mutants.

MATERNAL EFFECT MUTANTS: COORDINATE MUTANTS

The mutants that have proven information in elucidating the determinative system in the early *Drosophila* embryo have three features in common. They are all strict maternal effect mutants. They are all homeotics, that is, they cause the substitution of recognizable parts or regions for other distinct parts or regions. Finally, the homeotic embryonic changes produced are all *global graded shifts in pattern elements.*

The effects of these maternal homeotics are best described with reference to the embryonic fate map of the blastoderm stage. A fate map shows which cells in the early embryo give rise to particular structures in the later embryo; it is a map of prospective potencies. Because the cells of the *Drosophila* blastoderm embryo comprise a monolayer, its fate map can be drawn as a two-dimensional projection by simply splitting open the sheet along a line and flattening it. The embryonic fate map shown in Figure 3.20 is based on the original study by D. F. Poulson (1950). Note that segmental fate is a strict

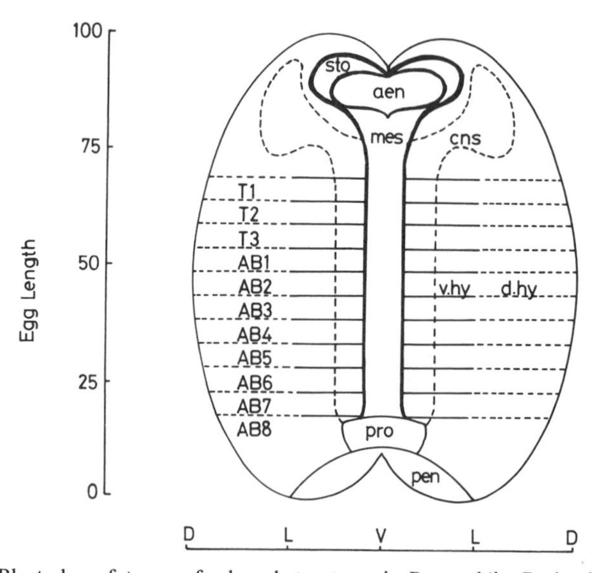

Figure 3.20 Blastoderm fate map for larval structures in *Drosophila*. Projection drawn for an egg cut down the dorsal midline and folded out. The ordinate shows the relative distance in egg length (EL) from the posterior end (0 egg length) to the anterior end (100); the abscissa indicates the position along the d-v axis. AB1–AB8, abdominal segments; aen, anterior endoderm; cns, central nervous system; d.hy, dorsal hypoderm; mes, mesoderm; pen, posterior endoderm; pro, proctodaeum; sto, stomodaem; T1–3, thoracic segments; v.hy, ventral hypoderm. (Adapted and reproduced with permission from Nusslein-Volhard, 1979b.)

function of the a-p position and that the d-v position determines the germ layer fate, that is, whether a region will be ectodermal, mesodermal, or endodermal. The delineation of segments does not begin until the first 5–7 hr of embryogenesis, but germ layer allocation begins immediately after blastoderm, when a central midventral strip of cells invaginates to become the mesoderm. (This process will be described in more detail in Chapter 6.)

The fate map shows that each region of the blastoderm cellular monolayer can be located by a pair of Cartesian coordinates, an a-p coordinate to designate segmental fate and a d-v coordinate to designate germ layer fate. The effect of each maternal homeotic is to shift the appearance of structures in a graded fashion along one axis or the other; each mutant thus alters the coordinates, along one axis, of the fate map. Sander (1976) has termed such mutants "coordinate mutants."

The various coordinate mutants are listed in Table 3.6. They are grouped in terms of the axis whose polarity they affect: the mutants affecting a-p polarity (segment identity) are listed in the top half of the table and those affecting d-v polarity (germ layer identity) in the bottom half. For purposes of illustration, one mutant from each category will be described. To represent the maternal a-p transforming mutants, *bicaudal* (*bic*) will be discussed, and to represent the d-v transforming mutants, *dorsal* (*dl*) will be described.

bic causes the production of double-abdomen embryos from homozygous mutant mothers. In such embryos, the two abdominal regions are in mirror-image symmetry, an effect reminiscent of the induced reversed abdomen embryos of *Smittia* (Fig. II.2). A comparison of the integuments of first instar larvae from wild-type and *bic* mothers is given in Figure 3.21.

bic was discovered as a spontaneous recessive second chromosome mutation by Bull (1966), who showed that the transformation affected larval organs as well as the outer epidermal structures. Initially, the mutant proved difficult to study because of its low penetrance and variable expressivity. However, when *bic* is placed over a second chromosome bearing the deficiency vg^{B}, which deletes the wild-type allele of *bic*, its penetrance is increased to 100% (i.e., all females produce *bic* eggs) (Nusslein-Volhard, 1977, 1979a,b). Even under these improved conditions, the expressivity of the mutation is not 100% but varies according to the age of the mother, with younger females showing higher expressivity, and with temperature, with higher maternal growth temperatures favoring mutant expression.

A preliminary question about any mutant phenotype is whether the mutant effects stem from a qualitatively new genetic activity, a so-called neomorphic mutation, or simply from loss of normal gene activity. In general, dominant mutants are dominant because they are neomorphic in some degree, and recessive mutants are loss-of-function mutants.

The *bic* mutant is recessive, and its phenotype therefore presumably results from an insufficiency of maternal bic^{+} activity. However, as discussed earlier, the particular mutant phenotype observed is often a function of the amount of residual gene activity. If the mutant is a hypomorph, pos-

Table 3.6 Maternal Effect Pattern-Shifting ("Coordinate") Mutants of *Drosophila melanogaster*

Locus (Number of Alleles)	Map Position	Mutant Phenotype	References
		Mutants affecting a-p polarity	
bicaudal (1)	2–67.0	Abdominalization	Bull (1966); Nusslein-Volhard (1977, 1979a)
Bicaudal-D (2)	2–59.9	Abdominalization	Nusslein-Volhard et al. (1982)
Bicaudal-C (3)	2–51	Abdominalization	Nusslein-Volhard et al. (1982)
fs(1)1502 (2)	1–14	Abdominalization	Gans et al. (1975)
Torso (1)	2–57	Abdominalization	Nusslein-Volhard et al. (1982)
Dicephalic (1)	3–46	Cephalization	Lohs-Schardin (1982)
		Mutants affecting d-v polarity	
K10 (4)	1–0.5	Dorsalizing; egg shell affected	Wieschaus et al. (1978)
spindle (3)	3–(N.D.)	Ventralizing; egg shell affected	Anderson and Nusslein-Volhard (1984)
dorsal (11)	2–52.9	Dorsalized embryos	Anderson and Nusslein-Volhard (1984)
easter (13)	3–57	Dorsalized embryos	Anderson and Nusslein-Volhard (1984)
gastrulation defective (8)	1–37	Dorsalized embryos	Anderson and Nusslein-Volhard (1984)
mat (3)2 (3)	3–17	Dorsalized embryos	Rice, cited in Anderson and Nusslein-Volhard (1984)
mat (3)4 (3)	3–51	Dorsalized embryos	Rice, cited in Anderson and Nusslein-Volhard (1984)
Toll (3)	3–91	Partially ventralized	Anderson and Nusslein-Volhard (1984)

sessing some residual wild-type activity, the phenotype is often different and less severe than if the mutant is an amorph, possessing none. To classify a mutant in terms of its expression characteristics, it is necessary to construct lines with different dosages of the mutant and wild-type alleles (Muller, 1932). In *Drosophila,* such strain construction is comparatively easy to do because of the existence of a large number of small deficiencies and duplications that collectively cover almost the entire genome. Furthermore, the

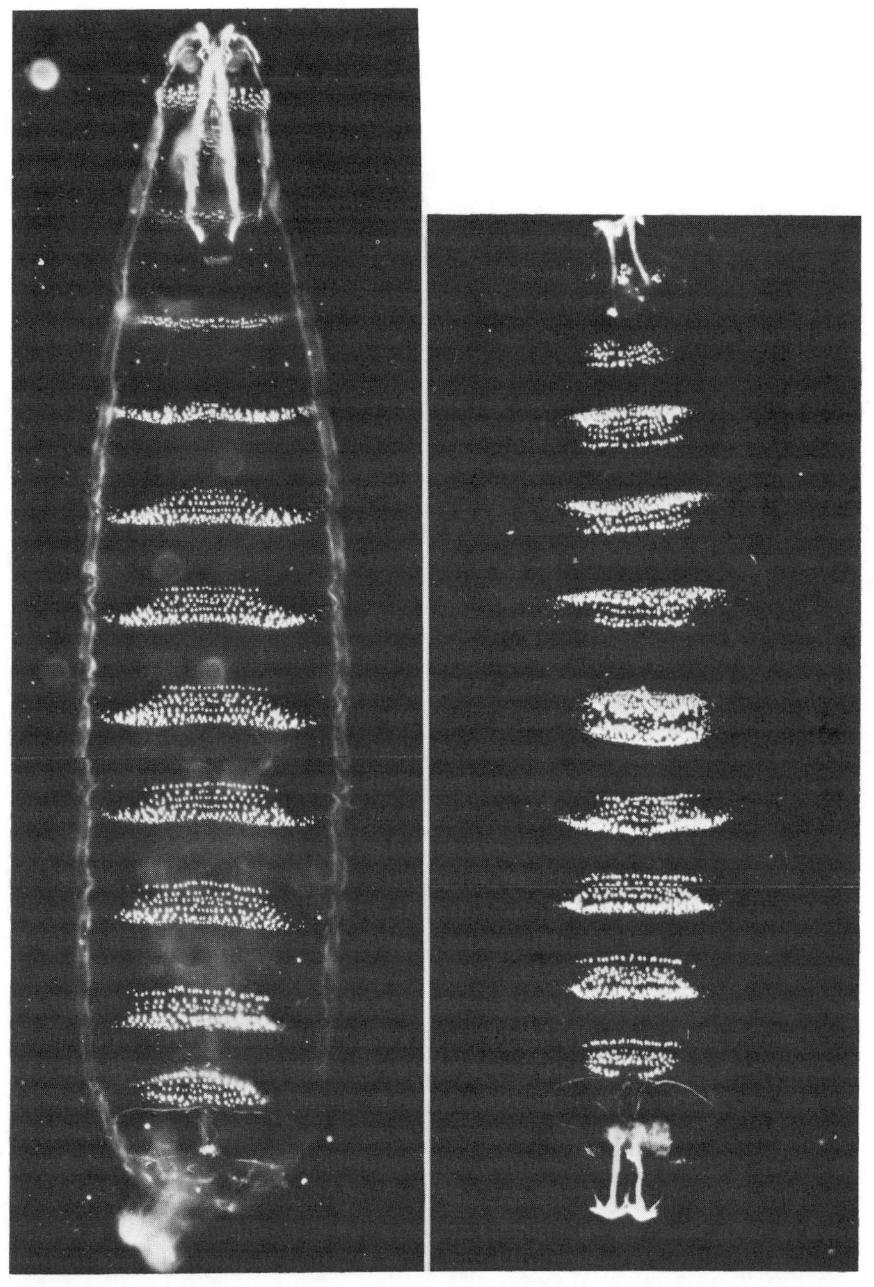

Figure 3.21 Segmentation pattern in a wild-type larva (left) and a larva from homozygous *bic* mother (right). Anterior at the top, posterior at the bottom. Abdominal segment phenotype shown by the presence of a thick denticle hook band. (Photographs kindly supplied by Dr. C. Nusslein-Volhard, 1977.)

Table 3.7 Genetic Tests to Determine the Type of Mutant Expression[a]

If *m*/*m* = *m*/Df, then the mutant is an amorph
If *m*/Df > *m*/*m*, then the mutant is a hypomorph
If +/*m* = +/+/*m*, then the mutant is a neomorph
If *m*/*m* > +/*m*/*m* (or *m*/Df), then the mutant is an antimorph
If +/*m* = mutant phenotype but +/+/*m* = wild-type, then the mutant is a dominant haplo-insufficient

[a] *m* = mutant gene; + = wild-type allele; Df = deficiency (deletion of + allele); > signifies more severe mutant expression.

positions of these deficiencies and duplications can be placed very precisely by examination of their polytene chromosomes. By constructing strains with the appropriate variations in doses of wild-type and mutant genes and comparing the phenotypes of these different strains, it is possible to establish unambiguously the character of the mutant. This set of comparisons and their possible outcomes and interpretations are summarized in Table 3.7.

The crucial comparison for *bic* is between the hemizygous (*bic*/Df) and homozygous (*bic*/*bic*) mothers. When the test is performed, it is found that the penetrance and expressivity of the mutant are much higher in the hemizygote (Nusslein-Volhard, 1977). The *bic* mutant allele evidently possesses some residual activity and must therefore be a hypomorph. Furthermore, the more extreme the reduction in maternal wild-type gene activity, the more extreme the resultant mutant embryonic phenotype. In fact, a single dose of bic^+ activity is insufficient to give full wild-type development: bic^+/Df mothers produce a small percentage of *bic*-type embryos. In other words, reduction of the bic^+ maternal gene product by a factor of 2 or more tends to reduce or abolish normal a-p polarity in the embryo, with the substitution of posterior for anterior structures. (The requirement for a full diploid dosage of maternal genes to promote normal embryogenesis may be a general one, as noted earlier.)

An interesting aspect of *bic* action is the production by *bic* mothers of embryos with a range of polarity defects. Because this variability in phenotype is observed in several different genotypic backgrounds, it must be an intrinsic feature of the expression of the mutant gene itself. These phenotypes range from complete and symmetrical reversed abdomens (from which the mutant takes its name), through a range of embryos with only partially reversed anterior halves, to embryos with normal polarity missing some anterior segments ("short embryos"), to fully normal embryos.

This range in polarity defects indicates that the a-p polarity setting mechanism has some lability. However, whatever the nature of this system, its metastable states appear to be the two extremes. Most of the embryos are either normal, showing the correct number and arrangement of segments, or are fully symmetrical *bic* types (Nusslein-Volhard, 1979a). Significantly, the latter are altered in their posterior as well as their anterior halves. In wild-type embryos, there are eight abdominal segments, all arising from the posterior half of the blastoderm; in symmetrical *bic* embryos, there is a range from one to five segments, with a mean of three, in the posterior half. Evidently, *bic* maternal action does not cause merely the subtraction of something in the anterior half of the embryo; rather, it produces a *global* alteration of the pattern throughout the embryo.

Most of the other strict maternals that affect a-p polarity also produce an embryonic phenotype comparable to that of *bic* (Table 3.6) (Nusslein-Volhard et al., 1982). In some instances, the effect is very weak, as with the hypomorphic mutation *fs(1)1502,* where the resulting embryos lack only the most anterior larval structures, such as mouth hooks (Zalokar et al., 1975). Interestingly, this mutation is allelic to another maternal effect mutation that is defective in the final cleavage division(s), a point that suggests the existence of a relationship between these divisions and determination along the a-p axis (see "Possible Mechanisms of a-p Gradient Formation and Action").

The exception to this pattern of "abdominalization" is the mutant *Dicephalic* (*Dic*) which, as its name signifies, creates "two-headed" embryos. The *Dic* syndrome seems to result from the relative disposition of oocytes and nurse cells within the egg chambers. In the mutant egg chambers, the oocyte does not occupy its typical posterior position but tends to be located between nurse cell clusters (Lohs-Schardin, 1982). Although few embryos develop from the oocytes of such egg chambers, those that do exhibit reversed anterior halves. The *Dic* syndrome confirms that one initial determinant of a-p polarity is the relative positioning of oocyte and nurse cells within the egg chamber.

The coordinate mutants that transform d-v polarity leave the embryonic segment number untouched but determine the relative extent of the germ layers and, in particular, that of the ectoderm and mesoderm. The effects of these mutants can be visualized in terms of the fate map (Fig. 3.20): mutants that create a "dorsalizing" effect reduce or eliminate the mesoderm, while maternal "ventralizing" mutants expand the mesoderm at the expense of the ectoderm.

dl, the most extensively characterized of these maternal mutants, produces a dorsalization of the entire embryo. In embryos derived from homozygous *dl* mothers, the early midventral invagination is abolished and the mesoderm never arises. The few embryos that form consist of segmented tubes of dorsal hypoderm (the ectodermal layer that secretes the larval cuticle) surrounding a yolk cylinder (Nusslein-Volhard, 1979a,b). The original *dl* allele, like several subsequently isolated alleles, is amorphic, and all of the affected eggs show the phenotype. However, like the *bic* locus, a normal diploid dosage of the wild-type allele in the mother is required for development. At 29°C, *dl* shows a partial dominant phenotype: *dl*/+ females grown at this temperature can produce embryos that are partially dorsalized (in certain background genotypes) (Fig. 3.26*c,d*).

This partial dorsalization is revealed in two ways. First, the embryos show a narrowing of the denticle hook bands, which are located at the ventral anterior edge of each segment. (These bands are the main ventral segment markers and are visible in Fig. 3.21.) Second, the fate map of the ventral and lateral hypoderm on the blastoderm embryo is shifted. In normal embryos, the ventral hypoderm of the larva is derived from blastodermal cells located on either side of the ventral strip that gives rise to the mesoderm (Fig. 3.20). With localized laser beam irradiation to the midventral line of wild-type blastoderm embryos, larval hypodermal defects are never obtained (Nusslein-Volhard et al., 1980). However, when blastoderm embryos from *dl*-dominant mothers are irradiated in the ventral midline, a high percentage develop ventral hypodermal defects. The detailed pattern of defects relative to embryos from wild-type others is most readily explained as a shift in the specification of ventral cells relative to that of cellular elements normally specified in more dorsal positions rather than as a deletion of a specific "mesodermal determinant."

Most of the other coordinate mutants that affect d-v polarity have similar dorsalizing effects. However, two have the opposite, ventralizing effect: the recessive maternal mutant *spindle* and the dominant *Toll*. Furthermore, the majority of the mutants affect the embryo without affecting the egg shell pattern. Again, there are two exceptions: one dorsalizing mutant (*K10*, described earlier) and one ventralizing mutant (*spindle*) affect chorionic structures in parallel with their effects on the embryo. Possibly these mutants act earlier than the others in establishing d-v polarity (Nusslein-Volhard, 1979a).

In addition to altering the early embryonic body plans, the coordinate mutant transformations produce corresponding shifts in the fate of the imaginal cell precursors within the embryo. Thus, when the anterior halves of *bic*-derived embryos are cultured in vivo, genital disc structures are frequently obtained after metamorphosis (Gehring and Nusslein-Volhard, cited in Nusslein-Volhard, 1979a); the anterior halves of embryos from wild-type mothers never give rise to genital structures (Fig. 3.8). A comparable polarized shift in imaginal cell determination has also been reported for early embryos derived from *K10* mothers. In implants derived from embryos produced by wild-type females, 80% produce disc structures typical of midlateral discs—eye, antenna, wing, and leg—with little contribution from the dorsal discs (labial and humerus) or the genital disc. In embryos from *K10* mothers, most of the implants are genital, with secondary contributions from the dorsalmost discs and none from the midlateral discs (cited in Wieschaus, 1979). (The predominance of genital disc material over labial and humeral material may reflect selective factors during culture.) The shifts in cell specification affected by the maternally acting coordinate mutants evidently affect *all* cells in the blastodermal cell layer, not just the precursor cells of the larval integument.

DETERMINATIVE GRADIENTS IN THE *DROSOPHILA* EMBRYO

The coordinate mutants suggest the existence of determinative gradients within the early *Drosophila* embryo. The transformations provoked by these maternally acting genes are systemic and apparently continuous. The simplest interpretation of the results is that the morphogenetic system of the embryo involves one or more monotonically increasing properties. The fact that each of the mutants transforms cell coordinates along either the a-p or d-v axis suggests that specification is carried out by orthogonal (and perhaps independent) gradient systems.

The simplest hypothesis is that there is a single gradient running in each direction. The actions of the *bic* and *dl* mutations can be interpreted, in this view, as affecting the shape of the two orthogonal gradients. Two schematized depictions are shown in Figures 3.22 and 3.23, respectively.

In Figure 3.22, the wild-type a-p gradient is shown as running from a maximum at the posterior end (0% egg length) to zero or minimal values at the anterior end (100% egg length). In this scheme, high values correspond to and induce "abdominalization" and low values "cephalization." The reason for positing a peak of posterior-inducing morphogen is that certain experimental findings suggest that posterior material has this abdominalizing

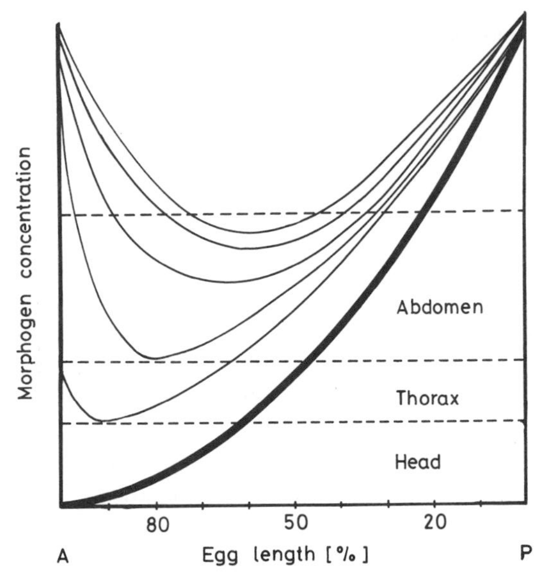

Figure 3.22 Hypothetical a-p determinative gradients in embryos from wild-type and *bic* mothers. The gradient in the wild-type (heavy line) runs from a high point at the posterior end to a low point at the anterior end. The several *bic* gradients (light lines) shown correspond to the various degrees of abdominalization observed; the top curve represents the symmetrical *bic* embryo. The dashed lines indicate threshold values for the different segment type determinations. (Adapted and reproduced with permission from Nusslein-Volhard, 1979a.)

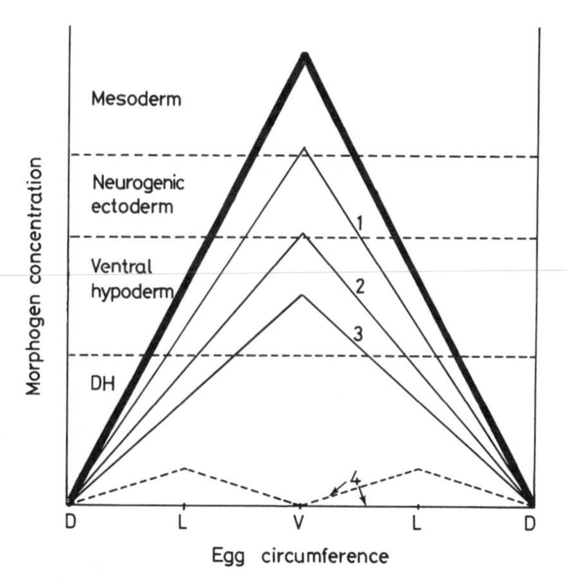

Figure 3.23 Hypothetical d-v gradients in the wild-type (heavy line) and in embryos with various grades of *dl* expression (light lines). D, L, and V represent dorsal, lateral, and ventral positions, respectively. Lines 1, 2, and 3 represent several degrees of *dl*-dominant (*dl*/+, 29°C) expression. The two curves labeled "4" represent alternative possible gradients for the *dl* recessive (*dl*/*dl*) phenotype. The dashed lines indicate the limits of gradient value for the several histotype determinations. (Adapted and reproduced with permission from Nusslein-Volhard, 1979a.)

effect (Sander, 1975, 1976). In this model, *bic* maternal expression redistributes morphogen in various degrees, producing increased levels of morphogen in anterior regions and hence abdominalization. Presumably, the processes of gradient maintenance and formation are complex; the other a-p altering coordinate mutants may affect either of these processes in a number of ways. However, the morphogen itself may be a simple molecule used in other processes in development or cell metabolism; the complexity of gradient formation need not reflect particular complexity of the morphogen itself.

Specification of cells along the d-v axis differs from specification along the a-p axis in two respects. It must be produced over a much shorter distance, that of a half-circumference of a segment, and it must take place symmetrically on both sides of the embryo. If dorsoventrality is engendered by a gradient, it must be produced by a bilaterally symmetrical gradient with its high point in either the middorsal or midventral line. Given that *dl* is amorphic, its dorsalizing action suggests a reduction in the slope of a ventralizing morphogen. The peak of the bilaterally symmetrical gradient in the wild-type would be at the midventral line. As gradient concentration is reduced by progressively reduced amounts of maternal dl^{+} gene product, specification of progressively more ventral structures is lost (Fig. 3.23).

Some support for the notion that *dl* lacks morphogen has been obtained. Santamaria and Nusslein-Volhard (1983) reported that injection of cytoplasm from wild-type but not *dl* embryos resulted in partial rescue of ventral specification. Rescue was limited to regions near the site of injection. Thus, the rescuing substance was either unstable or diffused slowly. If embryos from *dl* mothers made morphogen but could not distribute it correctly, wild-type cytoplasm should not have rescuing ability. This experiment opens up the possibility of a direct investigation of the molecular biology of d-v fate specification.

POSSIBLE MECHANISMS OF a-p GRADIENT FORMATION AND ACTION

The possible molecular basis of the a-p gradient system has received some attention from theoretical biologists in recent years. However, there have been few attempts to synthesize the proposed models with the genetic findings. The various considerations and possibilities will be presented here.

The simplest gradient model is that of a "source-sink": morphogen is elaborated at one end of an embryonic field (the source) and declines uniformly to a sink at the other end, where it is destroyed. In the context of the early insect embryo, it is most plausible to think of one pole of the embryo, either the anterior or posterior pole, as the source and the other pole as the sink.

Meinhardt (1977) formulated such a gradient scheme to account for fate specification along the a-p axis. The hypothesis envisages gradient formation as based on the relative concentrations of an activator (A) and an inhibitor (I). Both A and I can be made by all cells, and I is postulated to be the actual morphogen. If A stimulates both its own synthesis and that of I, and if I diffuses more rapidly than A, then an initial peak of A will give rise to stable monotonic gradients of I and A, with a peak at the initial point of synthesis of A. If the posterior pole is assumed to be the initial point of A synthesis, the gradient will run from a high point at this pole to a low point at the anterior pole (Fig. 3.24). Furthermore, at the low point, I will stably repress further synthesis of A.

In this model, double-abdomen production will take place whenever production or activity of I at the anterior end is reduced. This action leads to an increase in activator production, with a consequent rise in I (morphogen) concentration at this pole, to generate a reversed-abdomen embryo (Fig. 3.24). This mechanism can account for the induction of double abdomens in *Smittia* by various physical treatments and the genetic creation of double abdomens in *Drosophila* by *bic*, although the more variable response of embryos from *bic* mothers suggests that the two organisms differ either in the speed of reequilibration of the gradient or in the way the cells of the two embryos "read" the different concentrations of morphogen.

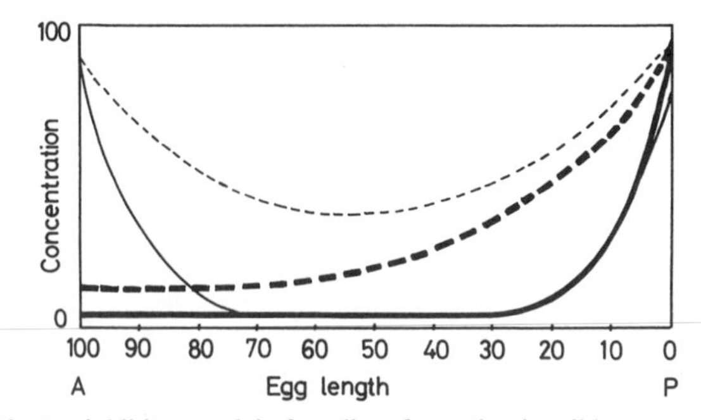

Figure 3.24 Activator-inhibitor model of gradient formation in wild-type and double-abdomen embryos. Details in text. Activator (heavy line) and inhibitor (heavy dashed line) distribution in the wild-type; activator (solid light line) and inhibitor (light dashed line) distribution in the double-abdomen embryo. (Adapted and reproduced with permission from Meinhardt, 1977.)

A central element of the hypothesis is that gradients are initiated and run from one end of the embryo to the other. However, several observations suggest that the assignment of determinative states along the a-p axis is more complicated than this, and may involve molecular signals that emanate from *both* poles of the embryo toward the center. This is a characteristic that is not expected from the Meinhardt model.

Indirect evidence for this signaling is that the embryo center is particularly sensitive to perturbations by certain mutations and physical treatments. The genetic observations are of several maternal effect mutants that produce offspring with two kinds of central segment imaginal defects: deficiencies for certain thoracic structures and homeotic transformations. The homeotic transformation most often seen with the mutants is similar to that produced by the zygotically acting mutation *bithorax* (*bx*). (The *bx* phenotype will be described in more detail in Chapter 6.) Briefly, it involves a transformation of anterior metathoracic structures to anterior mesothoracic structures. Mutant maternal effect conditions (which permit some progeny to survive) that are marked by the presence in the progeny of *bx*-type defects or deficiencies for thoracic structures are described by Rice et al. (1979) and Gans et al. (1980).

The *bx* phenotype can also be produced by certain physical treatments of early embryos. Such treatments that mimic particular mutant conditions are said to be "phenocopying" procedures; the mutant-like but genetically wild-type survivors are termed "phenocopies." When blastoderm embryos are exposed to ether vapors or heat shock, they transform with high frequency to *bx* phenocopies. The phenocopying can be obtained over a 4-hr interval (from 1 to 5 hr of development), but the peak period of sensitivity is at the blastodermal stage (Capdevila and Garcia-Bellido, 1974). The production of *bx* phenocopies again suggests that the center of the embryo (from which the thoracic segments later arise) is singular in some respect.

Direct evidence that information of some sort propagates from both poles comes from numerous embryological experiments. These experiments involve constricting or "ligating" embryos at successive stages, in different positions between the two poles, and measuring the capacity of the ligated fragments to differentiate larval structures or give rise to imaginal ones. In lower insect systems, the capacity to give rise to the complete inventory of structures develops only gradually during cleavage, and the development of central regions requires exposure to influences emanating from both poles (Sander, 1975, 1976). Recent evidence suggests that a comparable system may operate in *Drosophila;* specification of posterior and middle thoracic segments requires influences that spread from both poles (reviewed by Schubiger and Newman, 1981).

The precedent for a dual gradient system of this sort is the sea urchin. In this embryo, correct specification of structures that arise from cells between the animal and vegetal poles requires exposure to substances or processes spreading from both poles. Half-embryos containing just the animal pole become excessively "animalized"; those containing just the vegetal pole develop correspondingly exaggerated vegetal-half characteristics (Horstadius, 1973).

The fundamental question about the a-p determinative gradients in *Drosophila* concerns their molecular basis. There is no direct information on this subject and little speculation. The only firm evidence comes from *Smittia*. In this embryo, the sensitivity of the "anterior determinant" of UV irradiation is that typical of an RNP. Consistent with the idea that this molecule consists in part of RNA is the fact that double-abdomen embryos can be produced by anterior injection of ribonuclease (RNase) (Kalthoff, 1979).

It is not clear, however, whether the RNA affected by this treatment is itself the determinant or a substance required for the action or stability of this determinant. Because it is possible to induce experimentally both double-cephalon and double-abdomen embryos in *Smittia,* it seems that the information for both anteriorness and posteriorness is present at both poles in this insect embryo. Kalthoff (1983) has hypothesized that both poles have both determining specificities but differ in the relative "strengths" or concentrations of these molecules. The developmental outcome in both embryo halves would reflect the respective ratios of the determinants in these regions.

Another possibility is that the process of determination is in some manner connected to the final cleavage divisions. In *Drosophila,* the possibility that mitotic waves or gradients during the final cleavage divisions are connected with the determinative process was originally proposed by Kauffman (1973). Such mitotic waves had already been described in another dipteran, *Calliphora*. The final four cleavage divisions in *Calliphora* embryos consist of two anteroposterior anaphase waves followed by two ends-to middle anaphase waves (Agrell, 1964). In principle, gradients of such polarities could generate specification along the a-p axis. Furthermore, the existence of ends-to-middle waves could establish a singularity at the center of the embryo, a property that otherwise remains mysterious. In addition, the existence of determinative waves with these geometries could explain the symmetrical *bic* embryonic phenotype; this could be envisaged as the result of symmetrical ends-to-middle gradients in the absence of superimposed anteroposterior gradients.

Do mitotic gradients exist in *Drosophila* and, if so, are they of direct significance for determination in the blastoderm embryo? The first question has only recently been resolved. In a detailed study of living embryos, Foe and Alberts (1983) observed distinct mitotic waves in the final four divisions and concluded that probably all embryos experience them. In contrast to *Calliphora,* however, all four waves appear to be ends-to-middle progressions.

The question of significance is more difficult. Van der Meer et al. (1982) have argued that mitotic waves in late cleavage have no significance for determination. Their argument is based on the fact that the normal a-p polarity in the mitotic waves observed in late cleavage of the beetle *Callosobruchus* can be reversed by various treatments without a correlated shift in the fates of the treated embryos. The results imply that the relative timing of nuclear division per se cannot play a direct role in fate setting, just as the relative rates of cell division in the nematode embryo cannot be causally related to determination in that embryo (p. 76).

However, it is possible that determination is normally linked to mitotic progression, but that it can be uncoupled from nuclear division under certain conditions. As with the nematode embryo, the process whose rate might normally correlate with that of nuclear division is DNA replication. Under normal conditions, mitotic waves may mirror "replicative gradients"; the nuclei that first complete DNA synthesis would be the first to complete mitosis, followed in sequence by the progressively later ones. If replicative gradients exist in the late cleavage divisions, then lengthened replication times within each division should be reflected in lengthened interphases. The preblastoderm cleavage divisions, in fact, feature progressively lengthening interphases (see Table 1 in Foe and Alberts, 1983). (The mitotic data, however, do not show regional variations in interphase length.)

The following model is speculative, but it may help to illuminate the possible connections between the processes of replication and determination. The key premise is that the *speed* of replication within a nucleus is related to the *number and identity* of genes being replicated in that nucleus. A gradient of replication rates in a field of nuclei would then involve a spatial pattern of specific nuclear gene replication orders.

Replication rate in *Drosophila* is determined by the frequency of replicon initiation, which decreases from several initiations per genetic unit to fewer than one at the time of blastoderm formation (Blumenthal et al., 1974). A gradient of replication might therefore consist of a gradient of initiation frequencies. If there is some specificity in gene/replicon initiation for a given overall rate of initiation, then a replicative gradient would correspond to a gradient of gene-specific replication patterns (Wilkins, 1976). A simple gradient scheme for generating such an array of replication patterns is diagrammed in Figure 3.25*a*.

To convert an array of gene replication patterns to one of gene expression potentials requires only a protein that binds preferentially to one replicative state of DNA (either unreplicated or replicated) and affects transcription. One of the histone proteins might fulfill such a role; indeed, switches in histone synthesis occur around the time of blastoderm formation (Newrock et al., 1977). A similar scheme, employing differential replication and histone binding to create differences in transcription-potential, has recently been proposed by Brown (1984).

The consequence of such preferential binding within a field of differentially replicated nuclei would be a gradient of repression; the nuclei at the top of the gradient would be least restricted in expression and those at the bottom of the gradient most restricted. Successive gradients would superimpose their patterns of restriction to produce a composite pattern (Fig. 3.25*b*). Ends-to-middle gradients of this sort would produce a developmental sink in the center of the embryo.

A determination mechanism of this kind is an inherently imprecise, error-prone process. In fact, however, all gradient models of determination share this probabilistic element. If determination is by means of gradients, the requisite consequence is that cells must correct the initial errors created in the process, perhaps by some form of averaging of decisions. The evidence that neighboring cells can affect each other's determinative states will be discussed in Chapter 6.

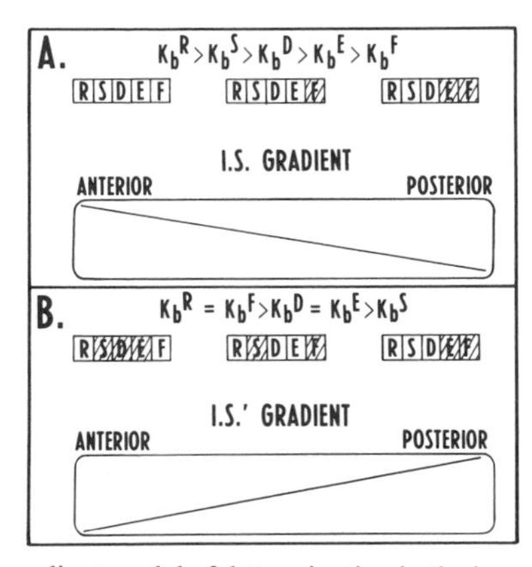

Figure 3.25 Replicative gradient model of determination in the insect egg. Gradients of replication are produced by gradients of replicon-initiator substances. Initiator substance (I.S.) binds to replication origins of different genetic units (R, S, D, E, and F) with the affinity values (Kb^x) shown. (*a*) The initiation pattern produced by an a-p gradient of I.S.; noncrosshatched genes are those that are initiated; crosshatched units indicate noninitiated ones. (*b*) The initiation pattern in a second replicative wave, produced by a second initiator substance (I.S.′), running from posterior to anterior. Noncrosshatched genes indicate those initiated in both replicative waves; crosshatched genes indicate those that were not initiated in the first or second waves and subsequently restricted in expression. (From Wilkins, 1976.)

GERM LINE DETERMINATION: IS IT A SPECIAL CASE?

There is one early cell assignment in the *Drosophila* embryo that appears at first sight to be independent of the determinative gradient system, that of the germ line. The pole cells, from which the germ line arises, are set aside before the final four cleavage divisions and develop from the distinctive posterior polar plasm. This specialness of the germ line and its cytoplasmic sources is also seen in other animal groups, including some amphibia and (as discussed earlier) nematodes.

In principle, pole cell formation could be the first phase of a single determinative process. However, several lines of evidence suggest that somatic and germ line determination are separate and independent events. In the first place, it is possible to eliminate germ line development by UV irradiation of the polar plasm in the early embryo without affecting the development of somatic structures (Okada et al., 1974). In the second place, numerous mutants exist that specifically alter either pole cell formation or somatic development. For instance, the embryos produced by the strict maternal effect mutant *mat(3)1* show a complete failure of somatic cellularization at the blastoderm stage, yet the pole cells form normally and even complete their normal dorsal migration (Rice and Garen, 1975). Furthermore, homeotic strict maternals that transform large regions of the soma of the blastoderm do not alter pole cell delineation; thus, *bic* embryos do not produce pole cells at their transformed anterior ends (Fig. 3.26*a,b*). Conversely, the elimination of pole cell formation without major somatic defects is produced by a wide variety of maternal effect mutants.

However, the strongest proof that germ line and somatic determination are independent is that polar plasm can specifically induce pole cell formation when transplanted to new locations within the embryo (Illmensee and Mahowald, 1974; Mahowald et al., 1979b). In this respect, the posterior polar plasm has exactly the expected properties of a qualitatively unique cytoplasmic determinant. Illmensee and Mahowald transplanted posterior polar plasm to anterior or ventral locations in early cleavage stage embryos and found that pole cells were produced at these locations. When these induced pole cells were transplanted into the posterior polar regions of genotypically distinct host recipient embryos, they gave rise to substantial portions of the germ line of the hosts, as shown by progeny tests. The experiment is diagrammed in Figure 3.27. The results show unequivocally that posterior polar plasm has some special quality that specifically induces germ line formation. Whatever the nature of this property, it can be shown to be present in stage 13 oocytes by the polar plasm transplantation procedure (Illmensee et al., 1976). This early time of development presents a further contrast to the somatic morphogenetic system, which comes into being during the cleavage divisions of the embryo.

The genetic analysis of germ line determination has centered on the group of mutants collectively designated *grandchildless* or *gs* mutants. These are

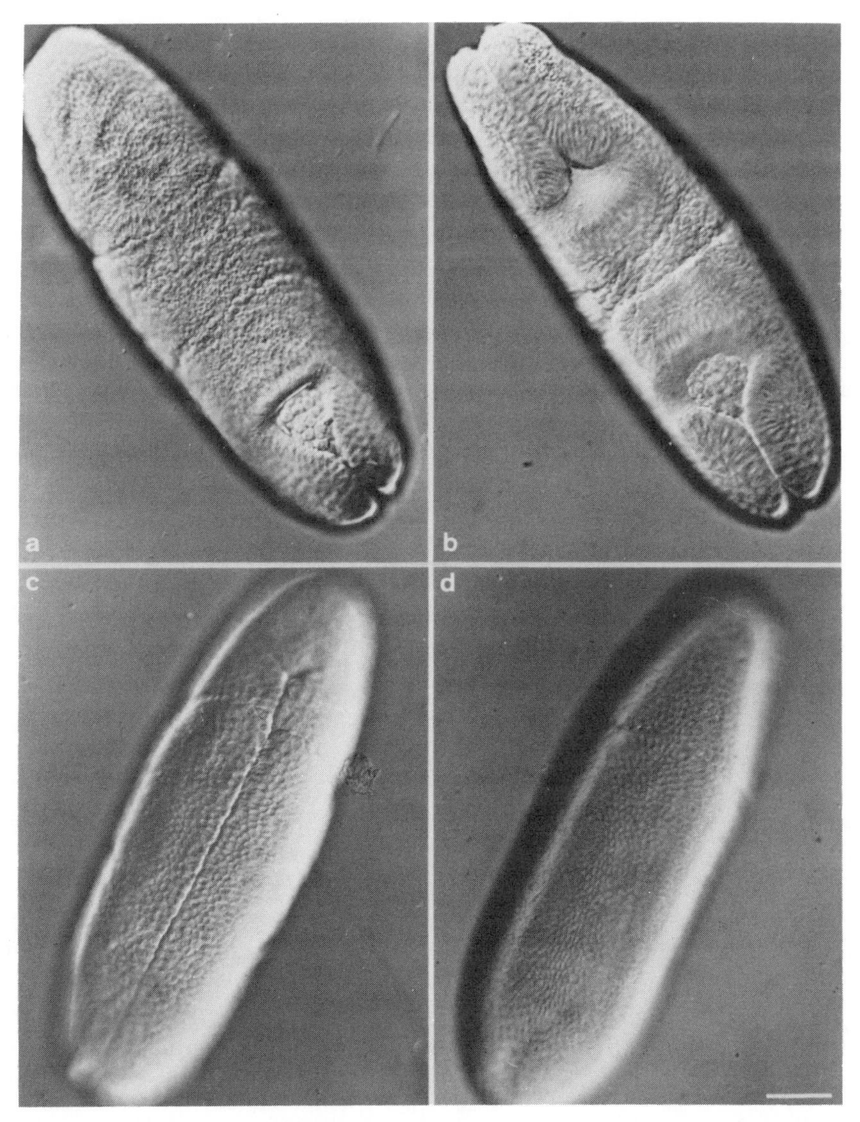

Figure 3.26 Gastrulation morphologies in wild-type and mutant embryos. (*a*) Wild-type; (*b*) *bic* embryo. Both embryos are seen from the dorsal side, with the anterior end in the upper left corner. Pole cells are carried forward by posterior midgut invagination. In the *bic* embryo, the comparable invagination occurs at the anterior end, but there is no cluster of pole cells; (*c*) wild-type embryo seen from the ventral side; (*d*) *dl*-dominant embryo seen from the ventral side. Note the absence of midventral invagination. (Figure kindly provided by Dr. C. Nusslein-Volhard; from Nusslein-Volhard, 1979a.)

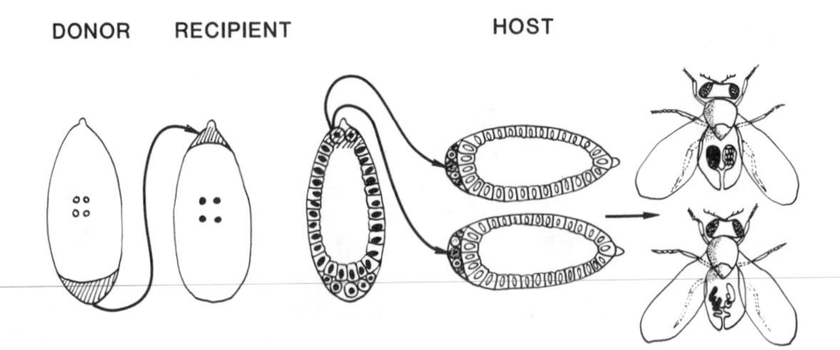

Figure 3.27 Procedure for inducing pole cells at nonposterior locations and using induced pole cells to create germ line chimeras. The genotypes of the donor, recipient, and host are all different to permit unambiguous delineation of the source of the induced pole cells. (Figure courtesy of Dr. A. Mahowald; from Mahowald et al., 1979a.)

recessive maternal effect mutants that produce sterile progeny; these progeny possess gonads, but the gonads are agametic. In all cases, the failure of gamete formation results from a failure of pole cell formation or function. The first *Drosophila* mutant of this type to be isolated was the *gs* mutant of *D. subobscura* (Spurway, 1948). The characterization of this mutant subsequently prompted a search for comparable *gs* mutants in *D. melanogaster,* of which a number have been isolated (Table 3.8).

Although all of these mutants were isolated on the basis of the canonical *gs* phenotype—the production of morphologically normal progeny that are sterile—nearly all of the mutants have been found to have some somatic defects. Furthermore, some temperature-sensitive *fs* mutants show a *gs* phenotype when they produce eggs at permissive temperature (Gans et al., 1975). These observations, in turn, suggest that the putative independence of germ line and somatic pathways may not be as total as was originally supposed. Might there be a connection that nevertheless allows some independence?

A scheme to unify the two determination pathways was proposed by Deak (1980). It was based on the Meinhardt hypothesis of gradient formation and on certain early observations by F. Seidel on lower insect embryos.

Table 3.8 ***Grandchildless*** **Mutants of** ***Drosophila melanogaster***

Locus	Chromosomal Position	Reference
fs(1)nas A	1–0.0	Counce and Ede (1957)
gs^{87}	1–20.0	Thierry-Mieg (1976)
gs(1)N26	1–33.8	Niki and Okada (1981)
gs(1)N441	1–39.6	Niki and Okada (1981)
gs(2)M	Autosomal	Mariol (1981)

Seidel's findings suggested that migration of nuclei into the posterior 20% of the egg activates a process essential for somatic development. The essence of the Deak hypothesis is that the polar plasm acts as a sink for an initial a-p gradient that runs from a high point at the anterior end of the embryo. Entry of cleavage nuclei into the posterior part of the embryo triggers the formation of a steeper morphogenetic gradient (by the Meinhardt mechanism). This gradient runs from posterior to anterior and is envisaged as the morphogenetic gradient that "instructs" somatic nuclei in their fates. However, the pole cells presumably remain unaffected by this event because they contain sink material. This removes morphogen and thereby promotes formation of the posterior peak of the second morphogen.

This hypothesis has the pleasing feature of providing a single mechanism of determination in the fruit fly. Nevertheless, several facts tell against it. Firstly, the model requires the formation of pole cells to sequester sink material in order to permit the formation of peak amounts of morphogen at the posterior pole. However, several *gs* mutants completely fail to form pole cells, yet undergo normal somatic development. Secondly, the model requires that cleavage nuclei enter the posterior part of the embryo to activate formation of the essential p-a morphogenetic gradient. However, ligated embryos in which the posterior parts receive no nuclei and fail to develop may give rise to the normal inventory of imaginal disc precursor cells in anterior halves (Schubiger, 1976). Finally, the hypothesis demands that sufficient posterior pole material transplanted to the anterior part of the embryo should induce posterior-half somatic structures, yet this transplantation induces only pole cells (Kalthoff, 1983). Whatever the connections between somatic and germ line determination may be, they cannot be obligate, as proposed by Deak.

Rather, the *gs* mutant phenotypes suggest that both somatic and germ line defects may result from aberrations in cleavage nuclei behavior. Many of the *gs* mutants, including the original mutant in *D. subobscura,* show delayed nuclear migration into the polar plasm (Mahowald et al., 1979a; Niki and Okada, 1981). Such delayed nuclei might therefore suffer the same fate as their blastoderm-destined sibs. Furthermore, these abnormalities of nuclear migration, which are not always restricted to the posterior pole region, could be responsible for the somatic structure defects seen in some of the *gs* mutants. Interestingly, two allelic temperature-sensitive *fs* mutants that show a *gs* phenotype at the permissive temperature produce embryos at the restrictive temperature with exaggerated mitotic waves and other cleavage abnormalities (Gans et al., 1975); this observation suggests that the permissive temperature (*gs*) phenotype reflects a less severe cleavage defect.

From this perspective, the function of the posterior polar plasm is not to actively *instruct* pole cells to become germ line cells but to *protect* pole cell nuclei from events of determinative restriction; as a result, the pole cells retain totipotency and hence the capacity to be germ line precursor cells (Kalthoff, 1979). We have already noted the apparently "derepressed" character of gene expression in the oocytes of sea urchins and *Drosophila*. In the fruit fly, at least, this characteristic may be established early by removal of the germ line precursor cells from the somatic determination system.

Whatever the fundamental character of germ line determination, the role of the polar granules in germ line determination remains an open question. In the original *gs* mutant of *D. subobscura,* polar granules of apparently normal morphology initially form but subsequently disintegrate; nuclear migration into the posterior polar region is delayed, and pole cells consequently fail to form (Mahowald et al., 1979a). (The disintegration of polar granule material in this mutant is not an autonomous property of the granules but appears to be related to abnormal cytoplasmic organelle segregation that occurs at both poles.) Although polar granule structure appears normal in certain other *gs* mutants, several observations suggest a general correlation between the amount of polar granule material present and the ability to form pole cells (Mahowald, 1983). It seems probable that the polar granules are essential for pole cell formation, but the nature of this requirement persistently remains obscure.

CONCLUSIONS

The distinctive contribution of genetic analysis to the study of the early development of *Drosophila* has been to define the general character of the morphogenetic system. The evidence is that determination of the somatic cells almost certainly involves gradient processes of some kind. At a minimum, one d-v gradient and one a-p gradient are required to specify cell fate in the blastodermal cell layer. The fundamental questions concern the nature of these gradients and their mode of action. Possibly, the two axial systems differ in these respects. Of the two, a molecular analysis of the d-v system is closer, with the discovery that maternal defects in embryonic d-v polarity can be partially repaired by injection of wild-type cytoplasm. The nature of the a-p axial specifying system is presently less approachable, although the geometries of the observed mitotic waves suggest that something keyed to mitosis, perhaps DNA replication, is involved.

Pole cell determination presents a separate puzzle, but illumination of the somatic determinative system may aid in its solution. By careful analysis of the *gs* mutants with delineation of the correlations between cleavage behavior, polar granule formation, and pole cell formation, the essential connections between somatic and germ line determination may emerge.

FOUR

MUS MUSCULUS

FROM OOCYTE TO IMPLANTING BLASTOCYST

Mus musculus, the house mouse of North America and Europe, is the experimental animal, par excellence, of modern biomedical research. By virtue of its manifold variations, convenient size, fertility, short gestation period, ease of maintenance, variable susceptibility and resistance to different infections, and exemplification of many diseases that affect mankind, it has found a major place in the laboratories of geneticists, developmental biologists, immunologists, cell biologists, and oncologists.

H. C. Morse (1981)

The laboratory mouse has the longest history of genetic experimentation of any animal. Unlike *Drosophila,* developmental hypotheses and phenotypic tests of gene effects accompanied the studies of inheritance from the beginning of genetic analysis. In consequence, more is known about the developmental genetics of the mouse than that of any other vertebrate.

The mouse as a model developmental system has a twofold interest. In the first place, it may be taken as a representative vertebrate. Although the early development of some vertebrates such as amphibians appears very different from that of mammals, these differences may be less significant than was once thought. Furthermore, the later development of body plans and organ systems in the five vertebrate classes have many clear similarities. In the second place, the mouse is a mammal, and as such is the most convenient model system for human development. Given this fact, it has special interest for us. Indeed, a number of human genetic disorders have their analogues in identified mouse genetic syndromes. The recently devised technique for genetic transformation of the mouse germ line (Chapter 9) may prove applicable to humans eventually and increases the relevance of mouse biology to human biology.

For mouse genetics, the primary genetic source material consists of nearly 300 highly inbred lines. These strains contain numerous genetic factors of interest. By making the appropriate crosses between them, it is often possible to identify the number and genetic map positions of the genes responsible for particular developmental conditions. To date, over 400 genetic loci in the laboratory mouse have been identified. A large number of these markers are biochemical, the structural genes for particular proteins. The use of such biochemical markers in developmental studies often allows genetic, molecular, and cytological analyses to be combined. An additional feature of the mouse that provides relatively easy experimentation is its relatively short generation time of 90 days.

In this chapter, we will look at the first stages of mouse development, from the beginning of oocyte development through the period of cleavage and independent embryonic growth. By the end of cleavage, when the embryo consists of about 120 cells, it contains two major cell populations, the outer trophectoderm (TE) layer and an enclosed group of cells, the inner cell mass (ICM). This preimplantation embryo is termed a "blastocyst." It is the blastocyst that implants in the uterine wall, beginning the phase of maternal-dependent growth and development. The embryo proper (the fetus) eventually develops from cells of the ICM; the later postimplantation stages of development will be taken up in Chapter 7.

The primary focus in this chapter will be on the nature of the morphogenetic system in the early embryo and its origins in the oocyte. The mouse oocyte has a special interest in this respect. In recent times, the mouse embryo has been viewed as a completely plastic system in which all early cellular commitments are impressed on essentially identical blastomeres by environmental forces. In effect, the oocyte has been interpreted as having no built-in morphogenetic instructions (Davidson, 1976). If this interpretation is correct and extends to the oocytes of other mammals, it would represent a major difference in the early developmental biology of mammals from that of possibly all other animal systems.

LIFE CYCLE AND EARLY DEVELOPMENT SUMMARIZED

Although mouse developmental biology is considerably more complex than that of either the nematode or the fruit fly, its life cycle is simpler, being the familiar one of eutherian mammals. In this life cycle, development takes place in one continuous phase, from the moment of fertilization until birth, the entire sequence taking place inside the mother's body.

Embryonic and fetal growth in the mouse takes approximately 20 days from fertilization to birth. The first 4.5 days are spent as an unattached, independent embryo, undergoing cleavage. When the embryo reaches the late blastocyst phase, it implants in the uterine wall, and the major phase of embryonic development begins. In contrast to most animals, the pace of the first cleavage divisions is very slow. However, the rates of cell division and developmental change increase following implantation, and accelerate further with the establishment of the placental connections to the mother's circulatory system and nutrient supply. During the first stages of postimplantation development, the three germ layers form within that portion of the embryonic structure that gives rise to the fetus, and by days 8–9, organ formation has begun. By 7.5 days, only 3 days after the 120-cell stage, the embryo proper consists of approximately 10^5 cells (McLaren, 1976b).

Because the embryo develops inside and attached to its mother's body, a special issue arises in mammalian development: the role of the mother in the developmental process. Is there a continuing "instructive" maternal role in development? This question has been investigated by culturing embryos in various nonuterine (ectopic) sites, particularly in males. Although such embryos show some abnormalities, the results indicate that the embryonic developmental sequence is governed by the embryo itself; the sole function of the maternal environment is to provide adequate nutrition to the developing individual (Graham, 1973; Johnson, 1979). Nevertheless, the question of maternal environment is pertinent in assessing putative maternal effect mutants, since early maternal effects on oocyte development must be distinguished from later maternal influences on embryonic development. This distinction can be made by transferring early embryos to surrogate mothers of a different strain, or, if preimplantation embryos are being tested for such effects, by comparison with control embryos in in vitro culture.

In general, preimplantation development is most readily analyzed in vitro. To obtain sufficient numbers of embryos for study, females are induced to "superovulate." This action is produced by injection with gonadotrophic hormones prior to mating. Normally, a female will release up to 10–20 eggs per ovulatory cycle; superovulated females can release several times this number of mature ova. The fertilized eggs can be readily removed from the female reproductive tract for further experimental analysis.

GENETICS AND GENOME STRUCTURE

The diploid chromosome number of *Mus musculus* is 40. Of the 20 chromosome pairs, 19 are autosomal and one is the sex chromosomes. Sex determination in the mouse is by an X-Y chromosome system, with XX individuals being female and XYs male, as in *Drosophila*. However, a fundamental difference in sex determination exists between the fruit fly and mammals. In the latter, sex is determined by the presence or absence of the Y chromosome, which establishes maleness, rather than by the X : A ratio, as in *Drosophila*. (The two sex determination systems will be described in Chapters 6 and 7.)

The genetic nomenclature is similar to that of *Drosophila*. Genes are given italicized symbols that are usually abbreviations of their names. In general, gene designations begin with a capital letter unless the gene was first identified by a recessive mutation (e.g., *qk* for quaking).

The development of chromosome banding techniques, G-banding and Q-banding, which distinguish specific homologous chromosome pairs, made possible the assignment of genes to particular chromosomes in the karyotype. This method uses translocations that move known genetic markers from one chromosome to another, in combination with an analysis of the changes in karyotype structure produced by the translocations, as detected by the banding techniques (described in Miller and Miller, 1975). (Until the advent of the chromosome banding techniques, genes could be assigned only to genetic linkage groups, but the identity of particular linkage groups with particular chromosomes could not be ascertained.)

A map of the mouse genome is given in the Appendix. All chromosomes are acrocentric and numbered from largest to smallest. The sex chromosomes are simply designated X and Y. The Y chromosome is similar in size to the smallest autosome pair (chromosome 19) but appears to be empty of genes, aside from male fertility genes and the locus for the H-Y antigen (a surface male-specific histocompatibility antigen). The nomenclature for chromosomal rearrangements is similar to that of *Drosophila* (see Lyon, 1981).

The molecular organization of the mouse genome raises many of the same fundamental questions encountered in connection with the genomes of the nematode and the fruit fly. The major gross difference between these genomes is in size. At 3×10^9 base pairs, the haploid mouse genome is nearly 20 times larger than *Drosophila*. Of this total, 8–10% consists of highly repeated satellite DNA, principally a 140-bp, adenine-thymine (AT)-rich repeat, reiterated 10^6 times in the centromeric heterochromatin. The single copy fraction is approximately 60–76% and the midrepetitive fraction is 15–25% of the genome (Church and Schultz, 1974; Ginelli et al., 1977).

The relative dispositions of single copy and repetitive sequences in the mouse genome are the subject of some dispute. The standard interpretation of eukaryotic interspersion patterns has been that there are two major types. In this interpretation, the most common pattern consists of single copy DNA sequences in 1000- to 3000-bp stretches, interspersed with short midrepetitive sequences about 300 bp in length (Davidson et al., 1975). The alternative interspersion pattern is designated the "*Drosophila* pattern," because it was first found in *D. melanogaster*. In this arrangement, long single copy stretches (averaging more than 13,000 bp) are interspersed with long blocks of midrepetitive DNA (5000 bp or more in length) (Manning et al., 1975; Crain et al., 1976).

In the original studies, all mammalian genomes were judged to follow the standard, short interspersion pattern. However, Moyzis et al. (1981a) have reevaluated these findings and concluded that much of the apparent difference between genomes in the interspersion pattern disappears when they are compared by identical renaturation methods. As these authors stress, the categories of "single copy" and "midrepetitive" are operational groupings, and the particular renaturation conditions employed can shift sequences from one category to the other. They show that under stringent reannealing conditions, the rodent genome shows the *Drosophila* pattern, with long single copy stretches (7200 ± 2000 bp) interspersed with long blocks of midrepetitive sequences. These midrepetitive blocks, in turn, consist predominantly of "scrambled clusters" of shorter (300- bp) blocks (Moyzis et al., 1981b). Interestingly, an analysis of the long blocks of midrepetitive sequences in *Drosophila* shows that many of these too consist of similar scrambled clusters (Wensink et al., 1979).

The functional significance of the sequence interspersion pattern remains unknown. For the reasons discussed in connection with the structure of the *Drosophila* genome, individual midrepetitive sequences are unlikely to have major, direct roles in controlling the expression of individual genes. However, it is not impossible that they have some general function in chromosome folding pattern or replication that indirectly influences gene expression. Conceivably, germ line transformation with engineered genes altered in neighboring midrepetitive sequences, followed by analysis of their expression patterns (see Chapter 9), may help to answer these questions.

OOGENESIS, FERTILIZATION, AND PREIMPLANTATION DEVELOPMENT

Like all animal development, the beginnings of mouse development are to be sought in the structure and properties of its oocyte. The development of the mouse oocyte takes place in two discrete phases. The first phase extends from midembryogenesis until 5 days after the birth of the female mouse. It lasts from the formation of oogonial stem cells by the primordial germ cells to the formation of immature oocytes, which are held in late diplotene of meiosis I. The second phase begins with the attainment of sexual maturity and involves oocyte growth and maturation. This takes place in small batches of oocytes within each estrous cycle (see Zamboni, 1972, and Wassarman and Josefowicz, 1978, for reviews).

In its general aspect, the initial development of the ovary in the mouse bears a certain resemblance to that in *Drosophila*. The primordial germ cells, which look identical in the two sexes, are distinguishable from other cells by their round shape and diffuse chromatin, characteristics similar to those of pole cells in *Drosophila;* however, they are most readily detected and tracked by their high alkaline phosphatase content. As in the fruit fly embryo, these primordial germ cells undergo a major migration from their posterior site of origin to the gonad rudiment. This migration, between 8 days postconceptus (pc) and 13.5 days pc, takes place from the caudal end of the embryo proper (the primitive streak) through the hindgut to the genital ridges (the somatic rudiment of the gonad). In female embryos, the arrival of these primordial germ cells stimulates the proliferation of the epithelial cells of the genital ridges; the primordial germ cells themselves also commence multiple rounds of cell division, to form oogonia. The oogonia subsequently undergo two waves of inward migration. Most of those in the first wave degenerate, but the cells of the second wave come to rest in the peripheral (cortical) region of the developing ovary. These cortical oogonia continue to proliferate, although remaining connected by many intercellular bridges. As we have seen, this syncytial arrangement is common in early germ line cells, and presumably serves to facilitate intercellular molecular transfer and synchrony of nuclear divisions.

During the migration to the ovarian cortex, groups of oogonia become surrounded by prefollicle cells of mesodermal origin. Eventually, each pro-oocyte is covered by a monolayer of these somatic cells to form a "unilaminar follicle." Pro-oocytes enter the first meiotic division but arrest at late diplotene until just prior to ovulation. Development of the unilaminar follicle is accompanied by the loss of syncytial connections between pro-oocytes. The proliferation of follicle cells and the secretion of extracellular material serve to isolate the follicles further from one another.

Oocytes remain quiescent until sexual maturation, when they are hormonally stimulated to commence growth and development in small groups. Although the estrous cycle takes 5 days, each preovulatory growth period lasts for 20 days, with most growth taking place between 20 and 6 days prior to ovulation. During this period, the growing oocyte increases in diameter from 20 to 80 μm and the surrounding follicle cells multiply to reach 50,000 per follicle. During these final stages, the oocyte comes to occupy a fluid-filled chamber, the "antrum," within the follicle. The follicle cells act as feeder cells for the oocytes; 85% or more of the metabolites required by the oocyte are passed to it through the follicle cells (Heller et al., 1981). The stages of oocyte and follicle development are shown in Figure 4.1.

The postnatal phase of oocyte development involves both growth and differentiation of the cell. A characteristic sequence of differentiative changes in the major cytoplasmic organelles—ribosomes, endoplasmic reticuluum, nucleoli, mitochondria—occurs, and marked protein synthesis takes place in both oocyte and helper cells. As in many other animal oocytes, synthesis and accumulation of histone proteins (Wassarman and Mrozak, 1981) and of tubulins (Schultz et al., 1979) occurs. These proteins are subsequently utilized in cleavage. The tubulin subunits may comprise as much as 1% of the total protein in the mature oocyte, thus constituting one of the most abundant protein fractions in the egg (as they are in the fruit fly oocyte). In common with other cellular differentiations, the oocyte growth phase features stage-specific alterations in the pattern of polypeptide synthesis, as assayed in two-dimensional separations (Schultz et al., 1979).

This period is also one of active transcription, of both ribosomal and informational RNAs (Bachvarova, 1974). The synthesis of poly A^+ mRNA is part of this transcriptional program. When [^{3}H]uridine is used as the radioac-

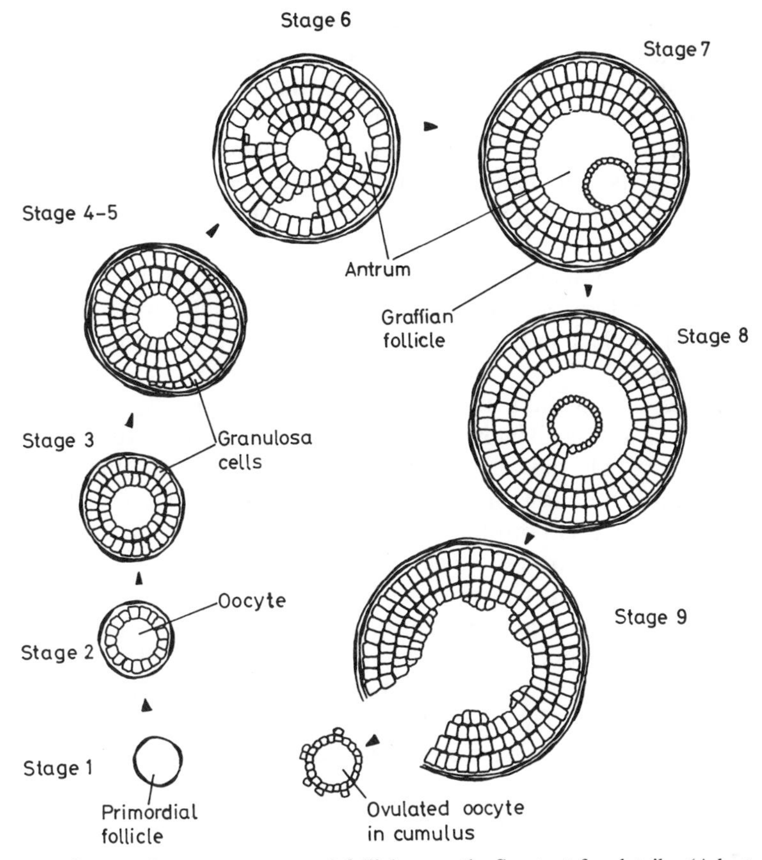

Figure 4.1 Stages of mouse oocyte and follicle growth. See text for details. (Adapted with permission from Baker, 1982, in *Germ Cells and Fertilization,* copyright Cambridge University Press.)

tive label, 9.6% of the total counts in the mature oocyte are found in poly A^+ mRNA (Bachvarova and De Leon, 1980). Indeed, the fraction of stored RNA that is in the form of mRNA is substantially higher in the mouse ovum than in the sea urchin or in the frog *Xenopus,* where the proportion is closer to 1–2% of total stored RNA (Davidson, 1976).

The period of oocyte growth and differentiation ends about 6 days prior to ovulation and is succeeded by a 5-day phase of follicle cell proliferation and antrum development. In the last 9–10 hr before ovulation, the oocyte reenters meiosis and completes the first meiotic division, giving rise to the first polar body. The oocyte nucleus then proceeds directly to meiosis II without re-formation of the nuclear membrane but halts at metaphase until fertilization. Ovulation involves release of the ovum from the follicle into the oviduct. It is surrounded by a polysaccharide-rich envelope, termed the "zona pellucida," and a group of follicle cells, the "cumulus oophorus."

Because receptivity to mating closely correlates with ovulation, the time of fertilization can be gauged fairly accurately under conditions of controlled mating. Fertilization involves penetration by the sperm through the cumulus oophorus, which is shed during passage down the oviduct, and the zona pellucida. Following entry, the sperm head is converted to the male pronucleus and loses its complement of chromosome-coating protamines, which are replaced by histones and acidic proteins from the maternal cytoplasm. Fertilization triggers resumption of meiotic progression in the oocyte nucleus; completion of meiosis is quickly followed by extrusion of the second polar body. The two pronuclei then move slowly into apposition. (DNA replication takes place in the pronuclei and is completed during this lengthy migratory phase.)

During the first 3 days following fertilization, the zygote progresses through four cleavage divisions to the 16-cell stage, the morula. In the following day and a half, the solid ball-like morula develops into the blastocyst. The blastocyst is a hollow, nearly spherical fluid-filled ball of cells, and, as mentioned earlier, it is both the implanting structure and the carrier of the precursor cells of the true embryo.

Cleavage in mammals is a slow process. In the mouse, the first cleavage does not take place until 24 hr postfertilization, after the prolonged migration and fusion of the pronuclei. Successive divisions take place at approximately 12-hr intervals. The eight-cell stage may be a developmentally critical juncture. It is at this point that the heretofore loosely knit blastomeres undergo the process of compaction. During compaction, the cells transform from spherical to columnar and develop a variety of intercellular junctions (Ducibella, 1977). The net result is the drawing together of the ball of cells. The eight-cell stage, before and after compaction, is shown in Figure 4.2. Compaction is almost certainly important in creating certain properties of polarity within the blastomeres of the eight-cell stage; the significance of such polarity will be discussed later.

From the eight-cell stage on, cleavage results in the formation of distinct inner and outer cell populations. By the 16- to 32-cell stage, the inner cells have begun to secrete fluid into the developing central cavity, the blastocoele. By the 60-cell stage, approximately 3.5 days after fertilization, the embryo is at the early blastocyst stage and consists of two cell populations. The outer layer of approximately 45 cells constitutes the trophectodermal (TE) layer and the internal 15 cells comprise the ICM. Formation of the early blastocyst is followed by reexpansion of the embryo to fill the zona pellucida; a picture of an expanding blastocyst is shown in Figure 4.3.

The final phase of preimplantation development involves the production of two further cell groups. The ICM gives rise to a group of endodermal cells that eventually cover the entire inside of the blastocoele. And the TE diverges into a cap of "polar TE" cells, which sit on top of the ICM and remain diploid, and the "mural TE" cells, which comprise the sides of the blastocyst. By the late blastocyst stage, some of these mural TE cells have

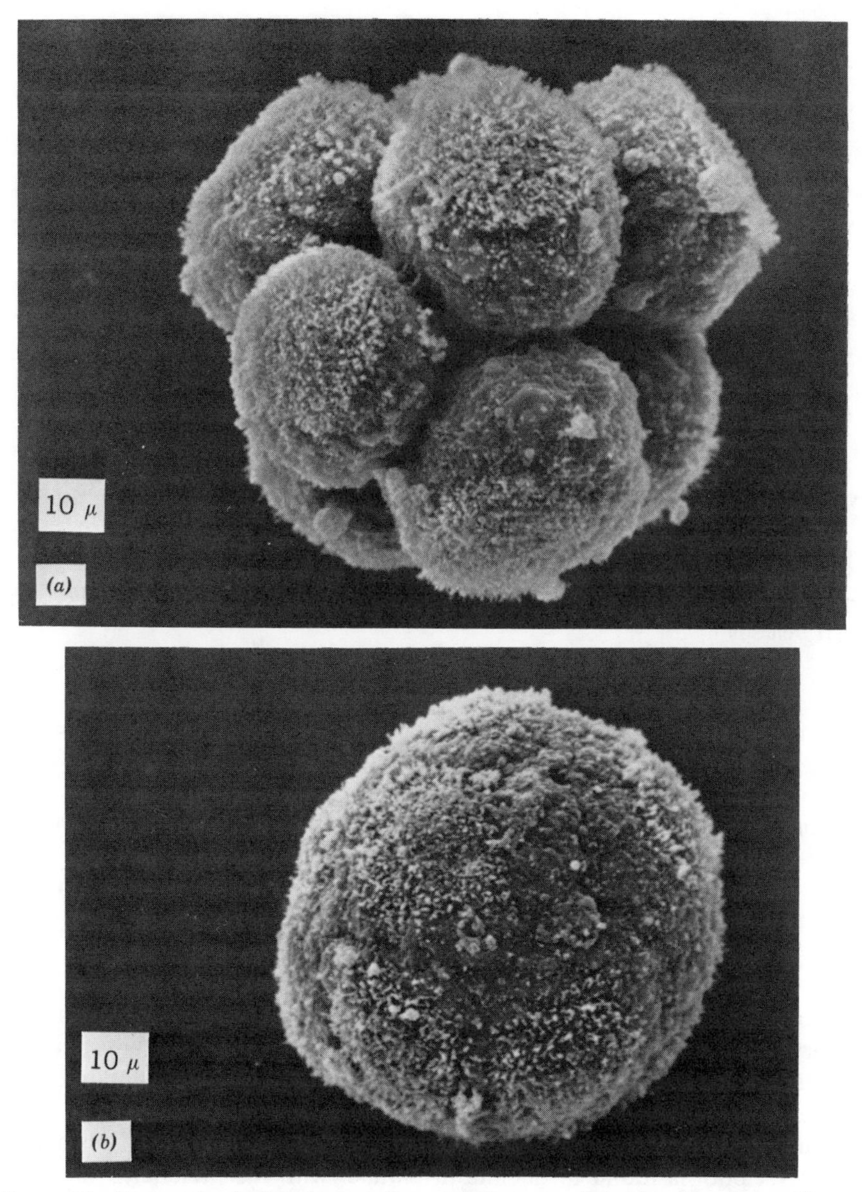

Figure 4.2 Compaction in the mouse embryo. (*a*) Uncompacted eight-cell embryo; (*b*) compacted eight-cell embryo. (Photographs kindly supplied by Dr. M.H. Johnson.)

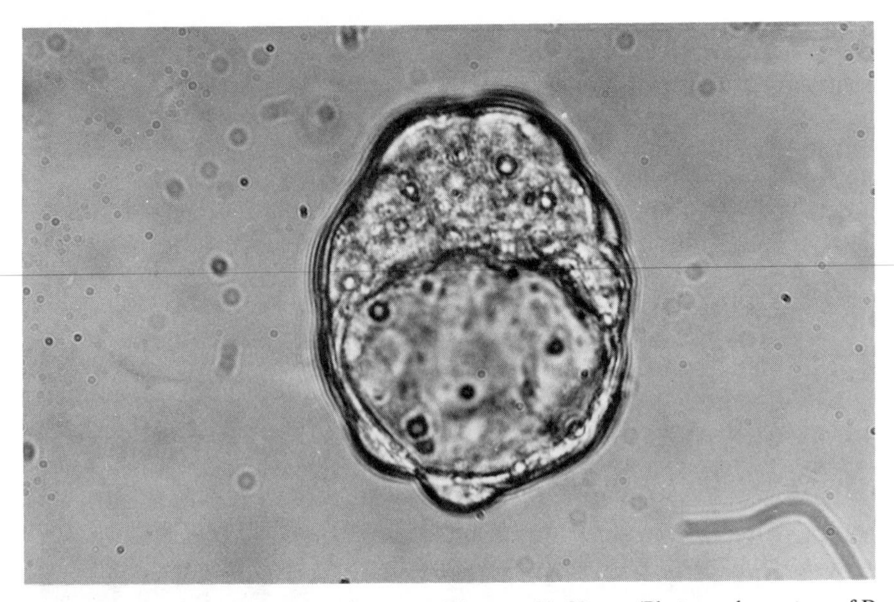

Figure 4.3 An expanded 3.5-day blastocyst. Diameter 70–80 μm. (Photograph courtesy of Dr. M.H. Johnson.)

begun to become polyploid; eventually, these large polyploid cells become the "primary giant cells" of the implanting blastocyst wall.

The sequence and timing of events in preimplantation development, from the one-cell stage to the late blastocyst, are summarized in Figure 4.4.

EXPRESSION OF THE ZYGOTIC GENOME DURING PREIMPLANTATION DEVELOPMENT

The changeover from maternal to zygotic genome control in the early mouse embryo has been tracked by several means. One of these is analysis of the stores of maternal mRNA and their depletion during preimplantation development. As noted above, the poly A^+ mRNA pool is unusually large, amounting to approximately 10% of the total RNA in the zygote. This maternal RNA pool is found to decline in parallel with the prelabeled maternal RNA. The measurements show an initial depletion of prelabeled mRNA to 60% of the prefertilization content during the first 24 hr of development (to the two-cell stage), followed by a more gradual decline between the cell-cell and early blastocyst stages to a final value of 30% (Bachvarova and De Leon, 1980). The actual decline may in fact be considerably greater; inhibitor experiments and tests of in vitro translation of extracted maternal mRNA indicate a rather rapid inactivation or depletion of these stores by the mid–two-cell stage (Flach et al., 1982; Pratt et al., 1983).

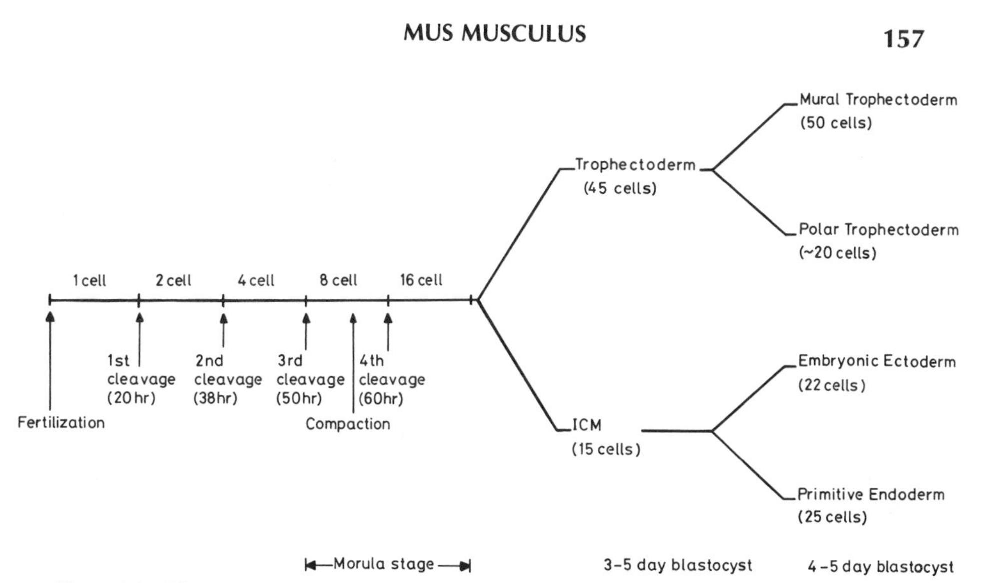

Figure 4.4 Time course and stages of preimplantation development. (Adapted with permission from Gardner, 1978.)

Despite the large starting maternal mRNA pool, the embryo becomes dependent on transcription from its own genome early in development. Blockage of transcription by either actinomycin D or α-amanitin stops development at any point after the first cleavage and reduces amino acid incorporation from the two-cell stage on. This process is significantly different from the early developmental programs of sea urchin and amphibian embryos. In these animals, actinomycin D treatment during cleavage has little immediate effect on amino acid incorporation or cell division. However, mRNA synthesis is shut off and development arrests at the postcleavage stage (Davidson, 1976).

In accordance with the inhibitor study results, labeling of new RNA synthesis by radioisotope incorporation shows that zygotic genome transcription begins early. Using [^{3}H]uridine incorporation, Knowland and Graham (1972) placed the beginning of transcription in the embryo at the middle of the two-cell stage. With the same label, Levey et al. (1978) similarly located the beginning of nonribosomal nuclear RNA and of poly A$^+$ mRNA at this point. However, when [^{3}H]adenosine, which is taken up more efficiently than uridine, is used, the synthesis of new species of nuclear RNA and poly A$^+$ mRNA is found to begin in the middle of the one-cell stage. Nevertheless, formation of labeled poly A$^+$ messengers at the one-cell stage involves primarily addition of poly A tails to preexisting maternal mRNAs (Clegg and Piko, 1982).

Given the early expression of the zygotic genome, when does the cytoplasmic composition of the embryo begin to change over from maternally encoded products to those characteristic of the embryonic genome? Biochemical experiments suggest that the transition begins no later than the two-cell stage. However, results of both inhibitor and labeling experiments are subject to various interpretations. Independent estimates for ascertaining when the zygotic genome begins to affect embryo cytoplasmic composition can be obtained by genetic means. Such tests involve detection and discrimination of paternal genome-encoded products from maternal ones; while maternal products may derive either from the oocyte or from the maternally derived chromosomes of the zygote, paternally encoded products in the embryo can come only from the zygotic genome.

In *Caenorhabditis* and *Drosophila,* it is possible to test for paternal gene activity by determining whether rescue from lethal maternal effects takes place in embryos with wild-type paternal alleles. In the mouse, where few maternal effect mutants are known (see the following section), one must resort to biochemical methods for detecting such paternal gene expression. One such method is to score for a quantitative increase in gene expression programmed by a paternal gene. An example is provided by a test for β-galactosidase expression in early embryos, an enzyme encoded by a gene on chromosome 9. During development to the blastocyst stage, this enzyme, like that of two other lysosomal hydrolases, undergoes a 50- to 100-fold increase in activity. This activity increase is mediated by a closely linked, *cis*-dominant regulator gene site, designated ***Bgl-s***. One allele of this *cis*-dominant regulator, $\boldsymbol{Bgl\text{-}s^{h}}$, produces a twofold higher level of activity than the other known allele, $\boldsymbol{Bgl\text{-}s^{d}}$. To establish the time of paternal gene expression for β-galactosidase in embryonic development, Esworthy and Chapman (1981) mated female mice homozygous for the $\boldsymbol{Bgl\text{-}s^{d}}$ (low activity) allele to males with the $\boldsymbol{Bgl\text{-}s^{h}}$ (high activity) allele and collected embryos at successive times after fertilization for measurement of enzyme activity. The strains employed were nearly identical in genotype ("congenic"), except for the region on chromosome 9 containing the structural gene and its linked regulator.

Figure 4.5 shows the plotted increase in enzyme activity in hybrids as a function of developmental stage when normalized to the activity increases seen in parallel $Bgl\text{-}s^{d} \times Bgl\text{-}s^{d}$ crosses. The results show a paternal allele-mediated increase in activity that extrapolates back to 36–45 hr, or shortly after the second cleavage division. That the increase mirrors a true turn-on of gene expression rather than an effect of heterosis is shown by the reciprocal cross (with a high maternal background) results, which did not give a comparable increase. The findings indicate that the cytoplasmic composition of the embryo begins to reflect zygotic genome expression by the two-cell stage. Comparable tests with the chromosome 5 structural gene for β-glucuronidase give a similar result (Chapman et al., 1976).

From the two-cell stage on, zygotic genome expression accelerates, its products increasingly dominating the cytoplasm. Ribosomal RNA synthesis cannot be detected in the one-cell embryo, but by the mid–two-cell stage it accounts for a significant fraction of the new transcripts. General transcription rapidly accelerates in the eight-cell embryo (Piko, 1975), the point at

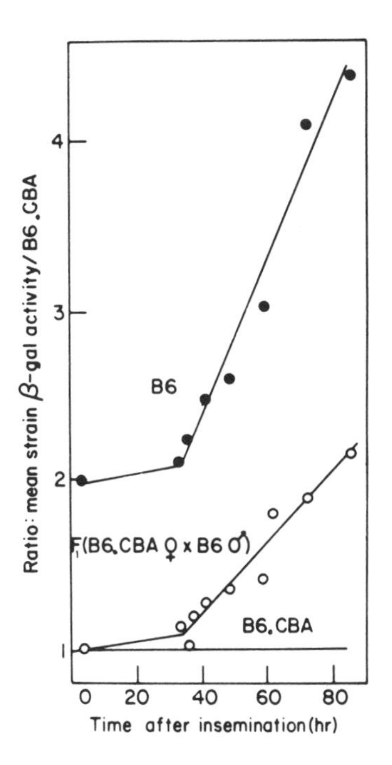

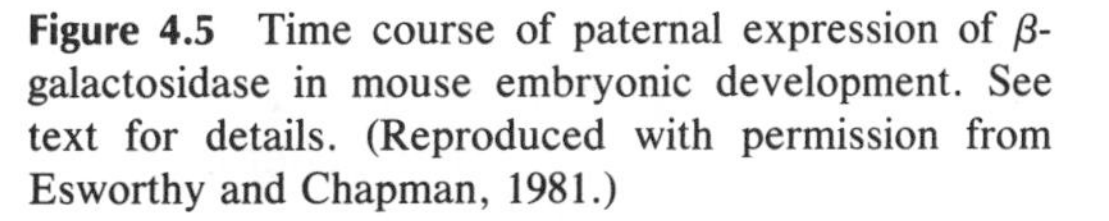

Figure 4.5 Time course of paternal expression of β-galactosidase in mouse embryonic development. See text for details. (Reproduced with permission from Esworthy and Chapman, 1981.)

which compaction occurs. By the early morula stage, rRNA synthesis is approximately 16-fold more rapid than in the two-cell stage (Clegg and Piko, 1982).

In contrast, protein synthesis is active throughout early development and even in the unfertilized oocyte. Much of this early synthesis is carried out on maternal templates. However, detailed analysis, employing two-dimensional gel electrophoresis, shows that zygotic genome-encoded protein products are detectable by the early two cell stage (Pratt et al., 1983).

One of the most interesting findings to come from such protein synthesis studies concerns the occurrence of posttranscriptional regulation in the maternal mRNA pool. These regulated messengers code for a group of proteins of 35 kd, which first appear during the late one-cell stage. The scheduled synthesis of these proteins continues unabated in the presence of the transcription inhibitor α-amanitin, at a drug concentration sufficient to inhibit all in vivo RNA polymerase II activity. Furthermore, when maternal mRNAs are extracted from unfertilized eggs and placed in an in vitro protein-synthesizing system, these proteins are made in quantities comparable to those of two-cell embryos (Braude et al., 1979). It follows that the responsible mRNAs must be held in some state that is unfavorable for translation, until the early two-cell stage, and are then released by some signal to the translation apparatus. One possibility is that this signal is a timed polyadenylation of these messengers, permitting transport out of the nucleus, because the onset of translation correlates with the approximate period in which long poly A tails are added to preexisting maternal mRNAs (Clegg and Piko, 1982).

The cellular functions of these posttranscriptionally controlled proteins and the significance of their controlled translation are unknown. We noted earlier the existence of a parallel phenomenon in *Drosophila* (Mermod et al., 1980). A third example of such posttranscriptional control has been reported in the sea urchin, involving delayed translation of a histone H3 maternal mRNA (Wells et al., 1981; Raff, 1983). One possible explanation for the phenomenon is that such posttranscriptionally regulated mRNAs might code for proteins required in abundance by the early embryo, and therefore must be supplied by maternal templates, but that their "premature" translation could cause some kind of damage to the embryo. This suggestion might be tested by examination of the effects on development of early injection of the purified proteins.

The question of mechanism is also unresolved but may involve different processes in different organisms or even for different mRNA species within the same animal embryo. *Drosophila* involves sequestration in cytoplasmic RNP particles, but the sea urchin histone mRNA seems to be retained by some selective mechanism in the nucleus (Raff, 1983). Delayed polyadenylation might, in principle, be a general mechanism of posttranscriptional control in the mouse, but the particular mRNAs that undergo delayed translation are stored as polyadenylated species (cited in Pratt et al., 1983).

Although changes in electrophoretic patterns of labeled polypeptides can be used as an index of changes in gene expression, they are not an infallible guide to changes in protein synthesis. About 50% of the changes occurring in early preimplantation development, as registered on two-dimensional separations, result from reproducible patterns of posttranslational modification—the addition of nonprotein moieties, such as glucosamine and phosphate, to preexisting polypeptides that cause a change in migration behavior (van Blerkom, 1981). Some of these posttranslational modifications are set in motion by oogenesis itself and occur in unfertilized eggs (Pratt et al., 1983). From the eight-cell stage, there is relatively less change in the overall distribution pattern of labeled polypeptides in two-dimensional gels from whole embryos. Thus, during the period of blastocyst formation, the overall protein synthetic pattern appears relatively stable.

Despite this relative constancy in protein synthesis between the eight-cell/compaction stage and the early blastocyst, a few significant changes in gene expression take place. The first is a key regulatory event in the late morula stage, as the embryo goes from a 16-cell structure to one of 32 cells, which is necessary for the secretion of fluid that takes place during blastocyst formation. This fluid initially accumulates between cells and then collects in the developing blastocoele, the entire process being referred to as "cavitation."

Smith and McLaren (1977) have shown that the onset of cavitation requires five DNA replication cycles. Presumably, the signal to begin cavitation involves either a direct "counting" of these replication cycles or the attainment of a characteristic nucleocytoplasmic ratio during cleavage. In investigating the molecular biology of this developmental event, Braude (1979) found that the onset of cavitation requires transcription and concluded that this transcriptional event occurs during or as a result of the fifth replication cycle. The nature of the link between the fifth replication cycle and transcription has not yet been elucidated. Since zygotic genome transcription has commenced well before this point, the control event cannot be a simple titration of transcription inhibitor by DNA as occurs in *Xenopus*.

The second set of events concerns the divergence of the first two distinct cell populations, the ICM and the TE cells. Van Blerkom et al. (1976) were the first to detect differences in patterns of polypeptide labeling between these cell groups. They dissected ICM and TE cells from 3.5-day (expanded) blastocysts, incubated the two fractions separately with [^{35}S]methionine, and then analyzed the patterns of labeled polypeptides with two-dimensional separations. Although the majority of spots were shared between the two cell groups, as expected from the general constancy of labeled polypeptides in whole embryos between the 8- and 64-cell stages, a few distinctive ICM and TE spots were detected. Thus, by the late blastocyst stage, the two cell populations, which are recognizably different in position and beginning to be distinct cytologically, can be distinguished by molecular markers.

Handyside and Johnson (1978) subsequently showed that the "ICM-specific" and "TE-specific" cells are first detectable between the early morula (12- to 25-cell) and late morula (25- to 30-cell) stages. When the inner cells from embryos of these stages were isolated by the technique of immunosurgery (described below), these inside cells were found to label just the shared and ICM-specific polypeptides and not the TE-specific ones. Some of these differences may reflect posttranslational modification rather than new gene expression per se, but the significant point is that the molecular events of differentiation are detectable before there are cytologically distinct ICM and TE cell populations.

The sequence of changes in gene expression and macromolecule composition that take place in the preimplantation embryo are summarized in Figure 4.6. The entire set of changes constitutes a complex picture. However, the salient feature is that zygotic genome expression begins early in this embryo, and zygotic genome products increasingly dominate the molecular composition of the embryo from the two-cell stage on.

DETERMINATION EVENTS IN THE PREIMPLANTATION EMBRYO: GENETIC ANALYSIS

In *Drosophila,* the analysis of early embryonic determination has involved two principal kinds of genetic technique. The first is the creation of genetic chimeras by removal of genetically marked cells from the blastoderm stage

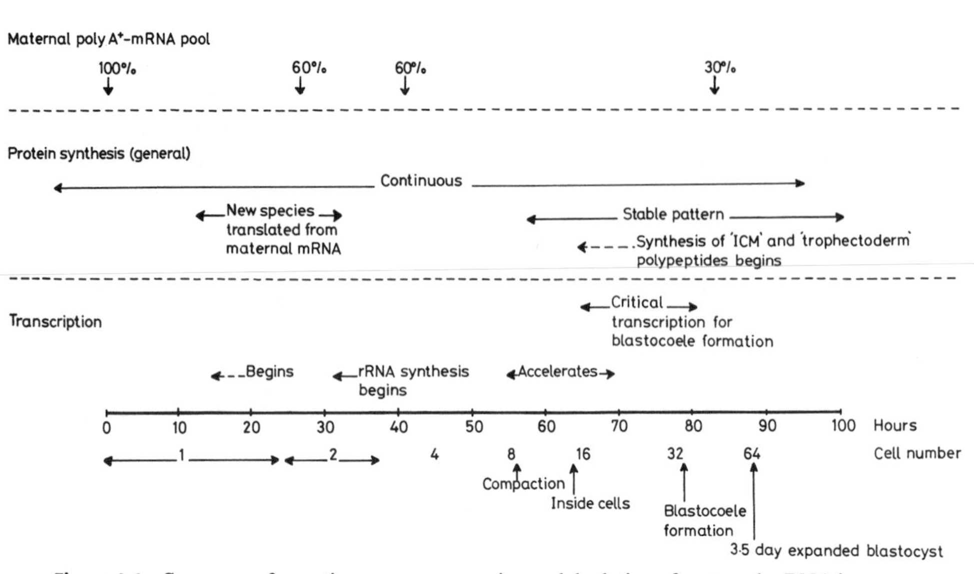

Figure 4.6 Summary of zygotic genome expression and depletion of maternal mRNA in mouse preimplantation development. See text for discussion.

and subsequent culture in recipient embryos or larvae. The fates of the donor cells in these chimeras are used to establish the timing of the first determinative events. The second genetic technique is the isolation of maternal effect mutants that produce aberrations in the morphogenetic system of the egg. These experiments have permitted a genetic characterization of the mechanism of determination.

Investigation of determination mechanisms in the early mouse embryo has relied principally on the first method, chimera construction. Extensive mutant hunts are impossible because of the limited number of progeny that can be screened; furthermore, the collection of existing mutants includes extremely few with maternal effects (McLaren, 1979, 1981b). However, the chimera experiments alone have proven instructive about the timing of determination events in the preimplantation embryo. This information, in turn, provides some clues to the mechanisms of determination in this embryo. We will first look at the methods of chimera analysis used in the mouse and the application of these techniques to analysis of early determination. In the next section, we will examine some of the hypotheses about determination in the mouse embryo that have been proposed.

Chimera Construction and Analysis: Methods

There are several methods of constructing chimeric mouse embryos; the two principal ones are illustrated in Figure 4.7. The first early embryo aggregation was invented by Tarkowski (1961) and Mintz (1962). In this technique,

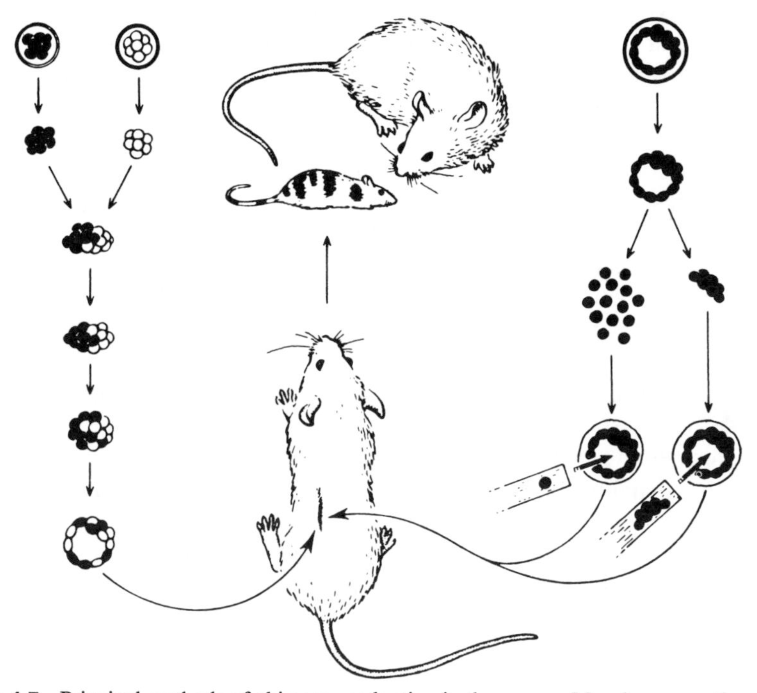

Figure 4.7 Principal methods of chimera production in the mouse. Morula aggregation shown on the left, blastocyst injection on the right. (Reproduced with permission from McLaren, 1976a, *Mammalian Chimaeras;* copyright Cambridge University Press.)

eight-cell embryos from two genetically distinguishable strains are stripped of their zonae (mechanically, or now usually by digestion with pronase) and then brought into contact; fusion normally occurs within a period of 2–4 hr at 37°C. The composite morulae are then transferred to the uterus of a pseudopregnant foster mother, where implantation and subsequent development take place. (Such foster mothers are typically prepared by prior mating to vasectomized mice.) A substantial proportion of such morulae develop and give rise to normal newborn mice. In experiments in which the two input genotypes differ in their coat color phenotype, the coat color pattern of the chimeras can have a striking appearance, as shown in Figure 7.9. (Note: the term "allophenic mice" is also found in the literature to denote chimeric mice.)

The second method, introduced by Gardner (1968), permits the assay of the developmental capabilities of cells of different stages, either singly or in small groups, without the major disruption of preimplantation development that embryo aggregation entails. In this procedure, single or multiple cells of the desired stage and genotype are injected into host blastocysts through a narrow incision. The recipient blastocyst is then placed in a pseudopregnant foster mother; those that have not been irreparably damaged by the injection procedure give rise to chimeric fetuses. The method allows the testing of developmental capabilities of any of the cells in the preimplantation embryo. The method can also be used to look at developmental potentialities of cells in postimplantation stages.

A potential drawback of the cell injection method is that the blastocoelic environment may affect the developmental paths taken by the injected cells. To avoid this complication, a third method has been invented. In this scheme, the cells to be tested are aggregated with eight-cell morulae to create chimeric embryos (Rossant, 1975a).

The informativeness of all chimera experiments depends on the sensitivity or resolution with which the cells of different genotypes can be resolved. When chimeric embryos are allowed to develop to birth, the use of coat color markers is adequate, at least for detecting respective contributions to the epidermal tissues. However, when internal tissues are to be traced, or when intermediate stages of development are to be examined, histochemically distinguishable genetic markers must be used. These are usually enzymatic or antigenic markers.

Biochemical markers are broadly classified as either *direct* or *indirect*. With direct cell markers, individual cells can be scored in tissue sections, permitting fine-scale tracing in situ of the location and relative contributions of the two genotypes. Plus or minus activity enzyme activity differences and surface antigenic markers are in this category. Indirect cell markers are those that can be scored only in bulk tissue preparations. The most usual indirect genetic marker is an electrophoretic difference in an enzyme, that is, electromorphic alleles. With indirect cell markers, the chimeric embryos are dissected into their component tissues, which are then assayed separately for the two genotypic contributions. The two most commonly used indirect markers are electromorphic differences for the enzymes isocitrate dehydrogenase (IDH) and glucose phosphate isomerase (GPI). For the latter, minor electrophoretic differences can be detected in proportions as small as 1–3% of total activity (Chapman et al., 1972). The separation of the two common GPI electromorphs, forms a and b, is illustrated in Figure 4.8.

In general, direct markers give more information with less chance of error. With indirect markers, there is always the problem of cross-contamination of tissue samples by cells of the other genotype. However, it is not always possible to find strains differing in alleles for a good direct marker. The recent development of techniques for discriminating cells of differing genotype at the major histocompatibility locus (H2) of the mouse may solve this problem (Ponder et al., 1983). Because of the lack of suitable direct markers, most experiments have relied on an indirect cell marker, principally the GPI electromorphic difference.

Chimeras and the Analysis of Early Determination Events

The analysis of determination events in preimplantation development by chimera construction is similar to that used in the analysis of *Drosophila*

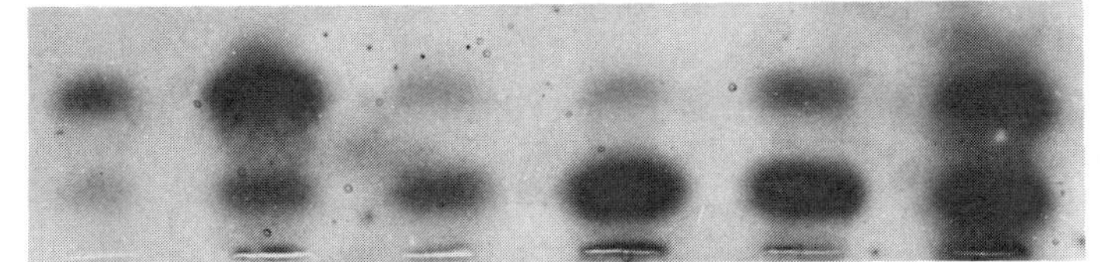

Figure 4.8 Electrophoresis of mixtures of GPI electromorphs. Kidney homogenates from strains homozygous for the two allelic enzyme forms were prepared, and mixtures were made in the proportions indicated; the samples were electrophoresed and stained for GPI activity. (Figure kindly provided by Dr. V. Chapman; from Chapman et al., 1972.)

determination in the blastoderm. One takes cells of genotype A, from a defined stage or location, and places them in host embryos of genotype B. Subsequent development is followed, and the relative contributions of A and B to the various tissues of the later embryo or adult are scored. If all tissue types have some donor cell (genotype A) contribution, then these donor cells must have been totipotent. If, on the contrary, they are found in only certain tissues, the original donor cells are presumed to have experienced some previous determinative events.

The chimera technique has been used to explore three fundamental questions about determination in the preimplantation embryo: (1) does the divergence of ICM from TE entail a restriction in developmental potency of either cell type? (2) If such restrictions do occur, when do they take place? (3) Does the formation of endoderm from ICM or the divergence of polar and mural TE involve determinative events?

The topic that has received most attention is the nature of the divergence between ICM and TE cells. It has long been known that these two cell populations differ distinctly in their embryological fates. The ICM of the late blastocyst contributes directly to the fetus and to certain of the extraembryonic membranes, while the TE contributes directly to tissues involved solely in the implantation process (Snell and Stevens, 1966). By 3.5 days pc, when the blastocyst consists of 60–64 cells, these two cell populations differ in distinctive cytological features (Johnson, 1979). These differences include the kinds of cellular junctions each displays, the presence of microvilli on ICM and their absence from TE, and recognizable differences in mitochondrial structure. Given the difference in fates between ICM and TE, is it based on an irreversible difference in developmental capability, and if so, when does this difference become established?

The experiments of Rossant (1975a) demonstrate that the cells of the ICM are already determined by the 3.5-day, expanded blastocyst stage. In these experiments, the donor ICMs were aggregated with recipient eight-cell morulae and transplanted to the uterine horns of pseudopregnant foster mothers, then allowed to develop for periods ranging from 8.5 to 18.5 days. The genetic marker used for tracing the ICM-derived cells was GPI, with the donor cells specifying the b electromorph and the recipient embryos specifying the a form. Following termination of development, the fetuses and associated embryonic structures were isolated and dissected into the ICM-derived embryo plus membranes fraction and the TE-derived fraction.

The analysis of the chimeric embryos showed a substantial and highly preferential contribution of donor GPI enzyme to the ICM-derived tissues, with little or no contribution to the TE-derived fraction. Of the 15 embryos showing a donor cell contribution, 12 showed an exclusive contribution to the embryo plus membranes fraction. The three apparent exceptions showed only a minor contribution, which probably reflected contamination with ICM-derived tissues. The results provide firm evidence that ICM cells from 3.5-day expanded blastocysts are already restricted in their developmental capacity. A similar result has been reported using a direct cell marker (an antigenic difference) in rat-mouse chimeras, in which rat ICM cells were injected into mouse blastocysts (Gardner and Johnson, 1975).

TE cells from 3.5-day blastocysts are comparably restricted in their developmental potential. When TE vesicles are emptied of their ICM cells before fostering, they give rise to trophoblast cells (the TE cells that mediate implantation) and undergo implantation. However, they cannot give rise to ICM or to those structures normally derived from ICM. Furthermore, injection of TE cells into 3.5-day blastocysts never yields embryos with a donor cell type contribution (Gardner, 1978). The results show that by the 60- to 64-cell stage, the cells of the ICM and TE are committed to distinct and restricted pathways of development.

When does this restriction in developmental potential arise? The experiment of Kelly (1975) provides a lower bound; it is diagrammed in Figure 4.9. She dissociated four-cell embryos in vitro and then allowed each blastomere to divide once, to give four "octet pairs." The members of each pair were then separately aggregated with individual eight-cell carrier morulae and fostered. Donor cells carried the *Gpi-1*^{a} allelic form, while host morulae carried *Gpi-1*^{b}. In addition, the donor strain was albino, so that chimeras that developed to term could be distinguished by coat color. Fourteen chimeras, derived from seven octet pairs, had donor cell contributions in the fetuses. In five of the pairs, donor cells had participated in the formation of both ICM- and TE-derived fractions. A substantial donor cell contribution was also seen in all chimeras allowed to develop to term. Evidently, the blastomeres of the eight-cell embryo do not show segregation of developmental potential for either ICM or TE. This work demonstrates that blastomeres at the eight-cell stage are totipotent. Therefore, the first determinative events occur between the 8- and 64-cell stages.

To pinpoint the time of developmental restriction more precisely, the technique of "immunosurgery" was used by Handyside (1978) to isolate and test the capabilities of inner cells at successive intermediate stages between compaction and the expanded blastocyst. Immunosurgery on mouse embryos involves the selective lysis of outside cells by exposure of the embryos to rabbit anti-mouse serum, followed by washing and treatment with complement (a substance that lyses antibody-reacted cells). When the treatment is

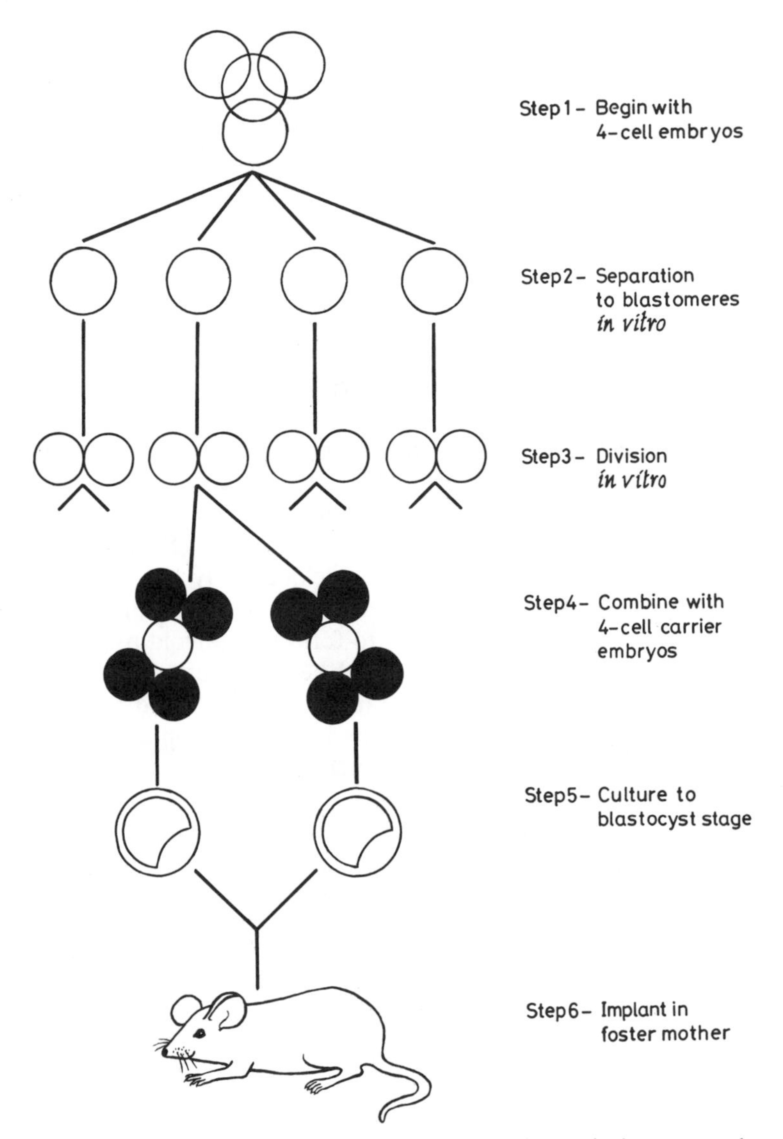

Figure 4.9 Test for totipotency of blastomeres at the eight-cell stage in the mouse embryo. See text for details. (Adapted with permission from Kelly, 1975.)

applied to embryos of successively more advanced stages and the isolated inner cells are tested in vitro for their differentiative capabilities, the inner cells of late morulae and early cavitating blastocysts are found capable of generating TE derivatives. However, the ICMs of 3.5-day expanded blastocysts do not possess this capacity and produce only endodermal layers.

A somewhat different result, obtained with the same procedure but with embryos from a different strain, was obtained by Hogan and Tilley (1978), who found some 3.5–4.0 ICMs capable of generating TE-derived structures. These authors suggested that the timing of the restriction event differs somewhat between different strains and that it is a progressive, gradual process, with some potential for reversal up to the 4.0-day stage. Such gradual loss of developmental potential could reflect either a decreasing plasticity of all cells or a progressive shift in the composition of the inner cell population from those with TE developmental capacity to those that do not have it.

Do these restrictions in cellular developmental capacity have nuclear state correlates? Two sets of published results suggest that they do. Illmensee and Hoppe (1981) compared the abilities of single nuclei from ICM and TE cells of expanded blastocysts to support development in eggs when substituted for the zygotic nuclear inheritance. They reported that ICM nuclei had a significantly higher capacity than TE nuclei to support development, even obtaining three mice that developed to term with substituted ICM nuclei. Modlinski (1981) performed a similar comparison without removing male and female pronuclei. He compared the ability of nuclei from morulae (12- to 18-cell embryos) to that of ICM and of TE nuclei to support embryonic development by transplanting nuclei directly into newly fertilized eggs without pronuclear removal. Under these conditions, the transplanted nucleus fuses with the zygote nucleus to form a tetraploid hybrid nucleus. When nuclei from either morulae or ICM cells are tested in this manner, approximately half of the surviving transplants develop to the blastocyst stage; in contrast, TE nuclear transplants invariably stop at earlier stages. The results of these nuclear addition experiments are thus similar to those obtained with nuclear substitution.

Less is known about the divergence of cell types within the ICM and TE cell populations in the final 24 hr of preimplantation development than about the initial separation of ICM from TE. However, the delamination of the internal layer of endoderm from the ICM probably also involves a determinative restriction. The endoderm cells of the late blastocyst differ from the ectodermal cells of the remaining ICM in three respects: they show more intense histochemical staining, they possess a more extensive endoplasmic reticulum, and their surface membranes have a "rough" character that contrasts to the "smoother" surface of the ectodermal ICM on dissociation in vitro under the appropriate conditions.

Gardner and Rossant (1979) purified rough (endodermal) and smooth (ectodermal) inner cells and injected them singly or in small groups into 3.5-day blastocysts, the latter marked with a different GPI electromorph. The results are summarized in Figure 4.10. Early endodermal cells contribute solely to the endodermal layer of the yolk sac, while ectodermal cells contribute solely to the mesodermal layer of the yolk sac and the fetus proper, but never to the endoderm of the yolk sac. (The structure of the early postimplantation embryo will be described in Chapter 7.) Evidently, by the time the two ICM-derived cell populations of the late blastocyst have separated, they have diverged both in fate and in their respective developmental potency.

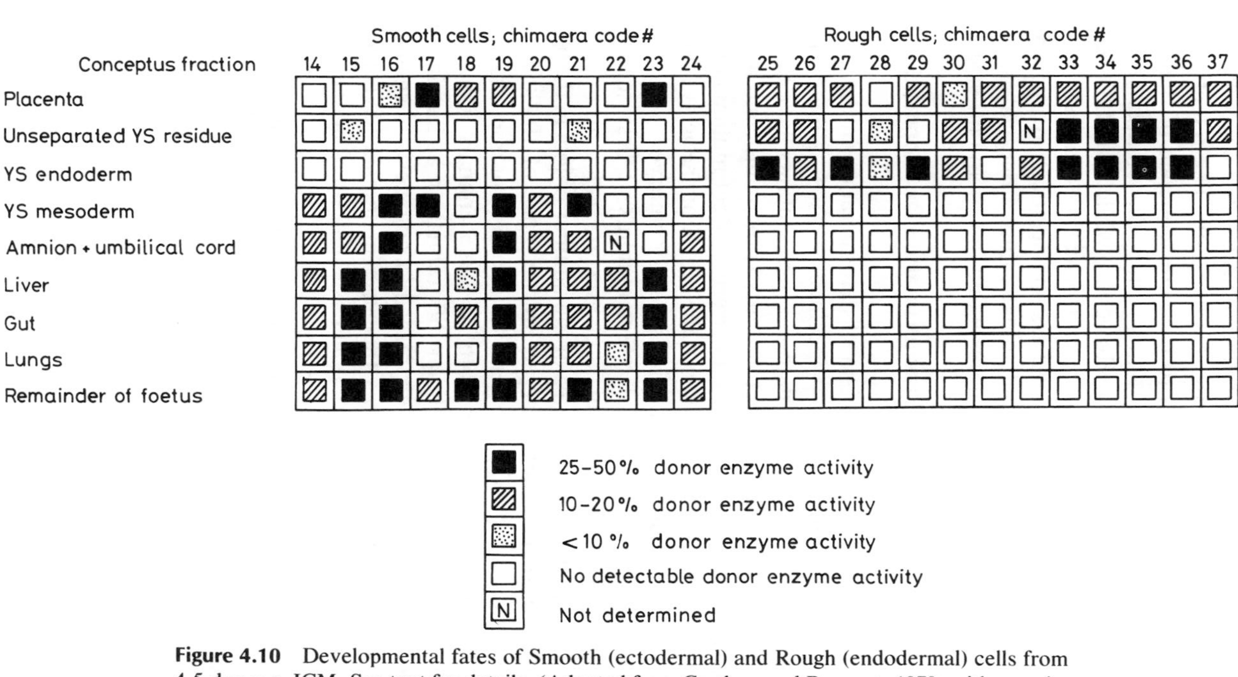

Figure 4.10 Developmental fates of Smooth (ectodermal) and Rough (endodermal) cells from 4.5-day *p.c.* ICM. See text for details. (Adapted from Gardner and Rossant, 1979, with permission.)

The case of the two TE cell populations may be subtly different. A principal difference between polar and mural TE cells is in their nuclear character. Polar TE cells remain diploid and retain their division capacity, while mural TE cells become polyploid and lose this capacity. The retention of division capacity by polar TE cells seems to be a consequence of their association with the ICM (Gardner, 1978); in some manner, this contact serves to maintain the polar TE as diploid cells. However, while mural TE cells cannot become polar TE cells, the reverse transformation can and does occur, as polar cells are pushed to the sides and assume positions away from the ICM. Determination in the TE cell populations is thus a one-way street, not the mutually exclusive pair of possibilities exhibited by the other divergences.

Mutant Analysis of Preimplantation Development

A genetic analysis of the mechanism of determination in the early embryo requires maternal effect mutants—specifically, those affected in the morphogenetic system of the egg. To date, no such mutants have been identified among the very few maternal effect mutants known in the mouse. However, among the few known and suspected maternal effect mutants of the mouse, none has yet proven informative about the morphogenetic system of the egg or about distinctive control events in the early embryo.

The dearth of mouse maternal effect mutants at first seems puzzling. Maternal effect mutants are abundant classes in *C. elegans* and *Drosophila*. As we have seen, minimal estimates of the number of loci in these organisms that can mutate to this phenotype are in the range of 10–20% of the total gene set; the real proportion of gene functions expressed in oogenesis is probably considerably higher in both of these animals. Oogenesis in the mouse involves as complex a cellular differentiation as any in this animal. Even if the proportion of genes expressed in oogenesis is smaller in mammals than in invertebrate systems (because mammals have more genetic functions), the absolute number of oogenesis genes and hence potential maternal effect genes should still be large. Indeed, there are very few early developmental mutants in the mouse at all, although zygotic genome expression begins early and is essential for preimplantation development from the early cleavage divisions.

The explanation is probably connected to the particular features of gene expression in early development, and specifically to the *relative demand* for certain gene products in oogenesis/early development in comparison to later stages. Many of the partial maternal effect genes in *Drosophila,* those that function in other stages of development, are first detected as maternal effect mutants precisely because the demand for their gene products is accentuated during early embryogenesis relative to later stages (Bischoff and Lucchesi, 1971). In the fruit fly, cleavage and early development occur very rapidly and make large metabolic demands on the stored materials in the embryo. In the embryo of the mouse, the situation could easily be the reverse. Cleavage and early development are leisurely processes compared to subsequent development (McLaren, 1976b). Under these conditions, a hypomorphic mutant of a gene expressed during both early and later developmental stages might show the more severe deficiency at the later stage and might therefore be classified as a late lethal.

It should eventually be possible to estimate the number of genes that are expressed in oogenesis and early development relative to later stages by molecular methods. As more cloned sequences of genes expressed in later stages become available, it will be possible to test for the presence of their transcripts in the late oocyte or early embryo by the technique of in situ hybridization (Chapter 9). Conversely, if and when cloned sequences are made by copying mRNAs of early embryos, the same procedure can be used to estimate the proportion of early expressed genes that are also expressed at later times. Such studies may help to illuminate some of the general features of the patterns of change in gene expression from early development to later stages, a prerequisite for framing a theory of gene expression control in development (Chapter 10).

Nevertheless, genetic methods must remain an adjunct to the molecular studies for ascertaining which functions are truly essential in early development and in which cells they are required. As methods of mutagenesis in the mouse improve, more early zygotic mutants should become available; their characterization can be expected to expand greatly our understanding of early development in the mouse.

An illustration of the genetic analysis of an early zygotic mutant is provided by a chimera analysis of the *yellow* (A^y) mutant, the classic mutant of mouse genetics. Early work established its time of action as within the period of preimplantation development. The experiments also showed that the effects on blastocyst development are delayed if the homozygous mutant embryos are allowed to develop in wild-type mothers, suggesting that part of the lethal effect involves the maternal environment of the developing embryos. However, the controversy over A^y has concerned the principal site of action of the gene. The mutant allele could be a general cell lethal, or it might act initially in either the ICM or the TE, with secondary effects on the other cellular component. Direct observation of developing homozygous mutant embryos has only produced conflicting interpretations and failed to settle the issue (reviewed by Pedersen and Spindle, 1981).

The controversy is potentially resolvable through chimera construction. The principle of such experiments is the same as that of the reciprocal pole cell transplantations in *Drosophila*. If one chimeric combination produces a mutant phenotype but the reciprocal combination does not, the locus of gene action is directly identified in the phenotypically mutant chimera. In practice, the main difficulty is similar to that in the *Drosophila* pole cell transplantations—knowing which embryos from the donor strain are the homozygous mutant ones. For any recessive lethal, the homozygous mutant individuals can be obtained only by intercrossing the heterozygotes ($m/+ \times m/+$), which comprise only one-quarter of the progeny.

In the particular case of *yellow*, the homozygous mutants are obtained by intercrosses such as $A^y/a^e \times A^y/a^e$, where a^e is the viable allele (it produces a black coat color). In 3.5-day donor embryos, before the lethal effect sets in, the three genotypes (homozygous yellow, homozygous viable, and heterozygotes) are indistinguishable.

Papaioannou and Gardner (1979) employed a statistical approach to avoid this problem. They used the progeny from two crosses as their sources of ICM and TE components: $A^y/a^e \times A^y/a^e$ (the experimental cross) and $A^y/a^e \times a^e/a^e$ (the backcross). The intercross produces one-quarter homozygous mutant embryos, while the backcross produces none. A difference in embryonic yields from ICM or TE components between the two crosses can signify the cell group in which the lethal mutation is creating its effect. In experiment 1, the two sets of ICMs were tested and compared by being placed into surgically emptied blastocysts of a control albino strain (CFLP). In experiment 2, the two sets of TE cells were tested by emptying the blastocysts of their ICMs and placing control (CFLP) ICMs inside the shells. The two pairs of reciprocal chimeric embryo groups were then transplanted to hormonally prepared foster mothers and analyzed for blastocyst development, measured as the ability of embryos to produce decidual swellings (implantation).

The detailed predictions for the various possible outcomes are given in Figure 4.11. If A^y/A^y are defective in either ICM *or* TE, then the two experi-

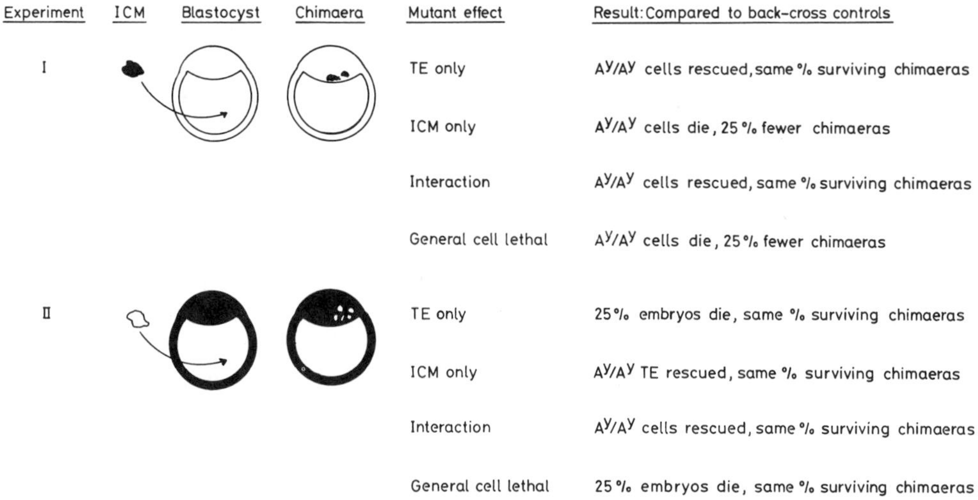

Experiment	ICM	Blastocyst	Chimaera	Mutant effect	Result: Compared to back-cross controls
I				TE only	A^y/A^y cells rescued, same % surviving chimaeras
				ICM only	A^y/A^y cells die, 25 % fewer chimaeras
				Interaction	A^y/A^y cells rescued, same % surviving chimaeras
				General cell lethal	A^y/A^y cells die, 25 % fewer chimaeras
II				TE only	25 % embryos die, same % surviving chimaeras
				ICM only	A^y/A^y TE rescued, same % surviving chimaeras
				Interaction	A^y/A^y cells rescued, same % surviving chimaeras
				General cell lethal	25 % embryos die, same % surviving chimaeras

Figure 4.11 Predictions of survival outcomes of chimeras involving A^y/A^y embryos from intercrossed $A^y/+$ parents. See text for discussion. (Adapted from Gardner and Papaioannou, 1975, with permission.)

ments should yield nonequivalent results: the chimeras with a defective mutant component should show a 25% deficit of chimeras from the intercross relative to the backcross embryos. As an additional control, the overall rates of chimera production (using donor GPI as a marker) within the surviving conceptuses were assayed.

The results are summarized in Table 4.1. Despite a slight reduction in overall chimera frequency in the intercross ICM transplantation (experiment 1), there was no significant difference in conceptus frequency between the intercross and backcross embryos as ICM donors. However, with a donor TE shell as the receptacle (experiment 2), the intercross embryos showed a distinct deficit not significantly different from that produced by a 25% loss. The results indicate that the initial action of A^y is in the TE of the lethal

Table 4.1 Reciprocal ICM/TE Transplantations to Test for the Cellular Site of the A^y Locus of Action[a]

Experiment 1: A^y/a^e strain ICMs injected into control blastocysts	Intercross	Backcross
Implantation		
Number of decidual swellings/number of transferred embryos	54/62 = 87%	70/80 = 88%
Proportion of chimeras		
Number of chimeras/total number of conceptuses	44/53 = 83%	62/66 = 94%
$Chi^2 = 4.02$; $P < 0.05$		
Experiment 2: Control ICMs injected into blastocysts of the A^y/a^e strain	Host blastocyst	
	Intercross	Backcross
Implantation		
Number of decidual swellings/number of transferred embryos	73/99 = 74%	89/98 = 91%
Proportion of chimeras		
Number of chimeras/total number of conceptuses	47/55 = 85%	68/82 = 83%
$Chi^2 = 1.4$; $P > 0.20$		

Source. Data from Papaioannou and Gardner (1979).

[a] The experiment is described in text. The null hypothesis tested was that the A^y/A^y component (obtained in the intercross) produces a 25% deficit of implantations. Employing the chimeric frequency in the backcross as the expected control level, experiment 1 reveals a significant departure from the expectation—no substantial loss when the A^y/a^e strain supplies ICMs—and experiment 2 reveals good agreement with the null hypothesis when this strain supplies blastocyst (TE) components.

mutant embryos. More significantly, this technique illustrates the utility of the reciprocal chimera approach in the mouse for examining sites of action of early zygotic lethals.

DETERMINATION IN THE PREIMPLANTATION EMBRYO: PROPERTIES AND MECHANISMS

A key question about the events that separate ICM from TE is whether they are initiated by some form of preexisting organization in the mouse oocyte. A specific hypothesis of this kind was initially formulated by Dalcq (1957) on the basis of fixed preparations of mature mouse oocytes. He reported that these specimens exhibited differentiated "dorsal" and "ventral" cytoplasm and proposed that these cytoplasms were differentially segregated to blastomeres by the eight-cell stage. In the Dalcq hypothesis, the dorsal cytoplasm eventually gave rise to the ICM and the ventral cytoplasm to the TE.

It now appears, in fact, that the regional differentiation of fixed eggs reported by Dalcq was an artifact of the fixation process (Wilson and Stern, 1975). Furthermore, the idea of a strict separation of ICM- and TE-forming potential was disproved by the experiments of Kelly (1975), described above, which demonstrated that blastomeres at the eight-cell stage are equally capable of giving rise to ICM and TE cells. Although the observations of totipotency at the eight-cell stage do not eliminate the possibility that a segregation of ICM and TE "determinants" occurs at a later stage, this too appears unlikely. The molecular and embryological evidence suggests that the developmental restrictions that accompany separation of ICM from TE take place relatively gradually and with a certain degree of reversibility or lability.

If the determinative process is not rigidly fixed by the cytoplasm that each blastomere inherits from the egg, then the divergence of ICM from TE must involve some form of differential cell–cell or cell–environment interaction. The obvious physical differences between these two initial cell groups is in their relative positions: the ICM cells derive from cells that are inside the blastocyst, while the developing TE cells are at all times outside and exposed to the environment. As first noted by Mintz (1964), this difference in physical position constitutes a difference in "microenvironment" that might shape the initial divergence in fate and developmental capacity between the two cell types.

The first experimental findings to support this "inside-outside" hypothesis of determination were presented by Tartowski and Wroblewska (1967). They examined the products of in vitro development of isolated blastomeres from both four-cell and eight-cell embryos and concluded that all cells at these stages had the capacity to give rise to vesicular (blastocyst-like) forms. Thus, there was no suggestion of a segregated ability to form ICM specifically from any of the blastomeres of these two stages. Significantly, however, the isolated one- and two-cell products of eight-cell embryos were found to form a higher percentage of purely trophoblastic vesicles (lacking ICM) than did blastomeres from four-cell embryos. This difference is explicable if fluid secretion from the outside cells, and hence blastocyst formation, tends to occur only after a particular number of cleavage divisions—for instance, after the fifth cleavage division (p. 161). Under these circumstances, cell clusters derived from one out of eight blastomeres (i.e., single blastomeres from eight-cell embryos) would necessarily be smaller at the time of fluid secretion than mini-embryos derived from one out of four blastomeres and would be less likely to have inside (proto-ICM) cells. In this view, the capacity to form ICM is solely a function of being inside at the time fluid secretion begins.

The strongest experimental support for the inside-outside hypothesis comes from chimera experiments in which cells from four- or eight-cell embryos were dissociated from one another and placed singly or in groups, either internally or externally, on carrier embryos. The fate of the donor cells, marked by prior labeling of their nuclear DNA with [^{3}H]thymidine or with a GPI-isoenzyme difference, was then followed in subsequent development (Hillman et al., 1972). The results showed a clear biasing of cell fate toward either TE or ICM development as a function of donor cell positioning on the recipient embryos. Cells placed externally contributed over 90% of their descendants to the outside region of the blastocyst, while internal placement produced a less dramatic but still substantial (40%) contribution of cellular descendants to the ICM. Furthermore, when 8- to 16-cell embryos were enclosed by six unlabeled embryos, aggregates were formed that developed into large blastocysts; of these, several showed labeling only within the inner cells and developed into normal appearing embryos by 13 days. In these embryos, cells that would have developed into TE had evidently been channeled by their position into forming ICM. Finally, chimeras with externally placed cells that were allowed to develop to birth showed reduced contributions from the donor cells to the development of the coat, as scored by coat color markers, regardless of their specific genotype. This result is the expected one if external cells are preferentially channeled toward TE formation and away from ICM formation. In sum, the results indicate that position rather than intrinsic blastomere cytoplasmic inheritance is a primary factor in influencing blastomere fate.

How might cellular position translate into a molecular mechanism for the differential restriction of cell fate? The first explanation proposed was that of Ducibella (1977), who suggested that inside cells find themselves within a qualitatively distinct microenvironment, created by the formation of apical zonular tight junctions between the TE cells that seal off the ICM cells completely from the external medium. The model is depicted in Figure 4.12*a*. The unique microenvironment of the internal cells then leads, by a series of steps, to the final differentiative and determinative steps.

However, two facts tell against this explanation. First, zonular tight junc-

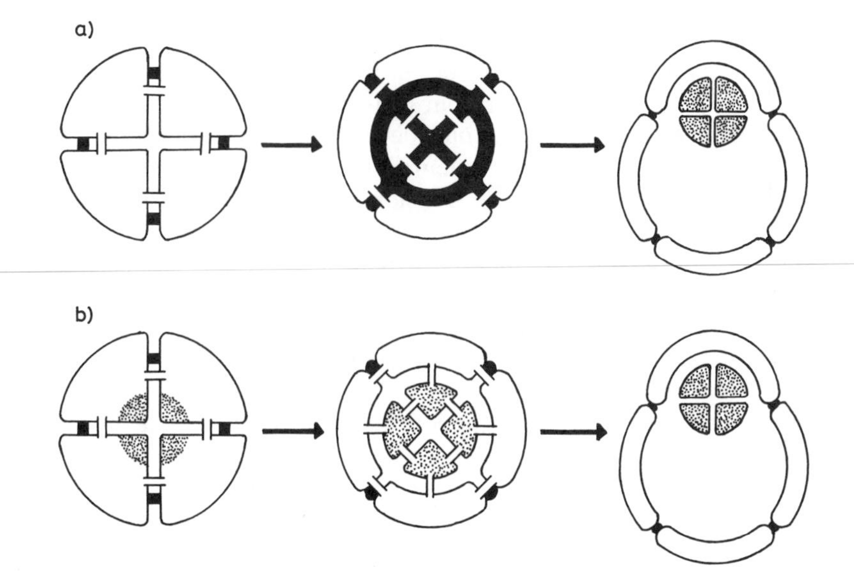

Figure 4.12 Two models of positional fate setting in the mouse embryo. (*a*) The permeability block microenvironmental hypothesis; (*b*) the polarization hypothesis. (Adapted and reproduced with permission from Johnson et al., 1981.)

tions only develop at about the 32-cell stage (Calarco and Epstein, 1973), *after* the initial divergence of the two cell types at the molecular level (Handyside and Johnson, 1978). Therefore, the first events in divergence take place without a communications barrier between the two cell types. Second, it has so far proven impossible to create by artificial means the postulated microenvironment. One observation, in particular, is fairly striking: injection of an entire eight-cell embryo into a fully expanded blastocyst failed to prevent the outer cells of this enclosed embryo from turning into TE despite the tight permeability block to the external environment provided by the host blastocyst (Pedersen and Spindle, 1980).

The microenvironment hypothesis of Ducibella might be termed an extreme ''instructionist'' one: the environmental difference is a complete, qualitative difference that automatically produces a corresponding qualitative cellular difference. However, if the initial environments of outside and inside cells are not different in kind but only in degree, the initial cellular differences may also be subtle ones of degree. In slightly different terms, the initial difference may be quantitative, involving a graded property of some kind.

Johnson et al. (1981) proposed a hypothesis along these lines. Their ''polarization hypothesis'' stated that inside-outside differences stem from a radial asymmetry within the embryo in which individual cells are polarized with respect to molecular and cellular properties at their (outer) apices and (inner) bases. In this view, cleavage in the compacted eight-cell embryo, occurring perpendicularly to the radial axis, separates cells with "inner" (presumptive ICM) and "outer" (presumptive TE) values (Fig. 4.12*b*).

This hypothesis makes compaction a key event in creating or amplifying a radially graded property. Compaction also draws the eight cells together into an organized unit in which communication is maintained through extensive intercellular contact and small intercellular channels at gap junctions. The importance of compaction in the separation of ICM and TE has been demonstrated experimentally using treatments that reversibly inhibit compaction during development. With one treatment that inhibits compaction without affecting cleavage, a delay in compaction beyond a certain time produces blastocysts that lack a genuine ICM component, yet the cells of these aberrant blastocysts synthesize the full spectrum of proteins characteristic of normal blastocysts (cited in Johnson et al., 1981). Under these conditions, the basic gene expression program is activated, but in the absence of cell compaction, none of the cells evidently perceive themselves as ICM.

An important second set of observations that support the hypothesis concerns the existence of individual cell polarization at the eight-cell stage. Whether in the intact embryo or in isolated cell pairs, cells at this stage become columnar and axially polarized, showing an outer cap of microvillous extensions that act as binding sites for a fluorescent ligand (fluorescein isothiocyanate-concanavalin A). When the cells of the eight-cell embryo divide in vitro, most divisions generate a polarized cell (possessing the ligand binding sites) and a nonpolarized cell (lacking these sites); the former correspond to outer cells, the latter to inner cells (Johnson and Ziomek, 1981). Some 2/16 cell couplets (pairs of cells at the 16-cell stage) of this kind are shown in Figure 4.13. In the normal embryo, these two types of cells tend to form two discrete lineages, polar cells dividing to give pairs of polar cells and apolar cells dividing to give apolar cells. The beginnings of the ICM-TE lineage divergences are foreshadowed in these cytologically defined lineages.

Does the suggestion of a radial gradient conflict with the discovery that cell fate can be biased in one direction or the other by the appropriate change of position within the early embryo? There is no essential conflict if the postulated gradient, with its high point at the center and decreasing along all axes out to the periphery, can "entrain" cells by means of cell–cell communication to assume the characteristic gradient value at any position. Hillman et al. (1972) point out that their results are consistent with a gradient hypothesis if the gradient source exists at the center of the embryo; transplanted blastomeres would tend to acquire the radial gradient value of their new position.

However, whatever the origins of polarity characteristics, whether they arise in the embryo or preexist in the oocyte, the maintenance of the two cellular lineages requires continuing interactions between polar and apolar cells. While isolated 2/16 (polar : apolar) couplets give rise predominantly to four-cell (4/32) clusters containing two polar and two apolar cells, synthetic

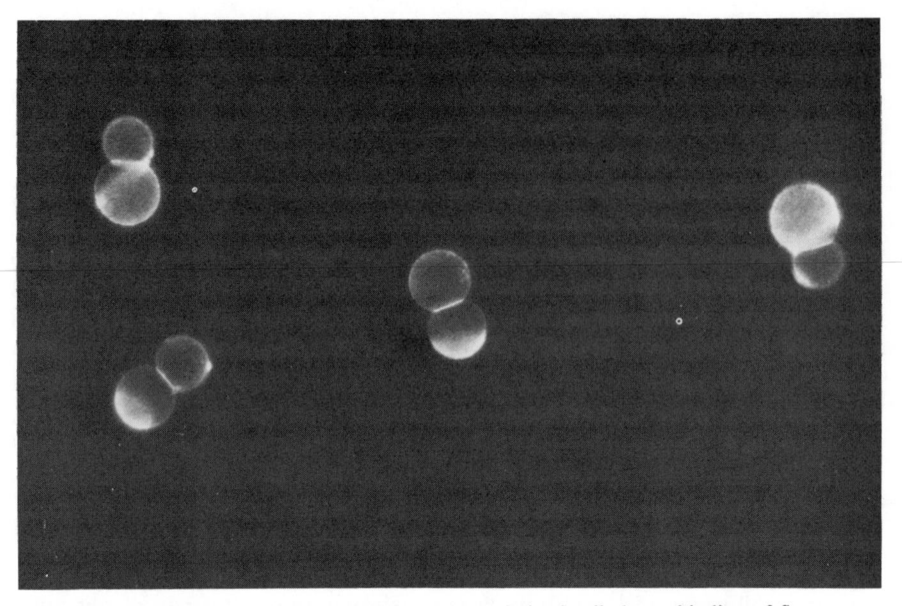

Figure 4.13 Polarity in 2/16 couplets. The outer polarized cell shows binding of fluorescent ligands; the inner cell is nonpolarized. (Kindly provided by Dr. M.H. Johnson.)

couplets consisting of two polar (TE-like) cells produce some ICM-like cells following division (Johnson and Ziomek, 1983). The results indicate that it is the continuing interactions between polar and apolar cells that tend to maintain the two cell type lineages.

The results indicate a complex origin of the "morphogenetic instructions" in the early mouse embryo involving both preexisting organization (presumably based on oocyte structure) and cell–cell interaction (Johnson and Pratt, 1983). Position *is* important in fate setting, as shown by the experiments of Hillman et al. (1972), but it is position with respect to the site of origin within the oocyte mass and to the other cells of the early embryo rather than to the external environment that is of primary importance.

There are two other features of the ICM–TE divergence that deserve note. The first is that there may be certain regularities in cleavage that affect cell fates. From a study of cleavage rates and spatial patterns in partially flattened embryos in vitro, C. F. Graham and his colleagues conclude that at the two-cell stage, one blastomere always cleaves first and its progeny tend to occupy internal positions preferentially (Kelly et al., 1978; Graham and Lehtonen, 1979). If the observations pertain to normal embryos, they suggest that a topogenetic bias leads to a biasing of cell fates. The other observation is that the first inner cells show more rapid cycles of nuclear replication than the outer cells (Graham, 1973; Herbert and Graham, 1974). In this system, as perhaps in others, it appears that cells with a greater range of developmental potential show more rapid rates of nuclear DNA replication.

The origins of the second pair of determinative events in the preimplantation embryo are more obscure than the ICM–TE divergence. However, they presumably also involve the intrinsic properties of the progenitor cells (established by the earlier events) and the interaction of these cells with each other and with the environment. The divergence of the two kinds of TE cells—polar and mural—seems to depend, as mentioned earlier, on the special interaction of polar cells with the ICM mass. In the absence of that interaction, TE cells invariably become mural TE. The divergence of endoderm from ectoderm in the ICM may, however, depend primarily on different relative exposures to the blastocoelic fluid. When ICM cells are placed into empty zonae, they invariably develop into endodermal cells (Rossant, 1975b). In intact embryos, the embryonic ectoderm, enclosed between endoderm and polar TE, remains totipotent and is the ultimate source of the fetus.

In both the TE–ICM and the endoderm–ectoderm divergence, it is the relatively enclosed inner cell group that retains the undifferentiated appearance of early morula cells and the greater set of prospective potencies. The high point of developmental capacity in the preimplantation embryo is thus at or near the center; cells formed from the periphery differentiate first and exhibit a smaller range of fates.

SOME CONCLUSIONS

The traditional method for evaluating the morphogenetic system of an oocyte involves the measurement of the developmental capacity in the first formed blastomeres. If the blastomeres exhibit typical "self-differentiation" when removed from the embryo, and if the remaining cells cannot substitute for blastomeres that have been removed, then the morphogenetic system is deemed to be a "mosaic" one. If, on the other hand, isolated blastomeres exhibit a range of developmental capacities in vitro or if manipulated embryos prove capable of recruiting cells to perform the roles normally assigned to other cells, the system is deemed to be a highly "regulative" one.

Based on such developmental criteria, the nematode and mouse eggs would appear to represent the two possible extremes of mosaic and regulative morphogenetic systems. Yet, as discussed above, the mouse oocyte is almost certainly not the tabula rasa commonly depicted but may possess a vestigial morphogenetic system that is instrumental in setting early blastomere fates. By extension, the mammalian oocyte in general may possess some form of morphogenetic system, perhaps also involving radially graded properties. The high degree of regulative ability displayed by early cells of the mouse embryo may reflect either a relatively late positioning of the morphogenetic system or its unusual lability. In contrast to the early embryos of many invertebrates, continuing cell–cell interactions may play an essential role in fixing the first cell diversifications and their conversion into determinative states in mammalian embryos.

PART THREE

CLONES AND POLYCLONES IN DEVELOPMENT

THE ROLES OF LINEAGE AND CELL INTERACTION

At some point in development, control passes from the maternal constituents sequestered in the egg to the gene products encoded by the zygotic genome. The transition may be fairly sharp, as in the *Drosophila* blastoderm, or relatively gradual, as in the early mouse embryo, but in all embryos it takes place relatively early in development. Most development occurs, therefore, under the control of the zygotic genome.

In the following three chapters, we will examine this phase of development in the nematode, fruit fly, and mouse, respectively. As in the previous chapters, the perspective will be dual: the individual and collective cell behaviors that comprise the developmental sequence and the gene activities that govern these behaviors.

Two major themes or questions will be discussed—the first about cells, the second about genes. The question about cell behaviors is: to what extent are particular tissues or organs founded by single cell lineages or clones, and to what extent are they founded by groups of cells? The principal techniques for exploring this subject are genetic: varieties of clonal analysis and the analysis of mutants that affect particular tissues or structures. Clonal analysis is used to determine whether at the time of tissue/organ foundation, or "allocation," a given tissue/organ arises from a single cell or multiple cells. The strategy is illustrated in Figure III.1. Marked clones are created shortly before or at the time of allocation, and the final composition of the tissue/

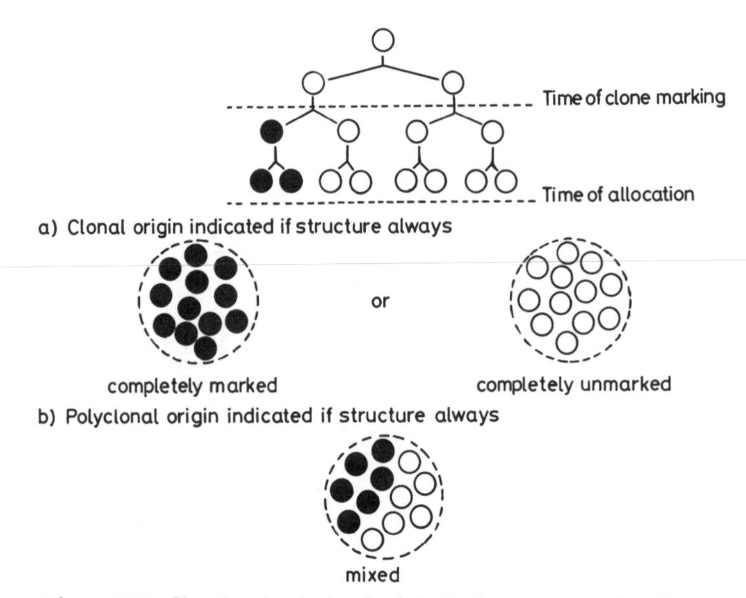

Figure III.1 Clonal and polyclonal origins in tissue or organ foundation.

organ of interest is assessed at the end of development. If this region is always completely marked or completely unmarked but never mixed, a clonal origin is indicated. If, in contrast, the great majority of the marked regions are mixed in composition, a multiple clone origin is indicated. When a region always originates from a group of cells, the founding cell group is said to be a "polyclone" (Crick and Lawrence, 1975). Polyclonal origins always indicate some form of cellular cooperation in the foundation of the region of interest.

The characterization of cell assignments defines the developmental phenomena of interest in cellular terms. The genetic question that arises from the phenomenological characterization is one of control: what gene activities are required for these cell assignments, whether of single clones or of polyclones, and how do they achieve their results? In particular, one would like to know which specific zygotic genes are first expressed under the direction of the maternal genome-encoded components of the egg. In systems such as *Drosophila* and *Caenorhabditis,* where mutant collection is easy, it is becoming possible to identify these initially expressed genes. For the mouse, in which the maternal–zygotic genome transition is gradual and the collection of large numbers of mutants is still impractical, the question cannot be readily approached. In asking questions about genetic control in the mouse, therefore, investigators still have to rely principally on interesting mutants that come to hand, although some of the new techniques for genetically modifying the germ line may soon change this practice (Chapter 9).

One important aspect of developmental systems that will be touched on only briefly in the following three chapters is the phenomenon of "pattern formation," the ability of cells to "know" where they are in autonomously developing regions (developmental fields) and, if necessary, following injury to the region, to reconstruct the spatial pattern. Pattern formation raises special and difficult issues; they will be described more fully in Chapter 8.

FIVE

CAENORHABDITIS ELEGANS

FROM EARLY EMBRYO TO ADULT

EMBRYOGENESIS AND POSTEMBRYONIC DEVELOPMENT

The basic body plan of the nematode is established during embryogenesis, well before hatching. The sequence of events in embryogenesis can be divided into two phases of nearly equal duration. The first consists of cleavage and is accompanied by extensive morphogenesis. During this period, all of the major organ systems except the gonads take shape. In the second half of embryogenesis, the animal, rotating within the egg shell, experiences a threefold elongation, acquiring the shape of the L1 larva, and completes the functional maturation of its cells. The newly hatched larva has fully functional digestive and juvenile motoneurone systems. Although all of the organ systems in the animal continue to develop and mature, the L1 has much of the behavioral repertoire of the adult.

The primary cellular lineages of the worm are founded during early cleavage with the formation of the six embryonic blast cells (Fig. 2.8). The E and P_4 blast cells are the precursor cells of the intestine and germ line, respectively; these tissues are the only ones to be founded clonally in the nematode. The MS and D lineages form the major portions of the pharyngeal and juvenile body muscle systems of the animal. The AB lineage gives rise to the neuronal system, part of the pharynx, and most of the external covering of the animal, the hypodermis; the C lineage contributes the rest of the hypodermis and the body muscle system of the juvenile. The AB and MS lineages, which furnish the greatest variety of cell types, also give rise to a critical group of somatic precursor cells termed the "postembryonic blast cells." These cells begin their divisions only at the L1 stage but are responsible for all of the somatic structures that distinguish the adult from the juvenile.

Morphogenesis

The task of morphogenesis is to convert the ball of cells produced by the first cleavage divisions into the longitudinally differentiated larval body. The opening movements of morphogenesis consist of a series of ventral invaginations and external cellular migrations. Cleavage divisions continue during these cell movements and continue to expand the various cell lineages. As described earlier in connection with the *emb-5* mutants (p. 69), the first event of morphogenesis is gastrulation, the ventral invagination of the daughters of the E cell. These two cells subsequently divide, and the four resultant E lineage cells give rise eventually to the 20-cell juvenile intestine.

Following the first invagination, the germ line precursor P_4 cells and the cells of the MS lineage enter through the ventral invagination. P_4, after completing its migration, divides to form two cells, Z2 and Z3, which remain quiescent throughout the remainder of embryogenesis and resume division only during larval development. From the eight MS cells originally present on the ventral surface of the embryo, all internal MS lineages increase in cell number during the first half of embryogenesis. Some contribute to the formation of the pharyngeal muscles in the future head region and the remainder form body muscle cells, coelomocytes (which may function as phagocytic cells), a number of neurons, and the ring gland of the head region.

The ventral invagination zone progressively widens and lengthens, after entry of the MS cells, first in a posterior direction, engulfing the C- and D-blast cell progeny, and then anteriorly, as AB-blast cell-derived pharynx muscle cell precursors enter. The complete sequence of cell invaginations take place 100–290 min after the first cleavage; the zone of invagination then closes. Throughout the entire period, the surface AB-derived cells that give rise to the juvenile hypodermis and ventral nervous system spread both posteriorly and ventrally. Concurrently, the C-blast progeny destined to give rise to hypodermis spread anteriorly and dorsally, covering the remainder of the external surface. (The embryonic hypodermis is composed of 11 syncytia and a number of discrete cells; the largest syncytium, hyp 7, comes to encircle most of the body and consists of nuclei contributed by both AB- and C-blast cell descendants.)

These external movements are paralleled by the formation of the central cylinder of the animal, consisting of pharynx cells anteriorly and intestinal cells medially. The body myoblasts, or muscle precursor cells, which are derived from the MS, C, D, and AB lineages, move into position between this central cylinder and the outer cells, the surface AB myoblasts entering last. The progeny of these blast cells form the four longitudinal muscle strips of the juvenile body.

Origins of the Neural System

The neural system of the juvenile is produced from AB-derived external neuroblasts. The ventral side of the animal, after the first cell invaginations, is largely composed of these cells; they divide and sink inward, being replaced externally by the sheet of hypodermis. The ventral neuroblasts form the juvenile ventral nerve cord. Dorsal and lateral neuroblasts similarly sink in from their external positions and are covered by hypodermis.

Later Cell Movements

In addition to these movements, a number of cells undergo extensive longitudinal migrations. These cells include the MS-derived postembryonic precursors of the somatic portion of the gonad, Z1 and Z4. These cells move from an anterior position to the posterior half of the embryo, where they assume positions proximate to the germ line precursors Z2 and Z3. A few circumferential cell movements also occur, as do extensive crisscrossing transverse nuclear movements within the syncytial juvenile hypoderm.

Programmed Cell Death

In addition to these invariant cellular and nuclear movements, another type of invariant event is programmed cell death. During embryogenesis, more than 100 cells die, their fates being written in their lineages. The function of cell death is illustrated by one of the few examples of sexual dimorphism seen in the embryo, the existence of six sex-specific neurons of the AB lineage. Four of these, the cephalic companion neurons (CEM cells) are produced in the head and constitute part of the male adult sensory apparatus. They probably serve to mediate chemotaxis of the male toward the hermaphrodite. The other two neurons are hermaphrodite-specific neurons (HSN cells), and are formed posteriorly and migrate anteriorly toward the midpoint of the embryo; they innervate the vulva of the adult hermaphrodite for the function of egg laying. All six neurons are formed in both sexes through identical lineage division patterns. However, only the neurons specific to the sex of the individual survive, the others undergoing programmed cell death. In a developmental sequence featuring great precision and economy of cell use, it is evidently less costly to retain division programs and expunge unneeded cells than to modify the basic division programs individually so as to save and utilize every cell.

Some General Considerations

The complete sequence of events in embryogenesis, summarized above, has been described by Sulston et al. (1983). The structure of the end product of the developmental sequence, the L1 hermaphrodite, is shown in Figure 5.1.

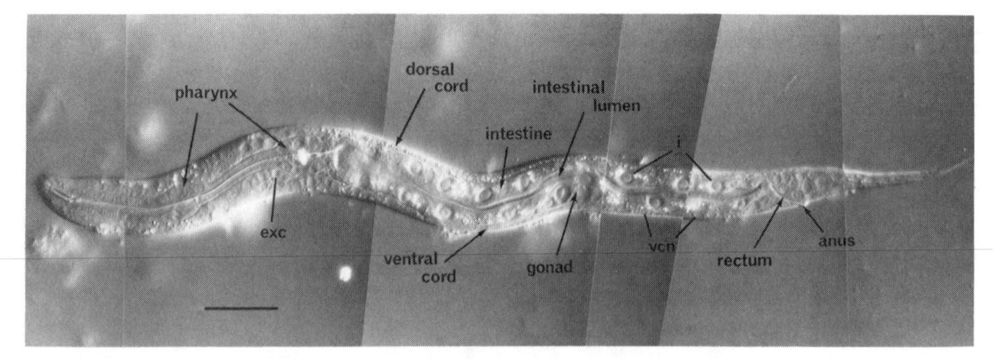

Figure 5.1 L1 hermaphrodite of *C. elegans*. (Photograph kindly provided by Dr. J.E. Sulston; reproduced with permission from Sulston and Horvitz, 1977.)

Although the formation of the juvenile worm is extremely complicated, three general features can be discerned. The first is invariance. From the first cleavage divisions through the final steps of functional maturation, the individual cells show highly reproducible behavior in the wild-type strain. Cells undergo fixed division programs and fixed movements (with a very few exceptions where cells of comparable lineage have interchangeable roles). The highly structured total program of embryogenesis is thus mirrored in and based upon highly structured cell lineage patterns. Furthermore, much of this invariance seems to be required for the normal execution of the individual steps of the program. From the 50-cell stage on, individual cells can be selectively killed by means of a highly focused laser beam. In the great majority of these embryonic cell deaths, there is no replacement by other cells or compensatory regulation by novel routes (Sulston et al., 1983).

The second point concerns the cellular origins of particular tissues. Although two major lineages, E (intestine) and P_4 (germ line), are clonal in origin, most tissues and defined structures arise from polyclones. In this respect, the embryonic and postembryonic programs for similar structures sometimes show differences. Sensilla, for instance, arise in both stages of development, each sensillum consisting of neuronal and supporting (glial) cells. In postembryonic development, however, a number of sensilla arise clonally from single mother cells, while in embryogenesis, each sensillum consists of two or more clones.

A consequence of polyclonality is that most cell lineages give rise to multiple cell types and structures. This is especially true for the AB and MS lineages. Within the AB ectodermal lineages, for example, there arise a few muscle cells in terminal embryonic cell divisions that also produce sister neuronal cells. Binucleate muscle cells also sometimes arise, resulting from the fusion of single AB-derived and MS-derived cells. It is apparent that ectodermal and mesodermal characteristics can be simultaneously present in certain lineages even at very late stages. Indeed, the derivation of mesodermal-type cells from ectodermal precursors is not unique to the nematode but also occurs in vertebrates, as instanced by the formation of migrating mesenchymal neural crest cells from the neural tube (Chapter 7).

The final point concerns the development of symmetry in the embryo. The L1 larva is bilaterally symmetrical, and much of this symmetry is derived from comparable division programs of lineally equivalent cells on the left and right sides. Nevertheless, there are also elements of threefold and sixfold rotational symmetry. The origin of some of these symmetries is intricate and is not reducible to simple rules. The case of the triangular pharynx is particularly complex (Sulston et al., 1983).

Different strategies of cell formation and recruitment are employed for the development of these symmetries. Evidently, specific ad hoc evolutionary modifications of basic cell deployment patterns have been incorporated into the developmental program as modifications of originally simple developmental mechanisms. In consequence, the original constructional mechanisms may often be obscured.

Emergence of the L1 larva from the egg shell marks the beginning of postembryonic development. The newly hatched L1 larva is similar in its general organization to the adult worm but is only one-quarter the length of the latter and is sexually immature, lacking a gonad and the associated secondary somatic sexual structures.

Like the adult, the immature worm is encased by a cuticle secreted by the underlying syncytial hypodermis. The body musculature consists of four longitudinal strips of muscle. Two of these strips are located on either side of the dorsal midline and two are in corresponding positions on either side of the ventral midline; each consists of 20 cells. The head of the L1 larva includes the pharynx and is fully formed. The intestine is functional but contains only 20 nuclei (14 fewer than the adult intestine, which is partly syncytial).

Postembryonic Development

Postembryonic development of the animal takes place during a series of four larval growth periods separated by intervening molts. Its most striking features are the large increase in body size and the development of the gonads and secondary sexual characteristics. The size increase results mainly from increased cell size and secondarily from an increase in cell number. In the gonads, however, a major increase in cell and nuclear number takes place, particularly in the germ line of the hermaphrodite.

In all of the postembryonic somatic cell lineages, as in embryogenesis, the pattern of cell divisions in the wild-type is highly precise. It involves oriented cell divisions, precise relative timing of cell division, oriented cell movements, and programmed cell death. This precision leads to precisely constant numbers of somatic cell nuclei in the adult. In the hermaphrodite, for instance, the L1 larva contains 558 cell nuclei (not counting those eliminated by cell death); the adult hermaphrodite contains exactly 959 somatic cell nuclei, of which about 200 are in various-sized syncytia (in the hypodermis, certain muscles, and the gut). The numbers are somewhat different in the male, but the same invariance of the somatic division program is observed. (The postembryonic multiplication of germ cells does not exhibit invariance; the development of the germ line will be discussed later.)

Like embryonic development, these highly specific division programs originate from specific precursor blast cells. These postembryonic blast cells, each denoted by a capital letter, are arrayed in specific positions along the length of the L1 larva; their positions are indicated schematically in Figure 5.2. Each of these blast cells undergoes an invariant number of rounds of division, ranging from one to eight. For some lineages and structures, the increases are correspondingly modest or great. For the head, which contains the majority of cells at hatching (305–310), there is virtually no increase in the content of its constituent cells (pharynx, sensilla, connecting neurons, etc.). Other structures, such as the secondary sexual structures (e.g., the vulva in the hermaphrodite, the elaborate tail structures of the male), come into existence only during postembryonic development.

The net result of postembryonic development is the building of an adult superstructure on top of the larval scaffolding. Part of this developmental sequence involves a remodeling of the larval foundation, with some major changes in preexisting neural and muscle connections for the innervation and movement of new somatic sexual structures. However, most of the events entail principally an addition of complexity rather than a dramatic reworking of the basic material. (In contrast, the metamorphosis of the fruit fly from the larva in *Drosophila* consists largely of a substitution of the adult structures for the larval ones.)

In certain respects, the program of somatic postembryonic development differs from that of embryonic development. In the latter, for instance, there are a few divisions that give rise to muscle and neuron sister cells; no such divisions "away from type" occur in postembryonic development. However, the general features of the two phases of development are similar:

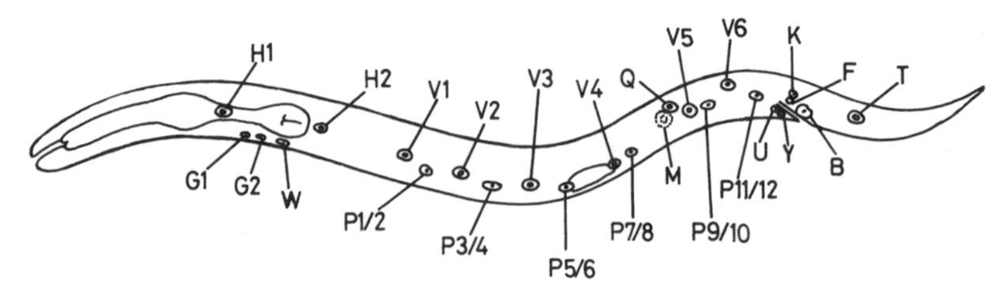

Figure 5.2 Positions of postembryonic blast cells in the young Ll larva. Blast cells G1, G2, and the male-specific B, U, Y, and F, which are found in both sexes but divide only in males, are located medially; M is located on the right side and K on the left; all others exist as pairs located on both sides (but with only the left representative shown). (W was formerly designated PO, U was originally E, and Y was C.) (Adapted from Chalfie et al., 1981, with permission; copyright MIT Press.)

oriented, precisely timed divisions in the somatic tissues giving rise to precisely constructed lineages, and the prevalence of polyclonal origins.

The invariance of postembryonic cell division and cell assignment necessarily requires explanation. Are all the events a function of the inherent capabilities of cells within lineages, or do cell–cell contacts play a part? Even where lineage plays a dominant role, is there any capacity for regulative substitution?

One way to answer these questions is to remove particular cells and examine the developmental consequences of this action. In *C. elegans,* such selective cell elimination has been carried out in two ways. The first is laser microbeam ablation of particular cells, followed by observation of subsequent developmental events. Where cells have been strictly programmed to perform certain tasks and cannot be replaced, such selective cell killing produces highly specific "holes" in the developmental sequence. The second method for eliminating particular cells involves the isolation and analysis of postembryonic defective mutants. In the next section, the use of these techniques to explore cell commitments in postembryonic development will be described.

POSTEMBRYONIC SOMATIC DEVELOPMENT: CELLULAR PHENOMENA AND GENETIC CONTROLS

Cell Equivalence Groups

Both the hermaphrodite vulva and the male preanal ganglion are ectodermal secondary sexual structures. Both are ventral structures derived from the central P-blast cells, which are situated initially in the lateral hypodermis. The hypodermis, a collection of surface syncytia and cells, consists of four longitudinal ridges in the L1—one dorsal, one ventral, a lateral right, and a lateral left. The four groups are connected by thin sheets of cytoplasm interposed between the internal musculature and the external cuticle.

The six pairs of lateral P nuclei (see Fig. 5.2) migrate ventrally through cytoplasmic connections to the ventral ridge. This migration takes place at about the middle of L1, the anterior nuclei moving first, followed in sequence by the others. Each P nucleus leaves behind a companion V blast nucleus, which contributes to the postembryonic lateral hypoderm. The 12 migrant P nuclei insert themselves between the 15 juvenile motor neurons that already exist in the ventral cord.

The member of a left-right pair of P nuclei that assumes the relative anterior position is variable, but having reached the ventral cord, the nuclei maintain their positions and produce a line of cells that then divide. The division patterns for the hermaphrodite and the male are diagrammed in Figure 5.3.

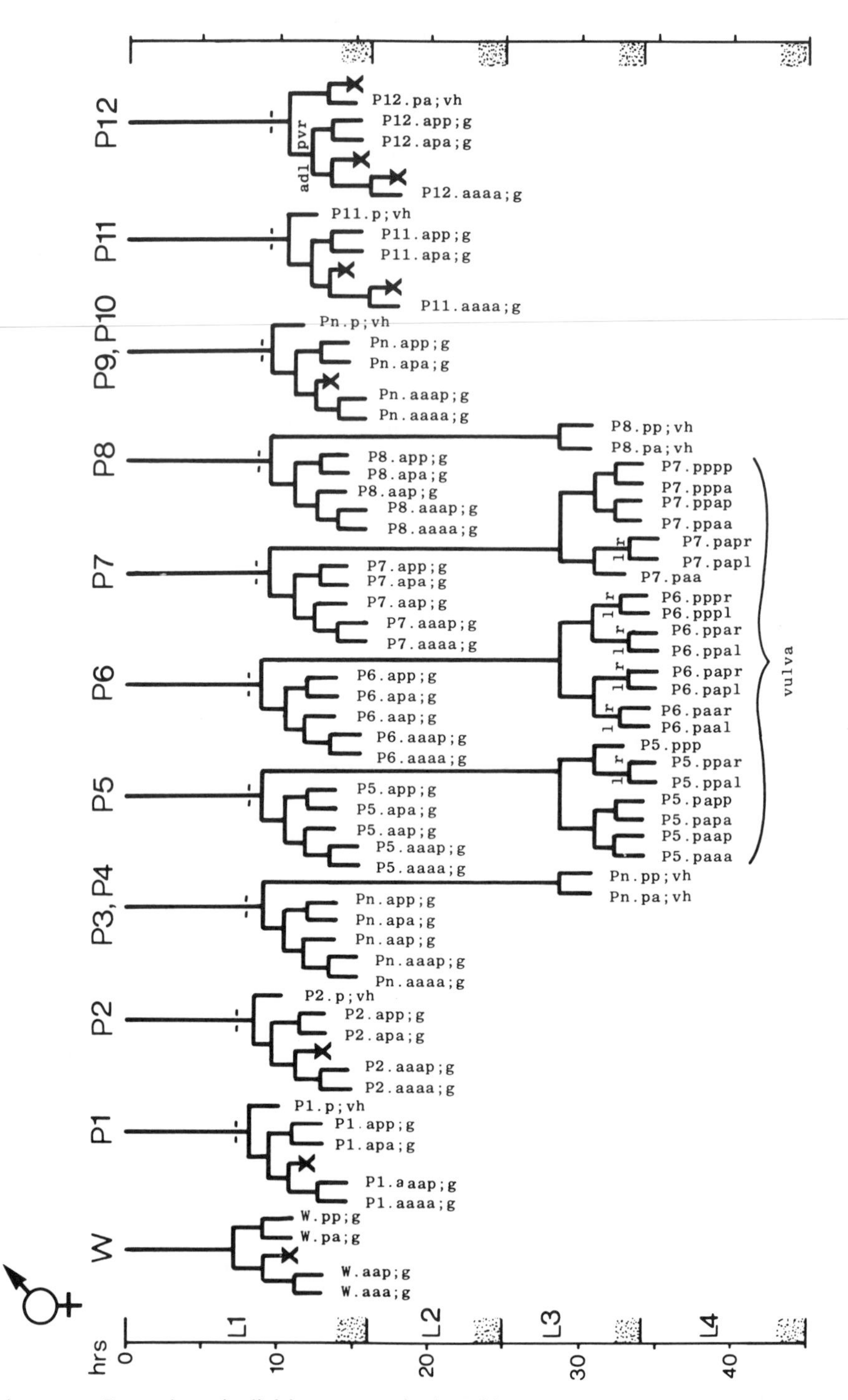

Figure 5.3 Postembryonic division patterns in the P blast cell lineages. Nomenclatural rules are given in the text. Anterior and left daughter cells are placed to the left, posterior and right

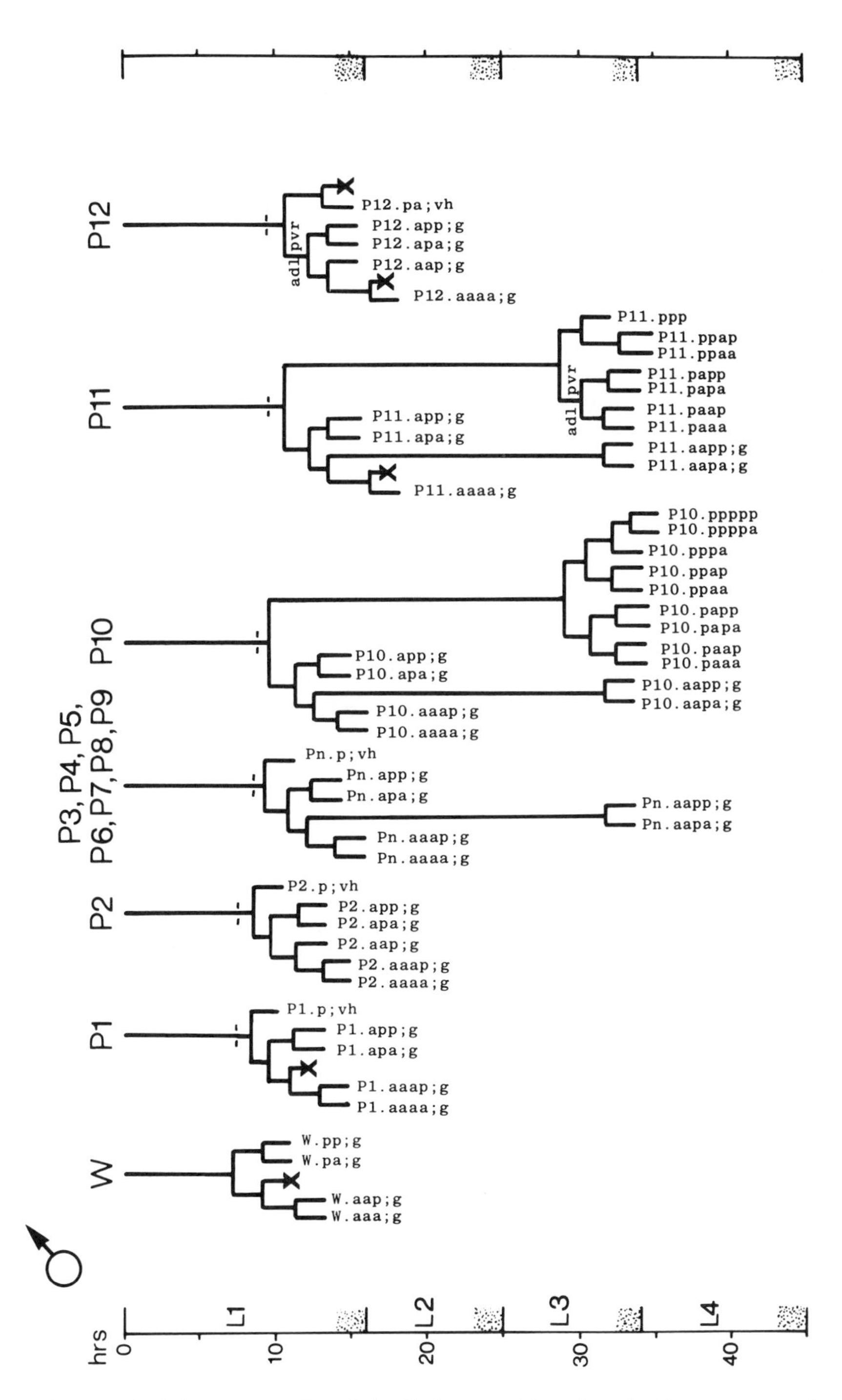

daughters to the right. g, neuronal or glial cell; vh, ventral hypodermal cell; x, programmed cell death. (Figure courtesy of Dr. J.E. Sulston; reproduced with permission from Sulston and Horvitz, 1977.)

In the hermaphrodite, each of the 12 P cells gives 5 neurons and 1 hypodermal cell. As in many of the lineages of the larva, the division planes are perpendicular to the a-p axis, producing an anterior and a posterior daughter. By convention, the division history of a particular cell is designated by the sequence of letters, each indicating the resulting position from a given division, following the name of the founding blast cell. Thus, P2.app is the posterior daughter of the posterior daughter of the anterior division product of the P2 blast cell.

It can be seen that the anterior daughter of the first P cell divisions gives rise to several progeny neurons; all of these neurons contribute to the ventral nerve cord. These divisions of Pn.a daughters are complete by the L1–L2 transition. Several of the posterior daughters, the Pn.p cells, however, have a different fate and contribute cells to the vulva in the hermaphrodite and the preanal ganglion of the male. In the hermaphrodite, the posterior daughters of P1, P2, P9, P10, and P11 all form ventral hypoderm (vh) cells that fuse with the syncytial hypoderm cord, but the midddle Pn.p cells (P3.p–P8.p) divide once more. This additional division is followed by further divisions of the P5.p, P6.p, and P7.p lineages. This last set of divisions produces the 22 cells that comprise the vulva (Fig. 5.4). The cell divisions of the vulva are complete by the early L4 stage, and all 22 cells have arrived in their final positions by late L4.

How does vulva development proceed if one of the three ancestral cells is destroyed? The precise outcome depends on which cell is destroyed, but significantly, there is some capacity for compensatory divisions (Sulston and White, 1980). Regulation is by means of cell recruitment from three Pn.p lineages that normally do not contribute to the vulva—namely, P3.p, P4.p, and P5.p. These recruitable Pn.p lineages are distinguished from the nonrecruitable lineages by the extra division of the ventral hypodermal cell during L3 (Fig. 5.3).

If the central vulval precursor cell P6.p is destroyed, then P4.p, instead of forming just two cells, forms seven, all of which contribute to the vulva, and P5.p., normally a source of seven cells, undergoes an extra division to give eight; the new result is a 22-cell vulva. If both P7.p and P5.p are destroyed, then both the P4.p and P8.p lineages are recruited and give a 24-cell vulva. If, however, all three standard precursor cells (P5.p, P6.p, and P7.p) are destroyed, then both P3.p and P4.p are recruited but manage to construct only a 16-cell vulva. Finally, if P3 through P8 (not the posterior daughters, but the blast cells themselves) are destroyed, there is no trace of vulva development, while ablation of P blast cells outside this group permits normal development of the vulva. The detailed analysis reveals that cells on either side of P6.p tend to replace it; the overall pattern of replacement may be described as one of recruitment to the center. In effect, the cells display a *hierarchy of fates,* with that of P6.p being primary.

It is evident that cell lineage plays a crucial role in vulva development. The group of Pn.p cells distinguished by extra divisions in the third larval stage are the only cells that can contribute to vulva development; cells outside this group cannot. At the same time, there is distinct regulatory potential to replace vulva cells within the special group of Pn.p cells marked by their extra hypodermal cell division. The group of potential (and actual) vulval precursor cells comprises a *cell equivalence group* (Kimble et al., 1979), which exhibits partial or complete interchangeability in developmental role. In the case of the vulval cell equivalence group, the lineage characteristics of the cells provide an obvious clue to their similarity of developmental capacity.

The vulval cell equivalence group is not unique; other cell equivalence groups can be found by the laser microbeam ablation technique. The preanal ganglion of the male provides a second example. In the male, the cells that correspond to the vulval equivalence group (P3.p–P8.p) plus P9.p all contribute to the ventral hypoderm (Fig. 5.3, lower half). However, P10.p and P11.p undergo a specific division program to form the preanal ganglion, one of the set of ganglia required for innervating the male tail (used in clasping the hermaphrodite). Cell ablation experiments show that P9.p, P10.p, and P11.p comprise a cell equivalence group in the male (Sulston and White, 1980). However, the replacement pattern is different from the recruitment-to-the-center pattern of the vulval equivalence group. For the preanal ganglion, recruitment proceeds from anterior to posterior. If P10.p is ablated, P9.p can replace it; if P11.p is killed, P10.p will replace it and P9.p will take over the role of P10.p. Nevertheless, P11.p never replaces P10.p, and P8.p is outside the equivalence group.

For the preanal ganglion, the hierarchy of fates is such that the anterior cells are subservient to the posterior cells. If any of the latter are removed, the more anterior cells drop their assignments and take up those of the missing posterior cells.

The second point of interest is that the equivalence group defines a boundary that is not apparent in the P lineages of the male, namely, that between P8.p and P9.p. In the hermaphrodite this boundary defines the posterior limit of the vulval equivalence group, while in the male these two cells normally have equivalent fates. This finding is consistent with a general conclusion about the relationship between the male and hermaphrodite somatic developmental programs based on cytological observations (to be discussed later); they are essentially modifications of the same scheme, with the male-specific features based on or superimposed on the hermaphrodite construction plan (Sulston et al., 1980).

In the examples presented so far, only P lineages have been examined. The lineages that form the rays of the tail in male worms illustrate the importance of equivalence groups in other lineages. The tail with its rays is shown in Figure 5.5; the function of the rays is sensory, and there are nine rays on each side of the tail. Each ray consists of two neurons and one supporting structural cell.

The cell division program that gives rise to the rays is shown in Figure 5.6.

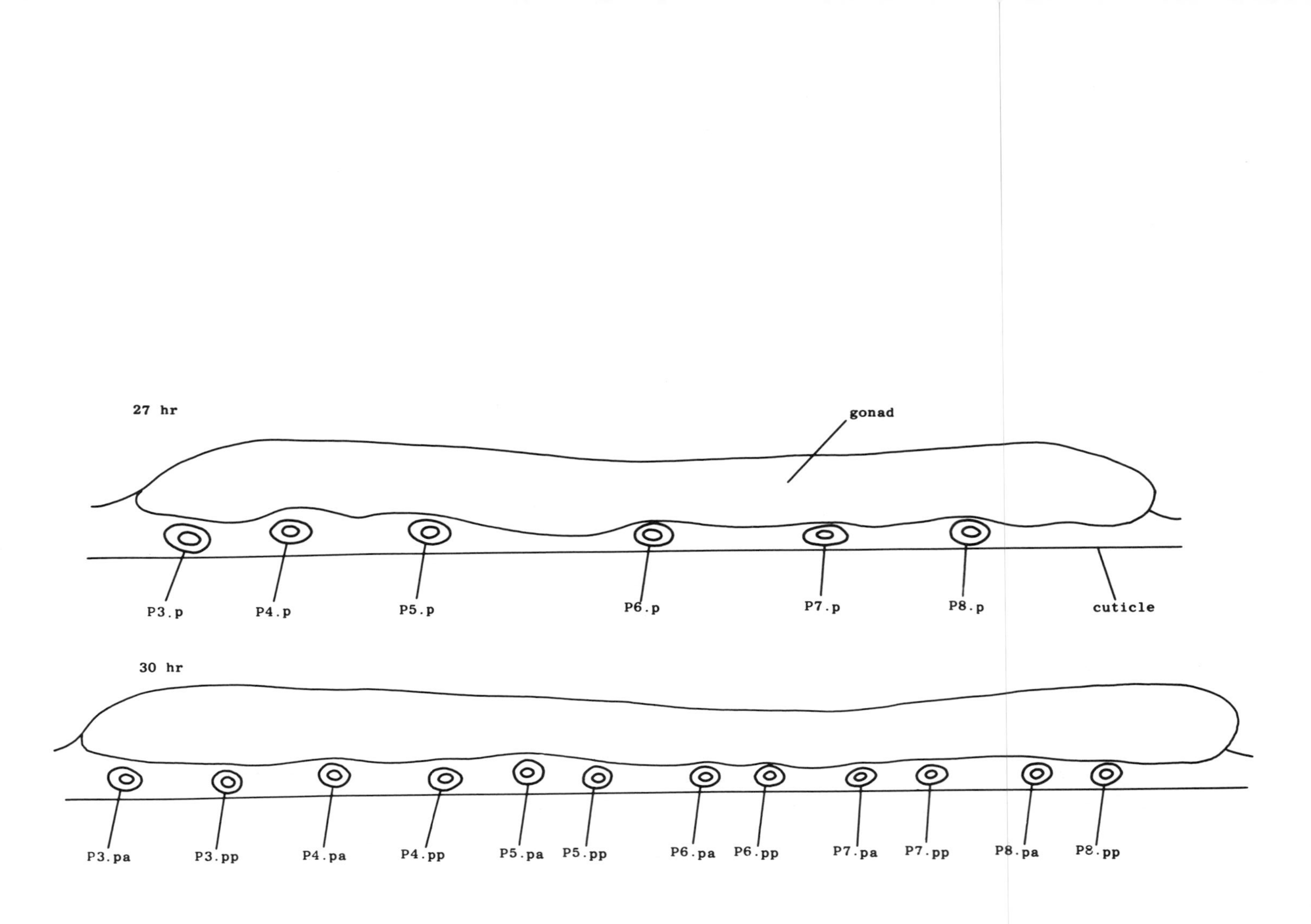
27 hr
gonad
P3.p
P4.p
P5.p
P6.p
P7.p
P8.p
cuticle
30 hr
P3.pa
P3.pp
P4.pa
P4.pp
P5.pa
P5.pp
P6.pa
P6.pp
P7.pa
P7.pp
P8.pa
P8.pp

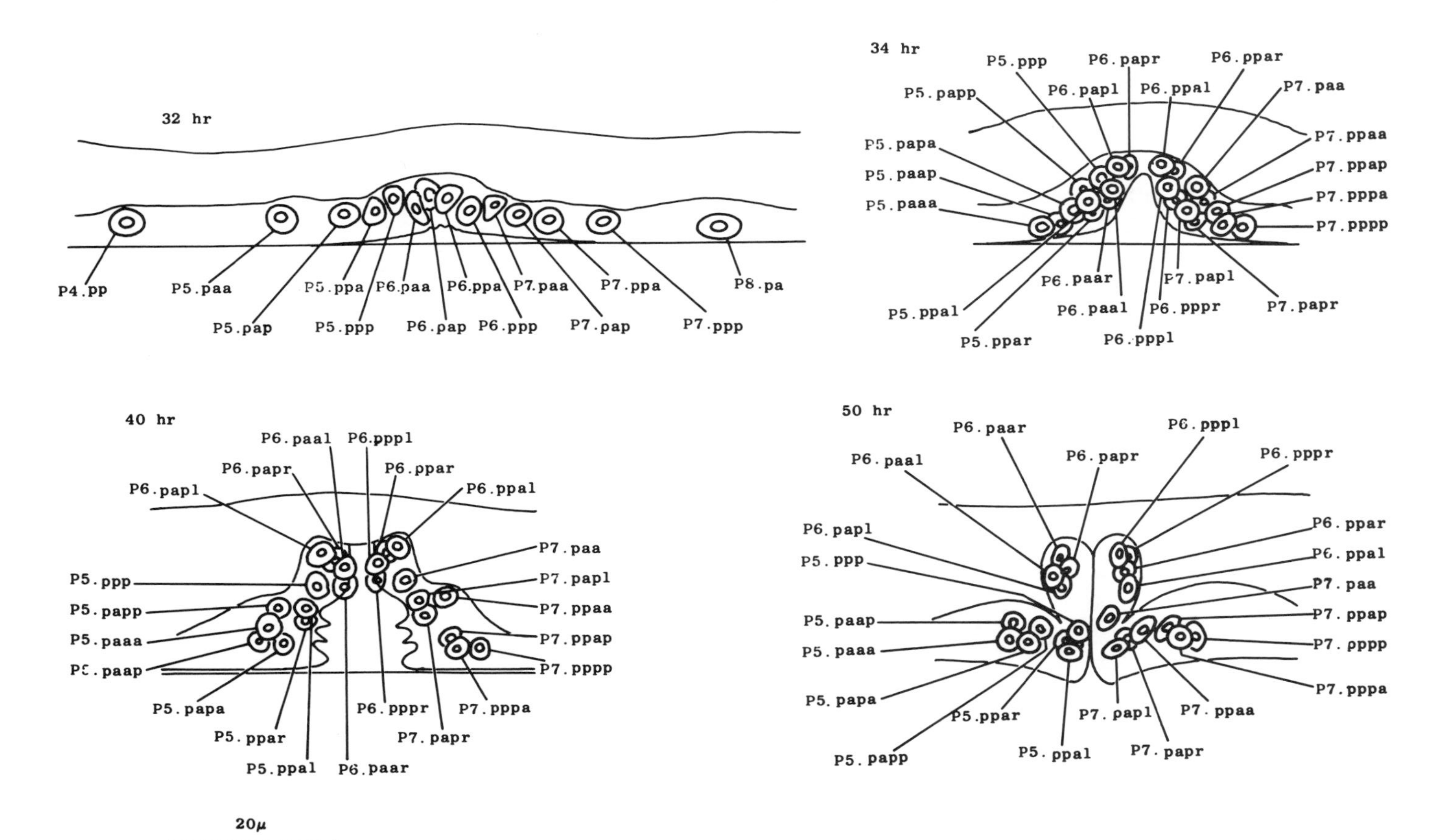

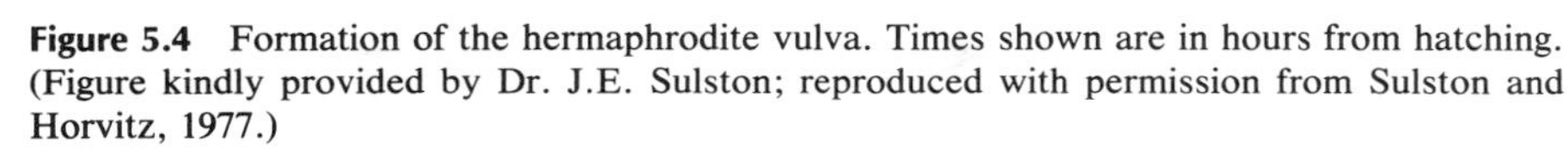

Figure 5.4 Formation of the hermaphrodite vulva. Times shown are in hours from hatching. (Figure kindly provided by Dr. J.E. Sulston; reproduced with permission from Sulston and Horvitz, 1977.)

Each cell group is derived from a ray precursor cell, R1 through R9, by three rounds of cell division. The ray precursor cells, in turn, are derived from each of one of three blast cells: V5, V6, and T. It can be seen from the figure that V5 gives rise to one ray group on each side, V6 to five, and T to three ray groups. Cell ablation experiments show that V4, V5, and V6 constitute a well-defined cell equivalence group and that V1, V2, and V3 are apparently outside this group. As in the preanal ganglion, recruitment is always from anterior to posterior, with V6 possessing the primary fate. V lineages never substitute for T, or vice versa, indicating that T is not a member of the cell equivalence group. (It is possible, however, that mechanical restraints prevent developmental substitution in this case.)

These observations demonstrate the existence of several cell groups related by ancestry and possessing similar developmental capacilities (Fig. 5.7). Each equivalence group is made up of contiguous precursor cells with similar cell division capacities. In each, the removal of a cell results in partial compensatory replacement, although the resulting structures are never morphologically normal. Not all similar-seeming precursor cells are clustered in equivalence groups; there are several groups of similarly behaving blast cells whose progeny have no developmental interchangeability. However, the equivalence group is a significant functional grouping in this organism. In

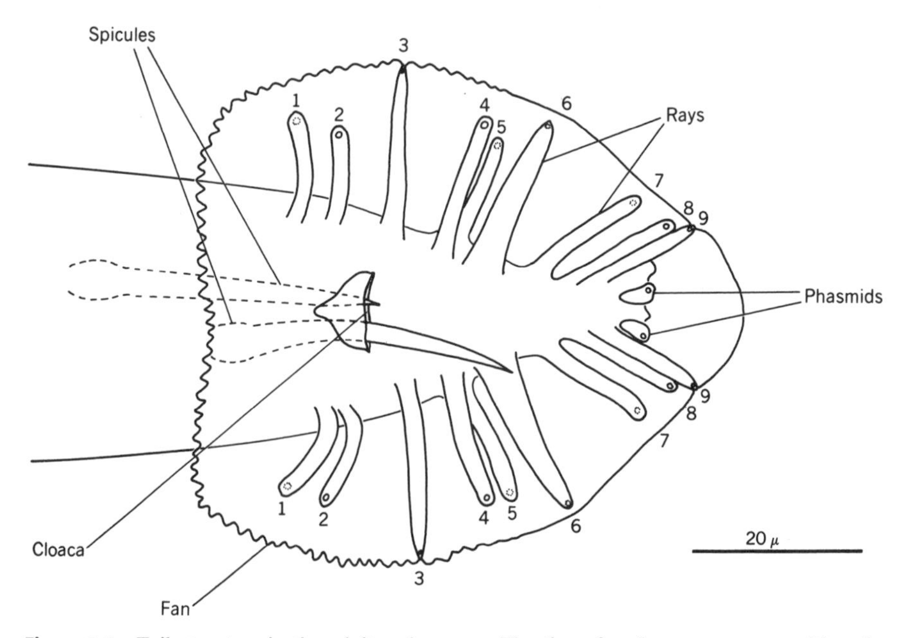

Figure 5.5 Tail structure in the adult male worm. The tips of each ray emerge on either the dorsal surface (dotted circle) or the ventral surface (solid circle). Phasmids are male-specific sensory elements; the copulatory spicules are also shown. The fan is an acellular cuticular product secreted by the hypodermis. (Figure kindly provided by Dr. J.E. Sulston; adapted and reproduced from Sulston and Horvitz, 1977.)

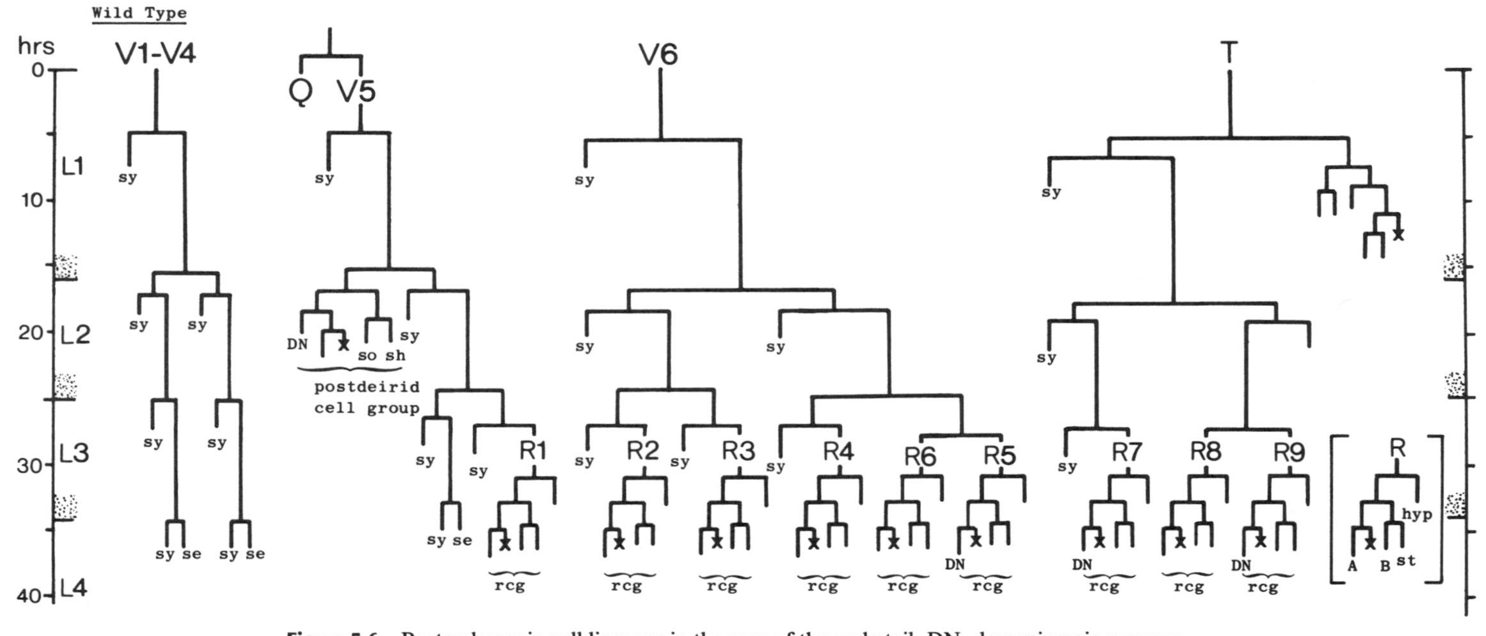

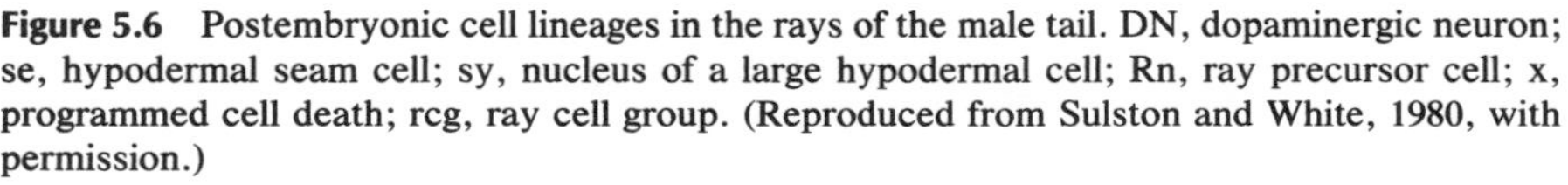

Figure 5.6 Postembryonic cell lineages in the rays of the male tail. DN, dopaminergic neuron; se, hypodermal seam cell; sy, nucleus of a large hypodermal cell; Rn, ray precursor cell; x, programmed cell death; rcg, ray cell group. (Reproduced from Sulston and White, 1980, with permission.)

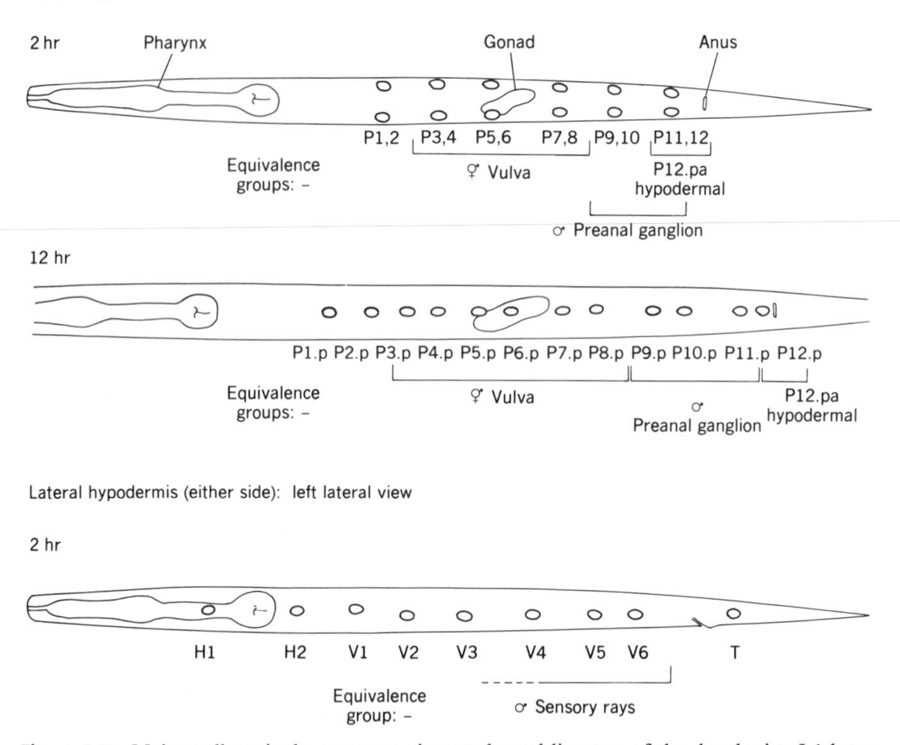

Figure 5.7 Major cell equivalence groups in ectodermal lineages of the developing L1 larva. The times given are from hatching. Note the changes in equivalence cell groupings of the P lineages between the early (2-hr) and late (12-hr) L1 form. (Figure courtesy of Dr. J.E. Sulston; reproduced with permission from Sulston and White, 1980.)

those pathways where there is some capacity for interchangeability, lineage largely sets the limits for the possible substitutions. (In those cell lineages derived from unique blast cells—for instance, the B, C, U, F, K, and M lineages—ablation of the precursor cell results in elimination of that lineage.)

It is clear that cell lineage history is very important in setting the developmental role of cells. Nevertheless, it is not the only factor in setting cell commitments. For cells of nearly equal initial capacity, positional setting can make a difference. One example is the set of 12 P lineages, where the specific assignment within each pair of cells (P1/2, P3/4, P5/6, P7/8, P9/10, P11/12), the two members of each pair arriving from the left and right sides, is determined randomly. However, once an anteroposterior order is established, these assignments are maintained.

Another factor in regulating developmental fate is cell induction. For example, if the gonad cell precursors are ablated, the vulval equivalence group does not produce its usual division pattern; instead, each cell P3.p–

P8.p) divides once, and all daughter cells join the ventral hypoderm (Kimble, 1981a). This interaction may involve a diffusible factor because the developing gonad and vulva are not in direct contract until the end of their division sequence.

In effect, cell lineage histories may prescribe the limits of developmental capacities, but other factors may often be necessary to actuate those potentialities. In principle, genetic analysis of mutants affected in postembryonic development can illuminate the requirements for the setting and expression of postembryonic developmental commitments.

Genetic Analysis: Nongonadal Tissues

The distinctive events of postembryonic development suggest the existence of distinctive programs of gene expression in the postembryonic period. If key control genes are expressed in postembryonic development, it should be possible to identify them through the isolation of postembryonic defective mutants. Some of these mutants might be affected in certain lineages, while others might have general late developmental defects. Genetic analysis should complement and extend the information derived from the cell ablation experiments. The genetic analysis of postembryonic development in nongonadal tissues will be examined first. In the next section, the special features of gonadal development will be inspected.

Postembryonic defectives can be isolated in either hermaphrodites or males. Most of the mutants described to date were detected as hermaphrodite abnormals. In the large mutant hunt described by Horvitz and Sulston (1980), the F_1 and F_2 progeny of mutagenized hermaphrodites were screened for the kinds of developmental and behavioral defects that could result from specific postembryonic cell lineage abnormalities. For instance, a distinctive part of the postembryonic somatic development of the hermaphrodite concerns the vulva and vulva-associated structures. Any early defects in the P blast lineages should produce vulva defects, and mutants affected in these lineages should be revealed as egg-laying defectives. Another class of postembryonic defect is Unc defects (although some of these reflect embryonic defects in the neuromusculature). The examination procedures include inspection of putative mutants by Nomarski optics (to determine if postembryonic lineages have abnormal cell numbers or arrangements), fluorescence staining after formaldehyde treatment (to detect the distinctive dopamine-producing neurons generated during larval development), and Feulgen or Hoechst staining (to detect aberrations in the nuclear DNA content of postembryonic cells).

Some of the loci identified by the various postembryonic mutant classes are listed in Table 5.1. The loci are grouped by mutant phenotype under three major headings: general defectives, vulva defectives, and homeotics. All of the mutants in the first category are defective in a fundamental nuclear or cell division property or behavior; the defects occur in homozygous

Table 5.1 Postembryonic (pe) Defectives in *Caenorhabditis elegans:* Hermaphrodite Mutant Phenotypes

Category	Locus	Mutant Defects	Reference
General defects			
Nuclear division	lin-5 (II)	No pe nuclear divisions; polyploidy in pe lineages	Sulston and Horvitz (1981)
	lin-6 (I)	No pe DNA replication in somatic cells; small somatic nuclei	Sulston and Horvitz (1981)
Nuclear migration	unc-83 (V) unc-84 (X)	Embryonic and pe P lineage nuclear migration delayed or blocked	Sulston and Horvitz (1981)
Cell division	unc-59 (I) unc-85 (II)	Defective cytokinesis; polyploid nuclei through nuclear division failure or nuclear fusion	Sulston and Horvitz (1981)
Vulva			
Vulvaless	lin-2 (X) lin-3 (IV) lin-7 (II)	Abnormal vulva precursor cell (P5.p–P7.p) division	Sulston and Horvitz (1981)
Multivulva	lin-1 (III) lin-8 (II); lin-9 (III)	Excess Pn.p divisions; pseudovulvae	Sulston and Horvitz (1981)
Homeotic			
Reiterative	unc-86 (III)	Reiterative cell divisions in Q, T, and V5 lineages	Chalfie et al. (1981)
Binary choice	lin-12 (III)	Affects cell choices in Z1 and P3.p–P8.p lineages	Greenwald et al. (1983)
Heterochronic	lin-4 (II)	Reiterated L1 division program	Chalfie et al. (1981); Ambros et al. (1983)
	lin-14 (X)	Skipping or reiteration of L1 and L2 division programs	Ambros and Horvitz (1984)

worms of both sexes. Perhaps the most interesting members of this broad grouping are the nuclear migration defectives. Nuclear migration occurs in both the embryonic and postembryonic hypodermal syncytia. *unc-83* and *unc-84* are defective in both sets of migrations; their postembryonic defect is in the migration of P blast nuclei and produces aberrant vulval and ventral cord development. Mutant males are less visibly defective but show abnormalities in the preanal ganglion, the structure derived from P10.p and P11.p.

The second two mutant categories, the vulva defectives and the homeotics, are the most informative about distinctive features of lineage specification in the assignment of developmental fate. The vulva defectives, consisting of vulvaless (Vul) and multivulva (Muv) strains, directly support the idea of individually programmed cell equivalence groups. In all Vul mutants, the P5.p–P7.p cells undergo the first round of L3 divisions (Fig. 5.3) but then fail to start or complete the second round of division. The two daughter cells of each simply join the ventral hypoderm, like the daughter cells of P3.p, P4.p, and P8.p. In these mutants, the division defect is specific for the vulval precursor cells; their division program has been reduced to that of the other members of the vulval equivalence group.

The Muv mutants produce a set of small vulva-like protrusions, "pseudovulvae"; like the Vul mutants, their effect is produced solely within the vulval equivalence group. In these mutants, *all* members of the equivalence group (P3.p–P8.p) undergo second stage proliferation, yielding pseudovulvae. In the Muv double mutant—*lin-8;lin-9*—the phenotype is absolutely dependent on both mutations.

The effects of these two Muv strains, however, are not restricted to the vulval equivalence group; both also effect the G lineages, derived from the G1 and G2 blasts, situated ventrally in the head (Fig. 5.2). All of the effected postembryonic blast cell lineages for these mutants are derived from the embryonic blast cell AB, and G2 is a sister cell of one of the two members of the P1/2 blast cell pair.

Most intriguingly, both Muv strains also produce an effect in the male preanal ganglion equivalence group, P9.p–P11.p. In both mutant strains, P9.p proliferates in a manner similar to that of P10.p and P11.p, the normal precursors of the preanal ganglion. Evidently, the wild-type gene products of *lin-1* and *lin-8;lin-9* are required for the normal behavior of two equivalence groups within the set of Pn.p lineages; the particular set of cells affected is a function of the sex of the individual. This finding is a further indication of the underlying similarity of the somatic developmental programs in the two sexes. Indeed, many mutants selected on the basis of their male phenotype reveal slight defectiveness in the hermaphrodite upon close inspection (Hodgkin, 1983a).

Two further observations indicate that cell equivalence groups are created by discrete gene expression programs. The first concerns a difference between the two Muv strains: in the *lin-8;lin-9* double mutant strain, the P3.p–P6.p cells undergo extra divisions in the male to produce pseudovulvae. In effect, the mutant condition reveals a cryptic cell equivalence group in males that shares a boundary with a manifest equivalence group in hermaphrodites. The second observation concerns a male-abnormal or Mab mutant (in gene *mab-5*), in which the V5 and V6 lineages are abnormal, failing to generate the rays of the tail, while the T lineage is normal (Hodgkin, 1983a). Clearly, this mutant "respects" the normal boundary between the V5–V6 equivalence group and the T-blast lineage in the male.

The last category of postembryonic defectives listed in Table 5.1 are the homeotics. They differ markedly in phenotype from one another and probably involve several different mechanisms of development alteration. Their shared characteristic is that they all transform several or many cell lineages, replacing particular cell types with others. (The Vul and Muv mutants could also be classified as homeotics, but their effects are restricted to the cells of the vulval equivalence group.) These nematode homeotics differ in character from the fruit fly homeotics discussed in Chapter 3 (*bic, dl*) in that they are zygotic rather than maternal effect mutations.

The first mutant listed, *unc-86,* produces a characteristic repeated reenactment of a particular cell division sequence in three specific cell lineages. In wild-type development, the precursor cells for these three lineages—the Q, V5.paa, and T.pp cells on both sides of the animal—participate in the production of specific neurons in a sequence of two divisions (Fig. 5.8). In each of these lineages, the posterior daughter of each of these cells gives rise to an anterior cell, which either becomes a neuron or gives rise in further divisions to neurons, and a posterior cell, which dies. In both *unc-86* mutants, the affected cell, instead of producing a posterior daughter that dies, undergoes the division pattern of its parent cell one or more times, thereby generating more neurons (Chalfie et al., 1981). In effect, the posterior daughter reiterates its parental cell's division program. In the case of the V5.paa lineage, the supernumerary neurons can be detected histochemically by the production of dopamine. In the wild-type hermaphrodite, there is only one

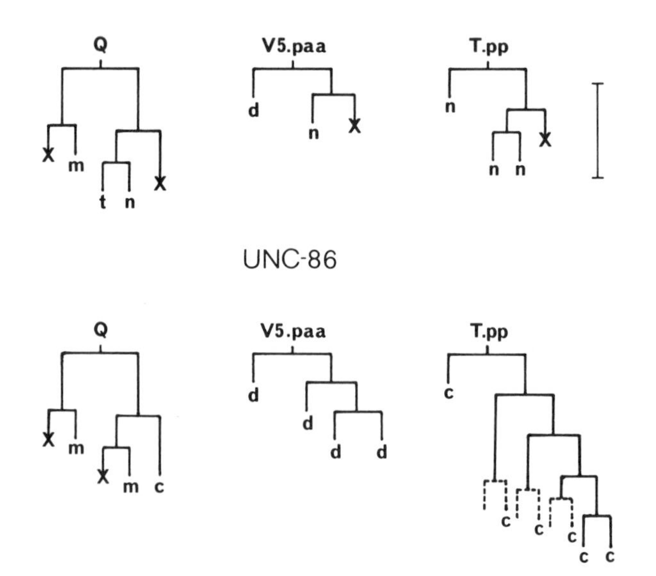

Figure 5.8 Reiterative cell divisions in the *unc-86* mutant. See text for details. n, neuron; m, migrating neuron; d, dopaminergic neuron; c, compact nucleus of a neuronal structural cell; X, programmed cell death. (Figure kindly provided by Dr. J.E. Sulston; from Chalfie et al., 1981; copyright MIT Press.)

dopaminergic neuron, the anterior daughter produced by V5.paa. In *unc-86* hermaphrodites, multiple dopaminergic neurons are produced, and each of these sends processes into the ventral nerve cord, just as the single cell in the wild-type does.

A second class of homeotics, represented by *lin-4* and *lin-14,* is the "heterochronics." In these mutants, major aspects of the developmental cycle are temporally shifted. In the recessive *lin-4,* the phenotype consists of a reiteration of the L1 cell division patterns. This repetition is even reflected in the cuticular composition of *lin-4* individuals. In the wild-type, the protein composition of adult cuticles differs from that of L4 cuticles, and the latter differ from L1 cuticles. In *lin-4,* the adult cuticle composition is that of the L1 stage (cited in Cox et al., 1981b).

Mutants of *lin-14* are also heterochronic but can have either of two effects. If semidominant, they mimic the *lin-4* defect, causing a reiterated L1 development with each molt. Recessive mutants, however, retain a normal L1 stage, but cause one or more of the following larval stage division programs (that of L2, L3, and L4) to be skipped in whole or in part. From the recessive phenotypes, it would seem that *lin-4*$^+$ activity is necessary for development from the L1 to the L2 stages, and that *lin-14*$^+$ activity is required to prevent precocious skipping of L2, L3, or L4. (The semidominant *lin-14* alleles are hypermorphs; evidently, too much of the gene activity prevents the transition from the L1 to the L2 stages, the same defect produced by *too little lin-4*$^+$ activity.) It is not known whether these gene activities are required autonomously in all of the postembryonic cell lineages, or whether the two genes exert a systemic effect by the production of diffusible products.

Like *lin-14,* the final homeotic gene listed, *lin-12,* can also mutate to two states of activity with opposing effects, but its actions are restricted to certain cell pairs rather than whole stages. In the wild-type strain, certain cells have alternative fates with respect to one another, and their choice of fates is a function of their mutual interaction. The set of vulval precursor cells, P5.p–P7.p, furnish an example. In this cell set, P6.p has the "primary fate" and both P5.p and P7.p have the "secondary fate"; the other members of the vulval equivalence group, P3.p, P4.p, and P8.p, effectively have the "tertiary" fate. Semidominant hypermorphic mutations of *lin-12,* designated *lin-12(d),* drive all cells of the vulval equivalence group to the secondary fate. In contrast, recessive *lin-12* mutants cause the three vulval precursor cells to go to the primary fate (and P4.p to produce a "hybrid" lineage with some primary and some tertiary characteristics). The semidominant and recessive mutants similarly produce opposite tendency effects in some of the other alternative-fate cell lineages. Greenwald et al. (1983) have proposed that in such alternative-fate cell pairs, cell–cell interactions determine cell fates by regulating the respective levels of *lin-12* activity in the two cells; high gene activity in one produces one fate, low activity in the other the opposite cell fate.

Cell Division Geometry, DNA Replication, and Fate Allocation

The genetic analysis of nongonadal lineages confirms that cell lineage is as important in postembryonic development as in embryonic development in setting cellular developmental capacities. Furthermore, in both stages of development, the geometry of cell division is correlated with the fates of the daughter cells. In symmetrical cell divisions, both daughters are similar in size and immediate developmental behavior. In asymmetrical divisions, the daughter cells differ in size and shape and in their immediate developmental fates. In postembryonic asymmetrical divisions, there is often a topographical difference, with the posterior daughter similar to its mother cell and an anterior daughter different from both (Sulston and Horvitz, 1977). The same relationship is seen in the first cleavage divisions of embryonic development: in the embryonic blast cell generating divisions, the division is always asymmetrical and each newly formed blast cell is always anterior to its Pn sister cell.

The correlation between cell division characteristics and cell fates is also illustrated by the postembryonic cell lineages affected by the homeotic *unc-86* mutants. In the wild-type, the posterior daughter cells in these lineages that are destined to die are always produced in an asymmetrical division, being the smaller of the two cells. In the mutants, the two sister cells are the same size.

These relationships raise the question of whether cell division per se is necessary for either the expression or the segregation of developmental potential in postembryonic development. We have seen that in embryonic development, neither cell division nor nuclear division appears to be required for the expression of such potentials. An analysis of the mutant *lin-5* indicates that neither cell division nor nuclear division is required for the expression of differentiated postembryonic characteristics but may be required for their segregation in sublineages.

In the *lin-5* mutant, the ventral cord is populated by 12 P-blast nuclei, but because nuclear division is blocked in this mutant (Table 5.1), the nuclei do not divide but become polyploid. Nevertheless, these blocked precursors show signs of the normal differentiation program (Albertson et al., 1978). For example, those Pn precursor cells that in the wild-type exhibit cell death in one of their descendants (W, P1 and P2, P9–P12) (Fig. 5.3) show partial or complete nuclear degeneration in the mutant. In contrast, the P3–P8 lineages, which do not exhibit programmed cell deaths in the wild-type, do not show nuclear degeneration.

The second sign of normal differentiation in the blocked cells concerns neuronal character. In the wild-type strain, the ventral nerve cord consists of five different classes of motor neurons (a motor neuron being one that innervates a muscle), and each kind is distinguished by its axon directions, the position of muscles innervated (dorsal or ventral), and the kinds of synaptic connections it makes (whether chemical or gap junctions, the types of interneurons with which it connects, etc.). Furthermore, each neuron type is derived, via a particular sequence of divisions, from the Pn precursor (Sulston, 1976). In the *lin-5* ventral nerve cord, the division-blocked P-blast precursor cells each develops 8–16 neuronal processes (each wild-type neuron has only 2), and these form neuromuscular junctions (NMJs) in approximately the right number and with the correct spacing. In addition, each Pn precursor develops characteristics similar to those that define the five normal classes of ventral cord motor neuron; each division-blocked Pn cell has a partially intermediate neuronal character, but only one or two types predominate for a given cell.

Thus, each kind of differentiation potential normally associated with a given member of a Pn ventral cord lineage can be expressed in division-blocked polyploid cells. It thus appears that the expression of differentiation potential is independent of the occurrence of actual cell division sequences, with their well-defined a-p polarities and times of occurrence. Similarly, Gossett et al. (1982), employing specific antisera to the muscle protein myosin and paramyosin, found that these two proteins, which usually appear in the muscle cell groups late in cleavage, are synthesized on schedule even when cleavage is blocked in two ts mutants under restrictive conditions.

However, the *segregation* of different potentials within a lineage may require cell division, as suggested by the occurrence of intermediate neuronal characteristics in the blocked cells. One can imagine a direct role for cell division asymmetry in this process. For instance, it is possible that some "regulatory" molecular influence is distributed unequally in each asymmetrical cell division, and that this quantitative absolute difference is amplified by subsequent events to give a strong differential bias in cell fate. In this situation, the change from an asymmetrical division to a symmetrical one should produce like daughters. Furthermore, reversal of the size difference between daughters should be accompanied by a reversal of daughter cell fates. A finding of symmetrization of cell fate following a symmetrization of division has been reported in early cleavage; the *zyg-2* mutant shows a symmetrical division plane and the production of two AB-like cells (Laufer et al., 1980). The second phenomenon, the reversal of cell fate following reversed placement of the division plane, has occasionally been observed in cell ablation experiments (Sulston and White, 1980; Kimble, 1981a). However, it must be emphasized that the cause–effect relationship between cell division geometry and cell fate specification is unknown; the placement of the division plane could be a consequence of a primary fate determination.

As in the case of E lineage and muscle cell expression in blocked blastomeres in embryogenesis, the expression of differentiated postembryonic cell characteristics in the blocked *lin-5* mutants apparently takes place on schedule. The process therefore appears to be governed by some kind of intracellular "clock," running independently of cell or nuclear division per se. One obvious possibility, suggested by the expression of AcE potential in ascidian muscle, is the number of rounds of DNA replication. As discussed earlier, the number of rounds of DNA replication may serve as either a counting or a timing device.

On the other hand, it is apparent that at least some events in postembryonic development occur independently of DNA replication. In normal development, certain of the juvenile motor neurons—those formed during embryogenesis and present in the ventral nerve cord at the beginning of L1—change their connectivity pattern from the innervation of ventral muscles to that of dorsal muscles and form new connections with postembryonic P lineage neurons. In the *lin-6* mutant, which is blocked in DNA replication (Table 5.1), yet whose cells continue to divide, similar changes in juvenile neuron connectivity occur, although these neurons fail to receive new connections from the late-developing ventral neurons whose formation is blocked in the mutant (White et al., 1978). Thus, certain differentiations can proceed independently of the formation or DNA synthesis of other cells of the wild-type program with which they are normally associated.

In several respects, the work on postembryonic development in *C. elegans* emphasizes the importance of questions about control raised by embryonic development. However, the greater accessibility of postembryonic development to direct observation and manipulation makes it the preferred stage in which to investigate these controls.

Gonadal Development: Patterns and Controls

The gonad in both sexes is the major organ system in the animal; in the hermaphrodite, for instance, the nuclei of the gonad comprise more than two-thirds of the total nuclei in the adult animal. Although the gonad in both sexes starts from what appear to be identical cells, the respective patterns of gonad development show some major differences (Fig. 5.9). The obvious structural difference between the hermaphrodite and male gonads is one of shape. The hermaphrodite gonad is bilaterally symmetrical, with twofold rotational symmetry and two gamete-generating arms that meet in the center, while the male gonad is asymmetrical, consisting of one gamete-generating arm that runs posteriorly. This difference is initiated early in the L1 stage, with an asymmetrical cell migration in the male gonad primordium that does not occur in the hermaphrodite. The consequence becomes manifest during L3, when the hermaphrodite gonad, which has been growing both anteriorly and posteriorly, undergoes a symmetrical reflexion in both arms; the male gonad, growing only anteriorly, makes a single bend and then continues growing posteriorly, eventually joining the cloaca.

Despite the pronounced morphological and growth differences, the two gonads share important similarities. First, that gametes mature from the most distal (outward) tips of the gonads toward the proximal (central) end(s). Second, the major somatic structures in both gonads are located proximally.

The developmental processes that underlie this similarity of organizational polarity are revealed by lineage analysis of the gonadal somatic cells

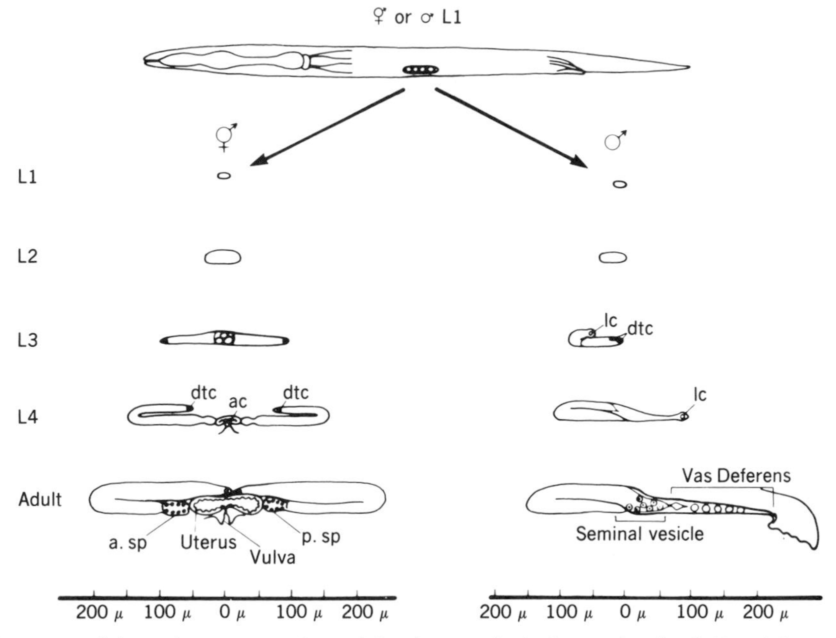

Figure 5.9 Schematic summary of gonad development in the hermaphrodite (left) and the male (right). See text for details. ac, anchor cell; dtc, distal tip cell; lc, linker cell; a. sp., anterior spermatheca; p. sp., posterior spermatheca. (From Kimble and Hirsh, 1979, with permission.)

(Kimble and Hirsh, 1979). Both types of gonad originate from just four precursor blast cells, designated Z1, Z2, Z3, and Z4, that are located mid-ventrally. In both sexes, the Z2 and Z3 cells (daughter cells of the germ line embryonic blast cell P_4) give rise to the gametes, while Z1 and Z4 (derived from the MS blast cell) give rise to the somatic cells of the gonad. The major differences in gonadal growth pattern between the sexes involve the activities of three key somatic cells, derived from Z1 and Z4, that are generated in the first divisions of the gonad primordia during L1.

Two are referred to as "distal tip cells" (dtc's) in both sexes, because of their location at the distal (outermost) point in each gonad, where the germ line precursor cells undergo premeiotic divisions. In the developing hermaphrodite gonad, each dtc leads one of the two arms of the developing gonad toward its final distalmost point. In the male, the two dtc's assume a central and distal position at the beginning and mark the stationary point of male gonad growth. Despite this difference in behavior, the dtc's are formed early in L1 in both sexes, and both occupy similar positions in the Z1 and Z4 lineages, being the anteriormost Z1 and the posteriormost Z4 descendants in both gonad primordia (Z1.aa and Z4.pp in the hermaphrodite, and Z1a and Z4p in the male primordium).

The third key regulatory cell is termed the "anchor cell" in the hermaphrodite and the "linker cell" in the male gonad. The anchor cell performs two functions: it provides a stationary point between the two growing gonad arms, and it acts to induce the vulva (Kimble, 1981a). The equivalent linker cell of the male, in contrast, provides the growing point of the male gonad: it leads gonadal growth throughout development, and its death at the end of its migration provides a junction with the cloaca (through which sperm will pass). Despite these marked differences in behavior, the anchor and linker cells are related in lineage. In the hermaphrodite, the anchor cell is one of 12 cells in the somatic gonad primordium by the end of L1, while in the male the linker cell is one of 10 cells. In both sexes, this critical third cell can be formed interchangeably by either of two cells in the somatic primordium (Kimble and Hirsh, 1979), each pair forming an equivalence group as defined by the cell ablation experiments (Kimble, 1981a). (These two cell choices are among those affected by *lin-12* activity.) Other resemblances in lineage program between the two gonadal soma are described by Kimble and Hirsh (1979).

These observations on the gonad substantiate the conclusion that the developmental programs in the two sexes employ many of the same elements, although in somewhat different fashion. The ultimate developmental consequences of the sex-specific modifications of cell utilization and movement, however, are large. The possible nature of the underlying controls in such program modulations will be discussed in the next section.

Like the somatic cell division programs in nongonadal tissues, those in the gonad are highly precise. In hermaphrodites, the gonadal soma consists of 143 cells, and in males 56 cells. In contrast, the germ cells, the descendants of Z2 and Z3, do not show this rigidity of division pattern. These cells begin multiplying without fixed orientations or precise schedules in L1 and continue to do so throughout development, becoming enclosed by the growing somatic primordium. Nevertheless, the germ line cells are under a form of division control. If the distal tip cells are ablated in either sex, the germ cells nearest the distal tips cease mitosis and become capable of entering meiosis (Kimble and White, 1981). Evidently, the distal tip cells exert some sort of meiotic inhibitory action on their germ line cell neighbors. One possible inference is that the whole polarity of gamete maturation *away* from the distal tips toward the proximal somatic structures is a function of this influence. This hypothesis receives support from the results of certain cell ablation experiments. When sister cells of dtc's are ablated, the dtc's sometimes assume abnormal positions. The resultant gonads grow in aberrant directions but gamete maturation is always polarized away from the dtc's, as occurs in normal gonads (Kimble and White, 1981).

Linear polarity in gamete maturation away from precursor mitotic stem cells is also seen in *Drosophila* oogenesis. The distal tip of the ovary is populated by mitotic stem cells that grade into meiotic stem cells down the length of the ovariole. Such linear polarity of gamete maturation is not universal, but it may be common (see the discussion by Kimble and White, 1981).

Although the absence of strict division control of germ line cells differs from the somatic cell division programs, it makes biological sense. In the nematode, somatic structures are constructed accurately and rapidly using a tightly regulated program of somatic cell division. Such accuracy is not required in the germ line cells, since each functions independently. Thus, germ line cells divide rapidly to provide many gametes and are organized in a polar array so that only mature gametes are discharged through the appropriate somatic structures. Beyond this, no accurate control of lineage is required. The developmental organization of the germ cells in the nematode fulfills these requirements.

Gene Expression and Genetic Control

A possible basis for the invariance of postembryonic somatic development is that a special program of gene expression exists to control these lineages. The isolation of mutants affected in specific aspects of postembryonic development might support this hypothesis. Although little direct molecular evidence on the nature of gene expression and its control during nematode development is available, the genetic observations permit some tentative inferences and conclusions.

The first such inference is that a large fraction of the postembryonic genes identified by the mutant set are fine-tuning genes in some sense. This belief follows from the fact that most of the mutants show highly variable penetrance and expressivity. Such variability cannot always be a function of leakiness because many of the mutations are true null mutations. If a mutant is completely deficient in a gene activity, yet shows a range of developmental phenotypes in the affected individuals, the gene function cannot be completely unique and indispensable, but is replaceable to some degree by related gene activities. This inference suggests, in turn, that the control of postembryonic development is a more subtle affair than the switching on of whole new blocks of genes in the different cell types. Because this point is important, the evidence that amorphic mutations can give variable phenotypes is reviewed here.

In *Drosophila,* the Mullerian gene dosage tests provide a means for detecting amorphic mutations. In *C. elegans,* where there are still relatively few duplications and deletions, these tests cannot be administered for many chromosomal regions. There is, however, an alternative test for assessing mutant expression character that is borrowed from prokaryotic molecular biology. It involves the isolation of a special class of suppressor mutations. These are suppressors that recognize and alleviate the deficiency of one category of amorphs, those produced by "nonsense" or terminator codons within the mutant gene.

The great majority of mutants are "missense" mutants, which cause the production of defective proteins through the insertion of an inappropriate amino acid at a critical site. Nonsense mutants, in contrast, cause premature termination of translation of the mutant protein. The consequence is a shortened and generally nonfunctional gene product, which in many cases is degraded by the cell.

Nonsense suppressor mutants are strains that can prevent this chain termination; they possess an altered cellular activity that allows the placement of an amino acid at the position of the stop codon, thereby permitting completion of synthesis of the polypeptide chain. Although the inserted amino acid is generally not the one present at the same position in the wild-type protein, the completed protein chain usually possesses much of the wild-type activity. In bacteria, yeast, and nematodes (Wills et al., 1983), the nonsense suppressor mutation is altered in a tRNA gene, and specifically in the anti-codon region, the triplet of nucleotides that recognizes and binds to the codon of the mRNA. The mutant tRNA pairs with the given nonsense codon and, in so doing, inserts the amino acid that it carries into the growing polypeptide chain. (The normal tRNA function is not lost, however, because every organism contains several copies of each tRNA gene in its genome, of which only one is altered in the suppressor strain.)

By the criterion of suppressibility by a known nonsense suppressor, several of the postembryonic mutants in the set described by Horvitz and Sulston (1980) are nonsense mutants. Yet these mutants exhibit variable penetrance and expressivity, similarly to the missense mutants. Evidently, phenotypic variability cannot be taken as an indication of mutant leakiness. In some instances, at least, it implies that a given gene product is not essential but can be partially replaced by one or more other gene products. Presumably, the alternative pathways or functions operate near the threshold of efficient substitution, and slight "developmental noise" can shift the result either way. In normal development, this background noise has little or no effect because of the efficient operation of the wild-type gene.

A second preliminary conclusion is that some, perhaps many, of the "postembryonic genes" are also expressed in embryogenesis. We have already noted that a large number of mutants isolated as temperature-sensitive *emb* defectives show postembryonic defects (p. 63), indicating a high degree of overlap. The existence of postembryonic defectives, isolated on the basis of their aberrant development in larval stages, might suggest that there are some genes that are utilized exclusively late. However, some of the genes identified by these mutant hunts are expressed early as well. Direct inspection of several of the mutants reveals embryonic defects; these include the *unc-86* mutants, which show additional dopaminergic neurons produced during embryogenesis, and the nuclear migration defectives (*unc-83* and *unc-84*) that are defective in the embryonic hypoderm at hatching. Furthermore, it is probable that maternal expression of many presumptive postembryonic-specific genes can supply an early embryonic requirement for zygotic genome expression of these genes. All of the extant mutants were isolated following self-fertilization of heterozygous (*m*/+) mothers; the homozygous mutant progeny may have enough gene product stored in the egg to sustain them through embryogenesis. From the genetic data alone, a reliable estimate of the overlap in gene expression between embryonic and postembryonic phases cannot be made, but it is likely to be large.

Indeed, a relatively small number of genetic controls could generate the various types and arrangements of the new somatic cells in postembryonic development. Each such control element might regulate the activity of a small number of structural genes, many or all of which might be active in embryogenesis; the properties of particular postembryonic cells would result from the particular *combination* of these activated gene batteries.

This idea follows from a comparison of the lineage patterns observed in wild-type development and following perturbations produced genetically or by cell ablation. The reiterative mutants of *unc-86* alter specific lineages through the repetition of a particular cell division event; the heterochronics produce their effects through reiteration of large parts of the entire cell division program. All lineages may be visualized, in effect, as consisting of combinations of basic patterns of cell division (Chalfie et al., 1981).

This statement requires amplification. Each cell division event can be viewed as consisting of either of two kinds of cell division: symmetrical (proliferative) divisions and asymmetrical (difference-generating) divisions. The former correspond to the simple cellular command "reproduce yourself without change," while the latter are the source of new cell types in postembryonic development. The basic asymmetrical division pattern itself can be altered in a small number of characteristic ways to generate further diversity. These are shown in Figure 5.10 and can be categorized as follows: (1) an

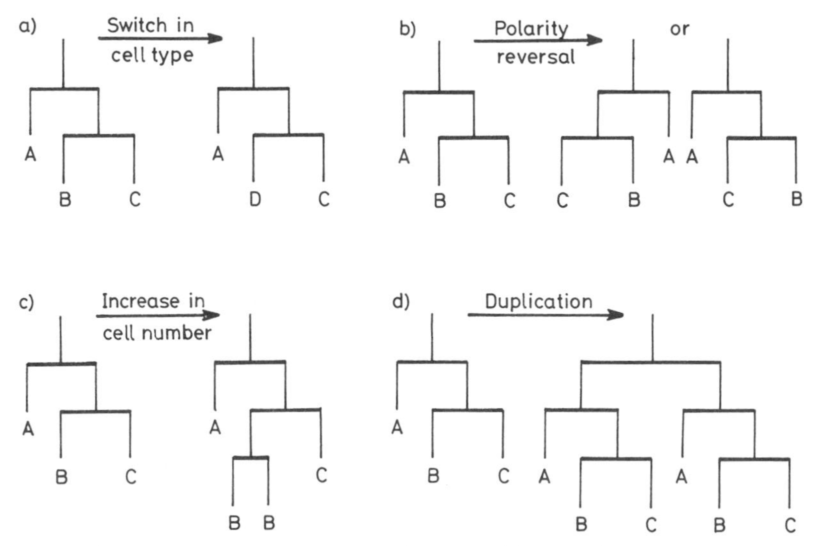

Figure 5.10 Four modes of lineage alteration through modification of the cell division program. A, B, C, and D represent four different cell types arising from a common precursor. See text for details. (Adapted from Kimble, 1981b; reproduced with permission from the Royal Society.)

alteration of cell *type* in a particular division; (2) reversal of *polarity* of a division, with consequent reversal of the lineage; (3) increases or decreases in cell *number* (e.g., insertion of a proliferative division); and (4) divisions that *duplicate* a sublineage through duplication of the precursor cell (Kimble, 1981b). Whenever the genetic programming incorporates one of these modifications of asymmetrical cell division, it may employ the *same* genetic command, irrespective of cell type. Complex lineages in the soma, consisting of sequences of symmetrical and asymmetrical divisions, can be built up through combinations of such simple genetic commands (Kimble, 1981a).

In such a developmental system, each new cell would acquire a unique "labeling" as the result of the particular sequence of commands given in its lineage history. There are many possible combinatorial schemes one can imagine, but the simplest is a binary coding scheme in which each event results in either of two alternatives, symbolizable as "0" and "1" states (Kauffman, 1973).

Assume, for instance, that only four basic cell types need to be distinguished. In principle, combinations of as few as two gene expression differences would be sufficient to differentiate these four types from one another. Letting the two states be represented by 0 and 1, the four gene expression state differences can be written as 00, 01, 10, and 11. The two state differences can be thought of as corresponding to "off" and "on" (without specifying which corresponds to which state of activity), but other molecular meanings might be attached. For the number of different kinds of postembryonic blast cell actually observed (Fig. 5.2), there might be as few as 16 basic cell types, whose distinguishing characteristics could be encoded by as few as four genetic differences in each of two possible states. Each four-letter sign (e.g., 1010) would serve as a group cell designation for all the progeny of that blast cell, while succeeding commands (e.g., "divide asymmetrically") could extend the binary labeling of each progeny cell accordingly through the addition of state differences in successive columns. The particular command for a given division might be an automatic consequence of the initial cell's signature or, more probably, a result of both the intrinsic cell properties and the influences exerted by neighboring cells.

The *lin-12* results fit into this framework comfortably. As described earlier, this gene's activity influences binary choices for those cells capable of alternative pathways, and the level of *lin-12* activity in the responding cells is determined, in part at least, by cell–cell interactions. Whether there are other key switch genes of this kind in *C. elegans* remains to be established.

Differentiation During Development: Does It Involve Somatic Genetic Change?

Until recently, a fundamental assumption about the control of differentiation has been that differentiative changes occur in nuclei without accompanying somatic genetic change. In an early discussion of the significance of the

Jacob–Monod model for eukaryotic development, Grobstein (1963) proposed that the principal significance of the bacterial regulatory models for eukaryotic development was that they permit the creation of stable developmental states without entailing genetic change.

In recent years, however, a number of programmed DNA sequence changes that provoke differentiative change have come to light. One set occurs in lymphocytes and is associated with the sequence of antibody production in these cells (reviewed by Tonegawa, 1983). A second example is that of yeast cells undergoing switches in mating type (reviewed by Herskowitz, 1982). The long-known phenomenon of somatic chromatin diminution in nematodes, which also occurs in certain insects and crustaceans, is a third example. Together with the classic work of McClintock on transposable regulatory elements in maize, these phenomena raise the possibility that directed DNA sequence arrangement may be a general source of differentiative change in eukaryotic development. Although detectable chromatin diminution does not occur in the somatic cells of *C. elegans,* it is of interest to know whether other changes in DNA and gene sequence arrangement occur during embryonic or postembryonic development in this animal.

The question has been approached by means of recombinant DNA techniques. The principle behind the experiment is straightforward: if a long DNA molecule is cut into fragments by a treatment that produces breaks only at specific sequences, then the same sequence that has experienced a prior DNA sequence rearrangement will show an alteration in the sizes of two or more of the fragments, depending on the position and size of the rearrangement and the position of the cutting sites (Fig. 5.11).

This test for rearrangement within the *C. elegans* genome was carried out by Emmons et al. (1979) using a set of 15 cloned DNA segments as the tester

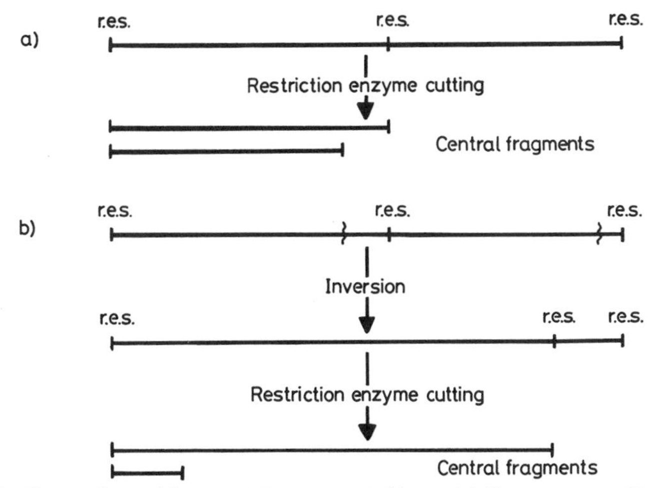

Figure 5.11 Comparison of fragment sizes generated by restriction enzyme cutting before and after a single DNA sequence rearrangement. r.e.s., restriction enzyme site.

sequences for rearrangement. In the first step of the experiment, DNA from L1 animals was digested with an endonucleolytic enzyme that recognizes and cuts at a specific short sequence, the Bam H1 site. This site occurs randomly about once every few thousand base pairs. The Bam H1 digest fragments were then "cloned" in an *Escherichia coli* plasmid, a small, self-replicating molecule found in many bacterial species, by the procedure that will be described more fully in Chapter 9.

To test for genetic rearrangement of chromosomal DNA sequences containing any of these 15 cloned DNA sequences, the recombinant plasmids were denatured and reannealed to the set of DNA fragments produced by restriction enzyme digestion of total genomic DNA from gametes (sperm cells) and genomic DNA from whole animals of several stages. The digest fragments were first electrophoresed through the gel matrix agarose, which separates fragments according to size, and then hybridized to the cloned sequences after transfer to nitrocellulose paper—a suitable medium for the DNA reannealing procedure. Reannealing was followed autoradiographically by the use of radioactively labeled cloned DNA sequences. Reannealed DNA shows up as a band of grains on the developed film, while nonhybridized, radioactive DNA is not retained. The positions of the "hot" bands thus reveal the positions and sizes of the genomic DNA fragments that have extensive homology to the cloned DNA sequence "probes." Rearrangement of sequences homologous to any of the DNA tester sequences should be detectable as a change in the position of a hybridized band.

To determine if rearrangement occurs in development, the pattern of hybridized, digested DNA from purified sperm (germ line DNA) was compared to the patterns of bands produced by digested L1 and adult DNA. The L1 DNA serves as a source of nearly pure somatic DNA following embryogenesis, while adult DNA contains both germ line DNA and somatic postembryonic cell DNA. No differences in band pattern between the three sources of DNA were observed for any of the 15 tester DNA sequences. For most, secondary bands were observed. These bands are caused by the presence of repeated DNA sequences in the cloned tester sequences that hybridize with related sequences in the genome; these secondary bands increase the proportion of the genome sampled.

Thus, rearrangements are not a common cause of developmental change. However, large (>20 kb) and small (<100 bp) rearrangements would not have been detected, nor would rearrangements affecting only a small proportion of the cells. These qualifications noted, it follows that the sources of developmental change probably lie in a regulatory circuitry imposed on a constant genome in this animal.

Analysis of Developmental Pathways: The Use of Epistasis

The development of a given cell type or structure requires not only a particular set of gene products but their expression and action *in a particular temporal sequence*. Furthermore, the ordered temporal expression of the genes required in a developmental pathway is expressed through a sequence of cell behaviors and functions; this cellular sequence is as critical to the attainment of the final result as the underlying gene expression program. The complete sequence of gene expression events and cellular changes constitutes the developmental pathway.

Part of the genetic analysis of development consists of the elucidation of these temporal sequences of gene and cell action within developmental pathways. The genetic procedure for pathway analysis requires establishment of the epistatic relationships between pairs of gene-controlled steps. To determine whether a mutation in one gene is epistatic to a mutation in another, the phenotype of the double mutant is compared to that of each of the single mutants; the epistatic mutant gene is the one whose phenotype the double mutant displays. However, the precise interpretation of the results depends upon one's assumptions about the fundamental biology. Given a reasonable hypothesis, the use of epistatic relationships can be highly informative about the sequence of gene-affected steps.

The simplest gene-controlled pathways are linear biochemical sequences in which each step is controlled by a single gene-encoded enzyme. In such pathways, the double mutant phenotype is that of the mutant that affects the *earlier* step of the pathway. Imagine a simple direct sequence of conversions of the following sort:

$$A \rightarrow B \overset{X}{\not\rightarrow} C \rightarrow D \overset{Y}{\not\rightarrow} E$$

in which each step is carried out by a single enzyme. In the sequence shown, mutant strain X is blocked in the conversion of B to C and mutant strain Y is blocked in the conversion of D to E. The X and Y phenotypes in this case are not distinguished by the absence of end product E, since both strains fail to synthesize E, but by the accumulation of two different precursors, B and D, respectively (or any metabolic products that these precursors are converted to by secondary pathways). In the double mutant strain X,Y, precursor B builds up to give the X phenotype, and the effect of the Y mutation is abolished since precursor D cannot be synthesized in the presence of the X block. The X mutation is therefore epistatic to the Y mutation; in this situation, and without knowing the biochemical specifics of the pathway, it is clear that the X step acts earlier in the pathway than the Y step.

The feasibility of analyzing biochemical pathways by genetic means was first demonstrated in a study of the formation of an eye pigment in *Drosophila*. In the 1930s, George Beadle and Boris Ephrussi performed a series of experiments on eye color mutants of the fruit fly. Their initial approach was to determine which mutant phenotypes could be phenotypically erased in mutant eye tissue by supplying wild-type gene product to mutant eye imaginal discs. The procedure was to place the undifferentiated eye imaginal discs of the various mutants into the bodies of third instar wild-type larvae and to score the eye color in the developed mutant discs after metamorphosis. (The discs differentiate during the pupal period along with their hosts.) For the great majority of mutants, the developed discs showed the eye color phenotype associated with the mutant; these mutants were thus autonomous in their behavior.

However, two mutants were found to be correctable to the wild-type, *vermillion* (v) and *cinnabar* (cn). Both of these mutants produce a bright red eye, reflecting the absence in both of the brown pigments, the ommochromes, found in the wild-type eye. (It is the combined presence of the dark ommochromes and the bright red pigments, the pteridines, that gives the *Drosophila* wild-type eye its characteristic dark red color.) v and cn were clearly nonautonomous mutants; wild-type flies could supply v^+ and cn^+ substance to the respective mutant tissues, allowing the blocked ommochrome pathway to proceed in the eye discs of both mutants. (The diffusible substances are not the enzymes themselves, but rather their metabolic products.) Since the v^+ and cn^+ activities are part of the same biochemical pathway, the question of interest concerns their sequence of action. To determine this sequence, v and cn eye imaginal discs were cross-transplanted between the two mutants and then scored after metamorphosis. cn eye discs transplanted to v larvae remained cn in phenotype. In contrast, v eye discs put into cn hosts became wild-type in color. Clearly, cn mutant hosts can accumulate and supply v^+ substance to genetically v discs. (The v mutants are genetically cn^+ and therefore capable of this reaction.) It follows that the v^+ gene product must act before the cn^+ gene product in wild-type development. The pathway itself can therefore be depicted as follows:

$$\text{precursor} \overset{v \text{ mutant}}{\not\twoheadrightarrow} \mathrm{v}^+ \text{ substance} \overset{cn \text{ mutant}}{\not\twoheadrightarrow} \mathrm{cn}^+ \text{ substance} \rightarrow \text{brown pigment}$$

(where each arrow represents one or more steps). This formal representation has since been verified biochemically, and the intermediates have been identified. Not surprisingly, this work on *Drosophila* eye pigments was highly instrumental in shaping Beadle's thinking about the relationships between genes and enzymes, which led on to the later work with Tatum and the one gene–one enzyme hypothesis.

The application of such logic to the determination of sequences of gene or cell action in development is broadly similar. Double mutants for two mutants that block different steps of the same developmental pathway are constructed, and the mutant phenotype that is epistatic to the other is determined. From the total set of results for many mutant pairs and a hypothesis about the nature of the mutant defects, one can construct a sequence of gene-mediated events for the pathway.

Two particularly interesting developmental pathways in *C. elegans* are the formation of the dauer larva and the events of sex determination. Although both developmental sequences are complex and remain incompletely understood, the use of epistatic relationships has contributed substantially to their elucidation; these analyses are described below.

Dauer Larva Development: The Sensory Perception–Development Link

The dauer larva constitutes a bypath of nematode development. It is a semi-dormant larval form exhibiting greatly reduced metabolic activity and no feeding behavior, and is formed at the end of the second larval stage under conditions of starvation or overcrowding. Entry into the dauer larva state is triggered by a response to the concentration of a small molecular weight, fatty acid-like "pheromone" that is secreted by the animals during growth. When the pheromone concentration is high relative to the food supply, as occurs under conditions of starvation or overcrowding, the animals begin dauer formation (Golden and Riddle, 1982).

The term "dauer larva" is derived from German and means "enduring larva." Lacking the normal pharyngeal pumping of medium that constitutes the nematodes' feeding behavior, the dauer larva does not feed and lies motionless unless prodded. It can persist for greatly extended periods at normal growth temperatures. In some parasitic nematodes, the dauer stage is an obligatory part of the life cycle, but in *Caenorhabditis* it is a developmental option, taken only under adverse conditions and available only at the end of the L2.

The structure of the dauer larva in *C. elegans* has been described by Cassada and Russell (1975). In form, it is slightly longer than the late L2 juvenile but only one-half as wide (Fig. 2.3). Consonant with the loss of feeding activity, the mouth is plugged and there is no open space within the pharynx and intestine. Dauer larvae also exhibit a higher density than L2 larvae, due to the shrinkage of the body that takes place during the developmental sequence. The outer covering is a thickened cuticle, possessing a characteristic striated inner layer not seen in the normal cuticle. This layer is not uniformly thick but grades down and vanishes at the lateral edges of the animal, being replaced there by a fibrillar layer typical of the basal layer of the adult cuticle. In combination with the absence of pharyngeal pumping, this thickened cuticle confers resistance to certain solubilizing agents added to the medium, in particular, the detergent sodium dodecyl sulfate (SDS). Normal worms exposed to 1% SDS for a few minutes quickly lyse, but dauer larvae can survive such exposure for hours.

The normal life cycle is resumed upon exposure to a new food supply. In the presence of food, pharyngeal pumping begins after 1–2 hr and increases in rate over the next few hours; molting takes place after approximately 14 hr at 22°C. During this recovery period, the animals swell to normal width but show no detectable increase in length, presumably due to the physical constraints of the cuticle. Bursting of the cuticle is followed by rapid growth, and after 6 hr, the normal length of L4 juveniles is attained. The recovery period thus replaces the normal L3 stage. Intriguingly, entry into or exit from the dauer state phenotypically circumvents some of the heterochronic mutant defects, restoring the match between cuticle phenotype and developmental stage in several mutants that either give reiterated juvenile cuticles or normally skip the L2 stage (Ambros and Horvitz, 1983).

The isolation of mutants affected in dauer larva formation is the first required step in a genetic analysis of the pathway. Because dauer larva development is a dispensable stage in development, it is possible to isolate mutants that affect dauer formation yet are fully viable. Three classes of mutants might be predicted: those that enter the pathway "constitutively" (under conditions of abundant food and no overcrowding); those that cannot transform to dauers under the standard inducing conditions ("dauer defectives"); and those that cannot readily leave the dauer state ("reversal defectives"). Of the three categories, the first and third are potentially genetic lethals, since animals that enter the dauer state but cannot leave it will not give rise to descendants. However, selection for hypomorphs or temperature sensitives facilitates recovery of these classes. All three categories of mutants have been found, although the distinctions between them are not absolute; many of the dauer constitutives, for instance, are slow or defective reversers.

The strong SDS resistance of the dauer larva permits a selective method for isolating constitutives and reversal defectives (Cassada and Russell, 1975). To isolate dauer constitutives, one exposes well-fed F_2 progeny of a mutagenized population to 1% SDS for a suitable interval (10–20 min) and then plates for survivors. (In normally growing populations, dauers are present at a frequency of 10^{-6}.) Since a survivor will be picked up only if it can subsequently emerge from the dauer state and give rise to a clone, the procedure can be modified to select for temperature-sensitive constituents. The worms are grown at 25°C prior to the SDS treatment and then, following treatment, plated at the permissive temperature (15°C), which allows reversal and production of the clones from the survivors. An analogous approach, using SDS selection, can be employed to isolate slow reversers. Dauer defectives, however, never achieve SDS resistance and can be identified only by direct inspection; they are identified as worms that have not been transformed into dauers upon exposure to normal inducing conditions.

The constitutives and defectives are, in a sense, opposite classes, and one might anticipate the existence of many differences between them. The constitutives clearly possess the capability for forming dauer larvae; they err in switching on the program when it is not called for. The most likely explanation of constitutive behavior is that sensory defects are involved, the animal being told by its nervous system that conditions are worse than they truly are. The dauer defectives, on the other hand, are each missing some essential component of the developmental equipment. For any given defective, the lesion might be part of the sensory apparatus, such that a high pheromone : food ratio cannot be sensed, or the defect might be in any of the components required to build the dauer once the sensory signal is received, such as enzymes required for making the dauer cuticle. The defectives would be expected to be a broader class, embracing a larger variety of defects and involving more gene functions than the constitutives.

This expectation was borne out in an extensive study of dauer formation (*daf*) mutants (Riddle, 1977; Riddle et al., 1981). It was found that mutation to dauer defectiveness occurs at a high rate, about 5%, following standard EMS mutagenesis, while mutation to constitutivity occurs with a frequency of only 0.25%. This difference primarily reflects the different numbers of genes that can mutate to produce these phenotypes. Dauer constitutivity can be produced by mutation in seven known genes, a number that is probably close to the maximum that can mutate to this phenotype. In contrast, dauer defectiveness can probably be produced by mutation in any of about 100 genes.

Both kinds of *daf* genes are found on all linkage groups and show no obvious clustering. One constitutive gene, however, shows hypermutability, and, like the highly mutable *emb* genes (p. 60), is near the middle of the third chromosome linkage group. All of the *daf* mutants are recessive, suggesting that each mutant defect results from a deficiency of the corresponding wild-type gene product.

The functional relationships between these two classes of mutant phenotype are shown by selecting "revertants" of dauer constitutives, by plating the ts constitutives at the restrictive temperature and selecting the survivors. When these strains are analyzed, the great majority are found to have a mutation in a second gene. Isolation of these second site mutations in homozygous form, by crossing out the constitutive mutations, reveals that most produce a dauer-defective phenotype. Furthermore, a number of these new defective mutants are found by complementation to be in genes previously identified as *daf* defectives on the basis of direct isolation of this type. Thus, some of the *daf* defectives are epistatic to and "suppress" some of the *daf* constitutives, showing that the two mutant classes affect a shared sequence of events in dauer formation.

This form of suppression must not be confused with the nonsense suppression described earlier. The latter is a form of "informational suppression," the correction of the translation of genetic information in the mRNA of the affected gene. Nonsense suppression is an *allele-specific* process; the suppressor works on only certain mutant alleles, specific nonsense codons, regardless of the genetic locus affected. Epistatic suppression of the kind described here is a *locus-specific* process, involving an alleviation of a deficiency in a particular gene activity, regardless of the nature of the mutant allele producing this deficiency.

The demonstration that the constitutives and defectives define a common functional pathway permits the delineation of the gene-controlled steps of that pathway by a series of tests for epistasis between gene pairs. The logic is similar to that in the biochemical pathway determinations but contains one critical difference. The simplest explanation of the constitutive response is that it involves a sensory defect, the generation of a false sensing signal in a particular cell type that thereby initiates dauer construction. Therefore, if a particular *daf*-defective mutation were found to be epistatic to the constitutive, the suppressor gene might act by canceling the neuronal instruction of the false signal. Viewed in terms of sensory transmission, the *daf*-defective suppressing mutation is interpreted as acting *after* the initiation of the false signal, presumably by interfering with a cell "downstream" from the signal-initiating cell. Conversely, if no suppression is obtained, the constitutive condition is epistatic to the *daf*-defective state, and this result signifies that the constitutive produces its false signal later in the neuronal transmission sequence than the step mediated by the *daf*-defective gene. Thus, the epistatic gene in this situation is the one that acts *after* the other mutant step, in contrast to the pattern for epistasis in a biochemical transformation pathway.

This difference in epistasis occurs because interference by one mutant gene with the effects of a second can occur in a variety of ways. The particular epistatic pattern observed depends upon the nature of the gene-controlled steps involved. If the developmental pathway consists of a series of substrate or cellular transformations mediated directly by the gene products in question, a deficiency of an early intermediary product will block the synthesis of a later one. However, most developmental pathways are more complex than this. Rather than direct, enzyme-mediated conversions of cellular type, they involve indirectly triggered patterns of cellular change. Such pathways can be seen as sequences of signals or commands rather than as product conversions. If these signals or commands can be uncoupled from their normal sequence, then it is usually only later defective events that count; in this circumstance, it is late blocks that are epistatic to early ones. The situation can be likened to a sequence of switches that must be thrown in order to permit the last crucial switch to be moved. If the latter is blocked, the positions of the earlier switches do not matter. The putative sensory pathway defined by the *daf* mutants has this character.

The pairwise mutant suppression tests are shown in Table 5.2, and translated into the pathway scheme of Figure 5.12, from Riddle et al. (1981). In the figure, the steps of the pathway are defined by the *daf* constitutives arrayed in two parallel paths; the points of blockade by the *daf* defectives are indicated by the dashed vertical lines. The basic premise is that the first gene-mediated constitutive steps will be suppressed by the greatest number of late (*daf*-defective) events and the last constitutive event in the pathway will show suppression by the fewest. The remainder of the *daf* constitutives should form a linear series between these points, the position of each defined by the number and identity of the *daf* defectives that suppress them. With the exception of *daf-2,* the constitutives can be arranged in a simple pattern. *daf-11* and *daf-8* are suppressed by the greatest number of *daf* defectives and

Table 5.2 Epistatic Interactions Between *daf*-Constitutive and *daf*-Defective Mutations[a]

Daf-Defective Mutations	*daf*-Constitutive Mutations						
	daf-11 *V*	*daf-8* *I*	*daf-7* *III*	*daf-14* *IV*	*daf-1* *IV*	*daf-4* *III*	*daf-2* *III*
daf-3 *X*	+	+	+	+	+	+	−
daf-5 *II*	+	+	+	+	+	+	−
daf-12 *X*	+	+	+	+	+	+	−
daf-20 *X*	+	+	+	+	+	−	+
daf-16 *I*	+	+ (60%)	+ (20%)	+ (25%)	+ (35%)	−	+
daf-6 *X*	+ (70%)	+ (30%)	+	+ (35%)	−	−	−
daf-17 *I*	+	+	+	−	−	−	+
daf-18 *IV*	+	+	−	−	−	−	−
daf-10	−	−	−	−	−	Not tested	−

Source. Adapted from Riddle et al. (1981).

[a] Suppression patterns are given or representative alleles of *daf* constitutives and *daf* defectives. + indicates suppression of the constitutive by the defective; − indicates the absence of suppression. In cases where suppression is incomplete, presumably due to leaky expression of the *daf* defective, the result is given in terms of the percentage suppression. All strains are scored as homozygous double mutants. Strains were constructed by means of closely linked visible markers; details are given in Riddle et al. (1981).

therefore mark the earliest constitutive steps of the pathway; *daf-4,* susceptible to suppression by the fewest defectives, defines the pathway's endpoint. The others, except for *daf-2,* form a consistent series inbetween.

An essential point is that frequency of suppression of a constitutive by the various defectives does not reflect degrees of leakiness of particular alleles but is locus specific (Riddle et al., 1981; Swanson and Riddle, 1981). The fact that *daf-2* cannot be fitted into the pattern suggests that its activity defines a second pathway in dauer larva initiation.

Many of the *daf* defectives have neuronal defects and altered chemotactic behavior. Chemotactic behavior is mediated in part by special head neurons,

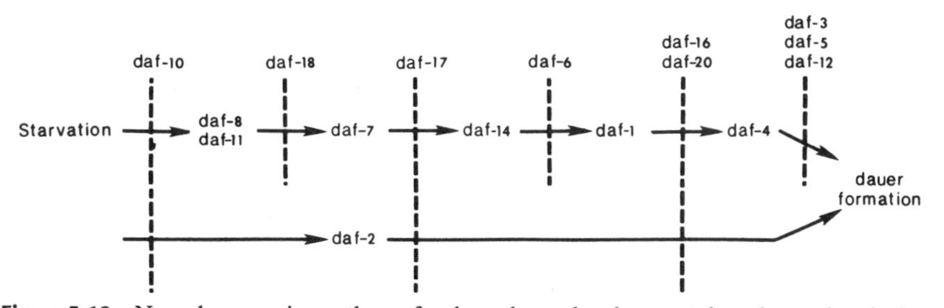

Figure 5.12 Neural processing pathway for dauer larva development, based on epistasis data. See text for details. (Reprinted by permission from *Nature,* vol. 290, p. 671; copyright © 1981, Macmillan Journals Limited; Riddle et al., 1981.)

the amphidial neurons, and in two of the *daf*-defectives these neurons are ultrastructurally abnormal. Such abnormalities, and others not visible as ultrastructural defects, presumably interfere with the registration of the pheromone : food ratio that normally acts as the initiating signal for dauer formation. Thus, it seems probable that the pathway of gene-mediated steps in Figure 5.12 represents a pathway of neuronal transmission affecting the first events in dauer formation, the chemotactic signaling that begins the developmental sequence.

The depicted pathway should not be taken as representing a sequence of gene expressions or gene activities. There is, in fact, little concordance between the order of steps in the pathway as drawn and the temporal order of gene actions suggested by the TSPs for some of the *daf*-constitutive mutants (Swanson and Riddle, 1981).

If the genes in Figure 5.12 are required only to initiate dauer larva formation, there must be other genes whose products are involved in its construction. Several known *daf* defectives (not shown) are probably in this category. A mutant in *daf-13* makes a dauer larva that looks normal but is SDS sensitive. This mutant is probably altered in some property of the cuticle that is essential for SDS resistance. When placed in a constitutive background, this *daf-13* mutation does not alter or suppress the constitutivity but causes the production of SDS-sensitive dauer larvae. Its action and that of all such mutants should be downstream from the Daf-defective steps shown. The final steps of the dauer development pathway can be considered those of dauer reversal. There seem to be few gene functions that are exclusively devoted to this purpose. Selection for slow reversers yields dauer constitutives and hypomorphic dauer defectives that are affected in both entry to and exit from the dauer state. From an analysis of the reversal defectives, it appears that the genes required for reversal are a subset of those required for entry (Riddle et al., 1981).

The analysis illustrates the usefulness of a genetic technique in exploring a complex set of cellular and developmental processes. It also illustrates the links that can exist between developmental and neurological or behavioral phenomena. The development–behavior connection can also run in the other direction. Investigation of a behavioral defect will often raise questions about the developmental origins of that behavior; two examples will be examined at the end of this chapter.

Sex Determination in C. elegans

The mechanism of sex determination is the oldest question in developmental genetics. T. H. Morgan and E. B. Wilson and their colleagues turned to it in the early years of this century and discovered that the sexes in several insect species differ in a particular chromosome or chromosome pair, the sex-distinguishing chromosomes being termed the "sex chromosomes." This finding simultaneously prepared the way for the modern chromosomal theory of inheritance through the discovery of sex linkage by Morgan.

Sex determination entails the sex-specific differentiation of many cell types and regions in the body. These include the germ line cells, the somatic cells surrounding the germ line, the accessory gonadal structures, and a variety of secondary sexual characteristics that are important either for gamete delivery or for sexual recognition. In some animals, the control of sex determination is purely physiological—for instance, in certain fish species in which sexually mature males transform into fertile females under the appropriate environmental stimulus. But for the great majority of animal species, sex determination is genetically controlled by a specific difference in chromosome composition between the two sexes, that of the sex chromosomes. The process involves a set of interactions between the sex chromosomes and a number of genes on the autosomes.

The chromosomal basis of sex difference in *Caenorhabditis* lies in a difference in X chromosome constitution. The hermaphrodite possesses two X chromosomes (the XX state) and the male a single X chromosome (the XO state). However, as shown by Madl and Herman (1979), sexual phenotype in the nematode is determined not by the absolute number of X chromosomes but by the ratio of X chromosomes to the number of autosome sets. Their proof involved manipulating the X/A ratio in polyploid strains and observing the resulting sexual phenotypes, the same approach used by Bridges (1925) to investigate the mechanism of sex determination in *Drosophila*.

The various X/A ratios and their associated sexual phenotypes are shown in Table 5.3. In particular, the observations that 2X;3A and 2X;4A individuals are male, not hermaphrodite, shows that sex is not a function of the absolute number of X chromosomes but of the X/A ratio. A ratio of 1 to 0.75 produces the hermaphrodite state, a ratio of 0.67 to 0.5 the state of maleness. By adding duplications of parts of the X to 2X;3A animals and determining which pieces shifted the phenotype toward femaleness, Madl and Herman

Table 5.3 X/A Ratio and Sex Determination in *Caenorhabditis elegans*

Sexual Phenotype	X Chromosomes	Autosomes	X/A Ratio
Hermaphrodite	2X	2A	1
Hermaphrodite	3X	3A	1
Hermaphrodite	4X	4A	1
Hermaphrodite	3X	4A	0.75
Male	2X	3A	0.67
Male	2X	4A	0.5
Male	1X	2A	0.5

Source. Data summarized from Madl and Herman (1979).

showed that there is no unique X chromosome site that promotes femaleness, but rather a cumulative effect of X chromosomal material in setting the X/A ratio.

The finding that X/A ratio plays a critical role in directing sexual differentiation raises the question of how the ratio is measured. The sensing of the chromosome ratio could take place in either of two ways (Fig. 5.13). In the first, the X chromosomes produce something in a dose-dependent manner that reacts with the autosomes. Low relative doses of X substance(s) turn on male differentiation or turn off female differentiation, while higher doses of the X chromosome gene products do the opposite. In the second mechanism, it is the autosomes that produce something that is titrated by the X chromosomes; in this case, a low X/A ratio would produce a relative saturation of X chromosomes by these autosomal products and provoke male differentiation, while a high ratio would titrate out these substances, triggering female differentiation.

The question of how the X/A ratio is sensed is connected to the equally important one of how X chromosome activities between the two sexes are kept in balance. In general, organisms are sensitive to large shifts in chromosome composition because such changes in gene dosage change the ratios of gene products to one another; if the "genetic imbalance" in chromosomal material is great enough, lethality is usually the result.

The difference in X chromosome composition between the sexes might be expected to pose just such a problem of genetic imbalance. In fact, various ways of equalizing expression of the shared sex chromosome have evolved. *Drosophila* has solved this problem through "dosage compensation," a

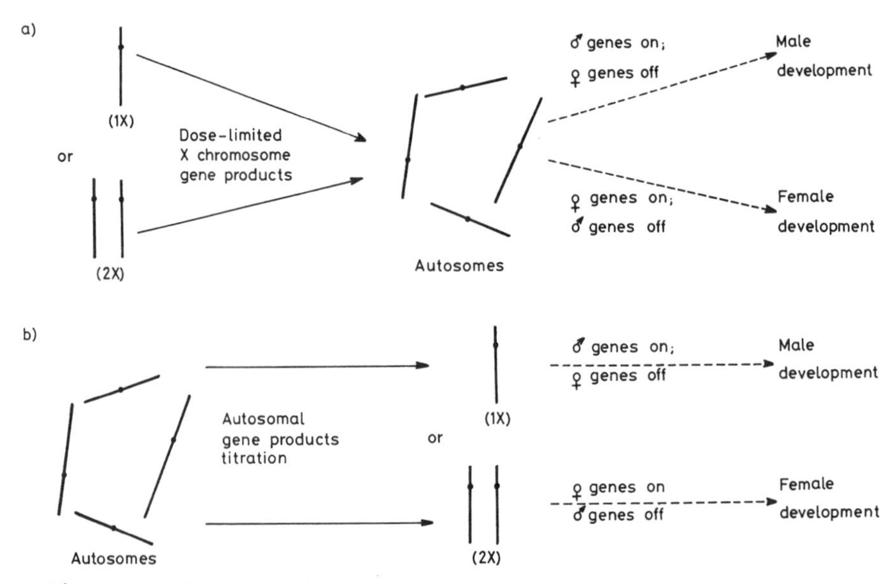

Figure 5.13 Two modes of measuring the cellular X/A ratio. See text for discussion.

mechanism that equalizes total X chromosome gene products between the two sexes by boosting the transcriptional activity of the single X chromosome in males to equal the summed activity of both X chromosomes in the female. Whether dosage compensation exists in the nematode is unknown, but it seems likely.

The phenomenon of dosage compensation is distinct from and potentially even antagonistic to the utilization of the X/A ratio as a sex-determining device. If equal amounts of *all* X chromosome gene products are produced in both sexes, then the sensing of the X/A ratio cannot involve differential titration of autosomes by X gene products but would seem to require the reverse, a titration of the Xs by autosomal gene products. On the other hand, even under dosage compensation, it is possible that one or more special genes on the X are not compensated but differentially expressed in the two sexes, with a concomitant effect on sex determination. *Drosophila* seems to have adopted this stratagem (see the section "Sex Determination" in Chapter 6).

The genes responsible for X/A sensing in *C. elegans* have not been identified, but one gene with an important role in regulating expression of the X chromosome in both sexes has been found. In mutant homozygotes for *dpy-21 V*, expression of the X chromosome is significantly elevated. One result is a partial transformation of XO animals into hermaphrodites (Meneely and Wood, 1984). The simplest explanation of the sex-transformation effect is that the higher level of X expression in XO embryos signals a hermaphrodite-like X/A ratio, producing a partial sexual transformation. The *dpy-21* gene was designated as a *dpy* gene because it mimics one of the effects of X chromosome hyperploidy, the transformation of 3X,2A wild-type animals into Dpy-phenotype animals. This developmental defect presumably reflects the hyperexpression of X chromosome genes. *dpy-21* homozygotes are Dpy only if they have two Xs, that is, in hermaphrodites, but not if they have one X, that is, males (Hodgkin, 1980). However, transformation toward both Dpy and hermaphrodite characteristics occurs in males if additional X chromosome material is added to the genotype (Meneely and Wood, 1984).

A chromosome balance mechanism of sex determination implies the existence of specific genes that are switched on in one sex in response to the X/A ratio and "translate" this ratio into a sex-developmental decision. It should be possible to identify these genes through the isolation of mutants that show the "wrong" sexual phenotype for a given X chromosome constitution. Precedents for such sex-transforming genes exist in both *Drosophila* and mammals.

Sex-transforming mutations in *C. elegans* fall into three categories: transformer (*tra*) mutations, which convert XX individuals into male or masculinized animals; hermaphroditization (*her*) mutants, which do the reverse, converting XO individuals into hermaphrodites; and intersexual (*isx*) mutations, which cause the formation of individuals of altered sexual phenotype in both chromosomal constitutions.

Since the *tra* and *her* mutants present an essentially or completely normal phenotype, a word of explanation about their isolation is required. The *tra* mutants were found as part of a search for individual hermaphrodites that produce a large number of male progeny. (Typically, hermaphrodites produce about 1 male per 500 progeny.) This screening for overproducers of males yielded two kinds of mutants: *him* mutants, which give large numbers of normal (XO) male progeny through increased nondisjunction of the X chromosomes, and *tra* mutants, which produce masculinized XX individuals. The two kinds can be distinguished genetically: hermaphrodites that are heterozygous for *tra* mutations (*tra*/+) produce 25% male (*tra*/*tra*) progeny through normal Mendelian segregation (Hodgkin and Brenner, 1977). The *him* and *tra* mutants can be distinguished further by tests based on the presence of two X chromosomes in the converted male progeny of the latter.

To isolate *her* mutants, which transform XO males into XO hermaphrodites, the expression properties of the *dpy-21* gene were employed. As noted above, hermaphrodite (XX) *dpy-21* homozygotes are Dpy, but male (XO) homozygotes are normal. A *dpy-21* strain that contains the *him-5* mutation (which increases the production of XO individuals) normally segregates just Dpy (XX) hermaphrodites and wild-type (XO) males. Any mutation that produces hermaphroditization of XO animals should lead to the production of wild-type XO hermaphrodite progeny (distinguishable from males under the dissecting microscope by their very different tails). Using this screening procedure, Hodgkin (1980) isolated eight allelic, recessive *her-1* mutants and a single dominant *her-2* mutant.

The various mutant classes are listed in Table 5.4. The single *isx* mutant was isolated fortuitously as a temperature-sensitive sterile; at the restrictive temperature, it produces spermless XX hermaphrodites (females, effectively) and converts XO individuals into partially feminized animals (Nelson et al., 1978). This mutant may be regarded as a "feminization" mutant rather than an intersexual one, since its direction of transformation is toward more female properties in both sexes. The dominant *her-2* is also a feminization mutant, transforming both XX and XO animals toward more female characteristics.

The extent of mutant conversion varies with the strength of the mutant alleles and with the individual loci involved. The strongest alleles of *tra-2* and *tra-3* produce an incomplete transformation, yielding males with only partially transformed tail structures and no mating behavior. In contrast, the strongest mutants of *her* and *tra-1* can be completely transformed in their somatic architecture, although *tra-1* mutants show a less than complete gonadal transformation and produce fewer sperm. All the mutants, with the exception of *her-2*, are recessive, suggesting that the transformations reflect the absence of particular wild-type gene products.

The development of a sexual phenotype is the result of two sequential processes, the initial sex determination decision and the execution of this decision in the differentiation of sex-specific structures. There are two rea-

Table 5.4 Sex Determination Mutants and Their Phenotypes in *Caenorhabditis elegans*[a]

Genotype	Linkage Group	XX Phenotype	XO Phenotype
Wild-type		Hermaphrodite	Male
isx-1	*IV*	Female (at 25°C)	Incomplete male (at 25°C)
tra-1	*III*	Male (gonadal defects)	Male
tra-2	*II*	Incomplete male	Male
tra-3[b]	*IV*	Incomplete male	Male
her-1	*V*	Hermaphrodite	Hermaphrodite
her-2	*III*	Female	Female
her-2/+		Hermaphrodite/female	Hermaphrodite/female
tra-1; tra-2		Male	Male
tra-1; tra-3		Male	Male
tra-2; tra-3		Incomplete male	Male
tra-1; her-1		Male	Male
tra-2; her-1		Incomplete male	Male
tra-3; her-1		Incomplete male	Male
tra-2; her-2		Hermaphrodite/female	Hermaphrodite/female
tra-3; her-2/+		Hermaphrodite/female	Hermaphrodite/female

Source. Adapted from Hodgkin (1980). Data for *tra* and *her* mutations are from Hodgkin and Brenner (1977) and Hodgkin (1980); data for *isx-1* are from Nelson et al. (1978).

[a] Results for *tra* mutations are shown for only the strongest mutant alleles. The single-allele notation in the first column designates a homozygous genotype for the given mutant gene.

[b] Maternal as well as zygotic effects.

sons for classifying the mutants described above as sex determination rather than sex differentiation mutants. The first is that two ts mutant alleles (*isx-1* and one of the *tra-2* alleles) show TSPs well in advance of visible sexual differentiation (Klass et al., 1976; Nelson et al., 1978). The temporal separation between TSP and differentiation is particularly striking for the *tra-2* mutant; the TSP ends at approximately 12 hr after hatching, long before the male and hermaphrodite gonads can be histologically distinguished but concurrently with the first divisions of the gonadal somatic primordium and the establishment of the decisive difference in progenitor cell position between the two gonads.

The second indication that the mutants are altered in sex determination gene functions is that the wild-type alleles of the *tra* genes and of *her-1* appear to be unexpressed in wild-type males and hermaphrodites, respectively. If these genes were involved in sex differentiation, then one would predict that the wild-type *tra* genes would be required for male differentiation and the wild-type *her-1* allele for hermaphrodite differentiation. In fact, XO males homozygous for mutations in any of the three *tra* genes are fully normal males and *her-1* hermaphrodites are fully normal hermaphrodites. If the wild-type alleles for these genes were important for normal differentiation, their dispensability in the sex whose differentiation they provoke (in the opposite chromosomal sex) would be difficult to explain.

When tested for their expression relationships, the mutants reveal a clear hierarchy of epistasis (lower half of Table 5.4). The existence of epistasis shows that the genes are involved in a common pathway of sex determination. However, the observations can be interpreted in different ways. In particular, the data on the order of action of the *tra* genes have been viewed in two different fashions. This difference stems from the ambiguity inherent in the term "pathway" as applied to development, and illustrates the important disparity between the view of a pathway as a sequence of substrate transformations versus a sequence of cellular "decisions."

The substrate transformation view of sex determination in the nematode stems from the initial observations that the *tra* mutations are without phenotypic effect in males. One possible interpretation is that gonadal differentiation in the nematode has a natural tendency toward maleness in the absence of sex-specific counter signals. In animals, it appears to be a general rule that the gonads in both sexes are initially "indifferent," with a tendency to develop as those of one sex unless instructed by physiological signals to develop into those of the other. The inherent developmental tendency is said to be that of the "neutral" sex and the imposed developmental tendency that of the dominant sex. If mature gonads of one sex are transplanted into an immature individual of the opposite sex, the gonad type that enforces its own differentiative pattern is that of the dominant sex and the one that responds is the neutral sex (McCarrey and Abbott, 1979). In mammals, the neutral sex is the female and the dominant sex is the male; in birds, the situation is reversed (although in both vertebrate groups, it is the heterogametic sex that is dominant). From the *tra* mutant effects, it can be argued that the male sex in *C. elegans* is the neutral sex and the hermaphrodite the dominant sex.

Based on this assumption of male neutrality and the observation that *tra-1* is epistatic to *tra-2* (shown by the complete *tra-1*-type transformation in the double mutant; Table 5.3), Klass et al. (1976) proposed that gonad development in XX individuals is toward maleness unless actively directed to differentiate toward the hermaphrodite state by the intervention of the *tra-1* and *tra-2* genes. Their proposed scheme of hermaphrodite development can be depicted as follows:

$$\underset{\text{(male character)}}{A} \xrightarrow{tra\text{-}1^{+}} B' \xrightarrow{tra\text{-}2^{+}} \underset{\text{(hermaphrodite character)}}{C'}$$

In this model, the absence of *tra-1*$^{+}$ leads to the A (male) state and a complete absence of hermaphrodite development, as exhibited by the *tra-1* mutants; failure of *tra-2*$^{+}$ leads to the B′ state, a masculinized version of the hermaphrodite—the incomplete or "pseudomales" produced by *tra-2* and *tra-3* transformation. The model treats the process of hermaphrodite development as a sequence of substrate transformations in which an early block (*tra-1*) completely abolishes the sequence and a later block permits partial expression of pathway character. Nelson et al. (1978) have offered a similar interpretation and, based on the observed partial epistasis of *isx-1* to *tra-1*, have suggested the following sequence of gene action: *isx-1* → *tra-1* → *tra-2*.

Hodgkin (1980) has constructed a different sequence and a different model of sex determination based on the alternative view of a developmental pathway as a series of determinative cellular decisions. From this viewpoint, an alteration in a later step in the sequence will always have the last word relative to an earlier step. The genetic consequence is that later steps are epistatic to earlier steps. Inspection of the results in the table shows that *tra-1*, *tra-2*, and *tra-3* are all epistatic to *her-1* and that *tra-1* is epistatic to both *tra-2* and *tra-3*. Under the view that late steps are epistatic to early steps, this leads to the sequence of gene-mediated actions:

$$\textit{her-1} \rightarrow \left.\begin{matrix}\textit{tra-2}\\ \textit{tra-3}\end{matrix}\right\} \rightarrow \textit{tra-1}$$

(where the relative order of *tra-2* and *tra-3* is unspecified).

What is the position of *her-2* in the sequence? *her-2* is epistatic to *tra-2* and *tra-3*, indicating that, like *tra-1*, it acts after these steps. Possible clues to the nature of *her-2* are that it is dominant, unlike the other sex transformation mutants, and that it cannot be separated by genetic recombination from the *tra-1* locus (Hodgkin, 1980). The last fact suggests that *her-2* and *tra-1* mutants may be allelic states of the same gene with opposite effects. If this conclusion is correct, then the *tra-1* phenotype corresponds to an amorphic or hypomorphic condition and the *her-2* state to an excess of hypermorphic condition. The *tra-1/her-2* gene, positioned at the end of the pathway, thus becomes the final and critical sexual state determiner, with the developmental outcome depending on whether or how much of its gene product is made. In this view, the early steps function only as indirect setters of *tra-1* activity. Indeed, segregation of a *tra-1* null and *her-2* in females (either XX or XO) mated to *tra-1* null homozygous males (either XX or XO) produces a self-perpetuating bisexual stock, consisting of males and females, in which X chromosome composition is irrelevant (Hodgkin, 1983b).

Hodgkin's model is based on these deductions and the observed epistatic relationships (Fig. 5.14). The model shows the sex determination pathway as a sequence of gene-mediated determinative decisions. In this scheme, the X/A ratio initiates the sequence, with a high ratio (as in wild-type hermaphrodites) turning off the first gene in the sequence, *her-1*. (As noted above, *her-1* hermaphrodites are phenotypically normal, indicating that this gene is off in this sex.) When *her-1* is off, $tra\text{-}2^{+}$ and $tra\text{-}3^{+}$ are derepressed (these

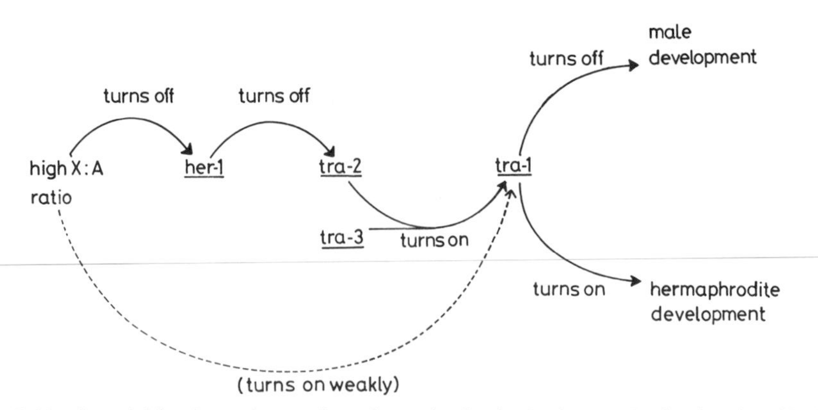

Figure 5.14 A model for the pathway of sex determination in *C. elegans*. Derivation described in text. (Adapted and reproduced with permission from Hodgkin, 1980.)

two genes being required for normal hermaphrodite development) and turn on *tra-1* (also required for hermaphrodite development, inessential in males). When on, this gene activity represses or antagonizes the neutral tendency to male development and actively promotes female development. The opposite starting signal, a low X/A ratio, leads to the reverse sequence of events, with consequent repression of *tra-1* and the promotion of male development. In this model, the *her-2* mutant is a constitutive; it does not require the action of *tra-2* and *tra-3* to be turned on and therefore automatically promotes female development.

Figure 5.14 also depicts a weak subsidiary pathway that may have no function in normal development. This pathway is independent of *tra-2*$^+$ and *tra-3*$^+$ activity but depends on a low level of *tra-1*$^+$ activity, perhaps induced directly by the X/A ratio signal. The existence of a subsidiary pathway was postulated by Hodgkin because the most extreme *tra-2* and *tra-3* mutants, including nonsense mutants, are only partially transformed into males; this partial transformation is dependent on *tra-1* activity, as shown by the epistasis results. The results suggest that some induction of the female phenotype, dependent on *tra-1,* can be achieved without *tra-2* and *tra-3* activity. Furthermore, boosting X chromosome expression (through the use of the *dpy-21* defect) can partially alleviate the male-transforming effects of a weak *tra-1* allele (Meneely and Wood, 1984), reemphasizing the importance of levels of *tra-1* expression as the determiner of sexual phenotype. Whether this *dpy-21*-mediated suppression operates through the normal intermediary genes or directly through the subsidiary pathway has not been established.

Genetic evidence alone cannot be used to decide between competing models derived from genetic analysis. However, the sequential determinative decision model shown in Figure 5.14 is more attractive than the substrate transformation model described previously. It is based on more complete data and accounts for male neutrality in a simpler and more realistic fashion. However, the scheme depicted in Figure 5.14 is incomplete. It does not incorporate the role of *isx-1*, and it makes no prediction as to when or where the sex determination events take place. The early TSPs of *isx-1* and the ts *tra-2* suggest that the sex determination genes act early, but more evidence is needed. In addition, it is unknown which cells these "decisions" are made in, whether they are required in every cell capable of sexually dimorphic fates or just certain cells. The development of methods of clonal analysis (see the section "Questions and Prospects") might help to answer these questions.

From Behavior to Development: Rol and Mec Mutants

The analysis of dauer larva formation illustrates the way a developmental problem can lead to questions about sensory function or behavior. Conversely, a genetically based behavioral anomaly can lead to an inquiry about its developmental basis. Much of the work on the genetic basis of various *Drosophila* behaviors has had this consequence (Hotta and Benzer, 1973). Similarly, researchers on nematode behaviors often become concerned with their cellular and developmental bases.

Some of the genetically based behavioral syndromes can be produced by cellular abnormalities in many different cell types. A prime example is the Unc phenotype, which results from any of a large number of serious malfunctions in the neuromuscular system. Not surprisingly, many genes can mutate to give the Unc phenotype; at present, more than 100 are known. Unc mutant defects include defects in the production of neural and muscle lineages, the synthesis and breakdown of neurotransmitters, muscle structural genes, and various aspects of neural cell differentiation.

Other behavioral abnormalities involve more specific defects, which sometimes involve only a few cells or cell types. Two examples are the roller (Rol) phenotype and the touch-insensitive (Mec) defect. The developmental genetics of these two behavioral abnormalities are discussed below.

The Roller Syndrome and Its Mutants

The Rol mutants are distinguished by a peculiarity of movement, as their name implies. Instead of the nearly linear path segments that wild-type worms trace while moving through the bacterial lawn, Rol mutants transcribe circular or near-circular craters. This feature makes isolation of rollers easy: one simply inspects the F_2 generation of a mutagenized population for those individuals that make circular craters.

Wild-type worms move on their sides in a sinusoidal motion in the d-v plane. Traction against the medium is provided by a three-ridged tread along either side of the body. These "alae" are cuticular specializations secreted by the hypodermis. Body movement results from rhythmic contractions of the longitudinal muscles working against the cuticle, which acts as a flexible external skeleton.

Rol mutants differ from wild-type both in the shape of their paths and in slowly rotating about the long axis of the body as they move. The relationship of this movement to cuticle structure was first explored by Higgins and Hirsh (1977). In isolating Rol mutants, they found two distinguishable types: left rollers (LR), who make counterclockwise circles, and right rollers (RR), who make clockwise circles. A key fact about these mutants is that their cuticles and inner longitudinal organs show matching helical twists of the same handedness. The left rollers display left-handed, helically twisted alae and neural cords instead of the linear alae and neural cords; right rollers show right-handed helices in their alae and internal organs. The helical path of the alae explains the rolling motion of the animal as it moves forward: gliding along the twisted alae automatically produces a rotation around the long axis of the body.

In a more extensive collection and classification of 88 Rol mutants, Cox et al. (1980) identified five distinct classes consisting of mutants in 14 distinct and nonclustered genes. Left rollers are all recessive and map to four genes. Left dumpy rollers (LDR), also recessive, are produced by mutation in any of six genes; the mutants are both Dpy and Rol in phenotype (but some alleles produce just the Dpy phenotype and others just the Rol phenotype). Right rollers, recessive mutations, all map to a single gene (*rol-6*). "Left squat" (LS) mutants, produced by mutation in a single gene (*sqt-3*), are semidominant, displaying left rolling in heterozygotes and a mild Dpy phenotype in juveniles. Finally, "right squat" (RS) mutants are produced by mutation in either of two genes (*sqt-1* and *sqt-2*) and exhibit right rolling in heterozygotes and a mild Dpy phenotype.

The mutants are all zygotic rather than maternal in character, and the behavioral phenotype is always expressed late in development, not in the early larval stages. The only major developmental events that occur late and that therefore might be the source of the defects are the completion of the gonad and the final larval molts. However, the gonad is unlikely to be the source of an imposed helicity since it effectively floats within the body. Cuticle formation is therefore the probable site of the genesis of the behavioral phenotype. If the cuticle is produced with a helical twist in it, then the muscles and neural cords, which appear to be anchored to the cuticle, will assume a corresponding helical twist.

The nature of the Rol defects can be assessed by comparison with the structure of the wild-type adult cuticle (Fig. 5.15). The wild-type cuticle can be seen to have an intricate ultrastructure consisting of three major layers. The basal layer itself consists of two layers of fibers that run in opposite helical directions around the body of the animal.

All of the mutant classes show visible abnormalities in one or more cuticular characteristics. In the LR and RR classes, the primary visible alteration is in the helicity of their alae, with little disruption of the internal cuticular

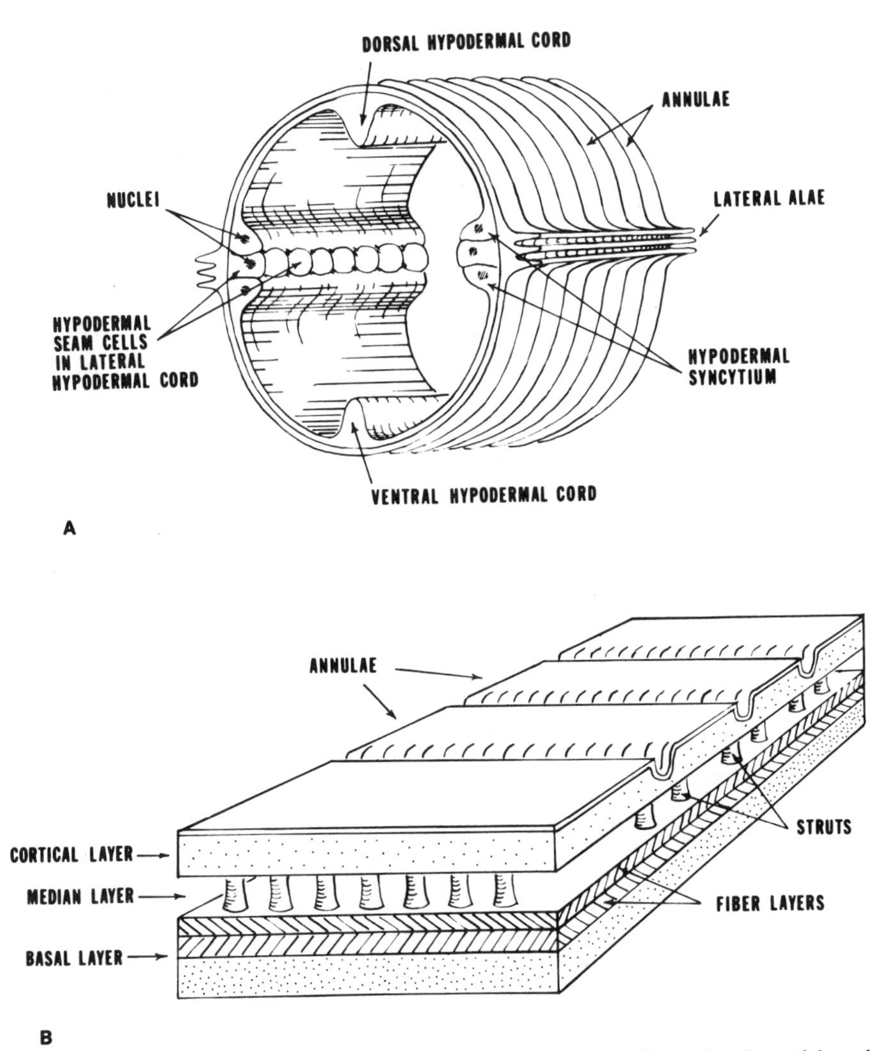

Figure 5.15 Cuticle structure in *C. elegans*. (*a*) A transverse section illustrating the cuticle and underlying hypodermis of the adult worm; (*b*) schematic of the ultrastructure of the adult nematode cuticle. (Figure kindly provided by Dr. G. Cox; reproduced with permission from Cox et al., 1980.)

architecture. The LDR, LS, and RS mutants, however, show various abnormalities of cuticle ultrastructure and various degrees of altered helicity.

The late development of the Rol phenotype in all mutants presumably reflects differences between the early larval cuticles and those of the L4 and adult. As noted earlier in connection with the discussion of the heterochronic mutants, passage through the four larval stages into the adult form is accompanied by specific changes in cuticle ultrastructure and composition. The details of these changes are described in Cox et al. (1981a,b). The cuticular defects associated with the Rol phenotype presumably derive from the aberrant development of features specific to the late stages.

The principal questions about Rol mutants concern the developmental origins of these cuticular defects. A probable site for at least the LR and RR mutants is the basal layers, which display counterhelical paths in the wild-type cuticle. A change in the orientation of one layer might produce an overall helical twist in one direction; a change in the construction of the second basal layer could produce the opposite change in helical twist. Aberrant placement of the topmost basal layer, to which the struts attach in the adult cuticle, might then transmit the imposed helicity "upward," resulting in the observable twist of the alae (Cox et al., 1980). A similar twisting could also be produced by an alteration in the deformability of either of the basal layers. Such an alteration could result from altered thickness in the degree of cross-linking or from the absence or alteration of a key cuticular constituent. Ultimately, the defects presumably all arise either from changes in cuticular composition or from changes in the cell biology of the hypodermis, which produce secondary effects on cuticular composition and structure. Further biochemical analysis of the constituents and of the hypodermis in the mutants should help to resolve this question.

The Mec Mutants: Touch Sensitivity and the Mechanosensory Neurons

If the head or tail of the worm is stroked with a fine hair, the animal moves in response. A gentle touch on the anterior half produces a backward movement; a stroking on the tail, a forward movement. The cellular and genetic basis of this touch sensitivity has been described by Chalfie and Sulston (1981).

Using laser microbeam cell ablation, they showed that the response is mediated by six distinctive mechanosensory cells located in defined positions just underneath the cuticle. Anatomically, the cells are grouped into an anterior and a posterior set of three each. The three most anterior cells and a posterior ventral cell mediate responsiveness to anterior stimulation, and the two most posterior cells mediate posterior responsiveness.

None of these cells have synaptic input from other neurons but seem to function solely as mechanoreceptors for tactile contact between external objects and the cuticle. All of these cells, and only these cells, are distinguished by bundles of darkly staining microtubules in their neuronal processes. These processes extend forward from the cell bodies, and one of the anterior cell group's connections extends to the nerve ring in the head.

Specific touch-insensitive mutants can be isolated by testing the progeny of mutagenized animals for the inability to respond to touch with a fine hair and eliminating general movement defectives from the set of nonresponders. (The movement defectives were identified by their inability to respond to prodding with a platinum wire; touch insensitives do respond to such prodding but not to gentle stroking.) Altogether, Chalfie and Sulston described 42 mutants, consisting of 39 recessives and 3 dominants. The mutants defined

12 new complementation groups plus *unc-86*. The new genes were designated *mec-1* through *mec-12*. (The report also describes a *mec-13* gene, but the defining mutation subsequently proved to be a dominant *mec-4* mutation.) The genes are not clustered in the genome and occur in all linkage groups. With the exception of one mutant allele of one gene, all mutants show touch insensitivity throughout the animal.

Mutant phenotypes for 4 of the 12 genes were found to be associated with visible cellular phenotypes, ranging from ultrastructural defects in the microtubule cells to the complete absence of these cells from the animal. One of the mutant phenotypes, seen in mutants of *mec-7*, was the absence of the distinctive microtubules of these cells. A second ultrastructural defect, associated with mutants of *mec-1*, involved the absence of a characteristic darkly staining "mantle" lying on one side of the processes of these cells. Mutants of a third gene, *mec-3*, seem to be affected in the maturation of these cells, causing a failure of neuronal process extension.

The last gene whose mutants produce a visible cytological defect is *unc-86;* the mutants lack all of the microtubule-containing mechanosensory cells. As we have seen, *unc-86* mutants cause a postembryonic reiterative cell lineage defect, with the resultant deletion of certain neurons from the Q lineages and the multiplication of other neurons in this lineage. Two of the neurons deleted in *unc-86* are the ventral microtubule cells, which develop postembryonically in the wild-type. The complete absence of the six mechanosensory cells from the mutant shows that the mutational lesion causes a corresponding defect in the four embryonic lineages that give rise to the remaining four cells.

Although the remaining *mec* genes do not give a visible phenotype, they seem to be affected in the specific functioning of the microtubule cells, since they are not pleiotropic. Their defects may be in specific biochemical properties of neuronal transmission. None of the mutants were altered in the *direction* of their neuronal processes, indicating that this property is dictated by the anatomical environment of the cells rather than by intrinsic and specific gene activities.

Altogether, the set of identified *mec* genes probably comes close to saturating the gene set required for the specific development and operation of the microtubule cells. Many other genetic functions are required for the development of these cells, but these genes must also be expressed and required in other cell types as well. The construction of double mutants for *mec* gene functions affecting these cells and the evaluation of the epistatic relationships might help to elucidate the pathway of development of these cells.

QUESTIONS AND PROSPECTS

Genetic analysis has proven to be a precise and informative approach to the understanding of developmental events in *C. elegans*. At the same time, it has helped to refine the questions that may be asked about the genetic control of these events. Although the questions are all related, it may help to group them in terms of category of phenomena. Within such a categorization, one can list the approaches that need to be developed for further progress in analysis and understanding.

Mechanisms of Cell Lineage Assignment. In the nematode, the fates of most cells are strictly specified in terms of their origins. Even for those cells that exhibit alternative fates, the choices are prescribed by their lineage histories. At the cellular level, one of the fundamental questions concerns the nature of the relationship between division geometry and subsequent daughter cell fate. Does asymmetrical cell division automatically produce divergent cell fates, or is the asymmetry a function of a preset divergence? The answer may emerge from a study of those mutants that alter both cell division geometry and fate as the molecular nature of the lesions is elucidated.

A necessary element of this exploration will be the development of additional biochemical markers to facilitate tracing cell divergences in terms of differential gene expression. These markers will include antigenic ones, for detecting cell surface components, and intracellular histochemical ones, to detect internal cytoplasmic macromolecules. Ultimately, understanding the mechanisms of cell fate specification will require knowledge of the chromosomal state changes that such specification must involve. Given the small quantities of cells that are available, this kind of description is a formidable task.

There is also the intriguing matter of the "combinatorial logic" of cell fate specification. If some form of modular binary coding of events does take place in postembryonic or embryonic development, what does each binary decision involve and how many such alternative choices can suffice to specify the complete animal?

The role of cell–cell interaction is not negligible in cell fate specification, but it is less central than lineage as a determiner of fate. In the case of *lin-12*, it appears probable that cell–cell interactions influence a binary choice of fates through differential settings of the expression of this gene. Are there other key switch genes of this character whose activities are influenced by cell–cell interactions?

Times and Cellular Locations of Key Gene Activities. The cytological analyses of mutant phenotypes show which cell types are affected in particular mutants but do not prove that the cellular defects arise autonomously within the visibly affected cells. Some of the defects might arise from lack of wild-type expression in other cells, thereby producing secondary effects on the visibly affected cells. Furthermore, it is often not clear precisely when the wild-type gene activities are required to prevent the observed defects.

In principle, clonal analysis can answer these questions. However, while induced mitotic recombination does occur in *C. elegans* (Siddiqui and Babu, 1980), its occurrence is too infrequent to permit analysis of the times and sites of gene expression. An alternative approach that may provide a form of clonal analysis has been described by Herman (1984). It relies on the fact that certain free duplications—pieces of chromosomes not attached to other linkage groups—are lost with measurable frequency during development. By coupling the wild-type gene of interest with a genetic cellular marker on such a duplication, it is possible (when the normal homologues bear mutant alleles for these genes) to discern the distribution of cells that have lost the duplication. By comparing the location of mutant patches with the expression or nonexpression of the mutant phenes, the cellular locus of action for the gene can be determined. The primary limitation at present lies in the limited set of cell-specific markers that presently allow unambiguous tracing of the duplications. Certain biochemical markers that produce neuron-specific staining exist, such as the dopaminergic neurons, but additional cell-specific markers or reagents need to be developed. An example of a problem in which clonal analysis is needed is that of sex determination. To understand the action of the various sex-determining loci, one prerequisite is to determine in which cells and at what times these genes must be expressed for development of the appropriate sexual phenotype.

Patterns and Molecular Biology of Gene Expression. The overall dynamics and patterns of gene expression in nematode development are only very sketchily known. Molecular approaches to measure these patterns are needed. One such method will probably be in situ hybridization with cloned genes to measure cellular transcript levels. The development of this and other methods will help to resolve the nature of the quantitative and qualitative differences in gene expression between embryogenesis and postembryonic development.

To elucidate the molecular biology of gene control, it will be necessary to identify particular gene products with the activities of particular key genes. However, searches for particular proteins associated with specific genes are generally unlikely to succeed, except for very abundant gene products such as muscle proteins. Even the very abundant gene products do not vary much during development. The gross profile of protein synthesis, as determined by means of two-dimensional separations of pulse-labeled polypeptides, remains nearly constant throughout normal postembryonic development (Johnson and Hirsh, 1979). Even when differences are discerned, as for instance in the protein patterns of a null mutant with and without nonsense suppression, it is difficult to discriminate primary from secondary effects. The development and application of sensitive in situ detection techniques, in particular the use of monoclonal antibodies, may make a substantial contribution in correlating genes with particular cellular components, although the problem of identifying primary defects will remain.

A different approach to the identification of developmentally significant molecules is that of bioassay: one injects extracts from wild-type animals into the oocytes or syncytial ovaries of mutant hermaphrodites and scores for rescue of the mutant defect in the progeny. In principle, any gene product that persists in the injected embryos can be assayed and ultimately identified through progressive purification by these means. In *C. elegans,* the potential usefulness of bioassay to rescue mutant phenotypes has been demonstrated by the successful suppression of a nonsense *tra* mutant through ovarian injection of suppressor tRNA (Kimble et al., 1982).

The final area that should contribute to an understanding of genetic control is that of gene analysis through recombinant DNA techniques (Karn et al., 1980; MacLeod et al., 1981). As more genes are purified and their flanking sequences analyzed, the common sequence features of coordinately regulated gene batteries should become apparent. This information, in turn, should provide clues in identifying the control mechanisms that regulate gene expression.

SIX

DROSOPHILA MELANOGASTER

EMBRYOGENESIS AND LARVAL DEVELOPMENT

THE DEVELOPMENTAL SEQUENCE SUMMARIZED

At the end of cleavage, the blastoderm stage *Drosophila* embryo consists of a single layer of cells surrounding a yolky center. The 6000 columnar cells of the blastoderm show little ultrastructural differentiation from one another, yet possess distinctive regional fates (Chapter 3). Embryogenesis converts this monolayer of cells into the segmented, differentiated first instar larva, which emerges from the egg case only 20 hr after blastoderm formation. Topologically, the problem consists of transforming a hollow ball of cells into a series of tubes within tubes. As in *Caenorhabditis,* the first steps involve multiple invaginations that remove cells from the surface and place them in characteristic positions within the embryo. Each of these invaginations alters the topology of the embryo and contributes to the formation of the three germ layers and those structures that arise within them.

The first invagination occurs almost immediately after blastoderm formation, at about 3.5 hr of development, and takes place along the midventral line (Fig. 6.1*a*); this invagination gives rise to the larval mesodermal tissue. Beginning about one-sixth of the way from the anterior end and extending nearly to the posterior end, a block of cells about 50 long and 20 wide, about one-sixth of the total surface moves inward to form a hollow tube just inside the midventral line. The sides of the tube soon become apposed, obliterating the lumen, and the closed tube flattens into a layer that gradually spreads laterally and dorsally on both sides of the embryo. This process of mesodermal spreading is not complete until the 11th or 12th hour of development. By the 7th hour, the mesoderm has become divided into an inner layer, the splanchnopleure, which eventually gives rise to the visceral musculature

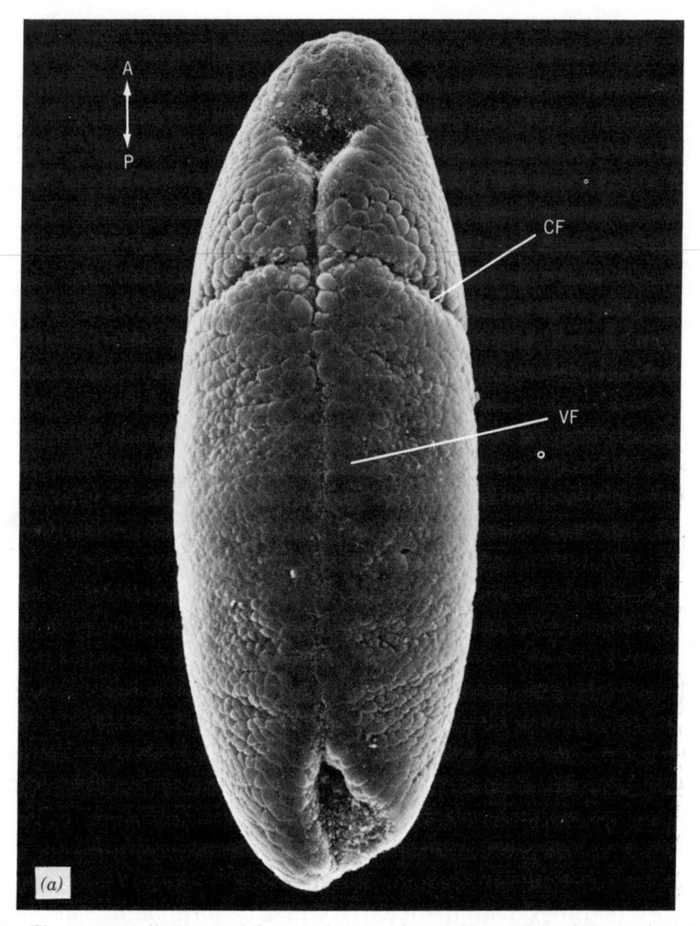

Figure 6.1 Three stages in *Drosophila* morphogenesis. (*a*) Formation of the ventral furrow; (*b*) extended germ band stage; (*c*) germ band shortening. AMG, anterior midgut invagination; CF, cephalic furrow; VF, ventral furrow; A, anterior; P, posterior; Hy, hypopharynx; Mn, mandibullary segment; Mx, maxillary segment; La, labial segment; T1–3, thoracic segments; AB1–8, abdominal segments; AS, amnioserosa. (Photographs kindly provided by Dr. F.R. Turner.)

surrounding the gut, and an outer layer, the somatopleure, which gives rise to all of the other mesodermal structures (the body musculature, the circulatory system, and the larval fat bodies). The midventral invagination thus lays the foundation for the development of one set of longitudinal tubular elements within the developing embryo.

The nervous system constitutes a second set of longitudinally distributed elements in the body. It arises from neuroblasts, detectable by their large size, flask-like shape, and basophilic cytoplasm. First observed in the future brain region, within the dorsolateral blastodermal layer, they become apparent along either side of the midventral invagination within an hour after

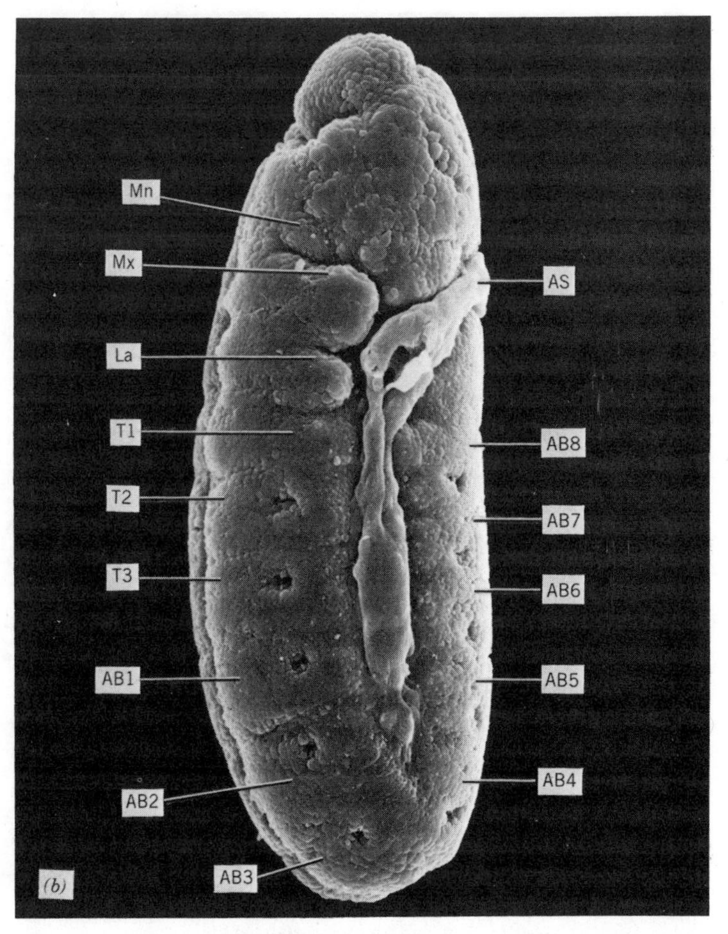

Figure 6.1 (*Continued*)

blastoderm formation. These progenitor cells always divide asymmetrically and perpendicularly to the outer surface, producing a large (outer) neuroblast and a smaller preganglion cell. The latter cells undergo further divisions to form the neurons of the larva; each initial progenitor cell generates approximately 18 ganglion cells in all during embryogenesis. Nerve fibers subsequently grow out from the ganglia to generate the body neural system, beginning in the 10th hour of development. The spatial relations of the blastodermal cell precursors of the mesoderm and body nervous system can be seen in the fate map of the larva (Fig. 3.20).

The third major set of longitudinal elements are those of the intestine and gut, which arise from a multiple set of invaginations occurring at several distinct locations. The primary component of the intestinal system is the midgut, which is derived from two invaginations near the opposite ends of

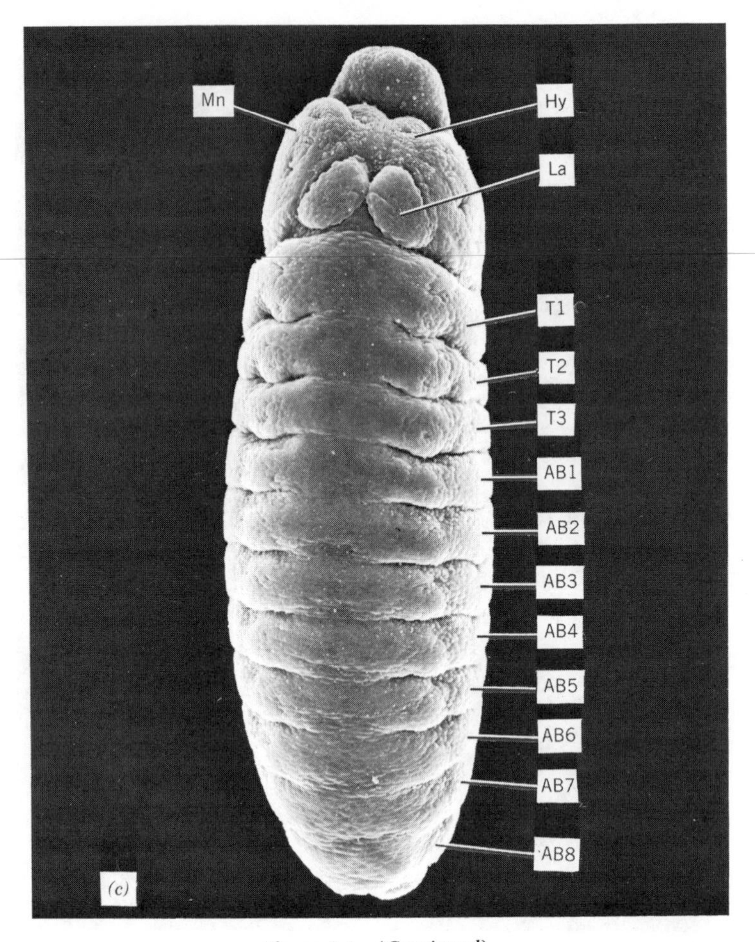

Figure 6.1 (*Continued*)

the midventral furrow; the anterior midgut (AMG), arising just anterior to the central furrow; and the posterior midgut (PMG), arising from an invagination posterior to it. A third very slight inpocketing, immediately posterior to that of the PMG, gives rise to the terminal portion of the intestine, the hindgut and the proctodeum. The mouth and foregut arise from a separate invagination just anterior to that of the AMG, the stomodeal invagination.

Originating at nearly opposite ends of the embryo, the AMG and PMG grow together to form the midgut intestinal structure. The stitching together of these two cell groups occurs as part of a complicated morphogenetic movement that involves the whole surface of the embryo, in the course of which the definitive segmented form of the larva emerges. The initial movement is led by the region of the PMG invagination. The PMG first moves dorsally, scooping up the pole cells (Fig. 3.26), and then anteriorly, eventually covering nearly two-thirds of the forward distance and reaching almost to the cephalic furrow. This latter invagination demarcates the future head region and encircles the anterior third of the embryo. The dorsal and anterior movement of the PMG involves both the ectoderm and the underlying mesoderm, the entire group of cells being termed the "germ band." The forward extension of the germ band takes place between 4 and 6 hr of development and is followed by the formation of visible segments which persist through the remainder of development.

The first internal stages of segmentation involve the development of segmental ganglia within the ventral neural cord and the segmentation of the body muscles. The first external sign of segmentation occurs 1 hr after the midventral invagination takes place, in the form of a series of ventral bulges at regular intervals (Nusslein-Volhard and Wieschaus, 1980). Between the sixth and seventh hours of development, regularly spaced clusters of large spherical cells appear. This action is followed by the development of the tracheal pits, one on each side of every segment. Distinct segmental folds develop between 7 and 8 hr. The disposition of the larval segments and the appearance of the tracheal invaginations in the fully extended germ band, 8-hr embryo are shown in Figure 6.1*b*. The internal branches of the T-shaped tracheal invaginations soon join up, and the external pits disappear. The fully developed tracheal system, terminating in lateral pairs of anterior and posterior spiracles, respectively, at the two poles, forms the larval system of respiratory tubes.

Germ band lengthening is followed by the reverse movement, shortening. During shortening, the germ band moves posteriorly and then ventrally around the caudal end. At the completion of shortening, the eighth larval segment comes to lie at the posterior end of the animal, an arrangement of segments that is retained throughout the rest of development (Fig. 6.1*c*). During the process of germ band development, the AMG and PMG extend, and as growth proceeds, the two regions come to enclose the internal central yolk mass. The fusion of the midgut extensions and the enclosure of the yolk are completed during the 11th hour of development and are quickly followed by enclosure of the gut by the visceral musculature. Externally, germ band shortening is succeeded by dorsal extension and closure of the ectodermal segments, a process that leads to the displacement and partial absorption of the amnioserosa membrane. (The amnioserosa is an extraembryonic structure formed between the cephalic furrow and the PMG.)

The last major morphogenetic change is the involution of the head structures. The maxillary, mandibular, and labial segments of the larval head and the imaginal disc cells that give rise to the adult head structures are still external at the end of germ band shortening. These structures all move forward into the ventrally located stomodeum. The consequence of head involution is that the first thoracic segment comes to occupy the anterior end of the embryo, preceded only by a region of external larval sense organs and the stomodeum. This arrangement persists until the pupal stage. Head involution is completed between 11 and 12 hr and is followed by rapid cuticle deposition over the entire external surface.

The remaining 8–10 hr of embryonic development are devoted to the further differentiation of the internal organs, and external cuticular processes of the larva further differentiate. The intermediate and final stages of development take place with very little additional cell division after 8 hr. The complete sequence of changes during embryogenesis has been described by Sonnenblick (1950), Poulson (1950), and Fullilove et al. (1978).

Beyond the striking morphological and differentiative changes that occur in embryogenesis, there is a further, subtler set of events that have great importance for subsequent development. These events involve the delimitation of the various groups of imaginal cells from the surrounding larval cells. From these imaginal cells, all of the external and most of the internal structures of the imago are formed. The imaginal cells are distinguishable from the larval cells by their smaller size, diploid chromosomal constitution, and retention of cell division capacity. Larval cells, having lost their division capability—some, such as the mesoderm, after an initial period of mitosis—become progressively larger during development and develop polytene chromosomes, as described earlier. An additional difference between imaginal and larval cells is that the former show little or no tissue-specific differentiation during embryogenesis or throughout most of larval development. The imaginal disc cell clusters thus appear as islands of small, undifferentiated cells within a sea of larger, differentiated (larval) cells.

Where do the imaginal cells arise? Some of the imaginal cells that give rise to internal adult organs derive from precursor cells carried into the embryo by the first set of invaginations (although the majority of these internalized cells give rise to differentiated larval cells). However, most of the imaginal cell precursors are located on the embryo surface, being set aside at characteristic positions within the sheet of cells that comprises the ectodermal surface of the larva.

Emergence of the fully formed larva from the egg at approximately 22 hr marks the start of postembryonic development. Postembryonic development consists of three consecutive larval instars followed by a prolonged period of metamorphosis, in which the animal finally transforms into the imago. The first two larval stages each take about 1 day, while the third lasts for nearly 3. During larval development, the animal increases greatly in size and many of the larval structures—mouth hooks, trachea, neural system—are further elaborated (Bodenstein, 1950). Accompanying these changes is continued division of the imaginal disc cells of the head and thoracic regions. In contrast, the imaginal cells of the abdominal histoblast nests begin their divisions only at the beginning of metamorphosis (when the imaginal disc cells stop dividing), rapidly increasing in number during the first half of the pupal period.

The onset of metamorphosis is triggered by a rise in the internal titer of the hormone 20-hydroxy ecdysone at the end of the third larval instar.

Metamorphosis itself is divisible into two broad stages. In the first, pupariation, the larval cuticle first darkens and the larva shrinks back from the surrounding third instar larval cuticle, secreting a new prepupal cuticle. The enclosing larval cuticle is termed the "puparium." At 12 hr from puparium formation, the developing animal, designated the "prepupa," begins the process of pupation. The larval mouth hooks and associated structures are ejected, the prepupal cuticle is shed, and the undifferentiated wing, leg, and haltere discs are pushed out ("everted"). The process of metamorphosis, lasting for 4 days from pupation, is extremely complicated. Some larval organs are completely replaced by imaginal ones; these include the salivary glands, fat bodies, intestine, and larval muscles. Nests of imaginal cells within each of the larval organs undergo rapid divisions and replace the histolyzing larval cells. Other larval organs, including the Malpighian tubules (the excretory system) and the brain, are retained, although with some "remodelling." Concomitantly, the external surface comes to be occupied by the unfolding and differentiating imaginal discs. During the final stages of the pupal period, the fully pigmented eyes and darkened wings can be seen through the puparium. At eclosion, the newly formed imago emerges through a slit in the puparium, the operculum, revealing a recognizably fly-shaped animal, although one with an extended, larvalike abdomen and still folded wings. The mature adult shape, accompanied by some body darkening, takes form within the first 2 hr after emergence.

A review of postembryonic development in *Drosophila* can be found in Bodenstein (1950), and a description of the hormonal physiology that drives these changes has been given by Doane (1973).

Gene Expression During Development: Molecular Surveys

The core subject of developmental genetics is the nature of the relationships between gene activities and observable phenotypes: whether these phenotypes are at the organismal, tissue, or cellular level. As exemplified by the oocyte, these relationships in general pose two different kinds of questions. The first concerns the *number* of genes required for the development of particular phenotypes and the temporal patterns of expression of these genes. The second concerns the actual cellular *functions* of these gene products and, in particular, the roles of those genes whose expression is absolutely essential for realization of the phenotypes of interest.

In this chapter, the focus is primarily on questions of gene function in relation to the divergence of cell types from one another. Questions of function belong primarily to the realm of genetics, and in particular involve the analysis of mutants to probe developmental changes. However, genetics is generally a poor tool for dealing with questions about the number of genes expressed and the temporal patterns of change in gene expression. Molecular analysis is much more suitable for approaching these questions.

As background to the genetics, the known molecular biology of *Drosoph-*

ila development will first be surveyed. Ultimately, the various genetic models of developmental control in *Drosophila* that have been derived from mutant studies must be evaluated in light of the molecular data on gene expression. The latter may even serve to eliminate certain formal models of genetic control and to discriminate between others.

Two broadly different patterns of gene expression change during fruit fly development might be envisioned a priori. One is that found in the sea urchin. In this animal, the early embryo starts out with a plentiful and diverse stock of maternal cytoplasmic transcripts, which becomes restricted in the course of development. In the final stages, the mRNA pools of the different cell types all share a modest number of housekeeping functions and differ in still smaller numbers of additional RNA sequences (Fig. II.1). A hypothetical alternative pattern would be the replacement of a relatively small set of maternal mRNA sequences by large, diverse mRNA groups in the various tissues. The actual pattern in *Drosophila* might be either of these or an intermediate one.

Attempts to elucidate the patterns of gene expression change have involved either surveys of protein synthetic capacities or the use of nucleic acid hybridization. The protein surveys have utilized either one-dimensional or two-dimensional separations of pulse-labeled polypeptides of various stages or imaginal discs. Although there are differences in the details of approach and result, the general consensus is that the profile of synthesized polypeptides shows a high degree of similarity between different stages and imaginal discs (Rodgers and Shearn, 1977; Seybold and Sullivan, 1978; Sakoyama and Okubo, 1981). In one detailed study of the changes in the polypeptide protein synthetic profile during the first 8 hr of embryogenesis, Trumbly and Jarry (1983) found that of 261 labeled spots on two-dimensional separations, 68 (26%) changed in a stage-specific manner. However, nearly half of the changes were losses of spots synthesized early and corresponded to the loss of specific maternal mRNA species. Most strikingly, the patterns of wild-type and *dl* embryos, derived from *dl* mothers, were found to be nearly identical over this period, despite the absence of mesodermal and endodermal tissues from the mutant embryos.

The large degree of constancy in the protein synthetic pattern in *Drosophila* development, assayed by these techniques, is reminiscent of *Caenorhabditis* postembryonic development and may reflect only the fact that most proteins abundant enough to be scored on gels perform general or housekeeping functions. However, the nucleic acid hybridization experiments to be described, which measure the diversity of total mRNA populations, reveal much the same thing: the total set of expressed genes is fairly constant throughout most of larval and pupal development and, insofar as the present data still show, gene sets between different cell types may be highly similar as well.

There are several different hybridization procedures for measuring mRNA pool complexities, which differ in accuracy and sensitivity. However, one common element in the more informative studies is the use of DNA enriched specifically for the expressed genes. This enrichment is accomplished by means of the tumor viral enzyme reverse transcriptase, which copies RNA into DNA. By copying an isolated mRNA pool into complementary DNA, or cDNA, with this enzyme, one effectively enriches for those gene sequences expressed in the cells that manufacture the mRNA. By eliminating the very large portion of DNA that is not expressed from the hybridization mixture, the ratio of RNA to hybridizable DNA in the reannealing mix is substantially increased, thereby enhancing the sensitivity and accuracy of the measurements.

Estimates of expressed gene number in different *Drosophila* stages, from several studies, are summarized in Table 6.1. Most of the estimates of mRNA complexity pertain to the poly A^+ mRNA pool only, because this fraction is particularly easy to isolate without heavy contamination from unprocessed nuclear transcripts (which lack the poly A tails). The majority of these results give estimates of about 5000–7000 different mRNA species for all stages tested. However, the poly A-tailed mRNA pool may account for only a fraction of the total mRNA complexity. The measurements of total mRNA complexity shown in the table indicate a figure corresponding to 15,000–17,000 different species for the different stages tested. If the mRNA used in these experiments were substantially contaminated with nuclear RNA, these estimates would be spuriously high. However, control experi-

Table 6.1 mRNA Complexities in *Drosophila*

Reference	mRNA Fraction	Stage or Structure	Number of Sequences
Levy and McCarthy (1975)	Poly A^+	Schneider line 2	~6,900
		Third instar larvae	7,600–7,900
Izquierdo and Bishop (1979)	Poly A^+	Whole embryos	3,500
		Larvae	≥4,900
		Pupae	~6,800
		Adults	≥4,900
Arthur et al. (1979)	Poly A^+	Embryos	~14,500
Zimmerman et al. (1980)	Total	Third instar larvae	13,000–17,400
	Poly A^+	Third instar larvae	~5,400
Levy and Manning (1981)	Total	Third instar larvae	~15,000
	Poly A^+	Third instar larvae	~5,400
	Total	Pupae	~15,000
	Poly A^+	Pupae	~6,600
	Total	Adults	~16,000
	Poly A^+	Pupae	~6,100
	Total	Adult head	~11,000

ments described by Zimmerman et al. (1980) suggested that nuclear contamination of this extent was not a factor.

The most significant feature of the data is the high degree of shared sequence relatedness between the different stages. For the poly A^+ mRNA fraction, about 85–90% of the sequences are shared between embryos, larvae, pupae, and adults (Izquierdo and Bishop, 1979). For the total mRNA pool, 15,000–16,000 sequences are found in third instar larvae, pupae, and adults, with the great proportion shared between these stages. Only a modest number, about 1600 sequences, are found exclusively in pupae and adults, and these are all in the poly A-tailed pool.

It might be that this near-homogeneity in gene expression patterns between developmental stages masks important qualitative differences in composition between different cell types. However, a preliminary result suggests otherwise. Levy and Manning (1981) measured the sequence complexity of isolated *Drosophila* heads, structures that are approximately 50% neural tissue and that lack numerous cell types found in the body (intestinal, reproductive, fat bodies, etc.), and obtained a figure of 11,700 different mRNAs. This number is approximately 70% of that determined for the entire adult body. The implication is that neural cells and probably many others will be found to have very high individual mRNA informational diversity. Indeed, complexity measurements on single vertebrate tissues reveal comparably high mRNA pool complexities (see Chapter 7).

The *Drosophila* findings are puzzling and interesting in two respects. The first puzzle is that different cell types may share such a large number of expressed gene sequences, perhaps on the order of 10,000–15,000. The existence of such a large common pool prompts one to consider all of the shared sequences as performing housekeeping functions. Yet, if one defines housekeeping functions as those necessary for cell metabolism and reproduction, numbers greater than 2000 appear to be excessive. The bacterium *Escherichia coli,* for instance, is an organism that devotes itself solely to housekeeping—it has no known developmental program—and does very well with 2000 genes. There are other bacterial cells, mycoplasmas, which get by with even fewer. Nor are eukaryotic cells, despite their greater complexity, necessarily more demanding. As we have noted, the estimated number of housekeeping functions in the sea urchin is in the range of 1000–1500.

Furthermore, there is an independent genetic estimate of the number of cell-essential housekeeping functions needed by *Drosophila;* the number is similarly small. Ripoll and Garcia-Bellido (1979) have scored the survival capacity of cell clones made homozygous for individual genetic deficiencies of varying length. Any deficiency for a cell-essential function that is cell autonomous will be lethal when homozygous and hence unable to give rise to a cuticular clone. From the distribution of chromosome bands whose deletion results in an inability to form a clone, Ripoll and Garcia-Bellido estimated that only 12% of all *Drosophila* bands are essential for cell viability and reproduction. With the number of genes in the fruit fly estimated at

6000–10,000 from genetic tests (see Chapter 3), the number of essential cell-autonomous functions would be 720–1000. Even taking 16,000 as the number of genes from the message complexity experiments, and assuming that essential cellular functions are distributed randomly among the 5000 different bands, a proportion of 12% still gives only 2000 cell-essential genes.

The biological function of potentially thousands of additional shared mRNA sequences thus is something of a mystery. Are they fine-tuning functions of some sort? Are they diverged duplicates of various essential genes, the copies being sufficiently different to register as discrete mRNAs in hybridization experiments but functionally equivalent? Or are they transcripts of genes that play defined developmental roles at discrete times and that are then left on at little or no cost to the organism?

The second puzzle raised by the data concerns the basis of cellular phenotypic diversity. The assumption that has informed the greater part of post-Jacob–Monod thinking about eukaryotic differentiation is that cellular qualitative diversity reflects underlying qualitative differences in gene expression. The *Drosophila* findings, supported by similar results in the mouse (Chapter 7), suggest that the qualitative patterns of gene expression between very different cell types are very similar.

If many of the mRNA sequences are without significant biological function, then the paradox disappears; a large number of qualitative, biologically significant differences could be hidden in the background of inessential shared mRNAs. However, if the greater part of the mRNAs are doing something useful for the cells that contain them, then qualitative cellular differences must spring either from a relatively small number of qualitatively different mRNAs between the different cell types, or from *quantitative* differences in gene expression, or from a combination of the two. Both kinds of regulation in gene expression occur during development: the mass hybridization experiments detect the former, while hybridization experiments with cloned genes reveal significant quantitative modulations for many individual genes (Biessmann, 1981; Scherer et al., 1981).

It is perfectly plausible that cellular differences should spring from either small numbers of differently expressed genes or from quantitative modulations of shared functions. The difficulties lie with the investigator, not the organism, and are both technical and conceptual. The technical problem is identifying those gene expression changes that are biologically significant. Comparisons of gene expression patterns between mutants and wild-type can provide some information, but pleiotropy often complicates the answer.

The conceptual problem lies in devising testable models of gene expression that derive qualitative cellular changes either from quantitative changes or small numbers of qualitative ones. The problem is not dissimilar to the one encountered earlier in *Caenorhabditis:* that a very large part of the genetic information appears to be expressed in different successive stages, but with obviously different developmental effects. For *C. elegans,* it was suggested that a relatively simple scheme of gene activations could account for the common base of shared gene expression while permitting the existence of qualitative differences. The same kind of regulatory scheme may similarly explain the large degree of apparent gene expression overlap in *Drosophila*. The analytical problem posed by such explanations is this: the greater the number of regulatory elements involved in producing the composite properties of a cell, the smaller the number of discriminating differences between cell types and the more difficult the assignment of particular roles to individual genetic regulatory elements. On the other hand, the manner in which quantitative modulations of gene activity (versus on-off switch-type regulation) might drive major qualitative change is a further complexity for which prokaryotic biology provides little precedent. The requirement for explanatory, testable regulatory hypotheses that relate the genetic/molecular observations to the developmental ones is becoming a pressing problem in developmental genetics.

CLONES, POLYCLONES, AND SEGMENTS IN *DROSOPHILA* LARVAL DEVELOPMENT

The history of the *Drosophila* embryo from the blastoderm stage on is an intricate and rapid series of cell and cell group diversifications. These changes involve assignments of cell fates and restrictions of cell potencies. Some of these cellular assignments become manifest quickly in the course of larval development and differentiation. Others are without immediate visible cytological effect, and some of these effects become manifest only during the final stages of development.

The specific approaches employed to investigate cell and regional assignments in *Drosophila* depend on whether one is examining larval or imaginal cells. Although some of the cell divergences that take place in the larval cell population are not immediately visible, many are and can be detected by direct examination. In contrast, most imaginal cells, both within and between clusters, look highly similar and continue to do so until the pupal stage. Despite this seeming homogeneity, each imaginal cell group experiences many hidden changes and assignments that eventually produce a large array of fates. For studying developmental pathways in imaginal tissues and structures, therefore, a much greater number of indirect investigative techniques is needed. Much of the recent investigation of *Drosophila* development has involved the use of genetic techniques to delineate the divergent fates of imaginal cells and those separations of cell potency that are cytologically invisible. The complementary effort has been to identify the genes whose activities are critically required for these divergences and to explain their mode of action.

Because the biological development of the larva and the imago take such different courses, it is simpler to review the developmental genetics of each stage separately, the approach that will be taken here. Nevertheless, the reader is advised that this separation is purely one of convenience, as an aid in keeping the distinctions between the observations and the interpretations clear. It turns out that both the larval and imaginal cells within a segment share many of the same genetic controls. The segment, and in some cases the half segment, appears to be the fundamental building block in the organism; the differences between larval and imaginal cells within the same segment are comparatively superficial. Interestingly, the underlying similarity of larval and imaginal cells is supported by a series of investigations that began with the opposite premise.

Larval and Imaginal Cells: How Different Are They?

In appearance, imaginal and larval cells differ dramatically. First distinguishable as clusters of cells in the newly hatched first instar larva, the cells of the imaginal discs are small, diploid, and essentially undifferentiated until late in development; only the eye imaginal disc, revealing the presence of ommatidia in the late third instar, shows signs of its future role (Ursprung, 1972). In contrast, most of the larval cells are large, have big nuclei with polytene chromosomes, and are differentiated with respect to one another from early or midembryogenesis on. Although most cells in the larval stages can be grouped in these categories, a small proportion cannot: some of the larval cells, such as those of the larval ganglia, remain diploid and continue to divide, while the histoblast cells, which give rise to the imaginal abdominal integument, are larger than disc cells and, although diploid, are partially differentiated, participating in the secretion of larval cuticle (Madhavan and Schneiderman, 1977).

One hypothesis to explain the broad differences between imaginal disc cells and larval cells is that there are distinctly different imaginal and larval developmental pathways involving the expression of unique sets of genes. If imaginal and larval development differ in this manner, then it should prove possible to isolate mutants that form normal larvae but that are deficient in the formation or structure of their imaginal discs. (The hypothetical complementary class of mutants, defective in presumptive larva-specific genes, would die in embryogenesis and therefore would be classified as embryogenesis defectives.) Recessive mutants of imaginal cell- or disc-specific functions should be viable as homozygotes throughout larval development but would die at or near the time of pupation, as the imaginal disc program commenced. A precedent for such imaginal disc mutants has long been known: the recessive *lethal giant larva (lgl)*, described by Hadorn (1961), forms functional larvae possessing severely defective discs. These larvae never pupate, but reach abnormally large sizes before dying. (Their large size reflects their impaired synthesis of ecdysone, which during normal development is secreted by the imaginal ring gland.)

To examine the genetic basis of imaginal disc development, Shearn et al. (1971) carried out a large screen for third chromosome, late larval lethal mutants following chemical mutagenesis. The search produced 134 mutant strains out of a total of 3167 mutagenized lines. Of these, 66 showed one or more disc types absent or defective in third instar larvae. One small class of mutants consisted of those defective in particular pairs of disc types; more will be said about these later. However, the great majority, 64%, proved defective in all of their imaginal discs (Shearn, 1977). The two major categories within this group were *discless* mutants, apparently lacking all traces of head and thorax discs, and *small disc* mutants, possessing rudimentary discs incapable of normal development. These mutant categories are not unique to the third chromosome; a comparable search for X chromosome-linked late larval lethals yielded 23 mutants with a similar range of disc-defective patterns, including some in which disc degeneration could be identified in late larval stages (Stewart et al., 1972). Indeed, the genome is liberally sprinkled with genes that can mutate to give a seemingly specific disc-defective condition. From the ratios of third chromosome complementation groups containing one, two, or more members, and assuming an equal density of such functions around the genome, Shearn and Garen (1974) calculated that 1000 genes are specifically required for general disc development and inessential for larval development.

Although the results support the notion of distinct imaginal disc and larval developmental pathways, closer inspection of the mutants combined with subsequent genetic results have undermined this notion. In the first place, many of the presumptive disc-specific defectives have larval defects. A simple criterion of larval defectiveness involves the time of death; any mutant strain that dies before puparium formation must be deficient in some larval-essential function. By this measure, 38% of the original 66 third chromosome imaginal defectives are also larval defectives (Shearn, 1977).

In the second place, whether or not a mutant appears to be a disc defective or a more general lethal is largely, and perhaps entirely, a function of the mutant allele rather than of the genetic locus itself. In the original study, Shearn et al. (1971) reported that one small disc mutant and one discless mutant were in the same complementation group; this result indicates that there is no fundamental distinction between these two categories of disc defectives. More strikingly, when additional mutant alleles of the gene identified by the small disc mutant *l(3)1902,* which itself shows no signs of larval defectiveness, were selected, many showed very different phenotypes from that of the original allele. Two were temperature sensitives that were zygotic lethals, with distinct maternal effects. The TSP of one stretched from early embryogenesis through larval development to the pupal stage (Shearn et al., 1978a). Presumably, the prototype allele was a hypomorph; more severely defective alleles fall into the category of general lethals. The converse also holds: some or many of the genes identified by single-mutant, general lethal mutant phenotypes may also mutate to alleles that give the disc-defective condition (Shearn et al., 1978b). The results illustrate a point noted several times previously: to assess the role of a gene in development, one either needs several mutant alleles whose effects can be compared against each other (as in the case of *dor*); alternatively, if only one mutant is available, one must be certain that it is an amorph.

It is nevertheless of interest to know what the characteristics of the late larval lethals are and why it is so comparatively easy to obtain mutants that seem specific for imaginal disc development. One clue has come from an unexpected source, an investigation of mutants involved in DNA repair mechanisms. All cellular organisms possess specific enzymes for removing and repairing certain kinds of damage to DNA. In *Drosophila,* as in several other organisms, one can isolate viable mutants of such repair functions that are characterized by high sensitivity to certain forms of induced DNA damage. Severely defective mutants of two such repair gene functions, first identified by hypomorphic viable mutants that exhibit a mutagen-sensitive (*mus*) phenotype, were found to give a late larval phenotype (Baker et al., 1982). Examination of the cytological defects in these lethals revealed a high frequency of chromosome breaks and aberrations in their dividing larval ganglion cells. The frequency is sufficiently high, reaching 0.6 breaks per cell per division for one mutant, to account for the death of all dividing cells, whether imaginal or diploid larval cells. Nor is this cause of late larval lethality exceptional; examination of a larger set of late larval lethals showed that a high percentage exhibit mitotic abnormalities (cited in Baker et al., 1982). It appears that isolation of imaginal disc defectives or late larval lethals selects preferentially for mutants of essential cell division functions. Since the imaginal discs undergo continuous division during larval development, while most differentiated larval cells do not, division defectivity is revealed primarily as a set of imaginal defects. The fact that such division-defective mutants can carry out cell division during embryogenesis probably reflects maternal storage of essential division components in the egg, provided by the heterozygous mother.

The hypothesis that division defectivity is the source of the disc-defective phenotype was confirmed in a detailed study by Szabad and Bryant (1982) on a group of five discless mutants. All five mutants show a strong inhibition of division in all cells that normally divide, both larval and imaginal. The larval cell defects show up as division blocks in the larval ganglia and reduced numbers of blood cells in the lymph glands. Careful inspection of the mutants shows that all, despite their initial classification as discless, possess vestigial discs containing the same cell numbers as those present in hatching wild-type L1 larvae. (One mutant, however, the X chromosome mutant *dl-1,* seems to lack even rudimentary wing and haltere discs, although possessing vestigial discs of the other kinds.) Thus, even in the most extreme class of imaginal disc defectives, the discless mutants, imaginal discs are normally established in nearly all instances; their defectiveness lies in an inability to carry out cell division. In those mutants showing only a partial cell division block, the defect may spur an increased rate of cell death, which would further reduce cell numbers in the discs (Murphy, 1974).

In conclusion, the imaginal disc cells as a whole do not require the expression of a gene set very different from that employed by larval cells as a whole. And, indeed, many imaginal cells cease to divide and their chromosomes become polytene later in development, in this sense acquiring a larval cell phenotype. The essential difference between the two cell types may therefore be one of timing in the shutdown of normal division and the onset of polyteny.

One implication of this revised view of the imaginal/larval cell distinction is the expectation that contiguous larval and imaginal cells may often have more in common with one another than do cells within each category that come from different blastodermal regions. This expectation is borne out in two different ways: the blastodermal fate maps of larval and imaginal precursors show a high degree of congruence for comparable features, and the larval and imaginal pathways within segments are governed by many of the same genes.

Larval Development: The Biology of Segmentation

One event that marks the emergence of the larval body plan is segmentation. The first sign of segmentation occurs 1 hr after gastrulation, approximately 4.5 hr after fertilization, with the appearance of a series of slight ventral bulges. The segmental pattern is fully apparent at 7–8 hr of development, when the embryo is in the fully extended germ band form (Fig. 6.1*b*). By the end of embryogenesis, each major body segment in the thorax and abdomen exhibits a distinctive cuticular arrangement of ventral denticle hooks (for grasping the surface of the medium), hairs, and, in the thoracic segments, certain sensory organs. This pattern is schematized in Figure 6.2. The replacement of the larval segment pattern by that of the imago during pupal development maintains segment identity. In the same sense that the imaginal discs and histoblast nests can be regarded as the building blocks of the adult's body surface, the segmental repeat of the larval body can be considered its fundamental element.

The process of segmentation raises several questions about the genetic controls that underlie the process and the actions of the relevant gene products in producing it. It is apparent, from the discussion of maternal effect mutants that affect the segmental pattern, that one component of the process is maternal, consisting of oocytic components active in the early embryo. However, the maternal instructions are only one component in the segment specification program. The second phase must involve the expression of zygotic genes in response to the maternal instructions. Faulty elicitation or expression of these genes leads to aberrant segment patterns.

In thinking about the nature of zygotic genome control in segmentation, it is helpful to view segmentation as consisting of two processes. The first entails the imposition of the characteristic regular and repetitive *spatial patterning* of the segments. The second aspect is that of *segment identity,*

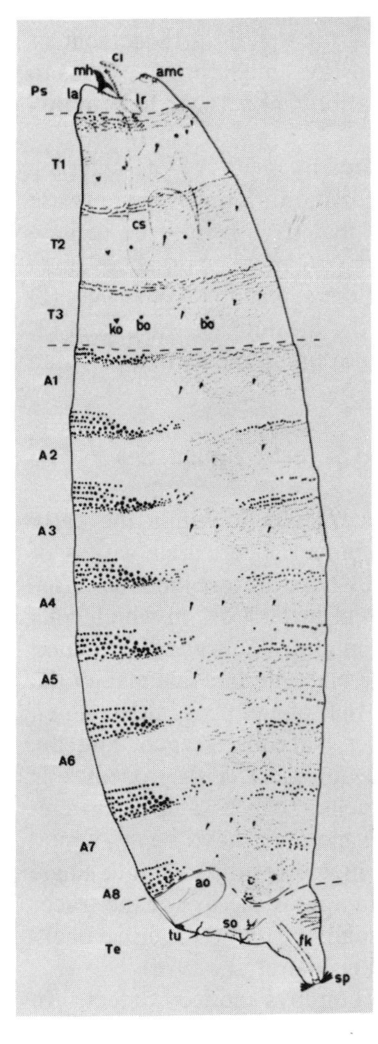

Figure 6.2 Cuticular pattern of the first instar *Drosophila* larva. Oriented with anterior at the top, ventral side at the left. Ps, pseudocephalon; la, labium; mh, mouth hook; ci, cirri; amc, antenno-maxillary complex; lr, labrum; T1–T3, thoracic segments; cs, cephalopharyngeal skeleton; ko, Keilin's organs; bo, black sensory organs; A1–A8, abdominal segments; Te, telson; tu, tuft; so, sensory organs; fk, Filzkörper organ; sp, spiracles. (Figure courtesy of Dr. M. Lohs-Schardin; from Lohs-Schardin et al. 1979b.)

the fact that each segment has a characteristic cuticular pattern. This distinction between the two features of segmental pattern control is not arbitrary: certain mutant changes can affect either aspect independently of the other.

To evaluate certain features of the genetic findings, one needs first to know something about the cellular field that participates in segmentation. Several findings, reviewed below, indicate that cell lineage restrictions develop between cells of neighboring segments, preventing substitution by cells of one segment for those of another. Such restrictions might indicate the existence of segment-specific states at the time the restrictions are laid down. When in development do the lineage restrictions and putative segment-specific states arise?

One would also like to know the *number* of cells set aside to form segments. If the restrictions are imposed early, how many precursor cells are assigned to a particular segment, and what is the spatial arrangement of these precursor cells? The answers to these questions might provide clues to the underlying processes involved in segmentation and perhaps help to discriminate between models.

To some extent, the questions can be answered by means of fate-mapping methods, through which the precursor cells for the different segments are identified on the early embryo. The methods that have been used include both surgical and genetic procedures. Collectively, these approaches have (1) localized the precursor region or *anlage* of the segmented epidermis; (2) identified the time of segment delineation as the blastoderm stage; and (3) refined the questions that can be asked about the commitments that partition the larval epidermis.

Fate Mapping of the Larval Surface and Segment Boundaries

Fate mapping is the process of locating precursor cells within the early embryo for structures that arise later in development. There are a variety of fate-mapping methods, and they produce somewhat different kinds of information. The most direct method is that of direct inspection, in which cells are tracked microscopically as they divide and migrate. This was the method employed by Poulson (1950) in constructing the blastodermal fate map of the larva (Fig. 3.20). A second tracking method is that of cell marking, in which specific cells are injected with a dye or other traceable reagent and the stained progeny cells are followed through development. This procedure is particularly suitable for embryos with large cells, such as amphibian embryos. Recently, such chemical fate-mapping methods have been supplemented by techniques involving the tracing of molecules injected into single cells that cannot diffuse between cells. An example is the use of the tracer horseradish peroxidase, whose activity can usually be detected above background peroxidase levels in early embryos (Weisblat et al., 1978).

A different method of direct fate mapping employs induced defects. In embryos possessing little capacity for regulatory replacement of cells, a selective ablation of early cells results in a specific deficiency or range of deficiencies in the surviving embryos. In contrast to the visual cell-tracking methods, which reveal the normal developmental potentials of cells in undamaged embryos, the induced defect methods establish the capacities of cells surrounding the defect to substitute for the deleted cells; a selective cell ablation that always produces a particular deficiency reveals an inability of the neighboring cells to change their developmental pathways in response to the injury. The cell ablation experiments using *C. elegans* are an example of this approach. The potential drawback in induced defect experiments is that the defects themselves may interfere with the developmental responses of the surviving cells.

Finally, there are genetic fate-mapping experiments. In these studies, one or more cells of the early embryo are made to differ from the rest of the embryo through possession of a distinctive genetic marker, either biochemical or morphological. Depending on the number and identity of the marked cells and the time in development when the genetic marker can (or must) be scored, the approach can yield a clear picture of early cell contributions to later developed structures.

For the *Drosophila* larva, all types of fate-mapping methods have been used. Some of the earliest embryological experiments using *Drosophila* involved the creation of localized damage in blastoderm-stage embryos and analysis of the ensuing developmental effects. It was these experiments that established the presumptive mosaic character of the *Drosophila* blastoderm; injuries induced in particular regions created characteristic regional defects. Nevertheless, these experiments produced damage on too large a scale to permit reliable fate mapping. Recently, newer methods for inducing highly localized defects have been employed; these experiments define precisely the region of the blastoderm surface that gives rise to the segmented larval epidermis.

Induced-Defect Fate Mapping

Two different procedures have been used in induced-defect fate mapping. The first is that of selective cell ablation with a laser microbeam (as in the nematode experiments). By focusing a microbeam of 15 μm in diameter, sufficient to kill about 15 cells at a time, Lohs-Schardin et al. (1979b) induced localized defects in blastoderm embryos and then mapped the sites of damage in the surviving first instar larvae. The second method is that of selective cell removal, in which a micropipette is inserted into the embryo surface and small cell patches are extracted. Using this method, Underwood et al. (1980b) deleted lateral cell groups of 15–30 cells at a time, at the cellular blastoderm stage and in slightly older embryos, scoring the sites of damage in 9-hr embryos, the stage of germ band shortening.

The two approaches give closely similar results. The mapping of first larval thorax and abdominal precursor cells along the a-p axis of the blastoderm by the laser beam cell ablation method is shown in Figure 6.3. Taking the posterior pole as 0% egg (length (EL) and the anterior pole as 100% EL, the entire larval thoracic and abdominal segment precursor regions are found to fall between 20 and 60% EL, that is, they extend over only 40% of the length of the blastoderm. Underwood et al. (1980b), employing cell removal and scanning electron microscopy of 9-hr embryos, found a similar blastodermal placement for the thoracic and abdominal regions. In addition, they localized the head segments to between 60 and 70% EL and the two vestigial abdominal segments, A9 and A10 (which are not distinguishable as separate segments in adult flies), to between approximately 20 and 15% EL. The entire larval epidermis is therefore apparently prefigured along 55% of the a-p axis, and blastodermal cells outside these limits cannot replace cells lost between them. Although multiple segment defects were found to be the rule

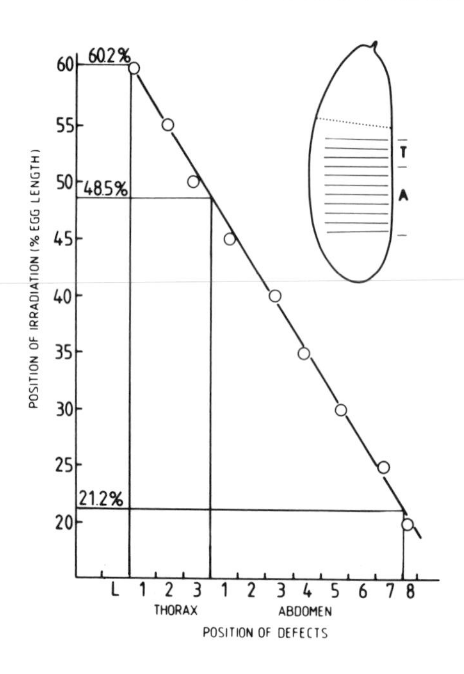

Figure 6.3 Induced defect map of segment position on the blastoderm stage embryo. From the observed centers of defects, for each position of irradiation, the map (inset) was constructed. See text for discussion. (Figure kindly provided by Dr. M. Lohs-Schardin; from Lohs-Schardin et al., 1979b.)

with both procedures, the defects always involved adjacent segments. Evidently, precursor cells for consecutive segments are arrayed consecutively within this region.

Lohs-Schardin et al. (1979b) also measured the d-v extent of the larval epidermal precursor region on the blastoderm. The results showed that damage can be induced throughout each half-circumference, except for the midventral 20% and the middorsal 4%. The latter areas correspond to those occupied by the cells that give rise to the midventral invagination and the amnioserosal membranes, respectively. Therefore, roughly 76% of the egg circumference, within the determined a-p limits, gives rise to the area of segmented larval epidermis. Taking the average EL as 500 μm, the portion of this distance involved in thoracic and abdominal segment formation is 200 μm (40% $\times$ 500 μm). With an egg diameter of 175 μm, the d-v circumferential distance occupied by thoracic and abdominal precursor cells on each side is also about 200 μm (76% $\times$ $\pi/2$ $\times$ 175 μm = 200 μm). Therefore, each lateral epidermal anlagen is approximately a square of 200 $\times$ 200 μm on the blastoderm embryo. From cell counts at this stage, a region of this size corresponds to about 1100 cells, or 2200 cells total for both sides, or just slightly more than a third of the total number of blastodermal cells. Part of the remainder forms the epidermis of the cephalic segments and the terminal abdominal segments (A9 and A10), but the remainder, approximately 3000 cells, goes to form internal tissue.

The information can also be used to calculate the distance along the a-p axis of the average segment anlage, which turns out to be surprisingly small.

With a total of 11 thoracic and abdominal segments and 1100 cells per side, each segment anlage consists of 100 precursor cells. Since each half-circumference is 36 cell diameters, the distance between segment precursors for different segments is only 3 cell diameters.

Because the entire larval epidermis of the newly emerged L1 form is approximately 7500 cells (Madhavan and Schneiderman, 1977), there are at most only two rounds of division between the blastoderm stage and the appearance of segmental boundaries. If the average longitudinal extent of each segmental anlage at blastoderm is 3–4 cell diameters, the assignment process must take place between this point and that at which each segment length is 12–16 cell diameters. Since the first signs of segmentation are visible at gastrulation, segment delimitation probably occurs at the smaller distance. Within these limits, cells are apparently instructed to contribute to one specific segment or another with a regular periodicity. Explanations of the segmentation process must ultimately account for both the regularity of segment placement and the small physical distances over which it is imposed.

Genetic Fate Mapping

Genetic fate mapping permits the localization of different precursor cells with respect to one another in the absence of physical trauma to the embryo. In principle, it can define boundaries between anlagen much more precisely than is possible with the surgical procedures. The partial drawback to genetic fate mapping is that, in the absence of fixed physical landmarks on the early embryo that can be related to the genetically mapped structures of later embryos, the mapping is purely relative.

To fate map the larval surface, one needs a suitable genetic marker that can be scored throughout the larval epidermis and a method for inducing clones in the epidermis. For the choice of marker, a histochemical activity that is expressed throughout all or most of the epidermis is best. One such enzyme activity is that of aldehyde oxidase, an activity controlled by the X-linked mal^+ locus, whose usefulness in tracing blastodermal cell fates was described earlier.

The choice of method for generating marked clones is more problematic. Mitotic recombination is unsuitable for two reasons. In the first place, even when mitotic recombination is induced by x-rays, the frequency of clones will always be low in embryos with small cell numbers. Furthermore, the small number of cell divisions between blastoderm and the first larval instar (about two to three) guarantees that any clones that are formed will be extremely small, consisting of only two or four cells. Searching for histochemically stained clones of these dimensions—for instance, white mal^-/mal^- clones on a dark blue background of mal^+/mal^- parental cells—would be a difficult task. In principle, this problem of clone size could be circumvented by obtaining larger clones through earlier, preblastoderm irradiation.

However, and this is the second problem, the induction of preblastodermal clones by x-rays is not feasible; the cleavage-stage embryo of *Drosophila* is too sensitive to x-rays to survive the doses needed for clone induction.

The fate-mapping method that is suitable for the *Drosophila* embryo is that of gynandromorph analysis, invented by Sturtevant (Chapter 1). The elements can be quickly summarized. If one of the cleavage nuclei in a female embryo loses one of its two X chromosomes, the embryo will develop as a mosaic of XX (female) and XO (male) tissue. The observed contiguity of the female and male sectors in most adult gynandromorphs shows that the descendant XO nuclei migrate as a group. If the positions of the anlage for the different structures of the larva or adult are fixed relative to one another on the blastoderm egg surface, then the boundary line between male and female tissue, which observation of the adult gynanders shows can fall anywhere, will invariably separate distant structures from one another more frequently than those produced from closer anlage.

Imagine that one has a ball (the blastoderm embryo) that is densely decorated on its surface with a variety of distinctive elements (larval or adult structure anlage), and that one neatly slices the ball through the middle in any one of an infinite number of possible but randomly chosen planes (the dividing line between XX and XO cells). Surface elements that are close together have an inherently much smaller probability of being separated by the plane of section than those that are far apart. In the absence of secondary complications, the frequency of separation of different elements should provide a *measure of the distance between them*. For the *Drosophila* blastoderm embryo, the distance between anlage for different structures is therefore indicated by the *frequency with which the mosaic border falls between these structures*. To perform the analysis, one needs only a genetic marker that unambiguously distinguishes all XO tissue from all XX tissue *and* a high frequency of gynandromorphs.

Sturtevant used the nondisjunctional property of the *claret* mutant of *D. simulans* (now termed ca^{nd} for *claret nondisjunctional*) to generate gynandromorphs. This mutant produces a relatively high rate of maternal X chromosome loss and thereby generates gynanders. Comparable gene-induced nondisjunctional systems are available in *D. melanogaster,* but the method of choice in this species employs an unstable X chromosome, the ring-X chromosome $In(1)w^{vc}$. This is a fully functional X chromosome that has lost the small, genetically inert right chromosome arm and has the tip of the major (left) arm joined back to form a ring. For postblastodermal development, the chromosome is apparently as mitotically stable as are normal (rod-X) chromosomes. However, during the first few cleavage divisions of the early embryo, the ring-X is unstable and tends to be lost.

Imagine that the loss occurs in the second cleavage division, as two nuclei are giving rise to four. The outcome can be diagrammed as follows, where only the X chromosomes are shown:

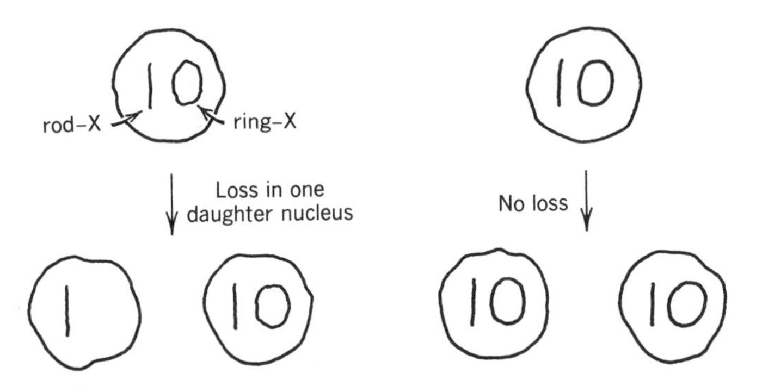

The resulting embryo will be one-quarter XO tissue, assuming that there is no preferential multiplication of male or female cells.

The ring-X system permits a fate mapping of the larval epidermal surface. If one constructs ring/rod-X heterozygotes, in which the ring-X carries a *mal*$^+$ allele and the rod-X carries a *mal*$^-$ allele, then ring-X loss produces XO *mal*$^-$ nonstaining regions on the larval surface. The distribution of non-stained tissue in a large set of gynandromorph larvae can reveal much of the early cellular history of these animals.

Szabad et al. (1979a) performed such an analysis on third instar larval gynandromorphs. Out of 2231 larvae obtained in a cross of *mal*$^-$ rod-X females with *In*(*1*)*w*^{vc}, *mal*$^+$ males, 152 mosaics were detected, of which 140 could be scored. Figure 6.4 shows the distribution of epidermal enzyme activity in 48 of these specimens. For the most part, there is a single male and a single female patch, but 18% of those analyzed had two male or female patches.

The frequencies with which different segments or cell groups are separated by the mosaic borders can be taken as a measure of the intervening distances of their precursor cells on the blastoderm surface, as explained above. The measure of distance is the percentage of mosaics in which the two segments (or cell groups) fall on opposite sides of the border. The unit of distance is taken as 1% separation and has been termed the "sturt," in honor of Sturtevant (Hotta and Benzer, 1973). Thus, if two cell groups fall on opposite sides of the male/female border 4% of the time, they are said to be four sturts apart. Given the waviness of the border in the larval gynandromorphs, Szabad et al. (1979a) divided each larval segment into subsections two cells wide by two cells long.

The analysis produced several points of interest. First, with the exceptions noted below, borders were found to separate different contiguous sections of the larval epidermis with nearly equal frequency. The measured distance between grid sections was 4.5 sturts, and the probability of a single cell interface being mosaic was 2.2 sturts. Such even spacing of segment precursor regions is consistent with the results of the induced defect experiments.

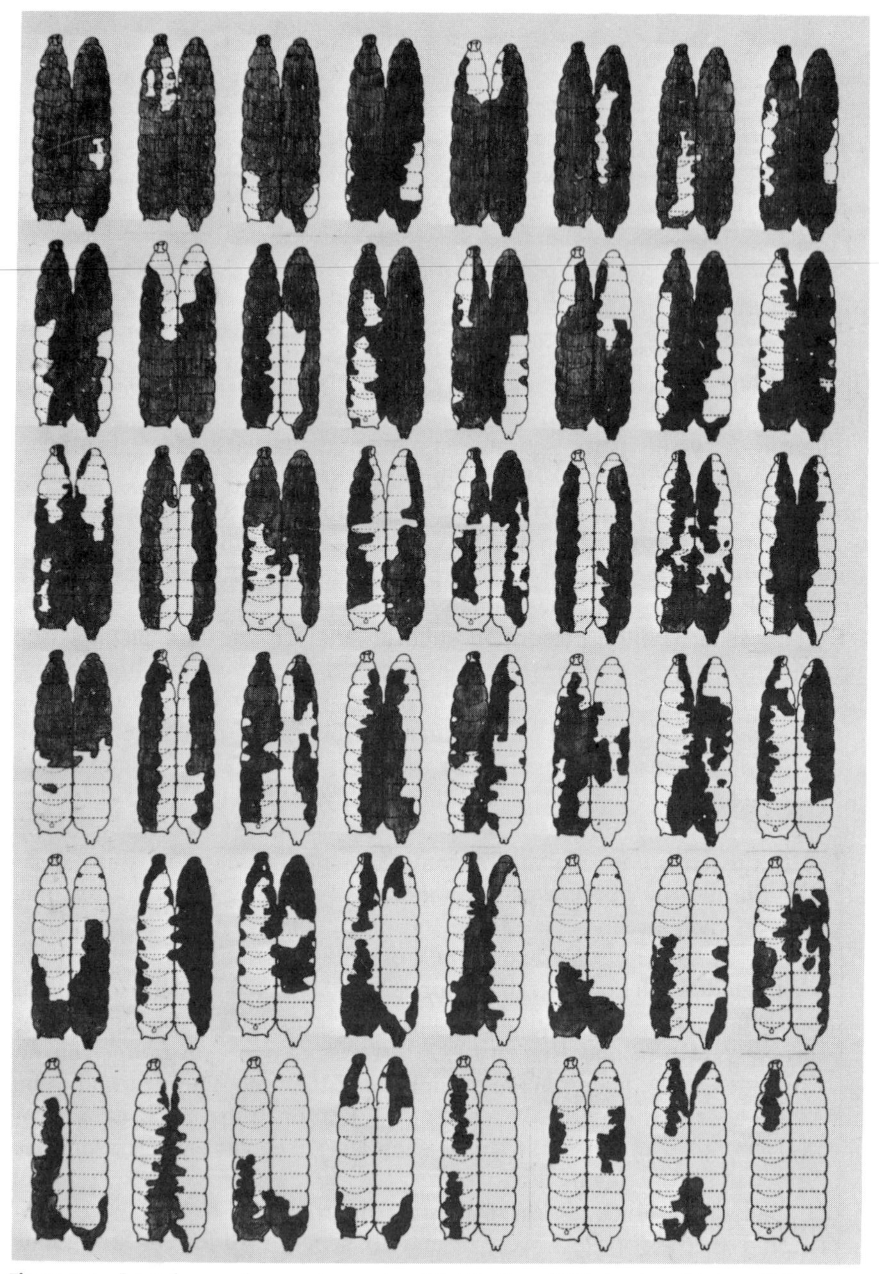

Figure 6.4 Larval gynandromorph mosaics. Distribution of XO (*mal*$^-$; white) and XX (*mal*$^+$; dark) cells in the epidermis of 48 third instar larval gynandromorphs. (Figure kindly provided by Dr. J. Szabad; from Szabad et al., 1979a.)

Second, the size of the blastodermal precursor area could be estimated, using the mosaicism of adult structures as a standard. Szabad et al. observed that a larval surface area of 46 subtended cells was mosaic approximately as frequently as an adult leg, or about 23% of the time. From the size of mitotic clones in legs induced at blastoderm, it has been estimated that about 10 cells give rise to leg disc primordia (Wieschaus, 1978), which reach a size of about 40 cells at the beginning of first larval instar (Madhavan and Schneiderman, 1977). Given the similar degrees of mosaicism, it follows that 7500 epidermal cells present in first instar larvae derive from about 10/46 as many precursor cells, or about 1630. Because estimates from clonal analysis of cuticular structures leave out internal tissues, the true number of leg precursor cells is somewhat larger than 10, perhaps 15 or so. In the latter case, the estimated size of the larval epidermal precursor region rises to about 2400 cells. In either case, the genetically derived estimate is satisfactorily close to the estimate of 2200 cells derived from the defect mapping experiments.

Third, contiguous regions on the larval epidermis that show higher than average sturt distances reveal something about their component cell histories with respect to one another. Either the precursor cells were relatively farther apart in the early embryo and intervening cells were physically removed by cell migration or death, or alternatively, the respective descendant cells simply moved with respect to one another following the demarcation of the mosaic boundary at blastoderm. The larval fate-mapping data provide possible instances of both kinds. A probable case of cell removal is signified by the sturt distances across the midventral line. Twenty-five percent of all mosaic borders that cut across a larval circumference lie in the most ventral subsections of the grid; this represents an excess of 16% relative to the lateral regions. A smaller excess is found over the dorsal midline. The simplest explanation for the difference is that these are regions in which precursor cells are removed. The 16% excess borders closely correspond to the 16% of cells invaginated in the midventral furrow; the excess middorsal border frequency corresponds to the loss of cells to the extraembryonic membranes.

A small but significant excess of mosaic borders is also found along the a-p axis at each segmental boundary. The difference is approximately twofold. These excess sturt distances between adjacent regions might represent a loss of cells through invagination. However, there is another possibility. If each segment, once it has been set aside from the other segments, constitutes an independent unit of development, then cells on either side of the segmental boundary should have some freedom of movement with respect to one another. Morphogenetic independence of this sort will result in shifted borders on either side of the boundaries, with a consequent increase in sturt distances. The inability of cells within one segmental repeat to replace those of another is one form of evidence that segments are such semi-independent developmental units.

However, the most interesting finding from genetic larval fate mapping pertains to the relationship between larval and imaginal cells in the abdomen. The imaginal abdominal cells are the histoblasts, cells that are quiescent during larval development and begin extensive proliferation only during pupation. In all of the gynandromorphs scored by Szabad et al. (1979a), the *mal* phenotype of the histoblasts was never found to be different from that of the surrounding larval cells. In those cases where the histoblast nest was cut by a mosaic border, the border was observed to run smoothly into the neighboring larval cellular region. These observations suggest that the histoblasts are singled out from a sheet of common precursor cells for both histoblast and larval cells. It appears to be not unlikely that the other groups of imaginal cells (of the thoracic and cephalic regions) are similarly derived from a precursor cell pool within each segmental region that also produces larval cells.

Gynandromorph fate mapping is not without its complications, and details of any fate map obtained are a function of the starting assumptions. This point will be amplified later, but one assumption will be noted here. The proportion of male (XO) tissue in adult (Garcia-Bellido and Merriam, 1969) and larval gynandromorphs (Szabad et al., 1979a) can vary from 3 to 90% but averages 50%. This fact is usually taken to signify that ring-X loss most commonly occurs only once in each gynandromorph, and then usually in the first division.

An analysis by Zalokar et al. (1980) suggests a different conclusion. They directly scored the distribution of rod-X and ring-X chromosomes in early cleavage and blastoderm embryos by employing two drugs that respectively freeze nuclear division in metaphase and relax the chromosomes to permit identification of the rod and ring Xs when stained. They found that chromosome loss can occur in either of the meiotic divisions of the egg nucleus, seemingly through failure of ring-X replication or of separation of chromatids, or during any of the mitotic divisions of the zygote nucleus. It is the mitotic losses that produce gynandromorphs. From the numerical distributions of XO nuclei among a group of 18 blastodermal gynandromorphs, Zalokar et al. concluded that most gynandromorphs arise from *two* successive ring-X losses during cleavage rather than single losses. One way to explain this result is to posit that the original two ring-X chromatids, produced during replication of the female pronucleus, are destabilizing entities that provoke mitotic chromosome loss following subsequent rounds of replication. In accordance with the idea of two chromosome losses per gynandromorph, 5 of the 18 blastoderms were found to have two or more discrete patches of XO nuclei. Furthermore, XO nuclei in the gynandromorph embryos averaged only 34% of the total, rather than the 50% male tissue seen in larvae and adults. This finding suggests the XO cells in the gynanders have a slight competitive advantage.

The cytological information does not invalidate the qualitative conclusions drawn from gynandromorph fate map studies, but it does suggest the need for caution in taking the numerical values as strict measures of the distance between precursor cells.

Pattern and Polarity in Segmentation: Mutant Analysis

The segmental body pattern of the first instar larva shows elements of both repetitiveness and segment individuality. The principal repetitive feature is that of segment spacing. From the anterior end of the first thoracic segment (T) to the posterior end of the eighth abdominal segment (AB8), the segment boundaries occur with well-spaced regularity. In addition, certain characteristic internal motifs recur. The ventral side of each segment is marked at its anterior edge by a band of small cuticular protuberances arranged in rows, termed the "denticle hook bands," and by a naked cuticle in the posterior portion (Fig. 6.2). In consequence, each segment possesses a well-defined a-p polarity. There is also a dorsoventral differentiation; numerous fine hairs cover the dorsal side, spreading laterally to differing extents on the different segments.

The principal external features that distinguish segments from one another are the size of the denticle hook bands, the extent of the dorsal hair region, and the occurrence of sensory organs. The broadest differentiation is that between thoracic and abdominal segments. The former possess relatively narrow denticle bands, each consisting of two to three rows, and small ventral sensory structures (Keilin's organs) and small black lateral sensory organs. The prothoracic segment (T1) differs from the mesothoracic (T2) and metathoracic segment (T3) in having an additional denticle band and pointier dorsal hairs, whereas T2 differs from T3 in showing slightly larger denticles and a wider denticle band. The abdominal segments differ from the thoracic segments in the greater width of their denticle belts and in the absence of sensory organs. The first abdominal segment appears intermediate between the thoracic and remaining abdominal segments in some respects (width of the denticle band, spread of the dorsal hairs), and the eighth is distinguished by the absence of a naked posterior ventral region. From the first through the seventh abdominal segments, there is a gradient of increasing width in the denticle band, permitting neighboring abdominal segments to be distinguished from one another.

It is the combination of repetitive and unique segment characteristics that permits a clear assessment of the effects of pattern-perturbing mutations. For zygotically acting mutations that derange the normal pattern of segmentation, it is possible to discern both general pattern disruptions and effects on individual segments.

A thorough genetic analysis of the zygotic genome functions that determine segment pattern and polarity has been carried out by Nusslein-Volhard and Wieschaus (1980). Large numbers of recessive embryonic lethal mutations were isolated, and these were screened for aberrations of larval segment patterns. Altogether, a total of 15 loci were identified in the search. From the average frequency of mutant alleles isolated per locus, it seems likely that these 15 loci represent the majority of functions in the genome whose deficiency drastically affects the segment pattern.

The mutant phenotypes associated with the 15 loci fall into three categories. The first group, consisting of six genes, is distinguished by a repeated internal deletion of part of each segment followed by a mirror-image duplication of the part that remains (''segment-polarity mutants''). The mutants of the second group of loci, also consisting of six genes, similarly display a repetitive aberration, but one that involves removal of integral, *alternate* segments (''pair-rule'' loci). The third group of mutants, totaling three loci, feature characteristic deletions of groups of segments (''gap mutants''). The names and chromosomal locations of the different mutants within each group are listed in Table 6.2, and schematic drawings of representative mutant phenotypes are shown in Figure 6.5. All mutants were isolated on the basis of their effects in homozygous lethal embryos. The degree or importance of maternal expression for these genes is unknown, except for *fused* (*fu*), one of the segment polarity mutants, whose lethality and segment

Table 6.2 Zygotic Genome Loci Essential for Normal Segmentation

Category	Locus	Map Position
Segment-polarity	*cubitus interruptus*^{D} (*ci*^{D})	4–0
	gooseberry (*gsb*)	2–104
	wingless (*wg*)	2–30
	hedgehog (*hh*)	3–90
	fused (*fu*)[a]	1–59.5
	patch (*pat*)	2–55
Pair-rule	*paired* (*prd*)	2–45
	even-skipped (*eve*)	2–55
	odd-skipped (*odd*)	2–8
	barrel (*brr*)	3–27
	runt (*run*)	1–65
	engrailed (*en*)	2–62
	ftz	3–48
Gap	*Krüppel* (*Kr*)	2–107.6
	knirps (*kni*)	3–47
	hunchback (*hb*)	3–48

[a] This locus has a known strong maternal effect on early development; several of the other loci listed may also have maternal effects, but these have not been fully characterized. (Because the mutants are zygotic lethals, tests for maternal effects can be done only by pole cell transplants of homozygous mutant pole cells or by induced mitotic recombination in the female germ line.)

Source. Data from Nusslein-Volhard and Wieschaus (1980) and Kaufman (1983) (for ftz).

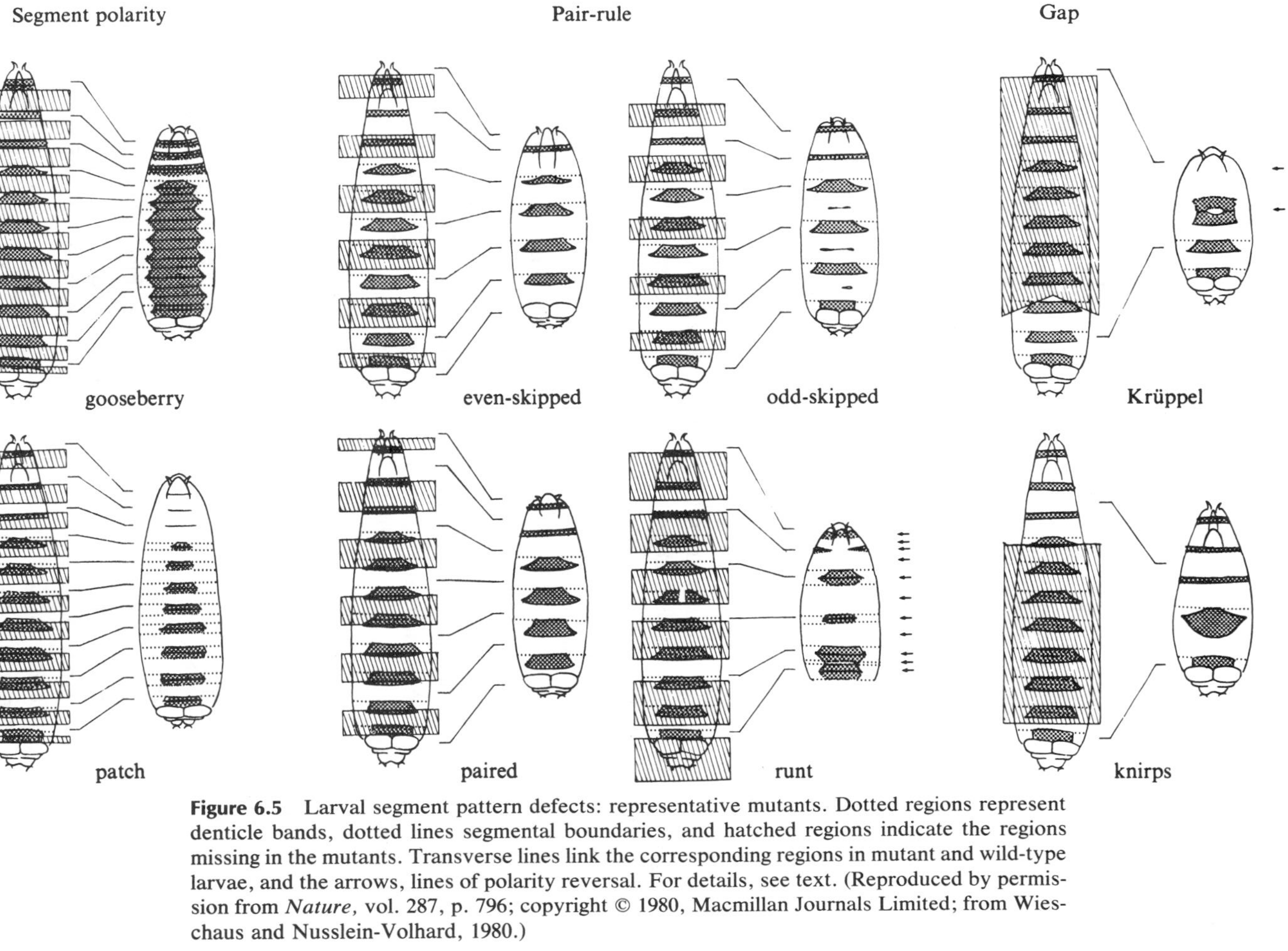

Figure 6.5 Larval segment pattern defects: representative mutants. Dotted regions represent denticle bands, dotted lines segmental boundaries, and hatched regions indicate the regions missing in the mutants. Transverse lines link the corresponding regions in mutant and wild-type larvae, and the arrows, lines of polarity reversal. For details, see text. (Reproduced by permission from *Nature*, vol. 287, p. 796; copyright © 1980, Macmillan Journals Limited; from Wieschaus and Nusslein-Volhard, 1980.)

phenotype defect can be rescued by maternal expression of the wild-type allele (King, 1970; Nusslein-Volhard and Wieschaus, 1980).

Although both the segment-polarity and pair-rule mutants produce a recurring, repetitive segment aberration throughout the thoracic and abdominal regions, these two general mutant phenotypes are strikingly different. The segment-polarity mutants are aberrant within every segment and, with only one exception, are deleted for the naked posterior cuticle in the ventral region of each segment. The duplicated portion of each segment is the anterior denticle band region, which is present in reversed orientation. (The polarity of the duplicated denticle band regions can be judged because the different rows of denticle hooks within each segment typically point either anteriorly or posteriorly.) The exception to this general mutant phenotype is *patch* (*pat*), which duplicates a bit of naked cuticle anterior to the denticle band. Because the duplication includes material from two adjacent segments, *pat* produces a doubling of the number of segment boundaries. The mutants *wingless* (*wg*) and *hedgehog* (*hh*) show the reverse effect. Because their duplication begins behind the position of the normal segment boundary, their mutant phenotype is of repetitive mirror-image duplications without any intervening boundaries. In only two of the mutants, *fu* and *gooseberry* (*gsb*), does the anterior margin of the duplicated region begin with a normal segmental boundary; these mutants therefore show duplications within the normal number of segment boundaries.

In contrast to the polarity mutants, the pair-rule mutants show effects in every other segment, being deleted for alternate segments. The *even-skipped* (*eve*) mutant is missing the even-numbered abdominal segments (AB2, AB4, AB6, and AB8) plus T1 and T3, while *odd-skipped* (*odd*) shows the opposite pattern (missing the AB1, AB3, AB5, and AB7 segments) plus T2. In both mutants, the deleted portion is the denticle band of the affected segment plus the adjoining naked cuticular region. In *paired* (*prd*), the deleted segmental regions are somewhat shifted. The missing region includes the naked posterior part of each odd-numbered segment and the anterior denticle band region of the following even-numbered segment. The phenotypic consequence of lesions in all three of these genes is the creation of bands of composite character, for example, anterior mesothorax and posterior metathorax bands for *eve* and *prd*. The segment deletions of *runt* (*run*) begin in alternate segments but, like the segment-polarity mutants, create mirror-image duplications of the denticle bands. The deleted portions in *run* encompass more than an entire segment, producing a greatly shortened embryo. The embryonic phenotypes associated with lethal *en* alleles are also complex. Although the embryos show a range of defects, one common category is that of embryos with deleted posterior portions in even-numbered segments. On this basis, it qualifies as a pair-rule mutant. However, in addition, every segment shows some alteration at the anterior margin and in the adjacent cuticle. The embryonic defects of *en* mutants are therefore spaced at both single- and double-segment intervals.

Gap mutants differ from the other two classes in showing specific regional aberrations of segment pattern. The *knirps* mutant, for instance, takes out all of the abdominal segments except for the termini; the thoracic segments appear approximately normal. In *hunchback* (*hb*), the defect is the complementary one; *hb* lacks a mesothorax and a metathorax while displaying a normal abdominal segmentation. *Kruppel* (*Kr*) is deleted for the thorax and most of the abdominal region, with abdominal segments 6, 7, and 8 being essentially normal. Anterior to the sixth segment in *Kr* embryos is a plane of mirror-image symmetry, producing a partial reversed abdominal phenotype. However, the position of the plane is variable. Although the *Kr* embryo phenotype is reminiscent of *bic, Kr* is strictly a zygotic genome function. The existence of loci that mutate to give the gap phenotype signifies the existence of some region-specific zygotic expression in the creation of a segmental pattern, an expression that is superimposed on the repetitive segment-generation signals.

In assessing the nature of wild-type gene action from the study of mutant effects in general, one may provisionally distinguish between *establishment* and *maintenance* functions. Does the mutant defect stem from an initial failure to create a particular developmental stage or to perpetuate that state once created? The segment pattern genes have not yet been definitively categorized in these terms. However, for several mutants, visible defects in segmentation become apparent very early in embryogenesis, suggesting that the wild-type genes may be important in establishing a segment pattern. For both *prd* and *eve,* the number of ventral bulges, the first sign of segmentation, is reduced in proportion to the reduction in segment number. For *Kr, run,* and *kni,* shortened germ bands, an indication of reduction in segment number, are apparent 15 min after the start of gastrulation (Nusslein-Volhard and Wieschaus, 1980). For a *prd*^{ts} mutant, the TSP was found to occur during the blastoderm stage, essentially coincident with the beginning of zygotic genome expression. For mutants of these five genes at least, the normal segment pattern appears never to be established. However, mutants of *en* appear first to establish and then to lose their normal segment pattern (Kornberg, 1981a). The embryonic function of *en* may therefore be one of maintenance. Nevertheless, it should be remembered that the distinction may be artificial: some or many gene activities may participate in both the establishment and maintenance of particular structures.

The significance of the genetic results is that they delineate a threefold level of spatial control in the process of segmentation when assessed in conjunction with the known maternal effects. The largest organizational unit in this process is that of the egg as a whole. In its capacity to evoke one or more a-p gradients during late cleavage that affect the segmental pattern, the egg is a single global system. The smallest unit of spatial organization is the individual segment or possibly the half-segment (in light of the half-segment duplications in the polarity mutants). It is within this unit that normal segment polarity is established and the segmental polarity mutants act. The third unit of organization is of intermediate size and was unsuspected until the genetic analysis revealed it: the double-segment unit. Its existence is suggested by the pair-rule mutants. The simplest interpretation of the pair-rule mutant phenotype is that certain signals act at spacings equivalent to double-segment widths.

Finally, superimposed on these basic single and double units of pattern organization, there appear to be region-specific zygotic genome activities whose existence is signified by the gap mutants. The pattern disruptions produced by mutations of these genes, however, probably have less to do with the process of segmentation per se than with the processes that confer segment character, both regional (thoracic versus abdominal) and individual (first thoracic versus second, etc.).

The formation of segments in the embryo is an instance of a general phenomenon, that of repetitive or periodic pattern formation. Some of the considerations relevant to the mechanisms of such periodic patterning, and the possible nature of some of the *Drosophila* mutants that affect the segmentation process, will be discussed in Chapter 8.

The Segment As Developmental Unit: Fields and Compartments

An embryonic ''field'' can be defined as an autonomous region of development within the embryo. Inside the field, cells can substitute for one another, to a greater or lesser extent, following removal or death of some members of the set; cells from outside the field cannot substitute for those within it. There is also a less precise definition of a field as the area of tissue that normally forms a particular structure. The functional and geographical definitions of field do not always coincide. For the sake of simplicity, the functional definition will be the one used here.

Although, in practice, the precise delineation of an embryonic field by experimental means is often hedged about with uncertainty, the concept of semi-independent cellular regions is a useful one and has motivated much research in experimental embryology. Even animals as small and simple as nematodes have embryonic fields. The cell equivalence groups, defined by cell ablation experiments, are the nematode equivalent of embryonic fields.

Does the larval segment qualify as an embryonic field? The physical defect fate-mapping experiments suggest that the cells destined for one segmental unit have an extremely limited ability to substitute for cells destined for another unit. However, because the commonly observed result in the induced damage experiments was multiple adjacent segmental defects, the results do not rigorously establish the individual segment as a field. In the gynandromorph fate-mapping experiments, the increased sturt distances between cells across segment boundaries may indicate some form of developmental restriction between segments and, hence, some degree of developmental independence. But again, the findings do not clearly establish the autonomy of segments as self-contained developmental units.

The strongest evidence that individual larval segments are such units comes from genetic experiments in *Oncopeltus*. In the milk weed bug, unlike *Drosophila,* it is possible to induce marked clones by X irradiation during cleavage stages. The genetic basis of these clones (distinguished by their color differences from the background cells) is obscure, since they can be induced in the wild-type as well as in heterozygotes for pigment marker genes. The basis of many of the clones is probably chromosome breakage and the formation of aneuploid cells, although in heterozygotes, mitotic recombination may make some contribution. The striking comparison in clone appearance is between that of clones induced in cleavage to those induced during or just after blastoderm formation. The difference is illustrated in Figure 6.6. Clones induced during cleavage often occupy or spread into adjacent segments; clones induced in the immediate aftermath of the

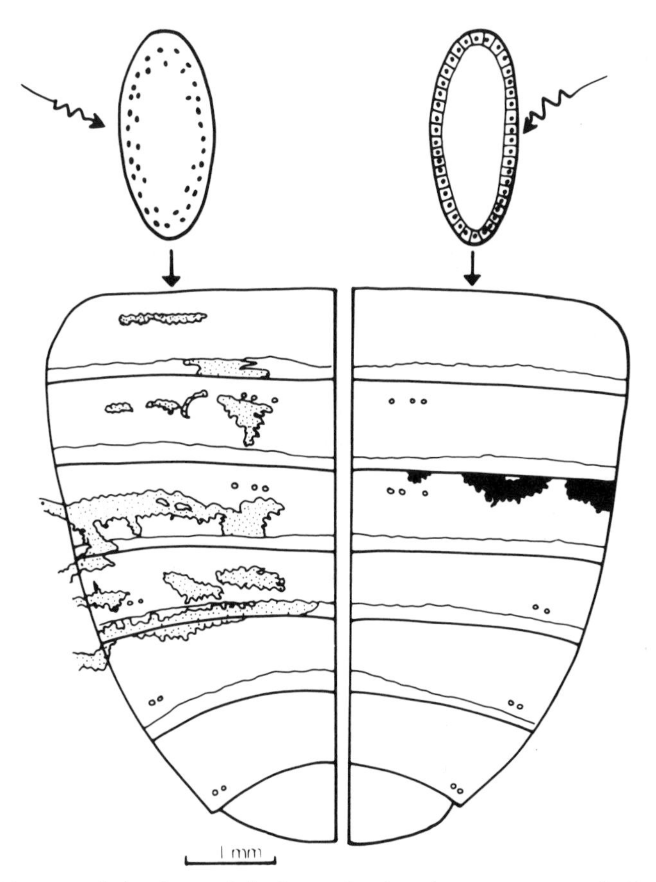

Figure 6.6 Patterns of clonal growth in *Oncopeltus* larval segments, ventral side, for clones induced before blastoderm (left) and after blastoderm (right). For discussion, see text. (Reproduced with permission from Lawrence, 1981; copyright MIT Press.)

blastodermal stage never do (Lawrence, 1973, 1981). This difference in clonal pattern between cleavage and postcleavage stages might in principle be a trivial consequence of cell growth dynamics. If the formation of the blastoderm is accompanied by a rapid burst of cell division, the postblastodermal clones would necessarily be much smaller than clones induced earlier. Occupying less space, they would be less likely to be large enough to straddle two segments. Nevertheless, clone size measurements rule out this explanation. The postrestriction clones are only about one-half the size of prerestriction clones, indicating that only one cell division intervenes between the two stages. Nor is the apparent restriction likely to be a consequence of cell loss at the boundary through infolding. If the latter were the case, the cleavage-stage clones would be expected to show marked discontinuities across segment borders. For the most part, they do not.

The observations imply that the *functional* demarcation of individual segments takes place before the morphogenetic events of segmentation. Cells formed on one side of the boundary cannot cross over and participate in the development of the segment on the other side. The boundary, in fact, makes a clean demarcation line; clones that abut the border are perfectly straight at this edge but highly irregular in outline where they project into the segment. This straightness of clonal borders along segment boundaries is also seen when boundaries are made to re-form. Mosaics can also be made by first transplanting pieces of *white* mutant integument into wild-type (orange) orange hosts in positions that straddle boundaries, followed by cauterization along both edges. In the initial stages of healing, the border between orange and white cells is convoluted. Regeneration of the segment border establishes a straight line between donor and host material (Wright and Lawrence, 1981b). Even when the segment border is considerably displaced from its usual position, the new boundary creates a perfectly straight demarcation between the two cell groups. Evidently, segment boundaries create or enforce separations of *defined cell lineage groups*.

The implication for genetic control is that initial separation of neighboring groups of cells creates adjacent but noninterchangeable embryonic fields. Within these discrete units, differences in gene expression occur that act to establish segment-specific differences in development. Initially, these differences must occur in the regions of the segment boundaries, at the initiation of segment separation, and presumably entail the creation of some form of cell surface differences between cells on opposite sides of the segment boundaries. These differences across boundaries presumably persist throughout development and are soon augmented by the differences that create individual segment phenotypes.

Viewed in this way, an embryonic segmental field is a domain of unique gene expression, although the number of distinguishing differences in expression may be small. Furthermore, each segmental field is composed of clones that contribute only to that segment. The segmental field is therefore a closed lineage group. From this genetic perspective, embryonic fields that are distinct lineage groups can be termed "compartments" and the process of separation of these groups designated "compartmentalization" (Garcia-Bellido et al., 1973; Garcia-Bellido, 1975). Furthermore, since the larval segments in *Oncopeltus* are never completely "filled" by one clone, each must be composed of several clones.

To sum up this line of thought: each insèct larval segment is a polyclone or compartment and, to some degree, a domain of unique gene expression. These differences are not great; indeed, biochemical and immunological tests have failed to disclose them (cited in Lawrence and Morata, 1983) to date, but subtle biochemical differences can be just as significant in development as major ones. The strongest evidence that segmental identities involve distinct patterns of gene expression is that mutations in certain genes transform all or part of different segments into one another in highly specific ways. These homeotic genes are the subject of the next section.

Genetic Control of Segment Character

Although mutations in many genes can affect segment character, two gene clusters in particular may play especially central roles. Both gene groups were detected through their mutant homeotic effects in the adult fruit fly. Only subsequently were the genes discovered to have profound effects on larval segment identity and structure. The first, the "bithorax complex" or "BX-C" (Lewis, 1978), is named after the original mutant mapped to this region, whose partial haltere-to-wing transformation was mentioned earlier (p. 138). The BX-C is located on the right arm of the third chromosome at 3–58.8 (bands 89E1–E4). The second gene cluster involved in the control of larval segment identity has been named the "Antennapedia complex" or "ANT-C" (Kaufman et al., 1980). The *Antennapedia* (*Antp*) mutants, which give the complex its name, are dominant homeotics that show degrees of transformation of the antennal structures of the adult fly into leg structures (Fig. 8.3). ANT-C is also located on the right arm of the third chromosome but is closer to the centromere, at 3–48 (bands 84B1–2).

The larval transformations produced by mutants of these genes have been most extensively documented for the BX-C. The initial description was given by Lewis (1978) and has been elaborated by Struhl (1981c). However, the history of recorded larval segment transformations in insects extends much further back. Transformations analogous to those wrought by BX-C mutants occur in the silk moth, *Bombyx mori,* and are mediated by the E locus in this species. These effects were first described in the 1930s; a review of the early literature of the E locus can be found in Tanaka (1953).

Although both gene complexes appear to play central roles in the delineation of the segmental phenotype, the BX-C has been more extensively characterized and is better understood. Nevertheless, before plunging into its description, some warning to the reader who plans to explore the original literature is in order. The interpretations to be presented are derived primarily from a synthesis of the combined imaginal and larval phenotypic transformations produced by BX-C mutations. For the most part, the effects of a given BX-C genotype are parallel for comparable larval and imaginal segments; however, some differences, which may reflect quantitative differences in expression or requirement between the two stages, are known. Furthermore, unless one reads the material closely, it is not always apparent whether the observations refer to a larval or imaginal segment. In addition, because imaginal segments are often more richly marked with landmarks than the larval segments, particularly in posterior half-segments, it is not always clear whether putative larval/imaginal segment differences for a given genotype are real or an artifact of scoring. Unless one is aware of these problems, the distinction between interpretation and observations can become lost.

These problems duly noted, it can be said that the study of the larval segment phenotypic characteristics produced by BX-C action yields a consistent and interesting picture of the operation of this group of genes. Four conclusions from this analysis have emerged: (1) the gene activities of the BX-C shift development away from mesothoracic (T2) development; (2) the greater the number of BX-C functions expressed in a segment, the more posterior in type is that segment's phenotype; (3) each abdominal segment seems to require a particular BX-C gene (an *infra-abdominal* or *iab*) function, plus all or most of those functions expressed in more anterior segments; and (4) there is a rough correspondence, at least, between the proximodistal order of the *iab* functions and the anatomical a-p sequence of the abdominal segments.

These conclusions are based primarily on experiments involving the addition of defined segments of the BX-C to background genotypes totally lacking the complex, and from assessment of the resulting larval phenotypes. However, several mutant classes, initially named for their effects in the adult, have contributed to the analysis; of these, two need to be mentioned. The first class is the *Ultrabithorax* (Ubx) group of mutants, recognized in adults by a slight haplo-insufficient phenotype of enlarged halteres. In late embryos, *Ubx* homozygotes show both AB1 and T3 transformed to a T2-type phenotype. Larval homozygotes for the next most distal gene, *bithoraxoid* (*bxd*), show a transformation of AB1 to a T3-like phenotype.

To assess the developmental effects of specific partial copies of BX-C, deletions of the complex must be induced. Although deletions can be induced directly by x-rays, most large deficiencies are inviable, creating problems of stock maintenance. In doing the analysis, it is preferable to create x-ray–induced translocations that split the complex into two parts. In such translocations, one part of BX-C is transposed to another chromosome. The process can be illustrated as follows, where A and B represent any arbitrary subregions of the complex, separated by the x-ray break:

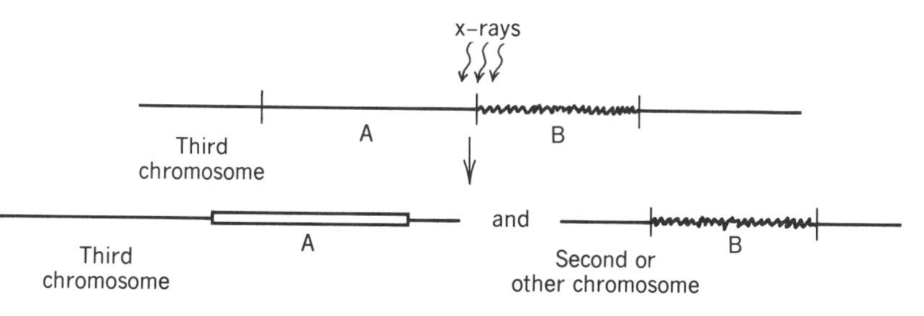

The position of the break is not under the control of the experimenter. However, by generating enough translocations, one can often obtain the separations desired. For many translocations, possession of both halves of the translocation by the animal permits viability. From these viable parents, the two halves can be segregated into progeny and analyzed for the embryonic phenotypic effects associated with removal of the other half of the translocation.

By placing the deleted third chromosome over a complete deficiency for the region, *Df(3)P9*, one produces progeny deleted for B and containing only A. Conversely, by placing the other chromosome, which now contains the B region, over a complete deficiency, one produces animals with the complementary deletion. By these means, one can assay the developmental effects mediated by any two complementary regions of the entire gene complex. To denote the two reciprocal deletion chromosomes, the original third chromosome from which material is removed will be termed the *deficiency* (*Df*) chromosome, and the transposed piece will be referred to as the *duplication* (*Dp*) chromosome.

A partial map of the BX-C is shown in Figure 6.7*a*, and includes a schematic representation of several deficiency and duplication chromosomes. Deleted regions are shown in white, the remaining regions in black. Thus, *Df P9* is deleted for the entire region, while *Df* Ubx^{109} is missing all but the portion most distal from the centromere. (In a heterozygous combination with a wild-type chromosome, this chromosome gives the imaginal *Ubx* phenotype (slightly enlarged halteres), by which the chromosome was identified as bearing a lesion in the BX-C.) The *Df* bxd^{100} chromosome is deleted only for the centromere-proximal part of the complex, while its complementary duplication, *Dp bxd,* possesses only that portion; *Df P10* and *Dp P10* are a similarly complementary pair, in which the break is somewhat more distal.

By placing these different deficiencies over the complete deletion chromosome (*P9*), one can assess the effects of adding back progressively larger pieces of the complex to the complete deletion background. The effects are schematized in Figure 6.7*b*. The most complete transformation is that seen in the *P9* deletion homozygote: all segments from T3 to AB8 are transformed to a T2 phenotype. The transformation can be seen in both surface views and in section (Fig. 6.8). Evidently, the T2 segmental state represents some form of "ground state" (a notion that will be examined more closely below): in the absence of any BX-C gene activities, all segments "fall" to this ground state.

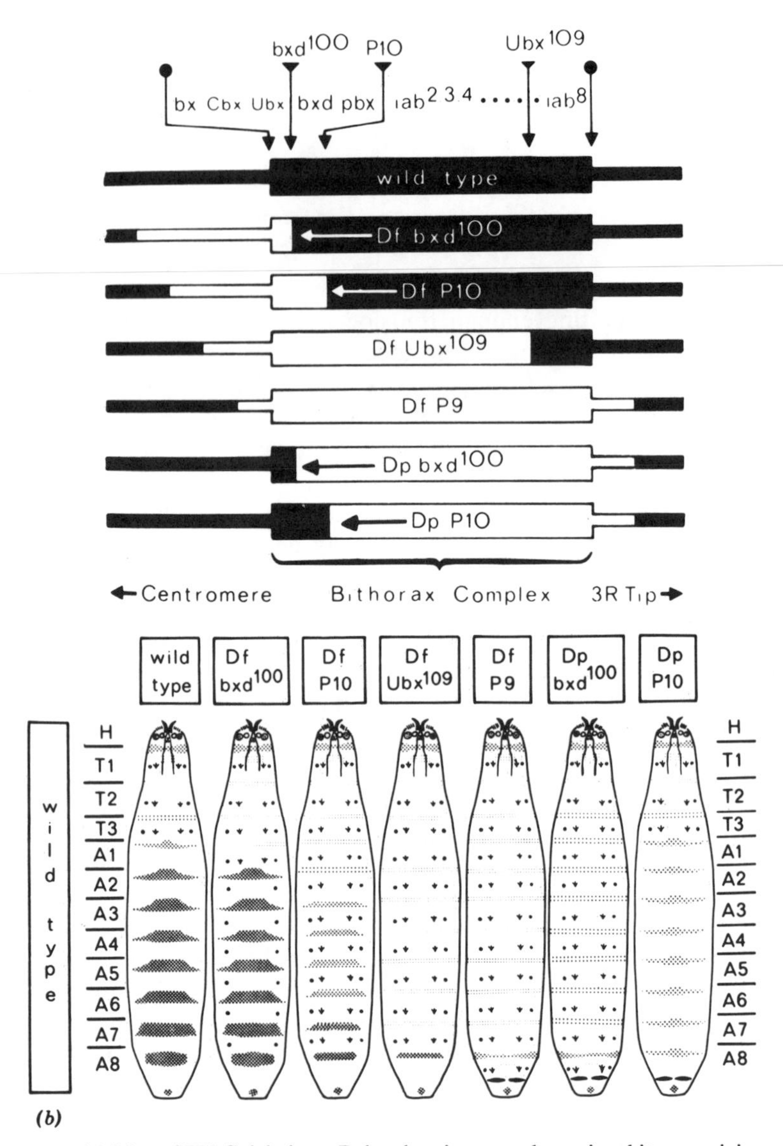

Figure 6.7 (*a*) Map of BX-C deletions. Deleted regions are shown in white, remaining regions in black. (*b*) Genotypes and corresponding embryonic phenotypes of BX-C deletions. H, head region; T1–3, thoracic segments; A1–8, abdominal segments. See text for discussion. (Reproduced with permission from *Nature,* vol. 293, p. 39; copyright © 1981, Macmillan Journals Limited; Struhl, 1981c.)

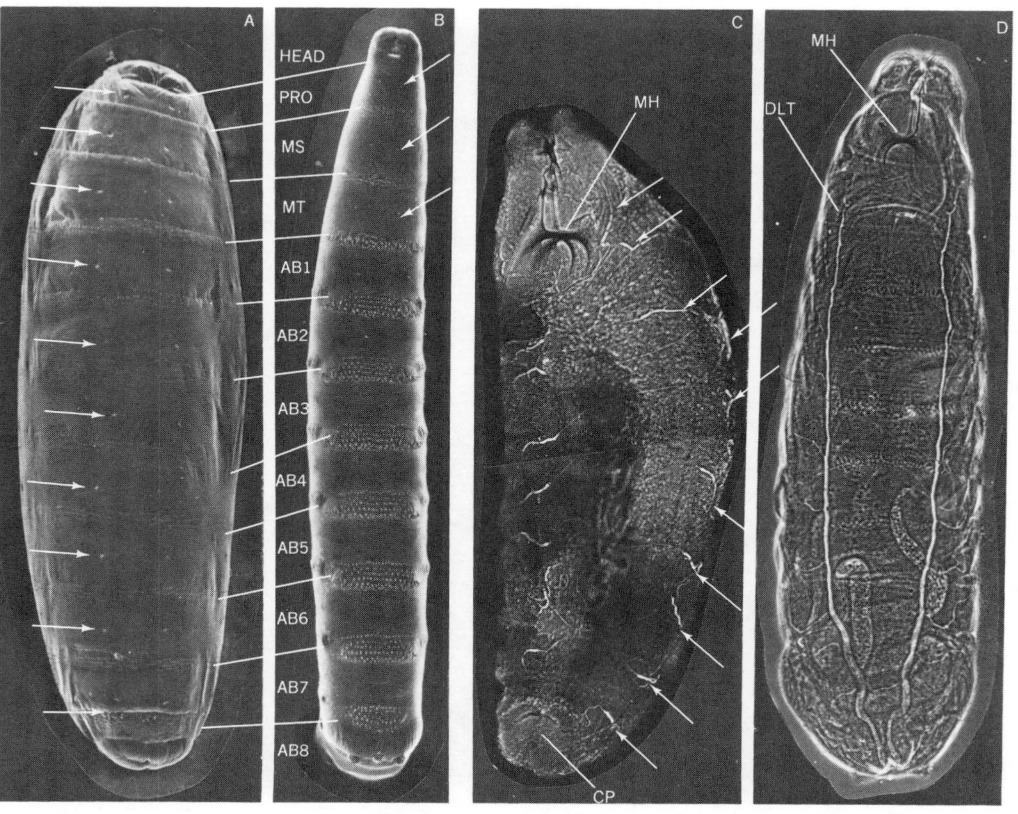

Figure 6.8 Comparison of *Df(3)P9* embryos (*a* and *c*) with wild-type embryos (*b* and *d*). (*a*), (*b*) Surface cuticular patterns; (*c*), (*d*) internal tracheal patterns. In (*a*), arrows indicate Keilin's organs (a thoracic structure); in (*c*), arrows indicate separate tracheal sections. DLT, dorsal longitudinal trunk; MH, mouth hooks; T1–T3, thoracic segments; AB1–AB8, abdominal segments. Magnifications: (*a*) 160X, (*b*) 60X, (*c*) and (*d*) 120X. (Reprinted by permission from *Nature,* vol. 276, p. 569; copyright © 1978 Macmillan Journals Limited; figure kindly provided by Dr. E.B. Lewis; from Lewis, 1978.)

By adding back just the most proximal portion of the BX-C (to the *P9* background)—the *Dp bxd*^{100} genotype—all segments are transformed to a T3-like state. This region contains *Ubx*$^{+}$ (plus a few genes whose inactivation affects the T3 imaginal phenotype). The result suggests two things: (1) only this most proximal portion is required for T3 development, and (2) *Ubx*$^{+}$ may be normally expressed in T3 and all more posterior segments. Conversely, the absence of this most proximal portion (genotype *Df bxd*^{100}) gives nearly normal AB2–AB8 development (although with thoracic ventral pits), suggesting that abdominal segment development is primarily governed by the part of the complex distal to *Ubx* and its nearby genes. (The abnormal development of AB1 in this genotype reflects the fact that the translocation breakpoint is in *bxd*, the gene required for AB1 development.) These results indicate that BX-C is roughly divided into a proximal-thoracic and a larger distal-abdominal section.

Beyond this gross division, the remaining genotypes suggest that progressively more distal pieces of BX-C within the "abdominal region" promote the development of progressively more posterior segmental phenotypes. Thus, *Dp P10,* containing slightly more distal BX-C than *Dp bxd*^{100}, produces an AB1 level of development in AB1–AB8 rather than a T3 level, while its complement, *Df P10,* gives less normal AB2–AB7 development than *Df bxd*^{100}. *Df Ubx*^{109}, containing only the most distal part of BX-C, shows a T2 level throughout, except for AB8, which has a recognizable abdominal segment phenotype (although not a normal AB8 one). The results suggest that there is a gene required for AB8 development specifically in the most distal part of the complex but which is unable to give full AB8 development in the absence of more proximal BX-C genes.

In sum, the findings indicate that each segment requires a particular function and all or most genes proximal to it (Lewis, 1978, 1981). When the larval effects of certain of the dominant mutants (to be discussed in connection with the imaginal phenotypes) are taken in conjunction with this analysis of deletion genotypes, the following proximodistal gene order of segment functions can be deduced: *Ubx bxd iab-2 iab-3 . . . iab-5 . . . iab-8*. This sequence parallels the sequence of thoracic and abdominal segments, in which each segment requires both its own defining function and most of those proximal to it. (AB8 may differ from the other abdominal segments in not requiring *Ubx*$^{+}$ gene activity.) Thus, in going from AB2 to AB8 within the larva, there seems to be a roughly corresponding step-like gradient of expression within BX-C such that with each additional posterior step (segment), an additional BX-C function is activated. In this construct, a given segmental phenotype results from the *summed* expression of the activated BX-C functions.

This idea is diagrammed in Figure 6.9. In this scheme, no BX-C functions are required for mesothorax (T2) development; perhaps all (except *Ubx*) are expressed and needed in AB8; and each segmental phenotype between T3

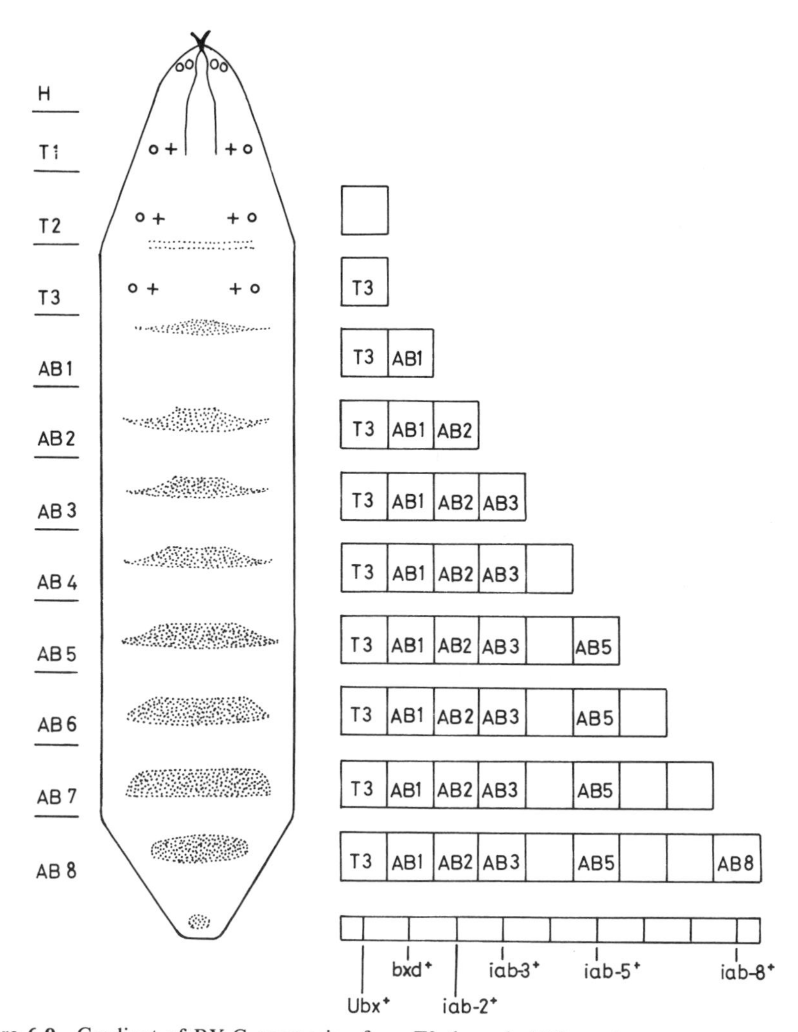

Figure 6.9 Gradient of BX-C expression from T2 through AB8; each segment expresses at least one unique BX-C function plus all those proximal to it. (Adapted with permission from Lewis, 1981.)

and AB8 is determined by the expression of a segment-specific gene function plus all those proximal to it in the complex.

How might this stepwise activation of gene functions be achieved? Lewis (1964, 1978) has proposed that a gradient of repressor is involved; the gradient is presumed to run from a high point in the center of the embryo (the T2 anlage) to a low point at the posterior end (the AB8 anlage). In principle, the shape of the postulated gradient should promote progressive derepression of gene functions from T2 to AB8. To account for the specificity with which genes are derepressed along this line, it is further posited that there is a gradient of increasing "operator strengths" from proximal to distal along the BX-C (Lewis, 1978). Weak binding constants of "thoracic gene" operators would allow derepression of these functions at below-peak repressor concentrations (from T3 posteriorly), while increasing repressor affinities of the *iab* functions would produce their sequential derepression from T3 to AB8 only as repressor concentration decreases. The gradient is postulated to be present at blastoderm, with the setting of BX-C gene activities produced at this time.

The action of the BX-C taken as a whole may be regarded as "raising" the development of all segments that are posterior to T2 from a mesothoracic "ground state." The notion of a mesothoracic ground state frequently recurs in the literature but is generally used without precise definition. The idea of a ground state is helpful only if it is carefully defined. Both the BX-C evidence and that of the maternal effect mutants and phenocopy experiments, reviewed in Chapter 3, suggest that the mesothorax is a developmental basal state of some sort. In purely developmental terms, the mesothorax may be a development "sink" or ground state, possessing the fewest prospective potencies of any other segment.

However, this developmental status is usually taken to signify that mesothorax is also a *transcriptional ground state,* that relative to other segments, the cells of T2 transcribe fewer genes. The model discussed on pages 140–141 interprets mesothorax in this way, but at the present time there is no independent molecular evidence to support this proposition. Also, the development of larval T2-type segments in homozygous *P9* embryos (Fig. 6.11) suggests that BX-C itself is completely unexpressed in T2. However, evidence on imaginal development, to be discussed, shows that at least one BX-C function is at least transiently expressed in the T2 segment anlage. Therefore, T2 is not a state of zero transcription for BX-C, although relative to other segments it is a transcriptional ground state for the gene complex.

In one further sense, the evolutionary one, mesothorax may be seen as a ground state. The two-winged fruit fly (and all Diptera) evolved from four-winged ancestors, and these progenitor insects most probably evolved from multilegged, millipede-like arthropods. In genetic terms, this evolutionary progression has consisted of the acquisition of wing- and leg-suppressing functions, and the BX-C can be regarded as a package of these genes, preserved as a genetic ensemble of related functions (Garcia-Bellido, 1977; Lewis, 1978). The mesothorax of *Drosophila* may therefore be the most direct lineal "descendant" segment of the basic segment in the original arthropod ancestor. In its own right, however, it is a complex and highly evolved unit; wings and legs, which are not possessed by the abdominal segments, are intricate anatomical structures.

Before leaving the Lewis model of BX-C action, one further point should be noted. The model stipulates individual segments as the domain of action of BX-C gene functions. The observed transformations that were the basis of the model are almost all of the *anterior* halves of segments; the posterior ventral portions of segments are so lacking in anatomical detail that transformations in them cannot be directly ascertained. Recent evidence, derived in part from imaginal phenotypes, suggests that many or most of the transformations affect predominantly the anterior halves of the segments (see Cole and Palka, 1982; Lawrence and Morata, 1983). Whether there are special genes embedded in BX-C that govern posterior halves specifically, or whether such genes are even required, is an open question (one of many) at this point.

While the domain of BX-C action is the group of segments from T3 to AB8, the ANT-C appears to govern the complementary region, namely, T2 and all segments anterior to it. The larval segment changes associated with mutations of the ANT-C have been described by Wakimoto and Kaufman (1981). Although many of the mutant embryonic phenotypes are difficult to classify, the defects in lethal embryos are primarily in segments anterior to T2. Thus, several of the complementation groups in ANT-C that produce a lethal response for noncomplementing members produce defective mouth parts and an abnormal pseudocephalon. The clearest segment transformations in embryos are those associated with lethal mutants of the dominant *Sex combs reduced* (*Scr*) locus and the *Antp* complementation group. The adult phenotype of *Antp* mutants, a leg-for-antenna substitution in heterozygotes, has been mentioned; the mutant imaginal phenotype associated with *Scr* is a diminution in the row of pronounced bristles, the sex comb, found on the prothoracic leg of the male. Lethal embryos deficient for these two complementation groups show nearly opposite segmental transformations. *Scr* embryos show a transformation of ventral prothorax (T1) toward a ventral mesothorax (T2)-like morphology (plus certain abnormalities of the head segments). Embryos homozygous for recessive lethal deficiencies in *Antp* show transformations of ventral mesothorax and metathorax (T2 and T3) to prothorax (T1). The possible basis of these reciprocal transformations will be discussed later in connection with the imaginal mutant phenotypes of mutations in this gene complex. The essential point in the present context is that mutations in ANT-C can produce recognizable larval segment transformations in segments anterior to T3.

"Regulation" of BX-C and ANT-C

Although the gene activities of BX-C and ANT-C are necessary for the correct specification of segments, they are only two subsets of these genes. A growing number of genes have been identified from their mutant phenotypes, which also play crucial roles. Mutations in a few of these genes seem to "drive" the segmental phenotype toward a T2-like state; the great majority drive the transformation in the opposite direction, toward an AB8-like state. All have effects in both imaginal and larval tissue, but their relative importance for the two stages is a function of the particular gene activity

Table 6.3 Additional Genes Regulating the Segmental Phenotype

Locus	Map Position	Zygotic/ Maternal	Embryonic Trans-formation	Imaginal Trans-formation[a]	Reference
Polycomb (*Pc*)	3–47.2	Yes/weak	→ AB8	→ AB8	Lewis (1978); Duncan and Lewis (1978)
Polycomblike (*Pcl*)	2–84	Yes/?	Weak or nil	→ AB8 (weak)	Duncan (1982)
extra sex combs (*esc*)	2–42	Yes/yes	→ AB8	→ AB8 (weak)	Struhl (1981c); Struhl and Brower (1982)
supersexcombs (*sxc*)	2–55.3	Yes/yes	→ AB8	→ AB8 (weak)	Ingham (1984)
trithorax (*trx*)	3–54.2	Yes/yes	→ T2 (weak)	→ T2	Ingham and Whittle (1980); Ingham (1983) Ingham (1981)

[a] As seen in heterozygotes for the dominants and homozygotes derived from heterozygous mothers for the recessives.

(judged from the relative severity of the transformation produced by amorphic or extreme hypomorphic mutations). A few of these genes are listed in Table 6.3.

There is little definite information on the nature of these gene activities, but a word of caution is in order. Most attention has been given to those genes that drive the segmental phenotype "caudad," toward an AB8 state. For several genes, such as *Pc* and *esc,* the transformation has been shown to depend on the presence of a functional BX-C; in the absence of the BX-C, produced by making the embryos simultaneously homozygous for the *P9* deficiency, the "derepressed" (AB8) phenotype of the mutant (in a BX-C^+ background) is replaced by a string of T2-like segments. Two interpretations have commonly been placed on this fact. The first is that one or more of these genes "regulate" BX-C, presumably at the transcriptional level. The second is that the gene in question may code for the postulated repressor in the Lewis model. Strictly speaking, neither conclusion follows.

The epistasis of *P9* to the derepressed phenotype, seen with *Pc* (Lewis, 1978) and *esc* (Struhl, 1981c), shows only that BX-C gene products are essential for the development of a string of AB8-like segments; it does not establish that the wild-type product(s) of the genes modulate the activity of BX-C at any level. As discussed in Chapter 5, the interpretation placed on a given epistatic response is only as plausible as the underlying model. The very fact that so many genes—and more than one constitutes a large number of postulated regulators—have been cited as possible regulators of BX-C, on the basis of similar transformation phenotypes, suggests that the application of prokaryotic repressor models in this instance is an oversimplification, at the very least. (The general problems of regulation by diffusible repressors in eukaryotes were discussed in Chapter 1.) Furthermore, clonal analysis has demonstrated that the wild-type allele of several of these genes must be present until late in the development of imaginal tissues to prevent the appearance of an AB8 segmental phenotype in segments anterior to AB8 (Struhl, 1981c; Duncan, 1982). In effect, *Pc* and *Pcl,* and perhaps many more, are primarily maintenance function genes. To posit that they are also establishment function genes is to stretch plausibility even further.

In this respect, *esc* seems to be different. It is a gene with a very strong maternal effect (Struhl, 1981c), and, judging from the TSP of one ts mutant that is centered on blastoderm, the gene is not required postblastoderm (Struhl and Brower, 1982). Of the various putative repressor-coding loci of those known, *esc* would appear to be the strongest candidate. However, one fact tells against it. When *esc* embryos from *esc* mothers are also homozygous for mutations of *trithorax* (*trx*), a zygotic function that drives imaginal phenotypes toward a T2 state, the normal segmental pattern is largely (although not completely) restored (Ingham, 1983). The result shows that the wild-type *esc* product is not required for the differential setting of BX-C activities. If *esc* were *the* repressor coding gene, the result would be inexplicable.

The cumulative results are intriguing and show that BX-C activities, and probably ANT-C activities as well (Struhl, 1981c), determine the segmental phenotype in combination with numerous other gene activities. However, the molecular basis is completely obscure. At this time, attempts to construct strict hierarchies of gene regulation are premature. And, indeed, reevaluation of the notion of a simple gradient of repressor is probably overdue. The complexities of expression of BX-C revealed by molecular studies have already undermined the concept of a simple gradient of "operator strengths" as the principal basis of differential BX-C expression (Chapter 9); without the postulated gradient of chromosomal repressor binding strengths, the concept of a repressor gradient is itself weakened.

Indeed, the observations on BX-C bring one back to the fundamental problem of the a-p determinative gradient(s) in the early *Drosophila* embryo and the fact that the gradient(s) begin or end in the middle of the embryo (Chapter 3). Although the nature of determination in the early embryo and the mechanism of segment assignment are usually treated as separate phenomena, they are undoubtedly two sides of the same coin. A conscious recognition of this point might facilitate progress on both problems.

Setting Segment Periodicity and Setting Segment Identity: Are They Separate Processes?

In the earlier discussion of the genetic basis of segment formation, a distinction was drawn between genes that affect the overall segmental patterning, in particular the spacing of segments, and those that govern individual segment character. Is this distinction valid? The answer is both "yes" and "no."

A general test for the independence of gene effects in creating a phenotype is whether the effects are additive. If they are, independence of action is established; if they are not, the two genes are deemed to affect the same process. If segment patterning and segment identity mutants control different processes, then double mutants, carrying both types of mutation, should show additivity. This is the observed result for several tested combinations. Double mutant *prd; Ubx* larvae show both double segments (as in *prd* alone) and a mesothoracic transformation of the *anterior half only* of the first double abdominal segment; the posterior half of this double segment (of AB2 provenance) is typically abdominal in its features (lacking mesothoracic sense organs). Similar additive and independent effects are seen for other comparable double mutant combinations described by Nusslein-Volhard and Wieschaus (1980).

However, it is also true that certain other genes can mutate to produce larval segment-spacing abnormalities and/or homeotic transformations of certain segments. Thus, lethal *en* mutations produce a variety of general segmental abnormalities, including the pair-rule defect, while the original en^1 mutation gives a distinctive intrasegmental transformation (p. 270). Two additional genes described by Kaufman (1983) similarly can mutate to give either effect. Deficiencies for both *fushitarazu* (*ftz*) in ANT-C and *Rg-pbx*, which is separable from and distal to the ANT-C, give dramatic segmental abnormalities, while rare hypermorphic or neomorphic mutations give a *postbithorax* (*pbx*) transformation; this mutant phenotype consists of a transformation of the posterior half of the haltere to that of the posterior wing. (*Rg-pbx* is not within the BX-C.)

The significance of such dual classes of effect produced by different mutations in the same gene(s) is unclear. One possibility is that in certain instances, initially correct segmental spacing or delimitation is a prerequisite for the correct establishment of segmental identity (Kaufman, 1983). However, the imaginal homeotic transformations produced by the rare dominant *Rg-pbx* and *ftz* mutations take place in the absence of a pronounced disturbance of the segmental pattern. There are at least three further explanations to be considered. One explanation is that certain gene products are needed for both the correct completion of segmental delineation and the establishment or maintenance of segmental identity; however, certain mutations can abolish one function without necessarily affecting the other. The second possibility is that at least one of the effects—either the segmental delimitation or the homeotic—is a nonspecific and secondary consequence of the primary mutational defect. The third is that the rare dominant in either or both cases actually competes with normal BX-C expression; some molecular evidence indicates that this is the case for the dominant *ftz* mutation (Laughan and Scott, 1984; Chapter 9).

Setting the Dorsoventral Larval Pattern

While the setting of cell fate along the a-p axis establishes segmental pattern, that along the d-v axis establishes the three germ layers and ultimately the organization of tissue types in the embryo (Fig. 3.20). Just as the creation of a segmental pattern involves a maternal "instructional" process and an "interpretive" zygotic genome response, the setting of tissue fates along the d-v axis similarly involves maternally synthesized factors and essential zygotic genome functions. However, as is the case in segment pattern formation, the distinction between maternal and zygotic genome functions is not always entirely clear; several of the key genes are expressed both maternally and zygotically, and have effects on d-v tissue determination.

The first visible event in embryogenesis marking the emergence of the d-v axis is the invagination of the cells along the midventral position to form the mesoderm. As discussed in Chapter 3, several maternal effect mutations abolish normal d-v fate setting, creating either a dorsalization of the embryo, with abolition of the mesodermal invagination, or a ventralization (the Toll mutants) that expands ventral pattern elements at the expense of dorsal ones. The original mutant of this series, *dorsal (dl),* gives an extreme dorsalized embryonic phenotype from homozygous mothers: embryos that consist of long, enclosed tubes of dorsal-characteristic cuticle. A milder phenotype, termed the "*dl*-dominant" phenotype, is seen in embryos produced from heterozygous mothers grown at high temperature; these embryos lack mesoderm but produce most of the characteristic ventral cuticular elements.

Two zygotically acting genes have also been identified that affect the d-v pattern. Mutant homozygotes of these genes produce embryos with a phenotype similar to that of the maternal *dl*-dominant effect. These genes have been designated *twist (twi),* at 2–100 and *snail (sna),* at 2–51, which is on the other side of the centromere from *dl* itself (at 2–52.9) (Anderson and Nusslein-Volhard, 1983).

The similarity between the maternal *dl*-dominant and zygotic *twi* and *sna* effects suggests a possible link between the action of these genes. If all three genes affect the same process (in setting d-v pattern polarity), then their similarity of embryonic mutant phenotype reflects the failure of completion of this process in all three cases. Simpson (1983) has examined the possibility of functional links between the three genes by testing for synergistic interactions between them. The test involves an examination for interaction between reduced maternal dosage of dl^+ activity and reduced zygotic genome dosage for *twi* or *sna*.

As noted earlier (p. 133), the temperature-sensitive maternal *dl*-dominant effect seems to reflect a reduced dosage of dl^+ product. Although *twi* heterozygotes from homozygous dl^+ mothers are fully viable at temperatures ranging from 18 to 29°C, such progeny from heterozygous *dl*/+ mothers show both temperature-sensitive viability and embryonic dorsalization; their +/+ siblings do not show the effect. A smaller degree of synergism was also found for *sna* heterozygotes. The effect is not a function of particular alleles but specifically of reduced maternal *dl* and zygotic *twi* and *sna* dosage. In effect, *twi,* and to a lesser extent *sna,* acts as an enhancer of the *dl*-dominant effect, indicating that all three genes affect the same process in d-v pattern setting at the midventral position.

Two further results of interest emerged from this study. The first was that an increased *zygotic* dosage of dl^{+} could produce some measure of rescue of *twi* heterozygotes. This finding shows that *dl is* zygotically expressed, despite the absence of measurable paternal rescue of $dl/^{+}$ embryos (i.e., in the absence of the *twi* effect). It confirms a point emphasized in Chapter 2: that standard paternal rescue tests provide only a crude measure of whether or not zygotic expression of a gene is occurring. The second point of interest is that no synergism could be detected between *twi* and maternal *K10* or *gad* effects. The failure of *gad* to show an interaction with *twi* indicates that mere similarity of phenotype is insufficient to indicate functional relatedness.

Simpson discusses two possible interpretations of the maternal–zygotic synergism. The first is that the dl^{+} gene product establishes a graded d-v pattern-setting instruction and that the *twi* and *sna* gene products interpret this morphogenetic instruction. The alternative possibility is that the *dl, twi,* and *sna* gene products physically cooperate in creating the morphogenetic instruction. If the second interpretation is correct, one would predict that both *twi* and *sna,* either separately or together, might cause a shifting of d-v positions in the fate map of homozygous embryos, as occurs in the *dl*-dominant effect (p. 134).

Following mesodermal invagination, the second observable event in the differentiation of the larval d-v pattern is the separation of neuroblasts from hypoderm in the ectodermal layer. In wide-type embryos, the neuroblasts arise in bands on either side of the newly invaginated mesoderm, and the surface ectoderm arises laterally to the neuroblasts (Fig. 3.20). The separation of neural from surface ectoderm has long been known to be under direct genetic control. In mutants of the X chromosome locus *Notch* (*N*) (named for its heterozygous dominant mutant effect on wings), the "neurogenic" region is greatly expanded at the expense of the epidermal precursor cells (reviewed by Wright, 1970). Effectively, *N* mutants ventralize a large part of the epidermal precursor region.

Indeed, mutants of seven genes are known to have this effect; they are listed in Table 6.4. Three have slight dominant effects (*N,Dl,* and *E*(*spl*)); the other four exert their neuralizing effects only in homozygous embryos. Although mutants of all seven loci create similar neural hypertrophy in the ventral and cephalic regions, they differ in the extent of this conversion, even for the "strongest" alleles of the several loci. Lehmann et al. (1981) list the degrees of embryonic neuralization for mutant homozygotes of five of the genes, as follows:

$$mam < bib < N < neu < Dl$$

Table 6.4 Loci That Mutate to Give Embryonic Neural Hypertrophy

Locus	Map Position	Reference
Notch (*N*)	1–3.0	Jimenez and Campos-Ortega (1982)
big brain (*bib*)	2–34.7	Lehmann et al. (1981)
master mind (*mam*)	2–70.3	Lehmann et al. (1981)
neuralized (*neu*)	3–50	Lehmann et al. (1981)
Delta (*Dl*)	3–66.2	Lehmann et al. (1981)
almondex (*alm*)	1–27.7	Lehmann et al. (1983)
Enhancer of split (*E spl*)	3–89.1	Lehmann et al. (1983)

One possible explanation for these differences is that they reflect differing extents of maternal wild-type gene product storage in the embryos, with the least extreme gene effects (*mam* and *bib*) being associated with large stores of wild-type products. Jiminez and Campos-Ortega (1982) tested this hypothesis by inducing homozygous germ line clones (see p. 122) for mutants of the relatively weak zygotic *mam* and the relatively strong zygotic *N* neuralizing functions, and measuring the extents of neuralization. It was found that homozygous embryos from these clones for both mutants showed a comparably strong neuralization. The equalization of neural hypertrophy shows that the difference in neuralization between these two "zygotic genome" loci is a function of differing degrees of maternal expression. Presumably, all of the differences in expression reported for ostensible amorphs of the different loci reflect such differences in the degree of maternal expression. The neuralization mutants provide another instance of a similar phenotype being produced by mutations in several different genes. Neuralization is not "controlled" by any one of them, but rather results from the absence of any of the respective wild-type activities.

The mode of conversion of epidermal to neural precursors in the neuralizing mutants is completely unknown, but the phenomenon exemplifies the difficulty of interpreting the nature of wild-type gene actions from mutant phenotypes. It may be that every one of the seven loci characterized by the neuralizing mutant phenotype produces something that is essentially for creating the neuroblastic state. However, it is equally possible that neural hypertrophy reflects some moderately nonspecific metabolic perturbation, with the neural cell state being some form of ectodermal sink; cells might fall into this sink when pushed by the metabolic derangement. In this connection, it may be pertinent that *N* point mutants are reported to have altered activities for four mitochondrial oxidizing enzymes (Thorig et al., 1981). Could the seemingly specific developmental aberration of the neuralizing mutants be a consequence of something as simple as an altered oxidizing capacity of the epidermal precursor cells?

DEVELOPMENT OF THE IMAGO: GENETIC ANALYSIS

The adult fruit fly takes shape during the pupal stage, when the imaginal discs unfold, differentiate, and join together. The entire sequence of differentiative and morphogenetic events is triggered by the hormone ecdysone, which is secreted at the end of the third larval instar. Yet ecdysone is just a releasing signal for imaginal development; it triggers preorchestrated but hidden programs of change within the imaginal cells. By the time of pupariation, all of the major imaginal developmental assignments have been made; metamorphosis only makes manifest the cumulative results of these hidden programs. (The ecdysone signal also triggers the proliferation of the partially differentiated but previously divisionally quiescent imaginal histoblast cells.) Because the developmental history of the imaginal discs is essentially invisible (apart from the increase in cell number), genetic analysis has assumed an even greater part in its investigation than in the comparable studies of cell commitments in larval development.

Genetic studies have contributed two different categories of information about imaginal disc development. The first may be termed "cell descriptive"; it consists of the delineation of precursor cell positions, numbers, and growth dynamics. This information is particularly important for delineating imaginal development in early embryogenesis, when the imaginal cells cannot be distinguished from the surrounding larval cells. The second group of observations concern the processes of allocation of developmental fate in imaginal discs development and the genes that govern these processes.

The cell-descriptive data will be discussed first. As in larval development, the fate mapping of the initial precursor cells of the imago is the logical place to begin.

Fate Mapping the Imago

The method of choice for fate mapping the structures of the adult fruit fly on the early embryo is that of gynandromorph mapping. Because of the wealth of surface detail on the adult fly, which can be partitioned by any one of innumerable dividing lines, this method is inherently more sensitive than the induced defect method. First invented by Sturtevant (1929), gynandromorph mapping was revived 40 years later by Garcia-Bellido and Merriam (1969). And of the various means for inducing X chromosome loss in the early embryo, the ring-X loss method is the most sensitive (Hall et al., 1976).

The localization of imaginal precursor cells on the surface of the early embryo is a more complicated procedure than the mapping of larval segments or segment sections discussed previously. Unlike the latter, which involves essentially one-dimensional mapping problems, mapping the surface of the imago onto the blastoderm embryo involves projecting a three-dimensional object onto a plane. It is accomplished by mapping structures initially two at a time, then expanding the map piecemeal by localizing the

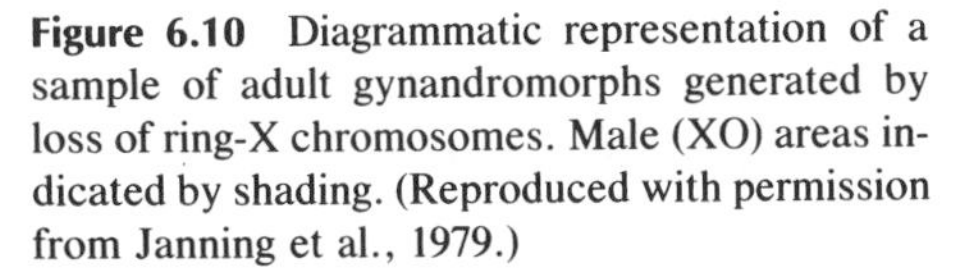

Figure 6.10 Diagrammatic representation of a sample of adult gynandromorphs generated by loss of ring-X chromosomes. Male (XO) areas indicated by shading. (Reproduced with permission from Janning et al., 1979.)

primordium for a third structure with respect to the first two, then adding a fourth by the same method, on so on. In effect, one localizes the anlagen by a sequence of expanding triangulations.

Figure 6.10 shows a set of schematic drawings of a number of adult gynandromorphs, the raw data for the analysis. There is a distinctly stronger tendency for mosaic boundaries to follow segmental boundaries and the longitudinal midline than in the larva, an effect of the separate development and subsequent suturing together of the different parts of the imago relative to larval gynandromorphs (see Fig. 6.4). As in larval mapping, however, the distance between each pair of structures is measured in sturts, the percentage of gynandromorph sides in which the structures are of different X-chromosomal composition. The mapping procedure can be illustrated. Imagine that three structures, A, B, and C, are found to be separated as follows: A and B by 5 sturts, B and C by 3, and C and A by 4. Taking the A-B line as defining one axis, there are two possible locations of C with respect to the second axis:

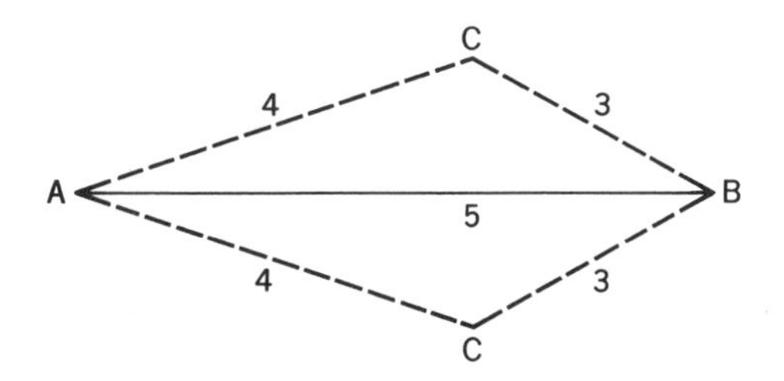

To fix the actual position of C vis-à-vis A and B, one adds a fourth point, D, then constructs new triangles, and establishes the most consistent set of distances between the points.

Sturt distances can also be used to map the relative size of different primordia. In general, the greater the number of precursor cells for a given structure, the more often that structure will be split by a mosaic dividing line. (For example, we have previously noted that mesothoracic legs are mosaic 23% of the time in ring-X loss gynanders.) The frequency of mosaicism is a function of the diameter of the anlage cell group (Hotta and Benzer, 1973) and therefore serves as a measure of the number of initial precursor cells. Structures that do not map as points are usually given as circles on gynandromorph fate maps.

This method is accurate only for mapping anlagen that are relatively close together. There are two reasons: the first is geometrical. On the assumption that one is mapping primordia to the cellular blastoderm stage, each measurement involves approximating an arc distance (the distance along the rounded surface of the blastoderm) to a linear distance. For short distances, the approximation is good; it becomes progressively less so over larger distances. The second reason for the greater accuracy in mapping short distances has to do with the observed shapes of the boundary lines. Although often following straight lines for short distances, as for instance along segment boundaries, the lines are often convoluted. With any curved boundary line that falls two or any even number of times between two structures, their anlagen will be registered as not being separated by the boundary line if one scores separations in terms of different X-chromosomal compositions. Compare these two figures.

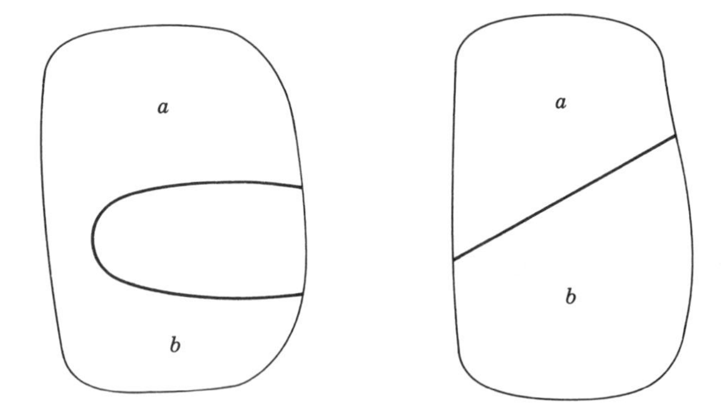

The greater the distance between *a* and *b*, the greater the probability of there being convolutions of the kind shown on the left, with a corresponding contraction of the calculated map distance between the primordia. One solution is to count the number of times that the boundary visibly passes between the structures of interest (Janning, 1978). An alternative is to map the internal structure of subregions of the embryo first, for example, the head or thorax, and then connect the subregions by further triangulation (Garcia-Bellido and Merriam, 1969).

In either approach, one obtains estimates of distances for well-separated primordia by summing all of the intervening small distances. The problem is not unlike the one encountered in conventional mapping of genes on chromosomes. The greater the distance between any two genes, the greater the number of multiple crossovers. Since any even number of crossovers produces chromosomes that appear nonrecombinant, the less accurate become the two-locus recombination frequencies the greater the true distance between them. Accurate genetic map distances are obtained by summing the smaller intervening map distances.

This analogy between genetic mapping and gynandromorph fate mapping suggest another solution: the construction of a mapping function. A mapping function is one that translates measured distances into actual ones by means of the appropriate mathematical conversion. Several fate-mapping functions have been constructed. An analytical solution has been given by Flanagan (1976). An empirically derived formula has been obtained by Janning et al. (1979):

$$f(x) = a(e^{x/a} - 1)$$

where x is the observed distance, $f(x)$ is the true map distance, and $a = 25$ (an empirically derived constant). The function permits the observed sturt distances between points taken two at a time to be fitted to the curve of estimated true distances, derived from summing the intervening component small distances.

To position the fate map with respect to the physical surface of the embryo, one needs some external reference point. A convenient one is the distance to the dorsal midline for the various imaginal primordia, obtained by dividing the sturt distance between equivalent structures on the left and right sides by two. The anteroposterior locations are obtained by normalizing distances along the longitudinal axis to the dorsoventral distance (Garcia-Bellido and Merriam, 1969). A modern gynandromorph map, constructed with the mapping function given below, is shown in Figure 6.11.

Fate mapping of the *Drosophila* embryo by gyandromorph mapping is a sufficiently precise and quantitative set of procedures to be a satisfying analytical problem in itself, and a large corpus of theory has been built up around it. Apart from the statistical and mathematical intricacies, however, it has produced several points of biological interest.

The first is that the fate map of imaginal structures bears a strong resemblance to that of the adult fruit fly. Head structure anlagen are anterior for the most part, thoracic structures are in the middle, and abdominal structures are posterior. The largest general difference between the fate map and the final adult plan is that all of the imaginal primordia occupy a dorsal position on the fate map. The ventral third, which is empty of imaginal

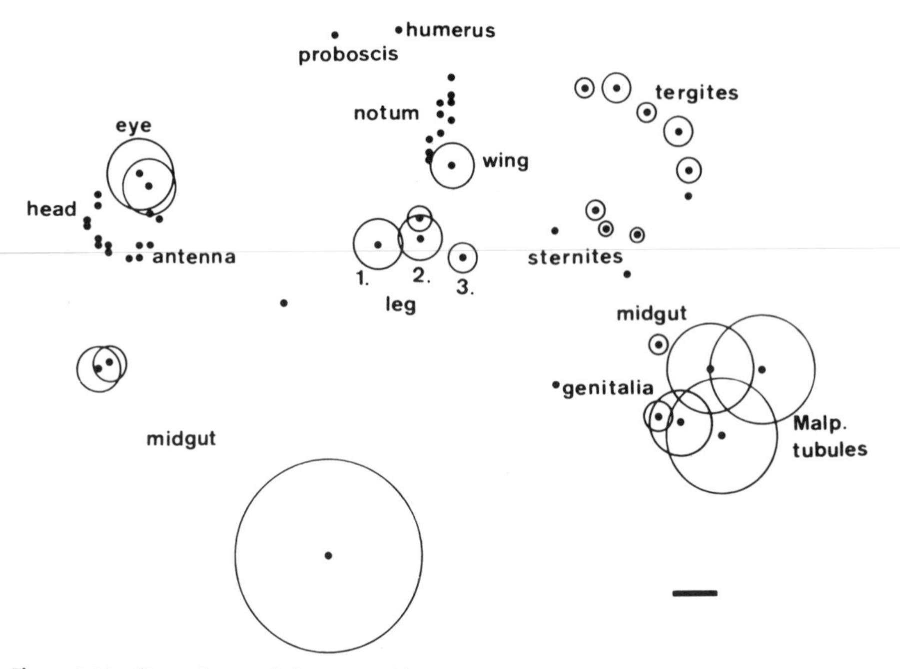

Figure 6.11 Gynandromorph fate map of imaginal primordia. The diameter of the circles is a measure of the size of the primordia at blastoderm. The positioning of the anlagen largely prefigures that of the adult structures. Scale bar: 10 sturts. (Reproduced with permission from Janning et al., 1979.)

anlagen, contains the mapped anlagen for the nervous system and the larval mesoderm; as we have seen, the precursor cells for these systems are indeed located ventrally at the blastoderm stage.

The second finding is that the larval and imaginal fate maps show a close relationship: comparable elements and segments for the larva and the adult fall in close proximity to one another on the two maps. The detailed larval maps are derived by creating gynanders for either X-linked enzymatic activities or pigmentation markers (*white, yellow, chocolate,* which distinguish particular larval structures). Thus, head structures for both stages are anterior, thoracic segments are in the middle, and so on. The fate mapping therefore strengthens the conclusion that imaginal and larval cells in proximity within the same segment initially have much in common.

(The concordance of larval and adult fate maps is also seen with fate maps constructed by the induced defect method (Lohs-Schardin et al., 1979 a,b). The centers of defects for the imaginal thoracic segments are between 55 and 50 EL, as in the comparable larval map, and the distances between the three leg centers—the ventral thoracic segments—are about three cell widths, the same distances seen between larval segments.)

The third general point concerns the stage of development that the fate map describe. The traditional assumption is that the maps pertain to the cellular blastoderm stage. In principle, the actual stage can be ascertained from the map position of any structure or cell group whose precise physical position at blastoderm is known on independent grounds and that subsequently changes position during development. The obvious candidate is the germ line, derived from the pole cells, which initially occupy the most posterior position in the blastoderm embryo and subsequently migrate dorsally and then internally. Gehring et al. (1976) were able to map the germ line, starting from the premise that gynandromorphs lacking mature cells possess a gonadal mesoderm of different sex composition from the germ line itself (see van Deusen, 1976). The mapping placed the germ line precursors at the posterior pole, as expected; a similar result, employing a different mapping strategy, was obtained by Nissani (1977). The results confirm that the blastoderm embryo is the stage whose fate map is described.

The fourth general conclusion from gynandromorph mapping emphasizes the probabilistic element in imaginal anlagen placement in the blastoderm embryo. Because restrictions on potency and assignments of developmental fate continue right up to pupation with an expanding imaginal cell population, it follows that the map itself cannot be a strict one-to-one localization of anlagen to particular invariant blastodermal cell positions. Each point on the gynandromorph fate map, or each center of a circle (for large primordia), is the *high point of the probability distribution* for the structure or cell type. This is a biological property (not a property of experimental error) and is precisely the expected property in any system of specification by gradient. As emphasized in Chapter 3, determinative gradients are inherently probabilistic specification mechanisms; the cells may possess checking or fail-safe mechanisms to reduce the degree of variance, but a perfectly rigid initial specification system embodied in gradients is impossible.

Beyond its usefulness in plotting locations of imaginal precursor cells, gynandromorph mapping can be used to estimate the number of cells in each primordium. The method is that of the minimal mosaic patch measurement. All imaginal disc primordia have been observed in one or more sets of gynandromorphs to be mosaic in at least some instances, with the male–female boundary dividing them into sectors. The necessary conclusion is that no imaginal disc anlage is strictly clonal; all derive from at least two cells. For primordia that are nearly half male and half female, nothing more than an estimate of two precursor cells can be derived. However, in most collections of gynanders, much smaller patches of male tissue can be found. In each case, this minority contribution was derived from at least one cell. If all of the precursor cells divide at the same rate, then the smallest male patch is presumably the product of one cell. It follows that the reciprocal of this fraction should provide an estimate of the total number of precursor cells set aside for that primordium. Using the minimal patch as a measure, different imaginal discs have been calculated as arising from 8–20 founder cells on the blastoderm surface. The potential pitfall is that the method relies on the tail of the distribution of patch sizes, with most of the data thrown away. If the smallest sectors grow more slowly than the other component clones in the primordium, the estimate of precursor cell number will be correspondingly inflated. An alternative approach is to count all the patch size classes, each class presumably corresponding to some initial fraction of the total set of founder cells; the number of classes minus one provides an estimate of founder cell sizes. The method works only if the number of precursor cells is small; for larger numbers, the discreteness of patch size classes is in question.

Mitotic Recombination and the Delineation of Imaginal Growth

The most versatile and useful of the mosaic-generating techniques for exploring imaginal development is that of induced mitotic recombination. By inducing marked clones during development, it is possible to investigate almost all of the features of imaginal cell growth. To do so, one employs genetic markers that alter a visible cell phenotype without affecting the underlying cellular growth pattern. Suitable "growth-neutral" markers include those that affect cuticular pigmentation [such as *yellow* (*y*)], bristle morphology [*singed* (*sn*) or *forked* (*f*)], or hair morphology [*multiple wing hairs* (*mwh*)]. (The essential differences between bristles and hairs are that the former have sockets and are the product of four cells, being an "organule," while the latter are socketless and formed by single cells.) Since most areas of the *Drosophila* imaginal cuticle are distinguished by their pigmentation or bristle or hair patterns, almost any region can be marked and its patterns of clonal development followed.

Several very different kinds of questions about imaginal growth can be answered by means of clonal analysis using mitotic recombination. These include questions about (1) the number of founder cells for given primordia (the answers providing a comparison to the gynandromorph-generated estimates); (2) the average and regional rates of cell division; and (3) the particular spatial patterns of clonal growth in imaginal discs or histoblasts during the period of larval development. In *C. elegans,* all of this information can be obtained by direct observation. In *Drosophila,* with its much greater cell complement, clonal analysis by mitotic recombination provides an invaluable substitute for such direct observation. Given the number and variety of observations reported in mitotic recombination studies, it is advisable to consider the findings under three general headings: founder cell numbers, clonal growth patterns, and growth kinetics. An extensive review of the uses of mitotic recombination in probing imaginal growth can be found in Postlethwait (1978).

Founder Cell Numbers

The procedure for founder cell numbers employs the principle illustrated in Figure 3.17. Using staged embryos, one induces homozygous clones by X

Table 6.5 Primordial Cell Numbers Estimated from Mitotic Recombinant Patch Sizes[a]

Disc	Average Patch Sizes Measured	Number of Precursor Cells Estimated	Cell Number Estimates from Gynandromorph Minimal Patch Sizes
Wing	1/11, 1/12	5, 6	8, 12, 40
Leg	1/20, 1/15	10, 7.7	20
Antenna	1/7	3.5	
Eye-antenna	1/12	6.2	>13
Tergite	1/9	4.5	8

[a] All estimates are for the early *Drosophila* embryo.
Source. Data summarized by and cited in Merriam (1978).

irradiation and measures the average fraction of total structure/tissue marked in the emergent imagos. The reciprocal of this fraction yields an estimate of the number of cells present at the moment of irradiation. If one is scoring both daughters (a twin spot), the reciprocal of the fraction of all marked cells provides the estimate directly; if one is scoring only one homozygous genotype, one multiplies that reciprocal by one-half.

Some estimates of imaginal disc primoridal cell number at the blastoderm stage are given in Table 6.5. The numbers are small but in all cases are greater than one, showing, in agreement with the gynandromorph findings, that imaginal primordia are always polyclonal in origin. Thus, as in the early mouse embryo, early cell assignments are cell group phenomena.

When one compares the founder cell numbers obtained by mitotic recombination with those obtained by the minimal patch method, however, a discrepancy emerges. The estimates produced by the former are almost always smaller than those produced by the latter. Why should this be? One contributory element has been mentioned, namely, that the minimal patch estimates may be distorted by reliance on slow-growing clones, the consequence being systematic overestimates of cell number. A second possible factor is that the x-ray treatment, which is used to induce recombination, causes some cell death in the primordia, with compensating overgrowth of the surviving cells to yield a larger mitotic patch, with correspondingly reduced reciprocals and hence estimates of founder cell number. Some evidence for this x-ray effect has been obtained, although contrary evidence has also been produced (Wieschaus and Gehring, 1976b). There is also a third possible reason for the discrepancy. If cells that *can* contribute descendants to imaginal discs/structures are not strictly *limited* to doing so at blastoderm, then clonally marked cells in the structure in question will make a disproportionately large contribution to that structure; this action, in turn, will lead to an underestimate of founder cell number (as discussed further below).

The question raised by this consideration is when cell "allocation" to particular structures takes place in development *relative to the time of cell marking*. The time of cell allocation, or tissue foundation, can be defined as that point at which cell exchange between different primordia ceases (McLaren, 1972). Although the traditional assumption in estimating imaginal primordium cell numbers from clones induced at blastoderm was that cells were restricted to particular imaginal discs at that stage, there was no independent evidence to support the proposition. The available data on determination of single blastodermal cells, referred to in Chapter 3, is too limited to have settled the point either way.

The effect of late cell allocation relative to the time of cell marking is illustrated in Figure 6.12. Imagine, for instance, that in a certain region there

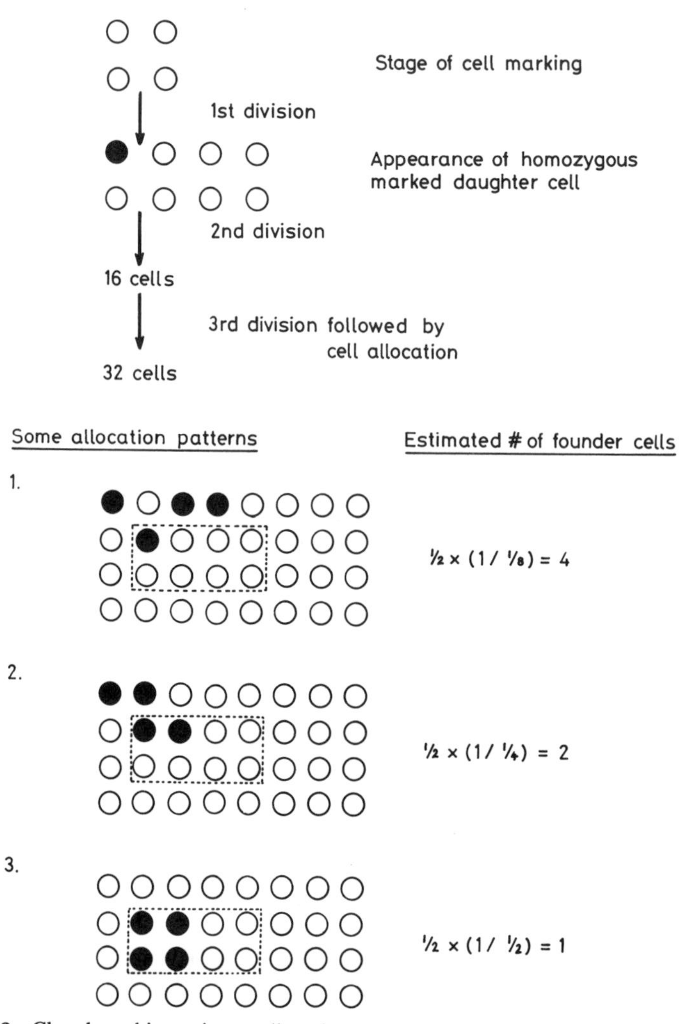

Figure 6.12 Clonal marking prior to allocation; some patterns of marked cell inclusion. See text for discussion.

are 4 cells that go through three rounds of cell division to produce 32 cells before fates are imposed, and that only 8 will form an imaginal primordium. Suppose further that during the period of growth, cells in this group can freely intermingle. If homozygosis was induced at the four-cell stage, then any marked cell may contribute one, two, three, or four cells to the primordium. Some of the corresponding estimates of primordial cell number are shown in the figure. In most cases, the observed contribution to the imaginal disc will be *disproportionately* large, although if one could score *all* of the descendant cells of the original four, the marked cells would always constitute four and (correcting for the descendants of the other daughter), hence one-fourth of the total.

The distinction between cell allocation and the process of determination should be emphasized. Allocation, or tissue foundation, refers merely to the process by which cells are provisionally assigned to form a particular structure and cease to mingle with other contiguous cells. However, these provisional assignments are not necessarily restrictions of developmental capacity. If placed in a different cellular environment, these cells might be able to develop differently.

Does cell allocation to different imaginal primordia truly take place after blastoderm formation, or is this just a hypothetical possibility? If disc-specific assignments of cells are made *following* the blastoderm stage, then some cells marked at blastoderm may be found to contribute to *two or more different discs*. If any discs do share descendants of the same blastodermal cell, they would be expected to be those whose primordia are close together on the blastoderm surface. In a detailed gynandromorph fate mapping of four pairs of thoracic discs—the three leg discs and the wing disc—Wieschaus and Gehring (1976a) estimated the size and relative location of their primordia using specific bristle landmarks within each. The results showed that all four primordia are contiguous, with many interdisc distances being less than many intradisc distances. One pair of wing and leg bristles were found to be particularly close; the anterior notopleural bristles (aNP) of the wing and the edge bristles of the second leg (IIEB) being only 6.2 sturts apart. (The leg preapical bristles, IIPAB, were also found to be about that distance from the aNP.) As can be seen in Figure 6.13, these bristles are not especially close in the adult (nor are the wing and leg discs contiguous in the larva). If cell assignments are not disc specific at blastoderm, then clones induced at this stage might cover both structures.

In a companion study, Wieschaus and Gehring (1976b) found this to be the case. In 13 clones marked at blastoderm that covered either the edge bristle or the preapical bristle, 7 were found to cover part of the wing, and then only in the region predicted from the fate mapping, embracing the aNP. Since only 1–2% of all wings and legs had any clones in these areas, the overlap could not have been produced by independently induced clones. By 7 hr of development, however, clones covering either of the respective wing or leg regions do not spread into the other disc. Evidently, some restrictive

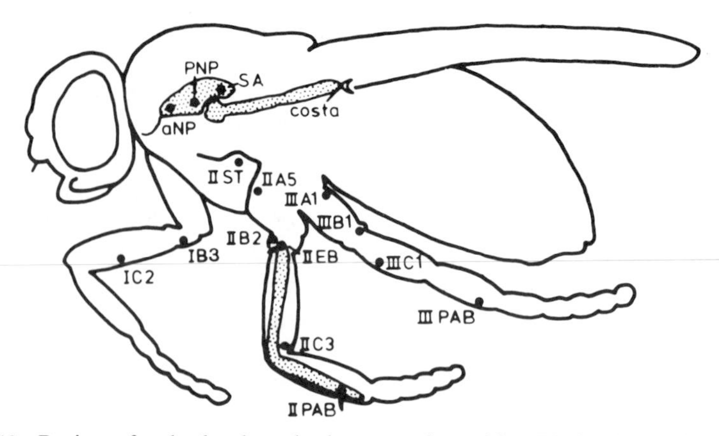

Figure 6.13 Regions of early clonal overlap between wing and leg. The key bristles illustrating shared wing-leg ancestry are: IIEB, edge bristle of the second leg; IIPAB, preapical bristle on the second leg; aNP and pNP, anterior and posterior notopleural bristles of the wing. (Reproduced with permission from Wieschaus and Gehring, 1976b.)

event intervenes between blastoderm (3 hr) and the seventh hour of embryonic development. Because clone size also decreases by a factor of two between these two time points, a cell division is inferred to take place between 3 and 7 hr. However, by itself, this decrease in clone size is not sufficient to explain the change. Evidently, either a disc-specific restriction in capacity or a disc-specific allocation occurs that separates precursor cells for the two discs.

The study also revealed that possession of common ancestral cells by discs is not simply a function of primordium distance. The closest precursor distance between landmarks of the second and third legs is essentially the same as that between the closest wing and second leg landmarks, yet mitotically induced clones are never observed to overlap both legs. This finding is reminiscent of that of *Oncopeltus*, where blastoderm marks a stage of clonal restriction between larval segments. It would appear that the anlagen of the thoracic segments in *Drosophila* are also compartments in the same way that larval segments of *Oncopeltus* are: the segment primordia consist of lineage-restricted polyclones by blastoderm. (However, the cephalic segments do not show such a restriction, and differ in some other respects from thoracic segment commitments to be discussed below.)

To sum up, it appears probable that the discordance between gynandromorph and mitotic recombination estimates of founder cell number are based, in part, at least, on the occurrence of disc-specific allocation after blastoderm formation. At the time of allocation, mitotic recombination estimates fix the approximate number of precursor cells for the visible cuticular structures of the different thoracic discs at about 10 per disc (Wieschaus and Gehring, 1976b). The total number of precursor cells per disc is, of course, somewhat larger because some precursor cells probably give rise to internal structures exclusively; for estimating total numbers accurately, additional methods, such as direct observation of early discs (Madhavan and Schneiderman, 1977) is needed. A review of the various techniques for estimating founder cell number can be found in Merriam (1978).

Clone Shape and Growth Patterns

The shape of marked clones on the adult cuticular surface can reveal much about the growth characteristics of the clone, and hence of the normal growth patterns of the constituent cells of each imaginal structure primordium. Several general features are apparent from the aggregate set of analyses.

In the first place, for most imaginal clones, the marked cells form a single patch of tissue. Thus, most individual descendant clones are coherent. This is true for all imaginal disc derivatives. A partial exception to the rule is provided by the abdominal histoblasts. Clones on the surface of the abdomen, as in *Oncopeltus* (Fig. 6.6), are often broken up into several patches. This difference between histoblast and imaginal disc clones reflects their different growth patterns. The histoblast cells, unlike those of the imaginal discs, begin their rapid period of multiplication only during the pupal period; rapid growth in this phase tends to disperse members of single-descendant clones (Madhavan and Schneiderman, 1977).

Furthermore, clones tend to have a shape characteristic of the region in which they arise. In the mesonotum (the proximal body portion of the mesothorax), clones are usually as wide as they are long. In the appendages—antennae, legs, and wings—clones are generally elongate, stretching along the proximodistal (p-d) axis. When clones are induced early, these elongate patches can extend over a large part of the appendage, frequently over many contiguous segments in the legs and antennae. In general, such clones are continuous, but in certain cases, reproducible discontinuities are observed for clones in particular regions. Such discontinuities reflect morphogenetic changes during the growth of the primordium. Thus, antennal clones stretching between the second and third antennal segments aways show a displacement of about 90°, reflecting a quarter-turn rotation between these segments in the development of the antenna.

Sometimes the discontinuity can be dramatic. The upper postorbital (UPO) and lower postorbital (LPO) bristles in the posterior part of the head, behind the eye, are contiguous regions, yet are never marked by the same clone except at the very earliest stage, blastoderm. When the eye-antennal imaginal disc is fate mapped by surgical means, using removal of bits to determine which region of the disc gives rise to which part of the head, the UPO and LPO regions are found to lie in opposite parts of the disc; they come together only during final morphogenesis. Thus, even when the clonal pattern appears odd, it can usually be explained in terms of the known growth or morphogenetic properties of the disc in which it arose. Within the discs of the appendages, the true shape of an early clone is generally that of a wedge-shaped, radial sector; eversion and stretching of the disc produces the elongate, p-d clones observed in the adult cuticular surface.

A third general feature of clonal growth is indeterminacy; collections of clones induced at the same time in different individuals and spanning the same region often differ in their precise shape and size. This is illustrated in Figure 6.14. Thus, single progenitor cells within a primordium can give rise to somewhat variable numbers of descendants, and specific delimited regions and structures can arise from any of several possible progenitor cells in the early primordium. (We have previously noted this fact in a different context, the mapping of progenitor cells for final structures in gynandromorph analysis.)

It is this property of indeterminacy in clonal growth that makes the strongest contrast to that in *C. elegans*. In the nematode, every division is

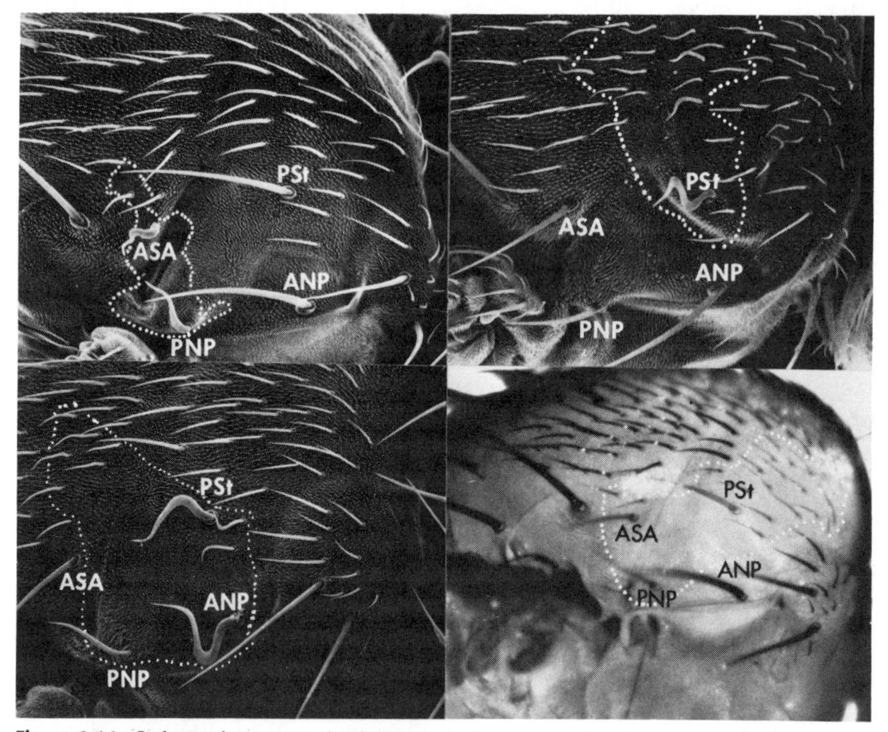

Figure 6.14 Indeterminate growth of clones on the mesonotum; dotted areas indicate clonal extent. Bristle landmarks: ASA, anterior supra-alar; ANP, anterior notopleural bristle; PNP, posterior notopleural bristle; PSt, posterior sternopleural bristle. (*a*), (*b*) f^{36a}; *mwh* clones; (*c*) a *y* sn^3f^{36a} clone; (*d*) a *y* clone. The clones in (*a*) and (*b*) were generated from induced mitotic crossing over in the genotype *Dp (3;1), y* mwh^+/f^{36a}; *mwh/mwh*. (With permission from Postlethwait, J.H., Clonal analysis of *Drosophila* cuticular patterns, in M. Ashburner and T.R.F. Wright (Eds.), *The Genetics and Biology of Drosophila*, vol. 2C. Copyright: Academic Press Inc. (London) Ltd. Figure kindly supplied by Dr. J.H. Postlethwait.)

determinative with respect to orientation and, except in the few instances of interchangeable roles, every cell has a fixed number of assigned divisions among its descendants. In *Drosophila,* which possesses many more cells, there is considerably more freedom in the division program. Within particular regions, the shape and relative size of each structure are fixed in the aggregate, but the division program of each clonal lineage is a statistical property with a measure of individual indeterminacy.

Imaginal Cell Growth Kinetics

Mitotic recombination is also used to measure the rates of imaginal cell growth. The principle is the same as that used in estimating primordial cell numbers. One induces clones at successive stages of development and takes the reciprocal of the measured average area marked by the clonal patches at each point, the growth curve then being constructed from these estimated cell numbers. In making the calculations, one needs to take into account whether twin spots (representing both daughters) or just single patches (one daughter only) are being used. For greatest accuracy, one should also correct for the fact that the cells in which mitotic recombination can be induced are in a restricted part of the age distribution of the growing cell population; the observed clones are, strictly speaking, derived from nonaverage cells. A recombination-inducible cell is always in the G_2 phase of the cell cycle, when chromosomes have replicated and each chromosome consists of two chromatids. These cells are therefore somewhat older than the average cell in an exponentially growing population, where young cells predominate. The progeny of a mitotically induced cell will therefore be relatively somewhat underrepresented. As discussed by Postlethwait (1978), the true numerical relationship between primordial cell number and clone size, when corrected for this age difference, becomes:

$$N_x = (N_f/C_f) \times 1.3 \qquad \text{for twin spots}$$

where N_x is the total cells in the primordium at the time of irradiation and N_f and C_f are the final cell numbers in the developed disc and the twin spot, respectively. For single spots, the relationship is

$$N_x = (N_f/C_s) \times 0.65$$

where C_s is the single clone size.

Growth curves for several imaginal primordia, constructed in this way, are shown in Figure 6.15. In each case, the estimate of final cell number was obtained either by counting, as is possible in the wing, where all cells produce a hair that can be counted directly, or from the relative areas of the disc surface and clone surface, using an average cell density measured over a small area. The curves represent average rates of growth throughout the

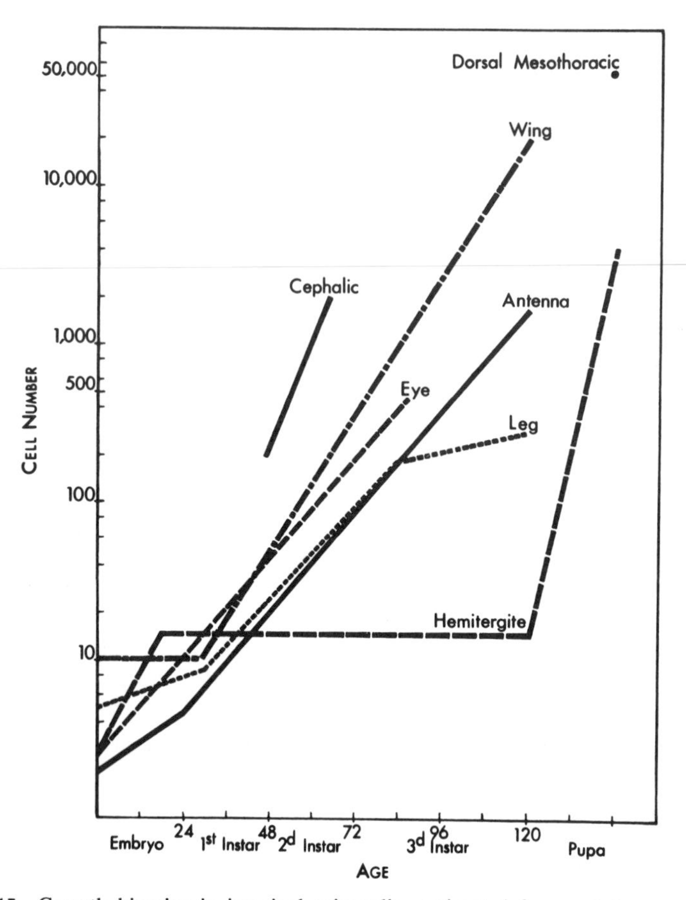

Figure 6.15 Growth kinetics in imaginal primordia, estimated from mitotic recombination experiments. Detailed references are given in Postlethwait, 1978. (With permission from Postlethwait, J.H., Clonal analysis of *Drosophila* cuticular patterns, in M. Ashburner and T.R.F. Wright (Eds.), *The Genetics and Biology of Drosophila,* vol. 2C. Copyright: Academic Press Inc. (London) Ltd.)

primordium. However, regional differences in precursor cell growth can, in principle, be measured in the same fashion.

The figure shows that (with the apparent exception of the leg) growth is exponential in the imaginal discs from the end of embryogenesis to the end of the third larval instar, with typical doubling times being 8–10 hr. The histoblast nests follow a different pattern: after a brief initial division period in embryogenesis, they do not divide again until the end of larval development. The results from clonal analysis thus perfectly parallel the direct observations on cell growth. (The leg appears to follow a third pattern, with an initial period of exponential growth followed by a later tailing off of growth, but this plot reflects the fact that only marked bristles were scored in the experiment; the asymptote of 440 reflects the average number of bristle cells in the legs.)

These depicted measurements were obtained before there were direct counts of cell numbers in the different discs. In a comprehensive study, Madhavan and Schneiderman (1977) provided the first complete set of cytological measurements of disc cell numbers. How well do the two sets of figures, genetic and cytological, agree? On the whole, very well, particularly with respect to rates of growth. Nevertheless, the genetic estimates consistently give 1.5 to 4.0 times fewer estimated cells than the direct measurements. There are at least two reasons for the quantitative discrepancy. The first is that x-rays, used to induce the recombination events, undoubtedly kill some fraction of the disc cells; the surviving cells, including the marked clones, may grow in compensation for these losses. The consequence is larger fractional marked areas and correspondingly decreased cell number estimates. The second and more significant factor is that the genetic experiments score only the outer cuticular cells, yet the disc possesses other cell types, including tracheal and mesodermal cells. Direct cell counts in discs include these cells, whereas clonal analysis of the cuticle perforce misses them. What the genetic approach can provide, in the absence of the more laborious direct cytological counts, is good *relative* measures of cell number and accurate determinations of average or regional growth rates.

Restrictions of Cell Fate

The most interesting observations from the mitotic recombination studies concern the apparent restrictions of cell fate that accompany growth. As development proceeds, induced clones cover fewer structures and occupy smaller areas. Three examples will illustrate this pattern. From the blastoderm stage until the middle of the first instar, clones marking bristles on the leg extend from the femur to the fifth tarsal segment, thus stretching nearly the full length of the leg; by the third instar, a given clone is restricted to only one leg segment (Bryant and Schneiderman, 1969). In the wing, some clones induced before the beginning of the first larval instar extend onto both the dorsal and ventral surfaces; from the first instar on, any clone is restricted to one surface or the other (Bryant, 1970). The third example involves cell type: until the early third instar (about 40 hr prior to pupariation), a clone can include both bristles and hairs; from that point on, all new clones consist only of marked bristles or hairs (Garcia-Bellido and Merriam, 1971a).

Other examples could be cited. Cumulatively, the results present a picture of a seemingly progressive restriction of developmental capacity, a process that has been referred to as "stepwise determination" (Gehring, 1976). However, if there are progressive restrictions in developmental capacity, they cannot be, except for the very late ones, to *specific cell types*.

The indeterminacy of clone shape necessarily indicates a certain indeterminacy in the derivation of particular structures from particular early cells, at least until late in development.

It could be that this progressive narrowing of clonal scope does not reflect an inherent restriction in cellular developmental capacity, but instead the trivial fact that later clones, being smaller, cannot include as many cells. If the distance between the two structures embraced by the large early clones is greater than that of the later clones, no single clone can possibly extend to both structures. Another facet of small clone size is that such clones are more likely to be truncated by any morphogenetic process such as infolding or regionalized cell death; larger early clones are more likely to extend through such local interruptions.

Imaginal Disc Developmental Commitments

Compartments and Selector Genes

To determine whether the restrictions on clonal extent reflect genuine restrictions in potential or the trivial constraints imposed by limited division capacity, one needs to uncouple growth restrictions from the developmental program. This can be done by generating clones that grow faster than the surrounding cells by virtue of some feature in their genetic makeup. If there are fundamental restrictions in developmental capacity that accompany normal imaginal development, these will occur as boundaries of clonal growth for the fast-growing clones. If, on the other hand, the apparent restrictions on clonal extent observed in the earlier experiments are a direct consequence of growth limits, then the previous boundaries should be readily transgressed by the fast-growing clones. The Minute technique permits precisely this kind of experiment to be performed (Garcia-Bellido et al., 1973; Morata and Ripoll, 1975).

The method depends on the fact that a class of dominant mutants, termed Minutes, when heterozygous, show a slower cellular growth rate relative to the wild-type. For the whole organism, this is manifested in a slower growth rate during postembryonic development. Altogether there about 60 Minute loci scattered around the genome. All strains heterozygous for a *Minute* mutation (the genotype being symbolized as M/M^+) exhibit delayed development, although the extent of delay varies between different mutants. Minute adults are characterized by shortened, thinner bristles and, in a few cases, by minor morphogenetic abnormalities. The biochemical basis of the mutant phenotype is obscure but may involve general defects in protein synthesis.

When embryos or larvae heterozygous for M mutations are x-irradiated to induce mitotic recombination, two kinds of homozygous daughter cells are produced: M/M, which are cell lethal and leave no progeny, and M^+/M^+, which are wild-type and grow at the wild-type cell rate. If the M^+ chromosome carries an autonomous cell marker, such as *y* or *mwh*, homozygosis for M^+ simultaneously produces cells homozygous for that marker, and these cells produce a distinctively marked clone against a wild-type background. The essential point is that because the clone has a wild-type growth rate, while the surrounding cells grow only at the Minute rate, the marked clone has a growth advantage producing a disproportionately large patch (Morata and Ripoll, 1975). The genetic scheme is illustrated and contrasted with that of standard twin spot analysis in Figure 6.16.

The Minute technique was initially used to generate large clones in the metathorax; these first results were striking. The size of the average M^+/M^+ clone is a function of the developmental stage at which the clones are induced, but the clone size for that stage is always severalfold larger than in wild-type animals because of the competitive growth advantage. Nonetheless, *the clones respect precise lines of demarcation* (Garcia-Bellido et al., 1973). Furthermore, clones become progressively restricted to given areas in at least two successive steps. The first demarcation line in the mesothorax is

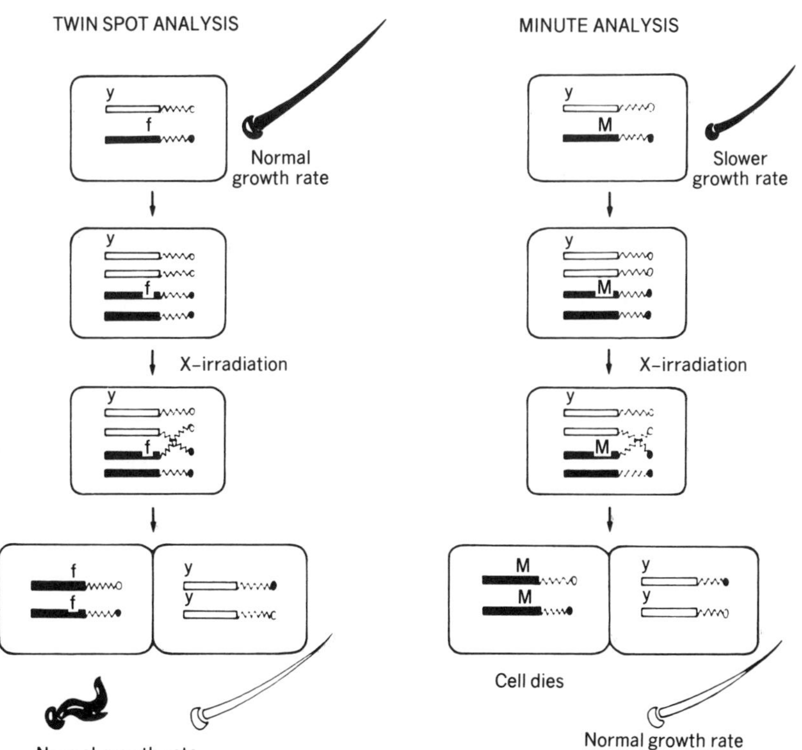

Figure 6.16 Comparison of mitotic recombination in non-Minute (M^+/M^+) background (left) and Minute background (M/M^+) (right). In the latter, the non-Minute clone (marked with *yellow*) outgrows the surrounding nonrecombinant cells, while its homozygous Minute sister dies. (Reproduced with permission from Wieschaus, 1978b.)

especially obvious; from the earliest times of clone induction at around the blastoderm stage, the marked area can occupy either the anterior part of the mesonotum (the body portion of the mesothorax) and the anterior part of the wing *or* the respective posterior sections of the notum and wing. The boundary line in the wing is perfectly straight for large clones and can extend for hundreds of cells; the boundary line defined by these clones is shown in Figure 6.17. Interestingly, this line of clonal restriction does not correspond to any obvious morphological feature of the wing (although it may have a special functional significance as the point of wing flexure during flight; see the discussion by Weis-Fogh in Lawrence and Morata, 1976b). The striking fact about the clonal patterns is that single clones apparently *never* occupy both the anterior (A) and posterior (P) regions. As development proceeds to the larval stages, clones become further restricted to either the dorsal (D) or ventral (V) surface of the wing. Within a demarcated region, early M^+/M^+ clones can occupy up to 60–90% of that region and later clones as much as 30–50% (Garcia-Bellido et al., 1976).

In the earlier discussion of the larval epidermis, it was noted that each segment's surface seems to be composed of all the descendants of a particular group of cells, a polyclone. These larval lineage groupings were termed "compartments." From the observations described above, it may be concluded that the imaginal mesothorax is similarly made up of compartments, each composed of the descendants of a limited group of founder cells. The process by which groups of cells are set aside as progenitors of particular regions has been termed "compartmentalization" (Garcia-Bellido et al., 1973). Within the wing, there appear to be at least two successive compartmentalization events that separate adjacent cells, an A-P and then a D-V division. (It should be noted that at least two further putative compartmentalization events in the mesothorax are described in the early literature, but there are reasons for believing that these reflect morphogenetic separations of cells rather than restrictions imposed between adjacent cells.)

Because each non-Minute cell in the heterozygous background can give rise to relatively more descendants than a cell of comparable age in the normal wild-type background, the progressive restrictions on clonal spread in non-Minute flies cannot be a trivial consequence of limited cell division capacity but must reflect genuine restrictions of developmental capacity. Furthermore, because the clones produced from cells generated early can

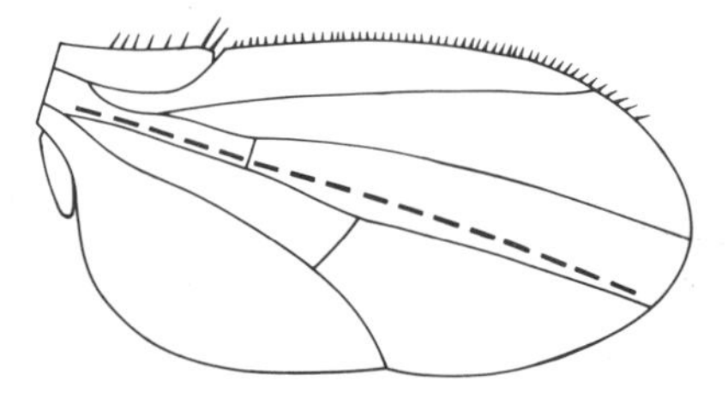

Figure 6.17 A-P compartment border in the wing. (Figure kindly provided by Dr. D. Gubb.)

occupy large compartments, but the clones of somewhat younger cells, including the daughters of the previous cells, only subcompartments of these, the exclusiveness of compartment occupancy is unlikely to be a simple consequence of the distance between progenitor cells of neighboring compartments. Rather, it must represent some kind of partitioning of neighboring cells into discrete groups, such that their surviving descendants can never mingle in the sheet of cells producing the adult cuticle. Furthermore, gynandromorph fate mapping of the A and P polyclones in the wing places the precursor cells in proximity on the blastoderm surface (Lawrence and Morata, 1977).

When do the compartmentalization events in the mesothorax occur? From the fact that single clones are invariably restricted to the A and P compartment, even when clonal induction occurs at blastoderm, it is usually assumed that this first separation occurs no later than the first division after blastoderm—the first division being required to generate the M^+/M^+ cell. The developmental stage at which the D-V restriction appears to occur in M^+ clones in M flies is during the second larval instar. However, in non-Minute strains, as we have noted, clones are restricted to the D or V surfaces by 24 hr, which suggests that the restriction takes place near the end of embryogenesis. The explanation for this discrepancy between the two results is unknown.

The potential significance of the findings is apparent when they are contrasted to more traditional notions of progressive specification of cell fate. The older view of such cell assignments, at least for vertebrate development, is of a series of specifications of cell *function,* usually with respect to a particular structure or tissue type. In mosaic systems such as *Caenorhabditis,* cell assignments are not always to tissue type (see Chapter 5) but are an intrinsic property of their clonal history. In contrast to both of these forms of cell assignment, the restrictions in cell fate in imaginal mesothoracic development consist of assignments to a future geographical area without respect to either future function or prior cell lineage. The cells of the outer surface of the mesothorax certainly differ in type and function—comprising hair-producing cells, bristle-forming cells, and vein-constituent cells—but these specific differentiative fates are fixed only at relatively late stages. All of the regions that are compartmentalized contain these different cell types, and any M^+/M^+ clone can give rise to all cell types within the region. What is restricted is the *area* to which the descendants of a particular cell can contribute.

The prospective significance of compartmentalization is enhanced by the finding that it is a general phenomenon for the cuticular derivatives of the imaginal discs. By means of the Minute technique, compartments have been sought and found in the leg discs (Steiner, 1976), the eye-antennal disc (Baker, 1978; Morata and Lawrence, 1979a), the proboscis which is formed by the labial disc (Struhl, 1977, 1981a), the haltere (Hayes, 1982) and the genital discs of male and female (Dubendorfer and Nothiger, 1982).

Each of these disc types, however, displays its own pattern of compartmentalization. In the wing, as discussed above, the initial event is the formation of the A and P compartments early in embryogenesis. In the eye-head capsule region, which is formed by the eye-antennal disc, the earliest subdivision may be into D and V compartments (Baker, 1978). An A-P separation occurs in the eye-antennal disc only in the second instar, long after the comparable event in the thorax; clonal segregations between eye and antennal tissue could not be detected even late in development (Morata and Lawrence, 1979a). The proboscis, like the wing, is divided into A and P compartments along a smooth boundary that follows no obvious morphological discontinuity, but further subdivisions could not be detected (Struhl, 1981a). In the legs, there is an early separation into A and P compartments, which may be followed by a further subdivision of the A compartment alone into D and V compartments (Steiner, 1976). However, the precise position of the A-P compartment border in both the leg and the antenna has a certain indeterminacy, unlike that of the wing (Lawrence, and Morata, 1979a,b). Some of the compartmental subdivisions of the discs are diagrammed in Figure 6.18.

Even though there are obvious differences between the various patterns of compartmentalization, the striking fact is the generality of A-P compartmentalization for those cell groups that give rise to the outer surface of the head and thorax of the fly. Furthermore, in all cases, the marked compartments and the metamorphosed discs as a whole have a normal shape and size, despite the overgrowth of the marked clones. This shows that there is some kind of compartment-wide constraint on and limitation of cell growth. At the very least, compartments must be, in the words of Crick and Lawrence (1975), "units for the control of shape and size."

There may, however, be a greater prospective developmental significance of compartmentalization, suggested by the finding that certain homeotic mutations transform particular compartments into recognizable alternative compartmental regions. These observations form the basis of a proposal on the nature of the genetic control of development in *Drosophila,* the selector gene hypothesis of Garcia-Bellido (1975).

Before describing this hypothesis, however, a few words about homeosis in general are needed. In the first place, there are many different instances of genetically produced homeosis in *Drosophila* (Ouweneel, 1976). We have already encountered several examples, the maternal effect homeotics such as *bic* and *dl* and the zygotically acting mutations of the BX-C and the ANT-C. Numerous other examples could be cited. One of the most dramatic is *Opthalmoptera,* which causes a partial replacement of eye tissue with wing tissue (Fig. 6.19); another is *spineless-aristapedia* (ss^a), which converts the terminal section of the antenna, termed the "arista," into distal leg structures (Fig. 6.20). The great majority of homeotics act zygotically rather than maternally and produce an intersegmental transformation, one involving the substitution of disc structures from a different segment. A few produce

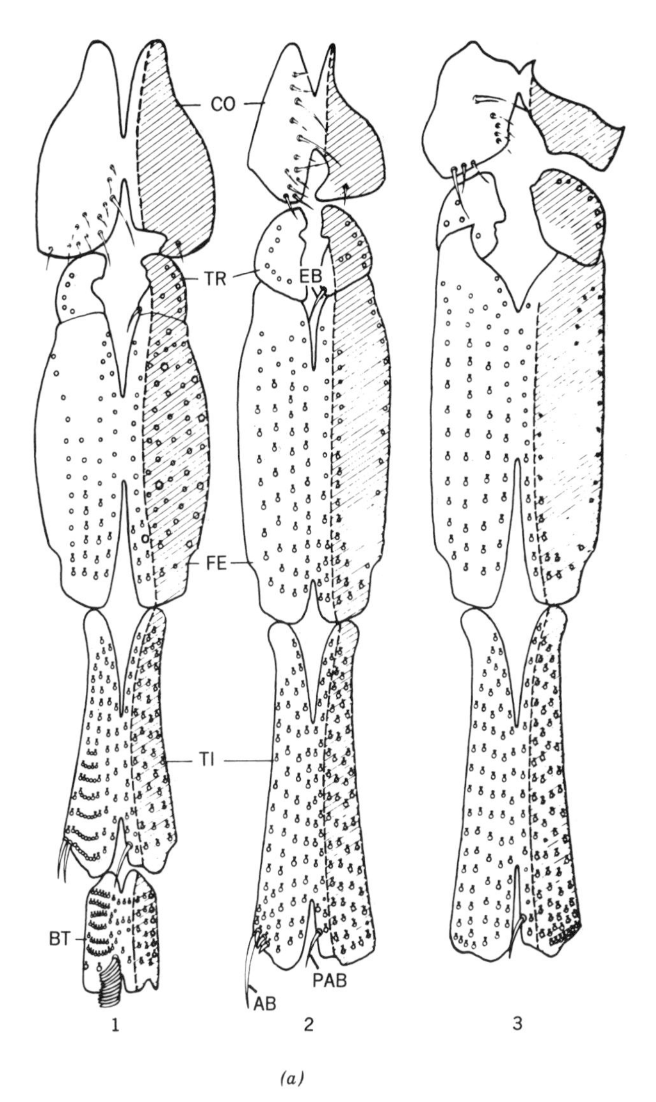

Figure 6.18 Compartmental divisions in three types of imaginal discs. (*a*) A (unshaded) and P (shaded) compartments in the legs. BT, basitarsus; TI, tibia; FE, femur; TR, trochanter; CO, coxa; AB and PAB, apical bristle and preapical bristle. (Reproduced with permission from Steiner, 1976.) (*b*) A (unshaded) and P (dotted) compartments in the eye-antenna disc. The antenna on the left side of the head (on the right in the diagram) is shown rotated to reveal the underside. I, II, and III, the three antennal segments; pr, prefrons area; v, vibrissae; cl, clypeolabrum; pa, maxillary palps. (Reproduced with permission from *Nature,* vol. 274, p. 473; copyright © 1978 Macmillan Journals Limited; Morata and Lawrence, 1978.) (*c*) A (unshaded) and P (shaded) compartments in the proboscis. Each proboscis consists of a left and a right half, each divided into A and P compartments. The numbers 1–6 designate the pseudotrachea of the labial palps; the smooth tubular area above the palps is the prementum. (Reproduced from Struhl, 1981a, with permission.)

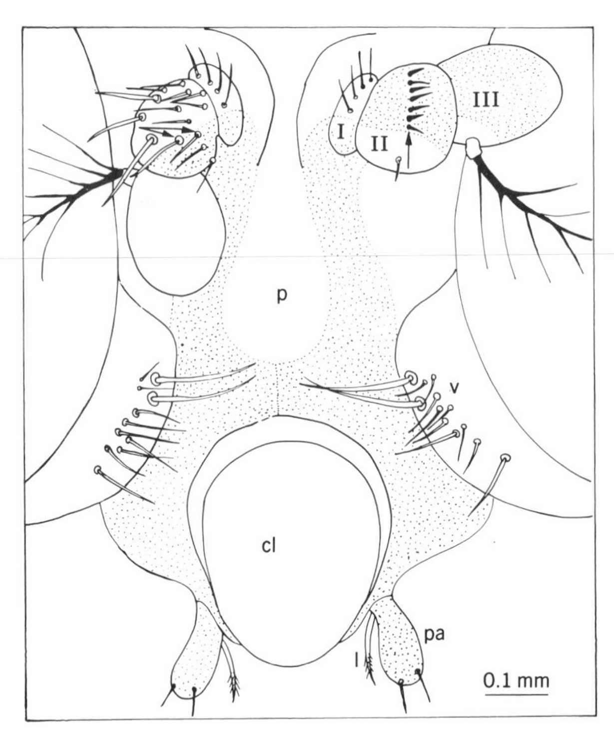

(b)

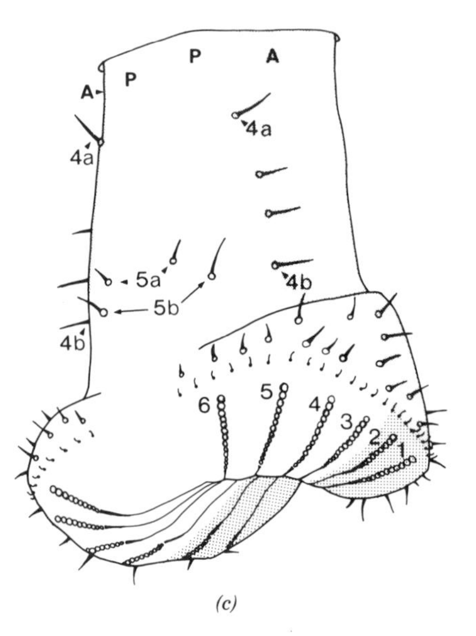

(c)

Figure 6.18 (*Continued*)

Figure 6.19 The *Opthalmoptera* transformation: eyes into wing tissue. (Photograph kindly supplied by Dr. J.H. Postlethwait; reproduced with permission from Postlethwait, 1974.)

intrasegmental transformation, causing the replacement of one subset of a disc's structures by that from another region of the same disc.

Despite their obvious interest for understanding the genetic basis of development, homeotics have proven difficult to interpret. There are several reasons for this problem. The first is that it is often unclear whether a particular homeotic response is a direct or indirect effect of the genetic lesion. In *Drosophila* and most other animals, there is a limited regenerative capacity for replacing dead or deleted cells with other cells that subsequently give rise to recognizable structures. When the replacing structures are from a contiguous region, the result may be a duplication of disc structures. This phenomenon will be discussed in more detail in Chapter 8. In other cases, localized cell death or removal may trigger replacement by cells of altered determinative state, so-called transdeterminations. Distinguishing such cases of indirect homeotic transformation, triggered by cell death or removal, from direct cellular transformations is not always easy. The second problem that often complicates interpretation is that the genetic basis of many homeotic mutants is complex, involving the presence in the genome of

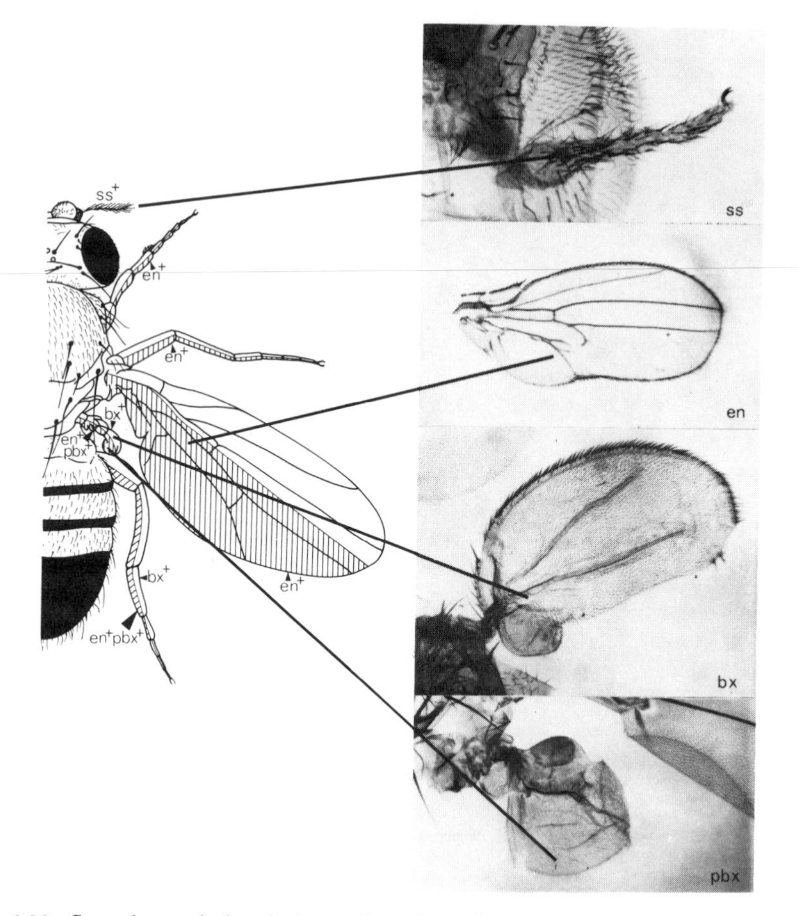

Figure 6.20 Some homeotic imaginal transformations. See text for descriptions of mutant transformations. The *bx* and *pbx* transformations shown were produced by placing strong point mutants over deletions. (Photograph kindly provided by Dr. P.A. Lawrence; reprinted with permission from *Nature,* vol. 265, p. 212; copyright © 1977 Macmillan Journals Limited; Morata and Lawrence, 1977.)

two or more mutations to create the effect. An example is *Opt* (Fig. 6.19), which requires two mutations for the transformation.

A final complexity is that the great majority of homeotics show variable penetrance and expressivity, giving a range of transformations, and in most cases, the mutations have not been characterized with respect to their degree of residual wild-type gene expression or the presence or absence of neomorphic properties. It is therefore often unclear to what extent the incompleteness of many of the transformations reflects fixed regional limits of the domain of gene action or only the hypomorphic nature of the mutations. Because it is essential to know the spatial realm of a gene's action in order to interpret its effects, this ambiguity often hinders reliable interpretation.

However, the discovery of compartmentalization has clarified this issue for one set of homeotics: the spatial limits of the regions of transformation of these mutants correspond rather precisely to compartments delimited by the Minute technique.

Mutants of the three loci that show this property are those of *bithorax (bx), postbithorax (pbx),* and *engrailed (en)*. The first two genes, *bx* and *pbx,* are part of the BX-C on the third chromosome (at 3–58.8), while *en* is on the second chromosome (2–64). All three loci, when mutant, affect thoracic segments. However, the BX-C mutants differ from *en* in causing an intersegmental transformation, the substitution of mesothoracic structures for metathoracic ones, while *en* produces intrasegmental transformations. Furthermore, *bx* and *pbx* act solely within the metathorax, whereas *en* affects imaginal structures throughout the imago (as well as in the segmentation of the early embryo).

The effects of *bx* and *pbx* are illustrated in Figure 6.20. *bx* transforms the anterior compartments of the dorsal (haltere) and ventral (third leg) metathorax into the equivalent anterior mesothoracic structures, while *pbx* produces the parallel posterior compartment transformations. Their effects on the haltere are shown in the figure.

These phenotypes suggest that the wild-type alleles of *bx* and *pbx* are required specifically in the anterior and posterior compartments of the metathorax, respectively. In the absence of the respective wild-type gene products of the loci, both dorsal and ventral metathoracic compartments transform into their mesothoracic equivalents, the anterior and posterior wing compartments. However, in the case of *bx,* the transformation of the anterior compartment is complete only in a fly that is also mutant for *pbx* and *abx* (another BX-C gene) (Cole and Palka, 1982).

In comparison to these BX-C genes, the changes provoked by mutation of the *en* locus are considerably less restricted in location within the developing imago. However, the direct effects of *en* action in imaginal tissue are either restricted to or occur preferentially within posterior compartments of imaginal tissue. Furthermore, the original mutant allele, en^1, produces varying degrees of transformation to the corresponding anterior compartments in those discs (Morata and Lawrence, 1975; Lawrence and Morata, 1976; Morata and Lawrence, 1979a); the partial transformation of P wing to A wing is shown in Figure 6.20.

The specificity of en^1 compartmental action is established by clonal analysis. Large homozygous en^1 clones, generated by the Minute technique, in the A compartment of the wing are morphologically normal and "respect" the A-P compartment boundary in the wing where they touch it for hundreds of cells. However, such clones arising in the posterior compartment are morphologically abnormal and, where they touch the boundary, transgress it (Morata and Lawrence, 1975). By several tests, the transformation effect is direct: there is no evidence of en^1-mediated cell death, and restriction of the effects to the posterior compartment is inconsistent with generalized cell death or cell loss as its basis. The simplest interpretation of the *en*[1] effects is that en^+ is not normally expressed in the A compartment—hence, the absence of effect of a large mutant patch in that compartment—but is expressed in the P compartment, where it is required for both the normal development of that compartment and the maintenance of the compartment boundary (Morata and Lawrence, 1975).

The existence of compartment-specific homeotic transformations by recessive mutations of these three loci (*bx, pbx,* and *en*) suggests that the wild-type development of these compartments is dependent on the respective wild-type alleles of the genes. In the absence of the wild-type gene product in the relevant compartment, the developmental sequence falls into a different channel. In the terminology of Garcia-Bellido, each such controlling homeotic gene is a "selector gene."

A selector gene is defined as one that sets a particular course of development within a compartment by governing the activities of numerous "realizator genes"; it is the activities of the latter that create the phenotype of that compartment. The realizator gene products presumably mediate such properties as division rate, mitotic cell orientation, cell surface properties, and the ability to differentiate such features as hairs and bristles. The hypothesis is that each compartmentalization step involves the switching on or off of a particular selector gene; it is the combined activities of the group of active selector genes within a compartment that create the properties of the compartment. In this view, compartmentalization is a basic sorting device for establishing which combination of controlling genes will be on and which will be off, and hence the development of the compartmental unit.

With the addition of a corollary, the selector gene model comprises a fundamental hypothesis of developmental control in *Drosophila*. The corollary is that the first differential settings of selector genes, which ostensibly occur at blastoderm, are mediated by the products of a third class of genes, "activator genes." These differential settings of selector genes are produced by the interaction of the products of the activator genes with differential concentrations of the fate-setting maternal substances (i.e., the putative determinative gradients). The activator gene product(s) "transduce" these differing cytoplasmic concentrations into a spatial pattern of initial selector gene activation/repressions throughout the early embryo (Garcia-Bellido, 1975).

The complete model has certain parallels with the earlier model of gene regulation proposed by Britten and Davidson (1969). An important difference is that the cellular domains of regulation are specified in the Garcia-Bellido scheme; they are compartments. Realizator genes correspond to the producer genes of Britten–Davidson (i.e., the structural genes in the still earlier Jacob–Monod formulation); selector genes are the equivalent of integrator genes (i.e., the regulator genes of Jacob–Monod). However, the activator genes are unique to the Garcia-Bellido model. The Britten–Davidson model was based on the facts then known about eukaryotic genome organization and gene expression, and has evolved as new facts have accumulated (e.g., Davidson and Britten, 1979). The selector gene model was based on the genetic analysis of cell behavior in the *Drosophila* mesothoracic discs, and has remained in its original formulation the key reference point for most of the subsequent work in *Drosophila* developmental gentics. In the past decade, how has it fared as a general explanatory scheme for *Drosophila* development?

The essential elements of the model are that (1) compartments are the basic building blocks of development and (2) the properties of each compartmental unit are the combined product of the activities of certain general compartment-class selector genes (A versus P genes, D versus V genes, etc.) and the expressed segment-specific selector genes. We will consider first whether there is evidence for selector genes in the first category, those that prescribe general compartment character, and then briefly look at the evidence bearing on the universality of compartments as building blocks of development. In the following sections, we will look at the properties and kinds of segment-identity selector genes and evaluate the current standing of the hypothesis as a whole.

The prime candidate for a compartment-class selector gene is *engrailed*. The original assignment of *en* to this role was made on the basis of the properties of the en^1 allele, a presumptive hypomorphic mutation. Homozygotes for en^1 show what appears to be a duplicated A leg compartment in place of the P leg compartment in the foreleg and a wing with a normal A compartment abutting a posterior compartment that is a partial mirror-image A compartment. Furthermore, the transformation is restricted to cells homozygous for the mutation, with en^1 clonal patches in the P compartment showing the P-to-A transformation characteristics (Garcia-Bellido and Santamaria, 1972).

Nevertheless, the P-to-A transformation produced by en^1 is not complete in the wings and legs, and the haltere remains wild-type (Hayes, 1982). This incompleteness of effect has been attributed to the presumed hypomorphic nature of the en^1 mutation, whose remaining wild-type activity was presumably sufficient to ensure some degree of normal P compartment development (Garcia-Bellido and Santamaria, 1972; Morata and Lawrence, 1975). If this explanation is valid, then more severely deficient mutations would be predicted to create a more complete compartmental transformation. When such mutants are isolated, on the basis of noncomplementation with en^1, and examined, the reverse is found. Numerous lethal *en* alleles—whose residual wild-type activity is presumably less than that of en^1, which is a viable mutant—have been tested, and their P-to-A transforming effects examined in various ways. It is found that the visible transformations associated with the *en*-lethal alleles are always *weaker* than those produced in en^1 homozygotes (Kornberg, 1981a; Lawrence and Struhl, 1982; Eberlein and Russell, 1983). Furthermore, for the wing transformation, en^1 homozygotes show a much stronger transformation than hemizygotes of en^1 (Eberlein and Russell, 1983). Thus, two doses of en^1 create a stronger transformation effect than a single dose, the property of an *antimorphic* mutation rather than a hypomorph (Table 3.7). (In contrast to the majority of antimorphs, however, en^1 is recessive to its wild-type allele.) However, there are important shared properties between the different *en* alleles: the similarities in phenotypic effect between en^1 and the lethal alleles lie in the ability of mutant clones originating in the P compartment to transgress the A-P boundary border and in the preferential occurrence of mutant effects in the P compartments.

The original observations on en^1 phenotypic effects were interpreted to mean that the difference between A and P compartments "depends critically on the selector gene *engrailed* which, when active, selects the posterior developmental pathway" (Lawrence and Morata, 1979a). The results described above suggest that the situation is more complex: parts of the transforming effects of en^1 are produced by neomorphic properties of that allele, and mutant alleles that probably have less en^+ activity than en^1 produce less dramatic or at least less obvious transformation effects of P to A compartments. The *en* gene may be *a* determinant of A-P compartment differences but it cannot be the *sole* determinant of those differences.

Similarly, other single-gene determinants for compartment-class characteristics have failed to emerge. Such single-gene selectors may remain to be discovered, but this seems increasingly unlikely, given the intensive genetic analysis of development in general and of compartmentalization in particular that is taking place in *Drosophila*. It is more probable that general compartment-class characteristics, such as A versus P or D versus V, arise from the cellular properties mediated by several gene products, possibly interacting in complex patterns, rather than from the activities of single selector genes. The work on *en* shows that its activity is particularly important for P compartment development but not sufficient to account for the A-P compartment differences. Other genes that may be important in this respect are those whose mutations create embryonic segmental disturbances (Table 6.2). The gene *fu*, in particular, is one possible candidate; mutations of *fu* interact synergistically with *en* mutations in several ways in imaginal development, suggesting functional interactions between their gene products (Fausto-Sterling and Smith-Schleiss, 1982).

A second major element of the selector gene hypothesis is the idea that the compartment is the fundamental, and perhaps the universal, module of development. This concept too requires revision (Gubb, 1985). Several homeotics, an example being ss^a (Struhl, 1982a), exert their effects over variable domains that are not compartments and that cross known compartment boundaries.

Furthermore, the idea of compartments as units of development seems to have less applicability to the internal imaginal tissues than to the external cuticular tissues. The observations of mesodermal and neuronal tissue, using a variety of clonal analyses and histochemically detectable gene activites, form a complex pattern, but none of these tissues shows a clear A-P compartment separation or any effect of *en*[1] or *en*-lethal alleles, either of transformation or in clone frequency (Ferrus and Kankel, 1981; Lawrence, 1982). A detailed study of the muscles of the thorax reveals an early separation into dorsal and ventral lineages. Using the Minute technique and succinate dehydrogenase inactivity (Sdh^-) as a marker, Lawrence (1982) has shown that early (blastodermal) clones can mark both the dorsal and ventral muscles of the mesothorax. By midfirst instar, however, clones can mark only dorsal or ventral muscles within any of the three thoracic segments. It is not clear whether this separation of lineages reflects an imposed restriction on cell mixing of primordia that continue to lie adjacent to one another or to a physical separation of the primordia.

The lineages of the thoracic musculature bear some similarity to those of the overlying cuticular lineages. In thoracic muscle, segment-crossing clones are never observed, and the ensemble of muscle tissue seems to comprise eight fundamental lineages by midfirst instar, comparable to the cuticular tissues. These consist of three pairs of dorsal lineages (one per segment), three corresponding ventral lineages, and two associated with the anterior mesothoracic spiracles (the openings of the tracheal system).

However, these resemblances do not reflect shared ancestry or compartments comprising the cuticular lineages and the underlying mesodermal tissue. Induction of clones homozygous for both a cuticular marker and Sdh^- reveals no correspondence between the cell lineages of the two tissues (Wilcox et al., 1981; Lawrence, 1982). Indeed, the conventional blastodermal fate map (Fig. 6.11) shows that mesodermal tissue derives from precursor cells that arise distinctly ventral to those that give rise to the thoracic cuticular anlagen.

The observed correspondences might signify instead the imposition of an ectodermal (cuticular tissue) pattern on the underlying musculature. Transplantation of Sdh^- wing disc fragments containing mesodermal precursor cells, the adepithelial cells, into larvae shows donor muscle cell contribution throughout the musculature of the imago (Lawrence and Brower, 1982). The transplanted adepithelial cells normally contribute only to the direct and indirect flight muscles of the wing, yet evidently retain sufficient plasticity as late as the third instar to contribute to virtually any muscle set in the fly.

Nevertheless, the development of the detailed pattern of muscle structure may require more than instruction by overlying ectoderm. In flies showing a strong homeotic transformation of the metathorax into a second mesothorax, there is no corresponding development of mesothoracic musculature (Fig. 6.21). Ectodermal instruction must be only one of several determiners of muscle development.

Mechanisms of Compartmentalization. How are polyclones segregrated from one another? Either cells of a given polyclone are formed in a semidispersed fashion and seek each other out by cell sorting, or polyclonal formation involves some form of geographical partition that allocates all cells and

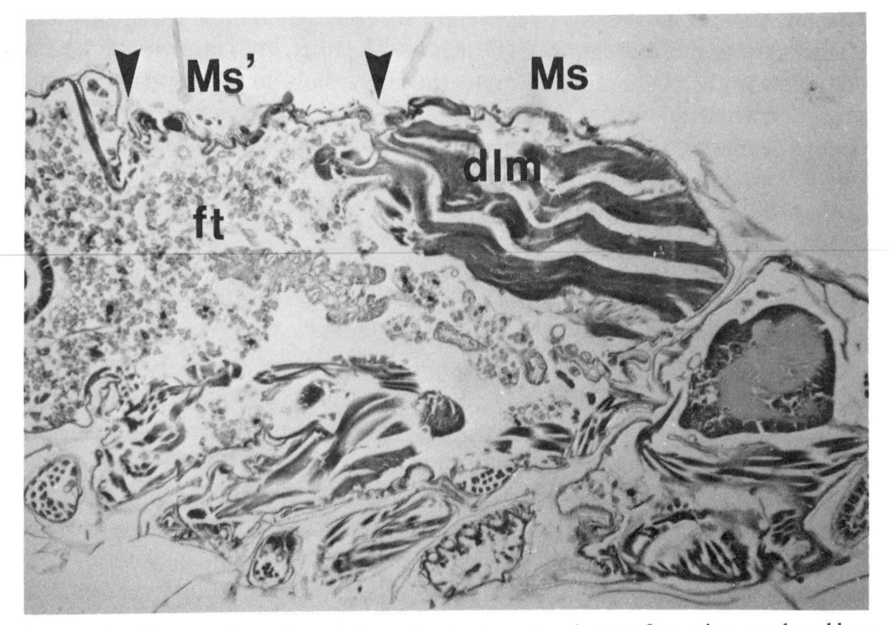

Figure 6.21 Absence of muscle transformation in strong *bx pbx* transformation, produced by a *bx pbx* chromosome over *Df(3)P9*. Ms and Ms′, mesothorax and transformed metathorax, respectively; ft, fat; dlm, dorsal longitudinal muscle. (Figure kindly provided by Dr. D.R. Kankel; from Ferrus and Kankel, 1981.)

their descendants by virtue of their initial position to one side of a boundary line or the other. The answer is unknown, but geographical partitioning seems the more likely method. If there were to be extensive cell sorting and reassorting, one would expect many clones to be broken up in the process; in fact, clones generally remain integral and possess smooth boundaries.

Evaluating the mechanism requires some information about the precise timing and hence the cell numbers involved in the time of compartmental segregation. Because clones in the mesothorax are never observed to cross the A-P border, the time of A-P separation is usually taken to be within one division postblastoderm, when the wing A and P primordial polyclones may contain as few as 10 and 5 founder cells, respectively (Lawrence and Morata, 1979b). The extreme earliness of this apparent segregation, indeed, seems to lend the event a special significance; one possible explanation will be discussed in Chapter 8.

Whatever the precise timing or mechanism of compartmental separations, one class of model that may be ruled out is one that ties the events to the growth dynamics of the disc. Kauffman et al. (1978) formulated such a model, in which the spatial distribution of a morphogen changes in a characteristic way as a function of growth of an ellipse-like disc. Their scheme generates a sequence of wave patterns whose minima, when projected onto the wing disc, mimic the observed lines of clonal restriction (including two described in the early literature that may not be genuine compartmental restrictions). The chief prediction of the model is that discs experiencing comparable increases in cell number should experience the same *number* of successive compartmentalization events. A secondary prediction is that successive divisions should divide the growing disc into *symmetrical* domains. Neither prediction has been verified. As Morata and Lawrence (1979a) remark, the eye-antenna disc exhibits size and growth dynamics comparable to those of the wing-mesothorax but shows a very different compartmentalization pattern with fewer and less symmetrical divisions. Although an increase in cell number may play some part in compartmentalization events, it cannot be the critical variable. Indeed, the connection between growth and compartmental organization may be the reverse; instead of total growth creating compartmental subdivisions, the extent of growth appears to be highly influenced by local factors, including compartment boundaries (Simpson, 1979; Brower et al., 1981; Kuhn et al., 1983). The fact that wounding can permit clonal transgression during the first 24 hr of the recovery period but not later (Szabad et al., 1979b) shows further that the process of maintenance and reestablishment of compartmental boundaries is also a complex matter.

Answers to the question about mechanism(s) in compartmentalization are still to be obtained. They will probably come only from a much more detailed characterization of the cellular properties and molecular biology of the growing discs than has yet been possible.

Segmentation

Like the larva, the imago is composed of a characteristic sequence of segments. These segments comprise the cephalic, thoracic, and abdominal regions. In contrast to the larva, whose basic segmental pattern can be discerned in early embryogenesis, the complete segmental pattern of the imago becomes visible only late in development, during the final events of morphogenesis. As the imaginal discs evert, differentiate, and fuse together, the characteristic head and thoracic segmental pattern emerges. In the abdominal region, the segments of the imago arise through a replacement of the outer surface of the larval segments as the histoblasts grow and differentiate.

The origins of the imaginal segment pattern therefore lie in the processes that set aside the imaginal primordia. As in the larva, the critical questions concern the mechanisms that establish the segmental divisions and the processes that create and maintain the differences between imaginal segments.

Formation of Segment Pattern. Does the imaginal segmental division arise separately from or as part of the same process that establishes the larval pattern? For the abdominal segments, the processes are clearly the same. The histoblast nests arise after the delimitation of the larval segments and probably share a pool of common precursor cells with the larval epidermal cells. Thus, in the abdomen, the establishment of the larval segments is the primary event, and that of the imaginal segment primordia is a secondary consequence.

The origin of the thoracic segments of the imago is undoubtedly similar. Firstly, the characteristic imaginal discs of the pro-, meso-, and metathorax can be directly observed within the corresponding larval segments from the time at which they can first be distinguished. Secondly, blastodermal fate mapping, using ultraviolet (UV) microbeam cell ablation, places the three imaginal thoracic anlagen at virtually the same positions as the corresponding larval thoracic anlagen (Lohs-Schardin et al., 1979a,b). In addition, irradiation at the later stage of the completed germ band (10 hr at 25°C) shows the respective target centers for maximum damage to the imaginal pro-, meso-, and metathoracic segments as lying within the corresponding larval segments.

However, the origins and exact number of the imaginal cephalic segments are contentious issues. Only three pairs of cephalic imaginal discs are found in the mature larva. These are the eye-antennal discs, which give rise to the eye and the major part of the head capsule; the cibarial discs, which form the central anterior structure, the clypeo-labrum; and the labial discs, which form the proboscis. The relationships between the discs in the third instar larva and the fully developed head are shown in Figure 6.22. In contrast, the preinvolution germ band embryo may possess an additional three segments that collaborate in the formation of the mouth parts of the larva. These larval segments may have no analogues in the adult. Struhl's (1981b) gynandro-

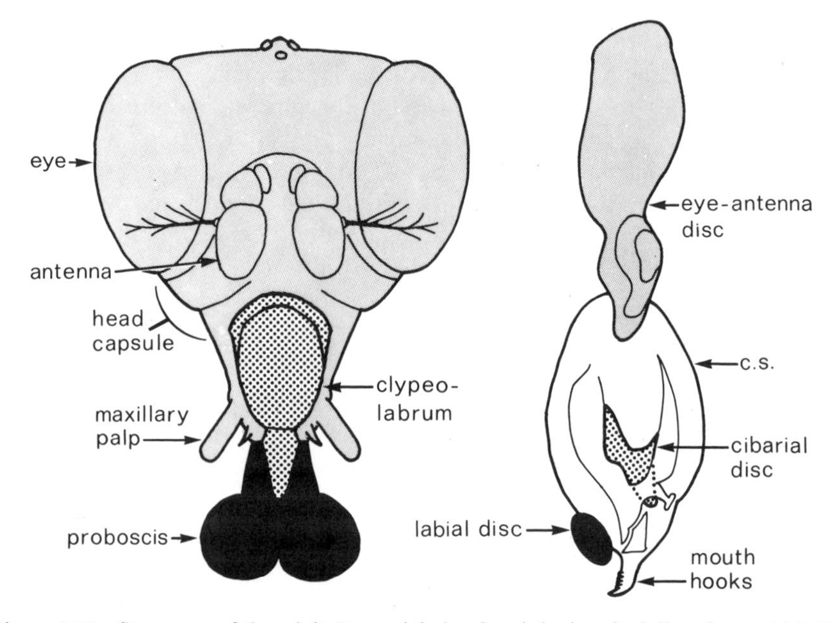

Figure 6.22 Structures of the adult *Drosophila* head and the imaginal discs from which they arise. (Reproduced with permission from Struhl, 1981a, based on a figure by Walter J. Gehring.)

morph mapping of the blastoderm revealed a distance between the primordia for the posterior compartment of the eye-antennal disc and the anterior compartment of the labial disc that is sufficient to accommodate three standard-sized segments. He suggested that the three "missing" segments in the adult are the premandibular, mandibular, and maxillary segments of the larva. Should this suggestion be confirmed, an interesting consequence is that the blastoderm or early gastrulating embryo is partitioned into equal-sized segment primordia, from the anterior (clypeo-labrum) end to the posterior (eighth abdominal segment) region of the germ band.

In summary, it seems probable that the precursor cells of the imaginal primordia arise following the partitioning of the embryo into larval segments. If so, then the imaginal anlagen might be stamped from the very beginning with some characteristics of the "host" larval segment. We have previously speculated that imaginal and larval cells arising in the same region might have considerably more in common with each other than their cytological differences suggest. The genetic analysis of imaginal homeotic transformations produced by mutations in BX-C and ANT-C, the two gene complexes that govern larval segmentation, confirms this relationship. To a large degree, the transformations in particular larval segments and their resident imaginal discs are parallel. However, it is the analysis of the imago in particular that reveals the full intricacy of the interactions between different genetic elements.

The domain of action of BX-C revealed by this analysis is from the mesothorax (T2) posteriorly to the terminal abdominal segment (AB8); that of ANT-C is from the mesothorax (T2) anteriorly to the cephalic imaginal segments. The expression of both gene clusters is seemingly regulated by loci external to themselves and, more surprisingly, by regulatory interactions between themselves; the final segmental phenotypes are the summed outcome of all of these gene activities. To keep the presentation relatively clear, we will first examine the behavior of mutants of the two complexes within an otherwise wild-type background and then the loci external to these complexes that affect their expression. The discussion will close with a second look at the selector gene hypothesis and the role of compartments in development.

Control of Segment Identity: BX-C. The genetic organization of the BX-C was described earlier (Fig. 6.7*a*). A more complete genetic map is given in Figure 6.23, along with a molecular map, which will be described in Chapter 9. The additional gene functions are those revealed by an analysis of the imaginal mutant phenotypes.

Before discussing the organization of the BX-C in relation to its actions, three genetic complexities should be noted. Firstly, many of the "point" mutations that identify individual BX-C functions are not single nucleotide changes, but are either small insertions or deletions. Secondly, many of these mutations were picked up in chromosomes that were found to carry

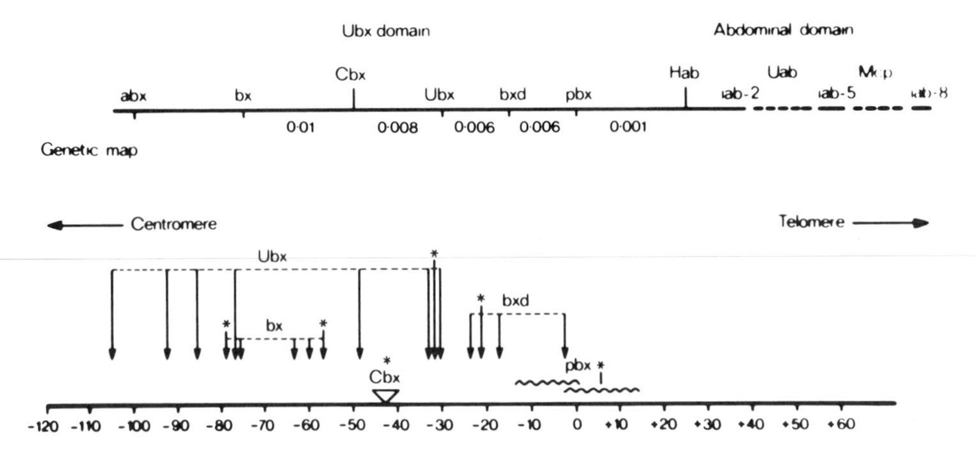

Figure 6.23 Genetic map of the BX-C and molecular map of the *Ubx* region of the BX-C. For gene symbols and a description of mutant transformations, see Table 6.6. The molecular map shows the positions of the *abx* and *bx* mutations (grouped under the small arrows, *abx* being the one on the extreme left); the *bxd* mutations (the intermediate length arrows); and a *Cbx* insertion (shown as a triangle). The molecular map is described in more detail in Chapter 9. (Reproduced with permission from Lawrence and Morata, 1983; copyright MIT Press.)

additional mutations within the complex. The origin of these coincidental occurrences is understood in only a few cases. Thirdly, several classes of mutant function that map as point mutants are associated with effects on the expression of adjacent or neighboring genes; the nature of these "polar" effects is also little understood.

As described previously in connection with larval segmentation, the map of imaginal BX-C functions roughly parallels the segmental organization, with key thoracic-segment genes being located relatively proximal and progressively more posteriorly abdominal functions situated progressively more distally. The list of BX-C functions, known and inferred, that affect imaginal development is given in Table 6.6, along with their associated mutant phenotypic effects. As in larval development, recessive loss-of-function mutations transform segment phenotypes in the T2 direction, either to the T2 state itself or toward that of a segment between T2 and the transformed one. Conversely, dominant gain-of-function mutations drive segmental phenotypes in the AB8 direction toward that of more posterior segments. The latter transformations may be provisionally interpreted as derepressions of posterior-type functions within more anterior segments.

At the level of gross morphology, a seemingly significant difference between the larval effects and those on the imago is that certain of the genes identified as affecting the adult are restricted in their action to specific compartments of the thorax. Thus, mutations of both *abx* and *bx* produce a haltere-to-wing transformation that is mainly limited to the A compartment

Table 6.6 *Drosophila* BX-C Mutants and Their Imaginal Phenotypes

Mutation	Phenotype[a]	Interpretation
Loss-of-function mutations		
postprothorax (*ppx*)	T2-A → T1-A	Loss of T2-A essential function
anterobithorax (*abx*)	T3-A → T2-A	Loss of T3-A essential function
bithorax (*bx*)	T3-A → T2-A	Loss of T3-A essential function
postbithorax (*pbx*)	T3-P → T2-P	Loss of T3-P essential function
bithoraxoid (*bxd*)	AB1 → T3	Loss of AB1 essential function
Ultrabithorax (*Ubx*)	AB1, T3 → T2	Loss of AB1, T3 essential functions
infra-abdominal-2 (*iab-2*)	AB2 → AB1	Loss of AB2 essential function
infra-abdominal-5 (*iab-5*)	AB5 → AB4	Loss of AB5 essential function
infra-abdominal-8 (*iab-8*)	AB8 → AB7	Loss of AB8 essential function
Gain-of-function mutations		
Contrabithorax (*Cbx*)	T2 → T3	Expression of T3 functions in T2
Hyperabdominal (*Hab*)	T3, AB1 → AB2	Expression of AB2 functions in T3, AB1
Ultra-abdominal (*Uab*)	AB1, AB2 → AB3	Expression of AB3 functions in AB1, AB2
Miscadestral pigmentation (*Mcp*)	AB4 → AB5	Expression of AB5 functions in AB4

[a] Denotes the direction of compartment- or segment-phenotype change; many of these transformations are not complete, even for the strongest alleles.

of the haltere, while *pbx* mutation effects are largely confined to the posterior compartment (Cole and Palka, 1982). However, it may be that the cellular domains of all BX-C functions are half-segment compartments (Lawrence and Morata, 1983) and that the whole-segment transformation effects discussed in the literature reflect the paucity of scorable cellular landmarks in the posterior abdominal compartments, preventing a clear discrimination between adjacent segments. The in situ hybridization methods for detecting transcripts (Chapter 9) may eventually resolve this issue.

A second difference between the larval and imaginal results is that the existence of at least one imaginal BX-C function has been deduced to be essential for normal imaginal T2 development. The observation is that when the proximal part of the complex is deleted by mitotic recombination early in development (between 3 and 7 hr), the A compartment of the T2 leg changes into an A compartment of the T1 leg. The postulated T2-essential BX-C function has been designated *postprothorax* (*ppx*) (Morata and Kerridge, 1981). Whether the complex contains comparable genes for either the posterior compartment of the T2 leg or of the dorsal compartments of T2 is unknown.

Leaving aside some of the intricacies (described in Lewis, 1978, 1981, and Lawrence and Morata, 1983), the pictures presented by the imaginal and larval transformations are comparable. There is a general, although not perfect, correspondence between the proximodistal order of the gene functions and the anteroposterior segments/compartments affected. Furthermore, the development of more posterior segments probably requires the additive expression of all genes in the complex expressed in more anterior segments. (The evidence for this conclusion, however, is based principally on clonal analysis and is less firm than for larval segment development.)

All of the effects described so far involve the outer cuticular surface. Is the BX-C active and essential for the development of internal tissues from T2 to AB8? Recent findings have established that BX-C is expressed in, and is necessary, for normal central nervous system (CNS) development. The strongest evidence has been provided by Jiminez and Campos-Ortega (1981), who studied the ventral neural cord patterns in several lethal embryonic BX-C genotypes. Although the various segmental ganglia, or neuromeres, fuse ("condense") in late embryonic development, they remain distinguishable by two structures. Jiminez and Campos-Ortega found that each neuromere contains a distinctive medial cell column and that the thoracic neuromeres are distinguished by an intracortical band of densely staining cells at the anterior edge of each neuromere. The number of these bands varies directly with the cuticular transformations produced in the mutants. Thus, wild-type embryos show three intracortical bands in the thoracic region; lethal *Ubx* embryos, which convert AB1 to T2, have four; *Df(3)P9* embryos, which have all segments posterior to T2 converted to a T2 phenotype, have 11; and *Pc/Pc* embryos, which convert all segments to an AB8 phenotype, have none.

In addition to these embryonic transformations, BX-C expression is also required in the imago. Green (1981) examined motoneurone patterns in wild-type and *abx bx pbx* flies, which exhibit a complete transformation from metathorax to mesothorax (Fig. 6.24). He observed a high percentage of partial or complete transformations from the metathoracic motoneurone pattern to that of the mesothoracic type. Because these neurons have their cell bodies in the ganglia of the CNS, the results demonstrate homeosis of the CNS.

Strictly speaking, neither set of findings establishes whether the effects occur autonomously in the CNS, independently of the surface ectodermal conversion, or as a consequence of the cuticular transformation. However, the evidence of Teugels and Ghysen (1983) shows that the CNS transformation can occur independently of the cuticular transformation. They determined the number of leg ganglia in adult *Hab* and *bxd* mutants, both of which show variable penetrance, and found that either reduced or additional numbers of leg ganglia could be produced by the two mutations respectively, independently of the occurrence of the cuticular transformation. The results show that the CNS can be transformed autonomously by mutation in BX-C, indicating that the gene complex is expressed in and necessary for normal CNS development. Molecular evidence has confirmed that BX-C is transcribed in neural tissue, and in mesodermal tissue as well (Chapter 9).

Figure 6.24 A four-winged *abx pbx bx* fly. (Photograph kindly provided by Dr. E.B. Lewis. From Lewis, E.B., Four-wing fly (Diptera produced by combining three mutants of the bithorax, etc.), *Science* Vol. 221, Cover, 1 July 1983. Copyright 1983 by the AAAS.)

Control of Segment Identity: ANT-C. The relationship between the genetic organization of ANT-C and its segmental domain—cephalic segments through T2—is less clear than that of BX-C. The initial mutations discovered in ANT-C were the dominant *Antp* mutants, distinguished by a transformation of the antenna to the second (mesothoracic) leg. Many of these mutations were associated with chromosomal aberration breakpoints. This kind of mutational change is intrinsically difficult to analyze because the aberrations themselves interfere with conventional genetic mapping. Delineating the numbers and kinds of gene functions in the region of ANT-C has therefore required a search for new mutants. The procedure is a two-step process.

In the first stage, "revertants" of the dominant mutants are isolated by screening heterozygotes possessing the *Antp* chromosome, marked with other distinguishing mutations, for those that do not give the homeotic transformation. The rationale for searching for such revertants is that many should have a second mutation that prevents the expression of the neomorphic phenotype; in the majority of these doubly mutant chromosomes, the second mutation is one that inactivates the primary gene function. In several cases, the inactivating mutations have proved to be deletions; these are viable over the standard wild-type third chromosome but inviable when homozygous. In the second step of the procedure, newly EMS-mutagenized third chromosomes are placed over such deletions and the heterozygotes screened for developmental effects. Chromosomes giving inviable, semiviable, or homeotic transformations must have a mutation in the same region as the spanned by the deletion. The complementation behavior of these new mutations, when placed in pairwise combinations with each other, is then ascertained to establish the number of complementation groups in the region (Lewis et al., 1980a,b).

The precise limits of ANT-C and the number of functions it contains remain to be established but a current map is shown in Figure 6.25. At least three, and possibly four, complementation groups are associated with a homeotic phenotype. These are listed and described in Table 6.7. The most proximal, *proboscipedia* (*pb*) is identified by recessive homeotic mutations that transform the proboscis into leg or antennal tissue to varying degrees. Mutations in the next most proximal gene function, *Deformed* (*Dfd*), give

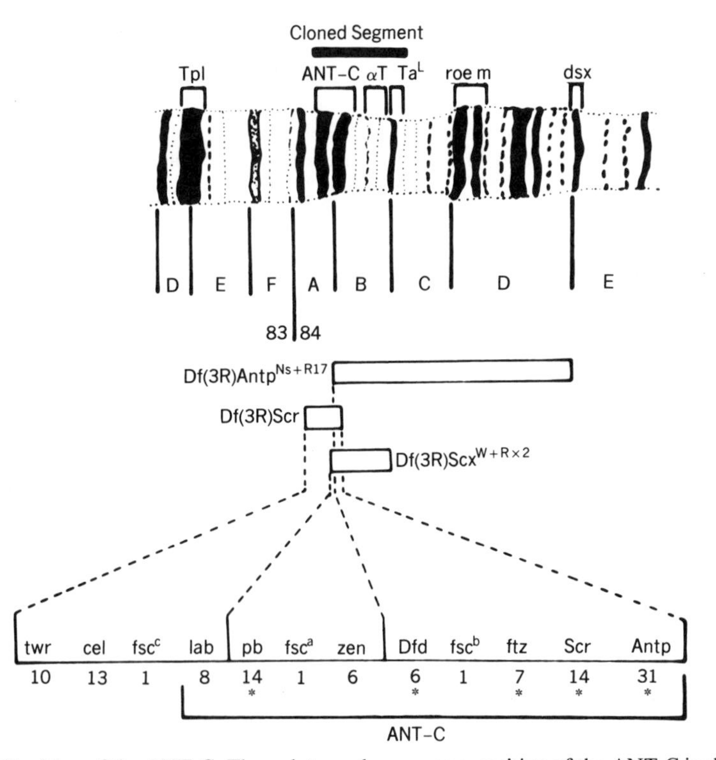

Figure 6.25 Map of the ANT-C. The polytene chromosome position of the ANT-C is shown at the top, and several deletions of the region are shown in the middle. The bottom portion is the genetic map of the region of the complex; the numbers below the gene symbols indicate the number of isolated alleles, and the asterisks indicate those loci having at least one allele associated with a rearrangement. For the principal mutant transformations associated with the ANT-C, see Table 6.7; for the other gene symbols, see Kaufman, 1983. (Reproduced with permission from Kaufman, 1983.)

Table 6.7 ANT-C Homeotic Complementation Groups

Complementation Group	Mutant Phenotypes[a]	Disc Required In:
Antennapedia (*Antp*)	*Antp* (dominant): antenna → second leg	Ventral meso-, pro- and metathorax
	Antp^{Scx} (dominant): second and third legs → first leg	
	Antp^{-} (recessive): second leg → antenna	
	Hu (dominant): humerus → mesonotum	
Sex combs reduced (*Scr*)	*Scr* (dominant): reduced sex comb (first leg)	Ventral prothorax
	Msc (dominant): second and third legs → first leg	
Deformed (*Dfd*)	*Dfd* (dominant) and *EbR11* (recessive): abnormal eyes, head bristles	Eye-antenna (?)
proboscipedia (*pb*)	Proboscis ↗ antenna Proboscis ↘ leg	Labial

[a] Denotes the direction of phenotypic change; many of these transformations are not complete, even for the strongest alleles.

abnormal eyes and certain aberrant head capsule bristles. This is followed by *Scr,* whose mutations produce a reduced sex comb on male forelegs indicative of a partial homeotic transformation of the first leg to the second leg. Like *Ubx* in the BX-C, the *Scr* phenotype reflects reduced gene activity for the locus and is produced by haplo-insufficiency for the locus. An interesting dominant mutation of *Scr, Multiple sex combs* (*Msc*), produces the opposite transformation, that of second and third (metathoracic) legs into first (prothoracic) legs. *Msc* appears to be a neomorphic derepression of prothoracic character in the legs of the meso- and metathorax. The simplest hypothesis for this gene's function is that *Scr*$^{+}$ activity is expressed and required in the prothorax for normal T1 development and is normally unexpressed in T2 and T3; the *Msc* effect would then reflect a neomorphic derepression of the gene in the latter two segments.

The function of the *Antp* locus, the most distal in the complex, is complementary to that of *Scr*: it is expressed and required in the mesothorax for normal second leg development, but can be derepressed in the prothorax, yielding forelegs with a second leg phenotype. The conclusion that *Antp*$^{+}$ activity is necessary for normal second leg development follows from the phenotype of clones homozygous for amorphic alleles (*antp*$^-$) of the locus; such clones are associated with the transformation of leg tissue to antennae (Struhl, 1981d). In contrast, *antp*$^-$ clones in antennae are perfectly normal, showing that the gene is not required for antennal morphology to be normal. If *Antp* is required for leg development but not for that of antennae, the dominant *Antp* transformation—antennae into legs—must reflect a neomorphic expression of *Antp*$^+$ activity in the antenna in these mutants. The locus may also govern some essential prothoracic function. The *Antp*^{Scx} phenotype, for the production of sex combs in the second and third legs, is reminiscent of *Msc*. It may signify that the *Antp* locus consists of several functional units, of which the prothoracic-essential function is normally repressed in all but the forelegs.

Although ANT-C, like BX-C, consists of a cluster of genes whose mutational alteration produces a set of related segmental transformations, there are some obvious differences between these two gene complexes. One such difference is that the linear sequence of genes in ANT-C bears a less clear relationship to the order of segments that these genes affect than in the comparable case of BX-C. Furthermore, it is not known whether any of the segments under the nominal control of ANT-C require the additive activities of the genes of this complex. An additional difference is that there is at least one gene embedded in ANT-C that plays no known role in setting segmental identities; this is *zerknullt* (*zen*), whose inactivation produces a block in gastrulation. BX-C may, of course, contain such additional genes, unrelated to segment phenotype control, but if so, they have not yet been discovered. The pair-rule gene *ftz* is also found within ANT-C, although, as we have seen, *ftz* can mutate to give a homeotic phenotype. A further complexity in ANT-C is that at least two of its gene functions, *Scr* and *Antp,* overlap to some extent in their control functions for the pro- and mesothorax.

The last point emphasizes the conclusion reached earlier, in connection with the BX-C gene *ppx,* that T2 (mesothorax) is not a state of nil transcription for segment-phenotype control genes, although it may be a ground state in evolutionary terms. Evidently, there are genes in both the BX-C and the ANT-C whose expression is required for the normal development of the imaginal T2 segment. [In larval development, there is a parallel requirement for at least one ANT-C gene, that of *Antp* itself (Wakimoto and Kaufman, 1981; Struhl, 1983).] The requirements for BX-C and ANT-C expression for normal development of the imaginal mesothorax may, in fact, be related. Clones in the T2 leg made homozygous for *Ubx* early in development show a transformation to the T1 leg; if, however, they are also *Scr*$^-$, the transformation is abolished (Struhl, 1982b). It appears that the early *Ubx* activity required in the mesothorax for normal leg development is needed only to prevent some *Scr*-dependent transformation. It is probably this same gene that keeps *Scr*$^+$ turned off in the second and third legs in normal development (preventing an *Msc*-type transformation). Although BX-C and ANT-C are regarded as clusters of selector genes, it appears that their activities may involve not just regulation of realizator genes but also some degree of cross-communication with each other. There is also some indirect evidence that ANT-C genes may affect BX-C gene expression in abdominal segments (Duncan and Lewis, 1982).

Establishment and Maintenance of Segment Phenotype. Apart from the genes of the BX-C and the ANT-C, there is a large and growing number of known genetic functions that, when mutated, produce striking transformations of the imaginal segmental phenotype. The great majority have corresponding effects on larval segment development and were referred to earlier in connection with the larval transformations produced by their mutants (Table 6.3). In a few instances, the larval effects are weaker than the imaginal effects, but for the most part, the degrees and domains of transformation seem comparable.

As noted earlier, a fundamental distinction may be made between establishment and maintenance functions. A striking generalization about the genes that affect the segmental phenotype is that all, with the possible exception of *esc,* are maintenance functions to some degree. If the wild-type alleles are removed from heterozygotes by mitotic recombination at any time until late in larval development, the mutant transformation is still produced. Indeed, at least some of the genes of the BX-C are maintenance functions by this criterion (Lewis, 1964), and probably many or all of those ANT-C as well (Struhl, 1981d, 1982b). The mitotic recombination test only establishes the latest time at which the gene(s) must be present; a late transformation produced by gene removal does not show that the genes must be present early in order to maintain the imaginal segmental phenotype, although this is often the simplest interpretation. For the larval segmental phenotypes, which are visible by mid-embryogenesis, it is apparent that the genes of BX-C and ANT-C are active and required from early embryogenesis on.

Indeed, the gene products may be needed even later than the times of gene removal seem to indicate, because the products may persist or "perdure" (Garcia-Bellido and Merriam, 1971b) for several cell divisions after the wild-type alleles that code for them have been removed. To determine the latest times at which the gene products themselves are required, one needs to employ ts mutants and the appropriate temperature shifts to obtain the approximate answers.

The failure to find "pure" establishment functions raises a question about the existence of activator genes, whose products are required to set up differential settings of the various selector genes. Presumably, activator gene products are needed as a sensing system to transduce a general signal, such as a morphogenetic gradient, into an informational signal that can read DNA sequences and turn on specific selector genes. As in the comparable case of compartment-class selector genes, they may have eluded detection only to date. However, the very intensive searches for new *Drosophila* mutants conducted during the past decade render this possibility unlikely. Although *esc* possesses some of the expected characteristics of an activator gene—a strong maternal effect on segmentation, an interaction with the bank of selector genes that comprise BX-C, and a requirement for its gene product around the time of blastoderm—its dispensability in the absence of a maternal *trx* product (Ingham, 1983) makes it unlikely that *esc* is an activator gene. If activator genes exist, their gene products must have additional roles in development. Mutational inactivation of such genes would produce pleiotropic effects and make the identification of their activator function difficult.

The other striking feature of these observations is the apparent genetic complexity involved in maintenance. The genes of the BX-C and ANT-C may be regarded as "positive maintenance" genes; their gene products are required to perpetuate specific segmental phenotypes. Most of the presumptive regulatory genes outside these two gene complexes, however, are "negative maintenance" genes. Such genes do not directly foster the development of particular segmental phenotypes but somehow ensure that the "wrong" genes remain unexpressed; in the absence of such maintenance gene products, inappropriate genes become turned on and segmental phenotypes "rise" to the AB8 level. The classical example is that of *Polycomb* (*Pc*), whose derepressed phenotype is dependent on the presence of one or more BX-Cs within the genome. The *Polycomb-like* (*Pcl*) gene (at 2–84) possesses much the same properties (Duncan, 1982), and several others with comparable properties have subsequently been identified.

The weakness of attributing such effects to repressors, ostensibly encoded by such loci was discussed earlier. It is more reasonable to posit that whatever the molecular basis of gene inactivity of BX-C and ANT-C functions within segments, it is susceptible to perturbation by a number of genetic influences, including some perhaps that are not necessarily directly involved in the normal regulation of the two gene complexes.

The heat shock phenomenon may provide an analogy that will illustrate this possibility. When cells of *Drosophila* or other organisms are exposed to a brief pulse of supraphysiological heat, they shut down normal gene expression and turn on the transcription and translation of a small set of "heat shock" genes. In *Drosophila,* these genes total more than a dozen, located at nine positions on the large autosomes (reviewed by Ashburner and Bonner, 1979). In fact, the heat shock response can be elicited by a wide variety of treatments. In *Drosophila,* the inducing stimuli include heat and recovery from anoxia, and treatment with such compounds as dinactin, dinitrophenol, hydrogen peroxide, hydroxylamine, and valinomycin. With respect to the question of mechanism (leaving that of physiological function aside), no one has seriously suggested—at least in print—that there is a common repressor of the heat shock genes that is antagonized by all of the inducing processes. Rather, it seems much more probable that the heat shock genes share some common feature in their regulation that makes them *sensitive* to induction by a variety of moderately nonspecific stimuli. This sensitivity might involve, for instance, some properties of the chromatin state, or shared noncoding sequences with special properties, or some combination of such features.

The maintenance of BX-C and ANT-C gene activity by *Pc* and *Pcl,* and by other genes, might have an analogous basis: some shared, unusual features of the genes of both complexes that render their nonexpressed states liable to a variety of perturbations, including reduced gene activity of certain other genes. This view, too, is just a hypothesis, but it is a simpler and more tenable one than the repressor hypothesis. The molecular biology of these genes complexes may help to identify such features (Chapter 9).

At the very least, the multiplicity of such "regulatory genes" illustrates the difficulty of inferring molecular mechanisms from observations of mutant phenotypes. More specifically, it suggests the need for caution in framing simple hierarchical models of gene control on the basis of a limited number of observations. The greater the number of genetic elements involved in the "control" of a specific phenotype, the more one is forced to move from hierarchical models to consider alternative models of regulation (Chapter 10).

The Selector Gene Hypothesis Revisited. As originally framed, the selector gene hypothesis had three central elements. These were as follows:

1. The compartment, as delineated by the Minute technique, is the fundamental and perhaps universal building block in development.
2. The phenotype of each region is determined by the combined activities of a small number of selector genes, consisting of two kinds, compartment-class specific and segment or half-segment specific selectors.
3. A special class of genes, activator genes, determine which selector gene(s) will be turned on or off in particular compartments.

Today, all of these propositions, in their simplest form, appear debatable. There is, for instance, little evidence for subsegmental compartments in internal tissues, with the possible exception of the thoracic musculature. Even in this case, however, the clonal restriction that develops may be produced simply by movement apart of the progenitor cells. Furthermore, there is little evidence for single compartment-class selector genes. It seems much more probable that general compartmental categories (A versus P, D versus V) are established by the activity of several, at least two and possibly many more, key genes. There is even less support for the existence of genes whose sole function is to act as activator genes, although of course, there may well be activator genes with additional functions (whose mutant phenotypes would be pleiotropic).

The element of the selector gene hypothesis that has been strongly substantiated is the postulation of segment or half-segment specific selector genes. The genes of the BX-C and ANT-C seem to have just this specificity, although their actions may involve the maintenance of a developmental pathway rather than its selection per se. Even with respect to these genes, however, a given (segmental) phenotype for many segments is determined not by a single selector gene setting that phenotype but by the combined action of several such selectors.

This combinatorial, additive action is most apparent for the abdominal segments, each of whose phenotypes depends on the action of a given BX-C function plus all (or most) of those proximal to it. But such additive effects may also pertain to thoracic half-segments. In their detailed electron microscopic classification of the different types of sensilla on wings and halteres in *Drosophila,* Cole and Palka (1982) found that sensillum character within the A compartment of the haltere was determined by the action of three BX-C functions, *abx, bx,* and *pbx*. The effect of *pbx* on sensilla in the A compartment is particularly interesting, given the presumptive specificity of this gene for the P compartment of the haltere. On an even grosser scale, however, the fact that both *abx* and *bx* mutations can independently transform a variable region of the anterior haltere to A wing (Table 6.6) shows that the development of even thoracic half-segments may require more than one selector gene. Rather than one gene–one phenotype relations, it is the combinatorial actions of genes that establish developmental pathways.

An additional complexity not anticipated in the original hypothesis is that the activities of two selector gene complexes, BX-C and ANT-C, exhibit some degree of interaction. In effect, these genes may govern not only groups of general cell function realizator genes but each other as well. In the literature, the term "selector gene" is coming to be used simply as a synonym for "regulator gene," without special reference to compartments.

Nevertheless, the hypothesis, unlike many in developmental biology, has been fruitful. Not least, it has stimulated much productive experimentation. The experiments have extended considerably our knowledge of compartments and of the genes that operate within them. Also, the hypothesis has served to focus attention specifically on the roles of the BX-C and the ANT-C, which is proving to be one of the most informative subjects in the molecular biology of development (Chapter 9). Finally, the results have served to sharpen the questions that remain.

One particular question of importance concerns the precise function of compartments in cuticular tissue. The simplest supposition is that the compartments are essential building blocks for each imaginal structure. In other words, for the fly to construct a leg properly, it needs to divide the imaginal anlagen of the legs into A and P compartments early in embryonic development. This hypothesis can be tested by asking whether the timing of events in compartmentalization is always a constant feature of the development of a particular structure. Morata and Kerridge (1981) performed this test by determining the time of the A-P segregation event in a leg formed in the head region, a so-called cephalic leg, produced by a strong *Antp* transformation. The time of A-P compartmentalization was found to be not that of the normal leg (postblastoderm) but that *typical of eye-antennal A-P compartmentalization*. Morata and Lawrence (1979a) and Struhl (1981a) have made similar observations using the less complete ss^{a} transformation of antenna to leg. It follows that the precise timing of a particular compartmentalization event within a primordium is not always crucial for the correct construction of that primordium.

It may be that the primary function of compartments is to act as units for the control of shape and size. The differences in compartmentalization patterns, in this view, reflect the different requirements of or constraints on growth and morphogenesis in the imaginal discs. As units of growth control, neighboring compartments may also serve to guide each other's growth in some fashion. Lawrence and Morata (1976) observed that en^1 clones at the margin of the posterior dorsal subcompartment of the wing cause a local englargement of the wing that extends beyond the clone on both dorsal and ventral surfaces. They concluded that the ventral surface/compartment is modeled in part on the dorsal compartment. Apparently, compartments may interact locally in morphogenesis even though their component cellular lineages are completely separate.

Nevertheless, it seems probable that compartments have a developmental significance that goes beyond their role as units of growth control. The A-P separation appears to be a fundamental property of most or possibly all imaginal discs (the picture in the genital disc is complex, perhaps reflecting its multisegmental origin). Given the numerous differences between the discs, the utility of this feature for purposes of growth control alone would be difficult to explain. Furthermore, the results of clonal analysis with *en* mutants—their preferential or exclusive effects in P compartments, the differences between A and P compartment *en* clones in transgressing the compartment border—imply rather strongly that there is a basic difference between anteriorness and posteriorness. This difference exists even within the histoblast nests, where *en* mutant effects seem to be restricted to posterior histoblast cells (Kornberg, 1981b). It may be that a primary function of the A-P separation is in the separation of embryonic/larval segment primordia, with the imaginal effects passively reflecting this embryonic function. The hypothesis is suggested by the difference between segmental and compartmental segregations between the cephalic segments and those of the thorax and abdomen. In the latter regions, both separations take place early while in the cephalic region, segmental separations are less distinct, and the A-P segregation is markedly late. Finally, compartments also appear to be units of distinctive gene expression, particularly in the combinations of segment or half-segment selector genes that they employ. However, their characterization in this respect is only just beginning.

Other questions that remain concern the mechanism(s) that generate compartments and, perhaps most importantly, the molecular and cellular activities of the two gene complexes, BX-C and ANT-C, that play key parts in the setting of half-segment and segment identities. Until the activities of these genes are identified, our understanding of the control of *Drosophila* development will remain superficial at best.

Determination and Transdetermination

One of the features of imaginal discs that has made them such favorable material for developmental studies is that they retain their characteristic developmental fate during prolonged culture. By this most stringent criterion, discs are highly determined pieces of tissue.

The standard culture procedure for discs is diagrammed in Figure 6.26. Discs are first removed from larvae (usually third instar larvae, which have the largest discs) and either cut into fragments or partially dissociated and reaggregated. The test pieces are then placed in the abdomens of females (females being used, since they have larger abdomens than males). The abdominal hemolymph serves as a culture medium. Because it contains little ecdysone, imaginal cells proliferate in it without differentiating. Breaking up the cell mass prior to implantation by dissociation/disaggregation promotes cell proliferation during culture. Over a period of 1–2 weeks, the implants grow, gradually filling up the host's abdominal cavities. At the end of this first culture period (or transfer generation, Trg), the implant is removed and cut into several pieces, and the procedure is repeated. At each stage, a small

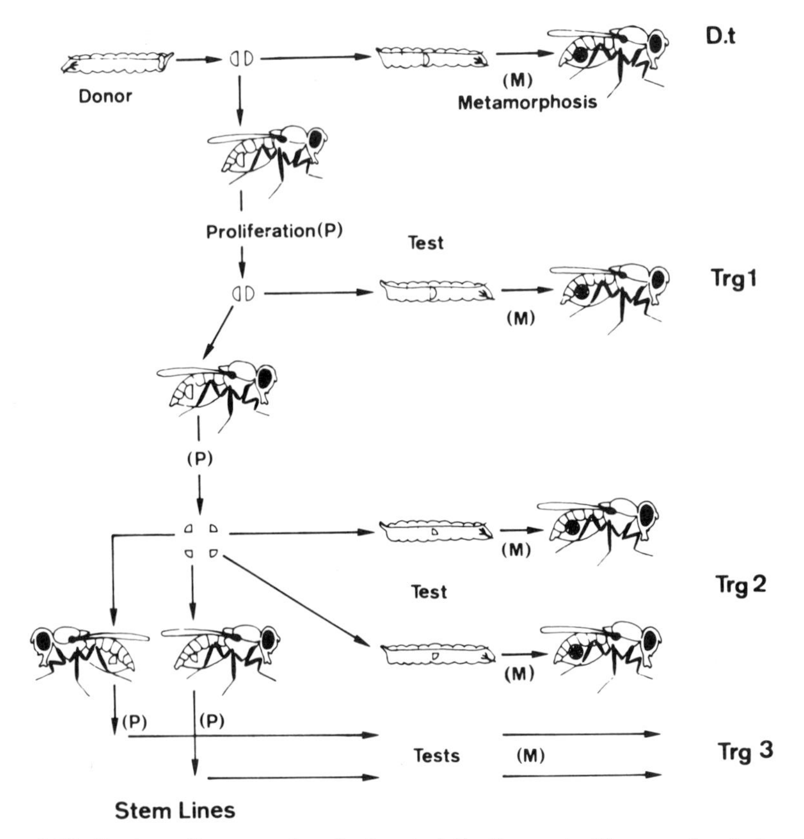

Figure 6.26 In vivo culture procedure for imaginal disc fragments. The procedure is described in text. D.t., direct test; Trg 1–3, transfer generations 1 to 3. (Reproduced with permission from Hadorn, 1978.)

piece can be assayed for developmental state by placing it within a mature host larva; the implant, exposed to the ecdysone of the metamorphosing host, differentiates along with its host, and can be removed from the abdomen after emergence of the imago. The adult structures generated by the tested fragment reveal its developmental capacities before differentiation. Thus, at each Trg, the state of determination of replicate implants can be tested.

The experiment reveals that disc-specific states of determination can be reproduced with great faithfulness. In one culture line started from a male genital disc, characteristic genital disc structures could be produced for a period of up to 55 Trg (Gehring, 1978). Each disc, in fact, appears to consist of a mosaic of regional capacities, each possessing great stability. Thus, in the genital disc experiment, there was an early segregation of two developmental potentialities, one for the anal plates and the other for male genital structures (clasper, penis, and ductus; see Fig. 6.31). This case may illustrate the segregation of two fundamental segmental potentialities because the *Drosophila* genital disc is a complex structure consisting of several structures that are found in distinct posterior segments in more primitive Diptera. However, other segregations during culture involve separations into true intrasegmental capacities, such as sublines from antenna or leg discs specific to regions of the antenna or leg, respectively (cited in Hadorn, 1978, p. 591).

Despite this great stability of disc-determined states during culture, it is not perfect; changes in state can and do occur. Intriguingly, the changes are generally to recognizable *alternative* disc types. (Only after very extensive subculturing do truly abnormal developmental states, shown by abnormal differentiated patterns, appear.) The genital disc experiment can again supply an example. After 6 Trg, implants differentiating into antennal or leg tissue were found.

To distinguish the original from the secondary disc states, the cell type characteristic of the original disc type will be designated as the ''autotypic'' cell type, and the new disc state will be referred to as the ''allotypic'' one. Strikingly, allotypic disc states show both the stability and types of (further) change characteristic of that of the disc type whose phenotype they mimic. By these criteria, the allotypic cells have acquired determined states indistinguishable from those of the discs they resemble. Hadorn (1965) termed these changes of state ''transdeterminations.''

Transdetermination: Character and Causes. The most important single conclusion about transdetermination is that it involves regulatory (''epigenetic'' in the older usage) rather than somatic mutational change. There are two grounds for this statement. In the first place, the frequencies of transdetermination are too high to be accounted for by mutational processes; in some experiments, frequencies of 70–80% of all of the implants exhibit transdetermination within a Trg. In the second place, a few of the reciprocal transdeterminations occur with comparable frequency; although preferential loss of a gene function might account for change in one direction, the nearly equal rate of acquisition would have no precedent in mutational phenomena.

Furthermore, transdetermination almost certainly entails a switch from one determined state to another (as the name implies) rather than the acquisition of a new determined state by uncommitted stem cells carried along in each implant. The evidence comes from an experiment by Gehring (1967), who argued that if transdetermined tissue derives from a pool of undetermined cells, then any genetically marked clone arising during culture should be either wholly autotypic or wholly allotypic; the presence of both autotypic and allotypic tissue within a marked clone in cultured material must signify a switch from the first to the second within the clone.

Gehring induced marked clones in second instar larvae, allowed further growth to the third instar, and then cultured antennal disc material that had been dissociated and reaggregated. The fully grown implants were then allowed to differentiate within host larvae and scored for the distribution of marked clones with respect to autotypic and allotypic structures. In nearly all of the 16 aggregates where marked cells were found in allotypic tissue, autotypic cells were also marked. Given the single-cell origin of each marked patch, as estimated from the frequency of induced clones, it follows that the transdetermined tissue was derived from cells that had the original (autotypic) determined state. [It is possible that the switch involves a transient "dedetermined" state; however, no evidence for the existence of such an intermediary state has been obtained despite attempts to detect it (Hadorn et al., 1970).]

The Gehring experiment reveals an additional fact about transdetermination: the process of switching often involves two or more cells at a time. This conclusion is based on the appearance of allotypic patches composed of both marked and unmarked tissue. In principle, such patches might result from two independent transdeterminative events—one within each of the two regions—followed by cell sorting movements to establish coherent territories of similarly determined cells. However, such reassortment would cause the breakup of the marked clone. In fact, the boundaries of marked clones run smoothly between areas of differentially determined cells. An example is shown in Figure 6.27; *mwh* homozygous tissue occupies both wing and occiput (head) tissue (both being derived from antenna in this experiment), and the boundary of the clone crosses without interruption from one tissue type to the other. Whichever change came first in these cells (whether from occiput to wing or the reverse), it must have simultaneously involved at least one *mwh* and one wild-type cell, with subsequent extensive growth in the implant.

In this aspect of cell cooperativity or entrainment, transdetermination resembles the action of certain homeotic mutants. An example is the change from antennal to homeotic leg growth within the antennal disc of an *Antp* mutant, which probably occurs in the early third instar stage. Yet, genetically marked cells induced before this period never solely occupy the ho-

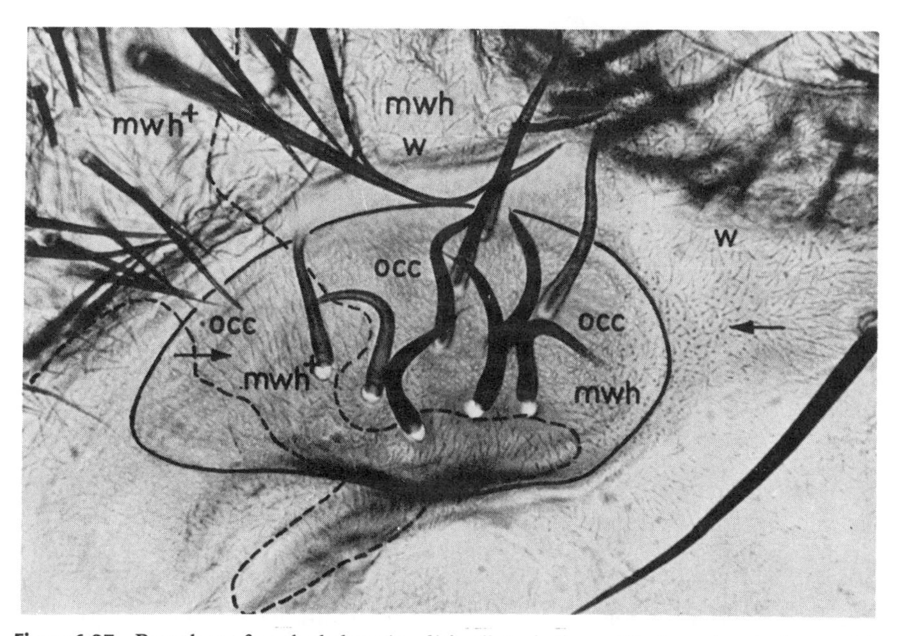

Figure 6.27 Boundary of marked clone (*mwh*) in allotypic tissue. The boundary is shown with respect to head (occiput; occ) and wing (w) tissue. Both the marked and unmarked areas smoothly traverse the region of allotypic tissue. (Reproduced with permission from Gehring, 1972.)

meotically transformed patch, as should occur some of the time if the *Antp* switch takes place in individual cells (Postlethwait and Schneiderman, 1969, 1971). Such homeotic changes within appendages must therefore entrain two or more cells at a time. Indeed, basic developmental assignments in *Drosophila* may always be given to small groups of cells. We have seen that compartmentalization always occurs in cell groups (polyclones) and that imaginal primordia always arise from more than one blastodermal cell.

Although the triggering factor is unknown, growth during the culture period is essential. Thus, treatments such as cell dissociation and reaggregation that stimulate cell division also promote transdetermination. The requirement for growth led Hadorn (1965) to propose originally that transdetermination might require a dilution of specific repressors; growth would diminish the intracellular concentration of repressing molecules, permitting the switch to occur. This hypothesis predicts an exponential increase in transdeterminative events as a function of the number of cell divisions experienced. Although precise measurements have yet to be made, the kinetics appear to be linear instead. An alternative hypothesis is that the growth requirement reflects the need for some form or degree of chromatin remodeling, something that might occur with a given probability only during chromosomal replication or division. At this point, the nature of the requirement for growth remains a mystery.

However, while growth is essential, it is not sufficient to provoke transdetermination. The stage of disc development at which culturing begins is also crucial, with late third instar material being the most favorable. Lee and Gerhart (1973) showed that fragments of early prepupal leg discs, which have begun to differentiate and evert, undergo growth in culture comparable to that of late third instar discs but show substantially less transdetermination. Furthermore, the background genotype can also influence the relative frequencies of transdetermination from different stages.

There is one last bit of phenomenology to be noted. The cells that undergo transdetermination are not a random sample of the cells of a disc but probably come from the regions with the largest regenerative potential. This fact, however, touches on the broader topic of the nature of disc patterns and the ways these patterns regulate following disc damage (Chapter 8).

Transdetermination: Pattern and Implications. The single most significant fact about transdetermination is that the entire sequence of observed changes comprises a single recognizable pattern. Each disc type undergoes primary changes to one or a few other specific disc types, with characteristic frequency and probability. With further growth, these changes are succeeded by characteristic secondary switches. Furthermore, autotypic and allotypic implants of the same disc type experience essentially identical patterns of change. Thus, allotypic antennal tissue derived from the genital disc transdetermines to the wing with the same relative frequency as an antennal implant derived directly from the antennal disc.

The existence of preferred types of change produces both an overall pattern and direction of transdetermination, shown in Figure 6.28. The general direction of transdeterminative change is toward mesothorax. Which-

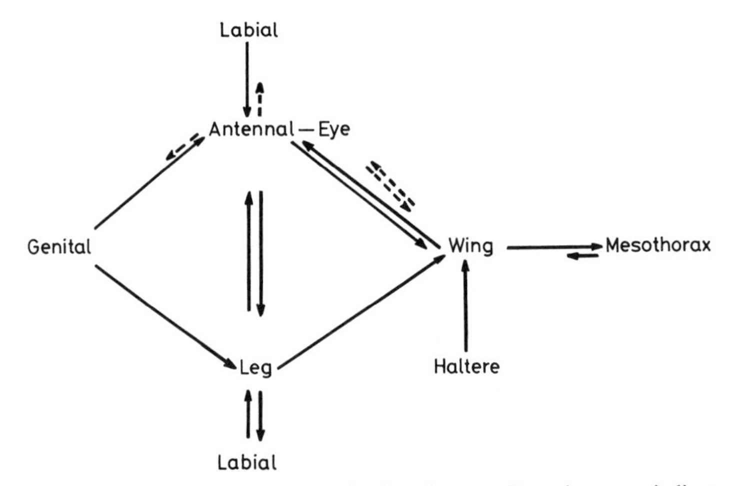

Figure 6.28 Global pattern of transdeterminative changes. Dotted arrows indicate rare or suspected changes. See text for discussion. (Adapted from Hadorn, 1978.)

ever disc type one begins with, the consequence of long-term culture is the eventual enrichment of mesothoracic (notal) tissue. Between the beginning of culture and the termination of a long sequence that ends in mesothorax, the culture line will usually produce many or all of the intermediary disc cell types predicted by the large arrows in the diagram, as well as minor contributions of tissue derived from the less frequent reverse changes. The transdetermination data confirm the genetic observation described earlier: that mesothorax is some sort of developmental ground state.

The interconvertibility of the various disc types by transdetermination emphasizes the underlying relatedness of the various disc states. The basis of this relatedness is not immediately obvious. A number of the conversions reflect neither the normal geographical proximity in the imago nor the standard developmental sequence of the discs in situ. In some instances, relationships between serially homologous structures are reflected, as in the conversion of haltere to wing or in the interconvertibility of antennal and leg cells; these interconversions are also found in the set of homeotic mutants. However, other changes, in particular that of genital disc to leg or antenna, entails a transformation of one cell type into those of other disc types that are neither geographically proximate nor obviously related in their normal development.

This pattern indicates that a fairly small number of connecting links differentiate the discs from one another. If one regards these switches as changes in activity of particular regulatory loci, it is reasonable to view each transdetermination as involving a switch of just one or two key activities. As noted above, the homeotics provide genetic parallels for many, although not all, of these switches.

Kauffman (1973) has formalized this picture of transdetermination. He has proposed that the activities of a small set of "control circuits," perhaps only three to five, underlie all the disc states. Each such circuit is posited to exist in either of two states, which can be symbolized in binary terms as 0 and 1. In his model, it is the combination of these control circuit states that establishes the particular disc states. Transdetermination events reflect the change of one, or at most two, such circuits from the 0 to the 1 state, or vice versa. To account for the overall direction of transdetermination toward mesothorax (notum), one need only postulate that one state, for instance, the 1 state, is more stable than the other. Mesothorax, the endpoint of transdetermination, would then be represented as a sequence of 1 values; for four control circuits, its representation would be 1 1 1 1. Another disc type, separated by just one control circuit state from mesothorax, might be 0 1 1 1 or 1 0 1 1, and so forth.

The molecular meaning of the 0 and 1 states is unspecified. One possibility is that each control gene is a bifunctional locus, with the two states representing alternative gene activities, each regulating a different set of downstream gene activities (sets of realizator genes). A simpler possibility is that one state represents a control gene in an on position, the other state the off position. The work on BX-C suggests that mesothorax is a condition of minimal expression. If one extrapolates from this to other disc states, the assigned 1 state in the Kauffman scheme can be provisionally taken as off and the 0 state as on; the direction of transdetermination can be viewed as a tendency for control circuits to switch into inactivity during development.

If each control circuit is either on or off in any particular disc, then any mutation that renders the on state defective, either completely or in part, will render all of those discs expressing that control circuit defective but will leave the remainder untouched. (Such defects could be either in the regulatory apparatus itself or in one or more genes whose activity is governed by the control circuit.) Such mutations should therefore divide the total set of discs into two discrete subsets, one consisting of defective discs and the other of normal discs. As Kauffman has stressed, one group of imaginal disc-defective mutants (Shearn et al., 1971) do just this. Eight of the disc defectives are affected in only a subset of the discs, and these fall into four complementary pairs. In one mutant class, haltere, wing, and leg discs are developmentally defective, but eye and antenna (taken as two separate discs) are normal; in the complementary class, eye and antenna discs are both affected, but wing, haltere, and leg discs are normal. The particular developmental abnormalities vary between the different mutants, but the partitioning into complementary classes is the noteworthy point. However, a further characterization of the mutants is desirable because several disc types were not scored in the study and the division of eye and antenna may be questionable, given their developmental links. Provisionally, however, the Shearn results support the concept of two-state control circuits underlying disc character.

If one were to speculate on the particular gene activities involved in transdeterminative switches, one would have to nominate the various homeotic loci. Most of the transdeterminative switches are paralleled by known homeotic mutations, although some are not. Despite the obviousness of the idea that most transdeterminative events involve gene activities identified in the homeotic mutants, the supposition has never been tested. In principle, it should be possible to do so, using a modification of the Gehring test of clonal origins. If, for instance, the change from antenna to leg involved in some cases the activation of *Antp*$^+$ activity—to repress antenna-specifying genes and activate leg-specifying genes—then mixing dissociated antennal discs from a strong *Antp* mutant with genetically marked, nonhomeotic antennal disc cells should increase the transdeterminative rate of the latter to leg. [Circumstantial evidence, in the form of an observed nonautonomy of *antp*$^-$ clones when surrounded by *Antp*$^+$ suggests that the *Antp*$^+$ gene product can be passed between cells (Struhl, 1981d).] Such tests could be carried out with any of the transdeterminative changes that mimic known homeotic mutants. Until the genetic activities that figure in transdeterminative changes are identified, the whole subject will remain simply a part of the phenomenology of developmental biology, without a satisfying linkage to the genetic data on developmental control.

Sex Determination

The numerous physical differences between male and female *Drosophila* all stem from a simple difference in sex chromosome composition; females are XX and males are XY. This difference actually decomposes into two differences. The first is the difference in X chromosome number between the two sexes; the second is that males possess a Y chromosome (whose primary function in single-X individuals is to permit normal sperm production). As in *Caenorhabditis,* it is the ratio of X chromosomes to the number of autosome sets that determines the sexual phenotype. Diploid individuals that have two X chromosomes, giving an X/A ratio of 1, are set from early embryogenesis on the path of female development; those with just one X chromosome, and an X/A ratio of 0.5, are directed onto the path of male development.

As in *C. elegans,* the difference in X chromosome number between males and females might, in principle, produce a lethal genetic imbalance in either males or females. The great majority of genes on the X chromosome have nothing to do with sexual differences per se, and the effect of such gene product differences could be a serious imbalance with respect to autosomal gene products in one sex or the other. In reality, this potential problem is circumvented because fruit flies have evolved a mechanism for equalizing X chromosome activities between males and females, the process termed "dosage compensation." Dosage compensation in *Drosophila* involves doubling the rate of X chromosome transcription in males relative to females in order to equalize X-gene transcription between the sexes.

In reviewing what is known about sexual development in *Drosophila,* we will first look at the visible differences between the two sexes, then examine what is known about the mechanism of sex determination in this animal, and finally discuss the relationships between sex determination and dosage compensation. An extensive review of sex determination and dosage compensation can be found in Baker and Belote (1983) and a description of the special problems of dosage compensation in Lucchesi (1983).

Sexually Dimorphic Traits

The effects of genetic alteration on sex determination can be assayed by inspection of three external regions of the fly body that are sexually dimorphic. Changes towards maleness are registered as a shift away from female-specific structures to male-specific ones and those toward femaleness by the reverse shift. The three sexually dimorphic regions are (1) the sex comb on the leg of the male, a group of 10–13 thick black bristles on the most proximal tarsal segment, the basitarsus (Fig. 6.29); (2) the fifth and sixth tergites in

Figure 6.29 Sex comb on the foreleg of male *Drosophila*. (Photograph kindly provided by Dr. D. Gubb.)

the abdomen, which are uniformly darkly pigmented in males but show only a posterior band of pigment in the females (Fig. 6.30); and (3) the genital disc structures, which show numerous external as well as internal differences between males and females (Fig. 6.31). Besides these features, the shape and size of the abdomen also differ, that of the male being smaller and more rounded.

Both the sex comb and major parts of the genital disc derive from precur-

Figure 6.30 Abdomens of wild-type female (*a*) and male (*b*) *D. melanogaster*. (Figure kindly provided by Dr. B. Baker; from Baker and Ridge, 1980.)

sor cells that exist in both sexes and take different routes, depending on their X chromosome composition. The sex comb derives from a group of bristle-secreting cells that, in the female, develop into the most distal row of transverse bristles in the basitarsus. In the male, these bristle precursor cells rotate 90° and develop into the thick "teeth" of the sex comb.

In the genital disc, the cellular homologies are complex, reflecting the multisegmental origin of the disc. The complete group of genital structures can be broken down into the analia (the anal plates) and the genitalia themselves; the collective set of genital disc structures are designated the "terminalia." Briefly, the analia derive from one set of primordial cells common to both sexes, but the male and female genital structures derive from spatially separate precursor pools.

The evidence for the respective origins of the analia and genitalia comes from an elegant gynandromorph analysis by Nothiger et al. (1977). Reasoning that the sexually dimorphic structures of the disc should develop in accordance with their X chromosome composition, these workers examined gynandromorphs in which the XX/XO boundary line was observed to pass through the terminalia. In all of these specimens, the sum of female plus male anal plates, which can be distinguished between the two sexes by size and orientation (see Fig. 6.31), was equivalent to approximately one whole set of structures; evidently, a common pool of precursor cells was partitioned in these gynandromorphs. In contrast, the summed male plus female genitalia ranged from practically zero to nearly two complete (one male, one

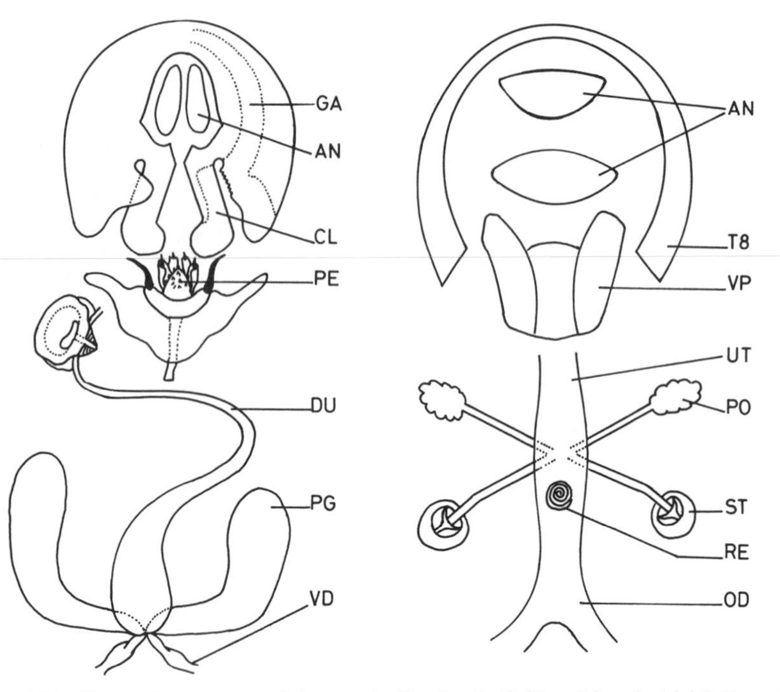

Figure 6.31 External structures of the terminalia of male (left) and female (right) *D. melanogaster*. Male structures: GA, genital arch; AN, anal plates; Cl, claspers; DU, ejaculatory duct; PE, penis; PG, paragonium; VD, vas deferens. Female structures: AN, anal plates; UT, uterus; PO, parovarium; T8, eighth abdominal tergite; OD, oviduct; VP, vaginal plate; ST, spermatheca; RE, seminal receptacle. Dotted lines are lines of clonal restriction. (Adapted and reproduced with permission from Dubendorfer and Nothiger, 1982.)

female) sets of structures. This latter result is explicable if there are two separate genital primordia and the development of either or both depends on their possessing the correct X chromosome number. In those gynandromorphs with nearly complete male and female sets, the male primordium consists largely or wholly of XO cells and the female primordium of XX cells, but in those with hardly any genital structures, both primordia must have received the opposite, and inappropriate, X chromosome compositions. By extrapolation from the segmental origins of genitalia and analia in lower Diptera, the female genital primordium is deduced to lie anterior to the male genital primordium and the analia posterior to both of these structures. A clonal analysis, employing mitotic recombination, has confirmed these conclusions (Dubendorfer and Nothiger, 1982).

Determinants of Sex: X/A Ratio and Autosomal Sex-Determining Loci

The evidence for the central role of the X/A ratio in sex determination in *Drosophila* comes from observations by Bridges in the 1920s on triploid flies.

The approach used was the model for the comparable analysis of X/A ratio effects in *C. elegans,* described in Chapter 5. Bridges first identified rare spontaneously occurring triploid females and then crossed these with normal males; the surviving progeny were found to have a variety of sexual phenotypes that could be correlated with their X chromosome and autosomal compositions. (Because of the need for chromosomal balance, all survivors had either diploid or triploid autosomal sets, except for some variability in the presence or absence of the tiny fourth chromosome, which is without major effect on viability when present in only one copy.) Bridges found that low X/A ratios favor maleness and high X/A ratios femaleness.

The complete set of observations for the different ploidy levels are shown in Table 6.8. An X/A ratio of 1, whether in diploids (2X/2A) or triploids (3X/3A), produces phenotypically normal females. Even haploid cells, with an X/A ratio of 1, formed in rare haplo-diplo mosaics (produced by the occasional double fertilization, in which the second sperm nucleus, carrying an X chromosome, gives rise to the haploid tissue), produce female tissue. Flies with an X/A ratio of 1.5 (3X/2A), so-called super- or metafemales (produced by rare nondisjunction events in normal diploids that result in eggs with 2X chromosomes), are also phenotypically normal females. In contrast, flies with an X/A ratio of 0.5 or less are morphologically normal males; thus, both normal X/Y flies (1X/2A) or triploid flies with just one X (1X/3A), so-called super- or metamales, are typically male in appearance.

The most interesting category are those with an X/A ratio between 1 and 0.5. As might be expected from the idea of chromosomal balance, these flies are neither wholly male nor wholly female, but intersexual. Thus, an X/A ratio of 0.67 (2X/3A) produces the intersexual phenotype. Indeed, there is no single intersexual phenotype, but rather a range extending from more male-like to more female-like intersexes. The placement of a particular individual on this scale may be determined by the number of small fourth chromosomes—which were claimed to have a feminizing effect—and by the pres-

Table 6.8 X/A Ratio and Sex Determination in *Drosophila melanogaster*

Sexual Phenotype	X-chromosomes	Autosomes	X/A Ratio
Metafemales	3	2	1.5
Female (triploid)	3	3	1
Female (diploid)	2	2	1
Haploid[a]	1	1	1
Intersex	2	3	0.67
Male	1	2	0.5
Metamale	1	3	0.33

[a] Scored in XO/XX haplo/diploid mosaics.

ence or absence of the Y, which has the opposite effect. Except for those triploid intersexes, the Y has no effect on the external sexual phenotype, but triploids seem less well buffered than diploids with respect to a number of factors that potentially affect the sexual phenotype (see the discussion by Laugé, 1980). Interestingly, the intersexes, whichever end of the male-female range they occupy, are a mosaic of female and male cells and structures. In nonmutant intersex genotypes, single cells are directed either toward a typical male path or a female one. A careful study of sex combs in intersexes demonstrates this point clearly. In these animals, the number of sex comb teeth is variable, but each sex comb tooth present has a completely normal male phenotype in length and width (Hannah-Alava and Stern, 1957).

The role of X/A chromosomal balance in setting the sexual phenotype led Bridges to propose that feminizing genes reside on the X and masculinizing genes on the major autosomes. A basic prediction of this hypothesis is that it should be possible to isolate recessive, loss-of-function autosomal mutants that create partial or complete sex reversal of males to females.

Of the four key autosomal loci that are known to affect sex determination, none fit the simple prediction. In fact, their mutant phenotypes are largely the reverse of that predicted; three loci are identified by recessive mutant functions that transform females toward maleness, and the fourth gene can mutate to produce changes in either direction. It follows that for at least the first three genes, their wild-type activity promotes femaleness. The existence of loss-of-function mutants in several loci that cause transformation to maleness suggests further that in *Drosophila,* males are the neutral sex (as they are in *C. elegans*) and females are the dominant sex—in other words, that the gonads and secondary sexual characteristics tend to develop male properties unless positively instructed to be female. As mentioned previously, the neutral sex in mammals and birds is the homogametic sex; in *Drosophila,* it is the heterogametic sex that is neutral.

The four autosomal loci that critically affect sex determination are listed in Table 6.9, along with a description of their mutant phenotypes. A complete description of the alleles and the genetic properties of the loci can be found in Baker and Ridge (1980). The standard reference alleles for three of the loci (*tra, tra-2,* and *dsx*) are amorphs by the Mullerian test, while *ix* has not yet been characterized with respect to its residual expression. The direction of transformation for *tra, tra-2,* and *ix* is from femaleness to maleness. *dsx,* however, causes both sexes to transform to intersexuality. (Three special alleles of *dsx* with different properties will be described below.)

In contrast to triploid intersexes with wild-type genomes, the intersexual phenotypes of both *ix*- and *dsx*-transformed individuals shows a true intermediacy at the cellular level, at least for certain features such as sex-comb teeth. This difference in types of intersexuality may be less surprising than it first appears to be. If the X/A ratio sets the sexual phenotype by regulating the activities of these key autosomal genes (see below), then a defective execution step involving either *ix* or *dsx* could give an intermediate cellular

Table 6.9 Autosomal Sex Determination Mutants in *Drosophila melanogaster*[a]

Locus	Map Position	Allele	Phenotype
transformer	3–45	*tra*	Transforms females into males; males normal
transformer-2	2–70	*tra-2*	Transforms females into males; males sterile
intersex	2–60.5	*ix*	Transforms females into intersexes; males normal
doublesex	3–48.1	*dsx*	Transforms males and females into intersexes
		dsx^{D}	Transforms females into intersexes; males normal
		dsx^{Mas}	Like dsx^{D}
		dsx^{136}	Transforms males into intersexes; females normal

Epistatic Relationships

Allele Combinations	Phenotype
dsx; tra	*dsx*-type; transforms males and females into intersexes
dsx; tra-2	*dsx*-type; transforms males and females into intersexes
dsx; ix	*dsx*-type; transforms males and females into intersexes
tra; ix	*tra*-type; transforms females into males
tra-2; ix	*tra-2*-type; transforms females into males
dsx^{D}/dsx*; tra-2*	*tra-2* type; transforms females into males
dsx^{136}/dsx*; tra-2*	Transforms females into intersexes (similar to dsx^{136}-transformed males)

Source. Adapted from Baker and Ridge (1980).

[a] The initial sex denoted under each transformation refers to the chromosomal sex: "females" = XX individuals; "males" = XY individuals.

phenotype; in the wild-type triploids, the execution step would be normal once the X/A ratio had set a cell toward male or female development.

All four autosomal genes affect all aspects of somatic sexual phenotype, including the internal genitalia. Their roles are therefore in the overall determination of the sexual phenotype rather than in the specification of particular structures. Nevertheless, in two key respects, the sexual phenotype remains untransformed in these mutants. In the first place, the mutants show no transformation of gamete production. Thus, *tra*-transformed XX animals are male in appearance but sterile, lacking sperm. The absence of sperm production stems from an incompatibility between the transformed mesodermal tissues surrounding the germ cells and the untransformed germ cells themselves. The absence of transformation of the germ line in *tra*-transformed females was proved by a pole cell transplantation experiment from X/X; *tra*/*tra* embryos. These cells were found to give rise to as many functional egg cells (as measured by progeny counts) when transplanted into female recipients as do control +/*tra* pole cells (Marsh and Wieschaus, 1978). Similar tests, and the use of mitotic recombination to generate germ line clones homozygous for *dsx, ix,* or *tra-2,* have shown that these sex-determining loci are also without transforming effect in the germ line (Schupbach, 1982).

The second respect in which the sexual phenotype remains untouched is that of dosage compensation. X chromosomal genes in wild-type males are twice as transcriptionally active as the same genes in females. If sex transformation included an alteration in dosage compensation, one might expect *tra* or *tra-2*–transformed X/X individuals to have twice the activities of X chromosome genes of their untransformed *tra*/+ or *tra-2*/+ female sibs. This result has not been observed: gene activities for the X chromosome are maintained at the same levels as those in untransformed individuals of the same sex chromosome constitution. The failure to transform the germ line and alter dosage compensation suggests that if there is a master genetic control of some kind in the sexual phenotype, then the somatic sexual phenotype must be regulated by genes that are on a separate pathway from those controlling dosage compensation and gametogenesis.

The genetic control of somatic, sexually dimorphic features bears certain resemblances to that governing the segmental phenotype. Like the homeotic genes controlling segmental character, the autosomal sex-determining genes are required until late in development. The particular time past which the gene is no longer needed is a function of both the individual gene and the tissue/structure involved. As in the comparable tests on BX-C and ANT-C function, one removes the wild-type allele from heterozygotes by mitotic recombination at particular times in development and then scores the emergent adults for the presence or absence of the homeotic phenotype. For all four genes, the retention of the wild-type sexual phenotype within individual clones is dependent on the continued presence of the wild-type allele until late in larval development or, in the case of the abdominal histoblasts, the pupal period (Baker and Ridge, 1980; Wieschaus and Nothiger, 1982). In effect, all four genes (*tra, tra-2, ix,* and *dsx*) are required in the sex in which they act until the last few cell divisions in all of the sexually dimorphic structures. The earliest times at which these genes are required to act is unknown; larvae show no sexual differentiation, except for the gonads, and only a slight difference in size.

The mitotic recombination experiments also do not establish the latest times of action of the products of these genes, since these gene products may persist (perdure) in the various tissues even after recombinational removal of the genes. In fact, temperature shift experiments with a temperature-sensitive *tra-2* mutant reveal a requirement for gene product in terminally differentiated cells of the adult female to ensure continued production of yolk proteins (cited in Baker and Belote, 1983). Surprisingly, these gene activities are apparently required for maintenance of sexual phenotype, even in adult flies.

Of these four sex-determining loci, *dsx* is the most interesting. Unlike the other three, it can mutate to produce at least three very different phenotypes. These are the amorphic and recessive *dsx* allelic type, which transforms both X/X and X/Y animals into intersexes; the dominant dsx^{D} and dsx^{Mas}, which masculinize females; and the recessive dsx^{136}, which feminizes males. Unlike *ix,* which does not affect males, *dsx* activity is needed in both sexes for the development of normal sexual phenotype. In the absence of dsx^{+} activity, individual animals of both chromosomal constitutions develop into highly similar-appearing intersexes (Hildreth, 1965). In both *dsx*-transformed X/X and X/Y intersexes, for instance, the sex comb develops into a structure almost exactly intermediate, in orientation and bristle morphology, between the normal sex comb and the homologous transverse bristle row of the female. Most strikingly, both male and female genital structures develop in all *dsx* intersexes; in wild-type animals, only one genital primordium develops, the selection being wholly dependent on the X chromosome constitution. Clearly, dsx^{+} activity is needed to discriminate in the choice between sexual pathways. However, as dsx^{D}, dsx^{Mas}, and dsx^{136} reveal, the gene can mutate to lock in development toward one sexual mode or the other.

The simplest hypothesis is that *dsx* codes for or controls two distinct activities—that it is a truly bifunctional locus; one activity would be required for female phenotype development, the other for male phenotype development (Baker and Ridge, 1980). These two activities can be provisionally designated dsx^{m} and dsx^{f} for the male-promoting and female-promoting functions, respectively. In wild-type animals, the activity selected would presumably be determined by the X/A ratio, but in sex-specific *dsx* mutants, only one activity can be expressed regardless of the X/A ratio.

The importance of *dsx* in the choice of somatic sexual pathway is given additional weight by its terminal position in the hierarchy of epistatic relationships (lower half of Table 6.9). The amorphic *dsx* allele is epistatic to *tra, tra-2,* and *ix*. In turn, *tra* and *tra-2* are epistatic to *ix*. The existence of an epistatic hierarchy shows that the genes control activities that are part of a single pathway of development. If the pathway is viewed as a sequence of determinative decisions, as in *C. elegans*, then the suppression of the three other mutant phenotypes by *dsx* indicates that the *dsx* locus has the final word in determining the sexual phenotype. In this sense, *dsx* may occupy a pivotal position in *Drosophila* sex determination somewhat comparable to that of the *tra-1* gene in *C. elegans* (Fig. 5.14). (An important functional difference between these two key genes, however, is that the nematode gene appears to be required only in one sex, hermaphrodites, while *dsx* is needed in both sexes of the fruit fly.)

One further similarity between the nematode and fruit fly sex determination mechanisms is that two extra activities are required to keep the critical control gene in the female mode in both animal systems. Thus, in *Caenorhabditis,* both *tra-2* and *tra-3* are needed to keep *tra-1* turned on. In *Drosophila,* the *tra* and *tra-2* genes may perform an analogous function in keeping dsx^{f} turned on. This inference is based on the *reversal* of the epistatic relationship between *tra-2* and *dsx* when the sex-specific *dsx* alleles are present. Thus, while *dsx,* the amorphic allele, is epistatic to *tra-2,* the latter is epistatic to dsx^D, as shown by the following: while dsx^D/dsx^+; X/X ($tra\text{-}2^+$) animals are intersexes, dsx^D/dsx^+; *tra-2/tra-2*; X/X are phenotypic males—like *tra-2*. If dsx^D is dsx^+ set in the dsx^m mode, this result shows that $tra\text{-}2^+$ activity is required for the dsx^f activity of the dsx^+ chromosome. Similarly, the dsx^{136} phenotype, which can be viewed as the expression of *dsx* set in the dsx^f mode, requires $tra\text{-}2^+$ activity. dsx^{136}/dsx, X/X animals are normal females, but when simultaneously homozygous for *tra-2,* they are transformed into intersexes closely similar to dsx^{136}, X/Y animals.

In both of these novel patterns of epistasis, the $tra\text{-}2^+$ activity has been interpreted as one of stabilization of the dsx^f mode of activity. (Alternatively, *tra-2* might mediate the reading of the X/A ratio, but this appears unlikely, since this ratio is read early in development, while $tra\text{-}2^+$ activity is required continuously; see below.) The function of the *tra* gene is unknown but is probably also one of stabilization of dsx^f. Possibly, the gene products of *tra* and *tra-2* act cooperatively or additively. A strain that is homozygous for a temperature-sensitive *tra-2* allele shows X/X animals transformed into males at the *permissive* temperature if it is simultaneously heterozygous for *tra* (genotype: *tra*/+). In a tra^+ homozygous background, the $tra\text{-}2^{ts}$ homozygous X/X animals remain female. Thus, a reduction in tra^+ gene dosage changes a conditional *tra-2* phenotype into an absolute block (Belote and Baker, 1982).

This set of observations has been synthesized by Baker and Ridge, as

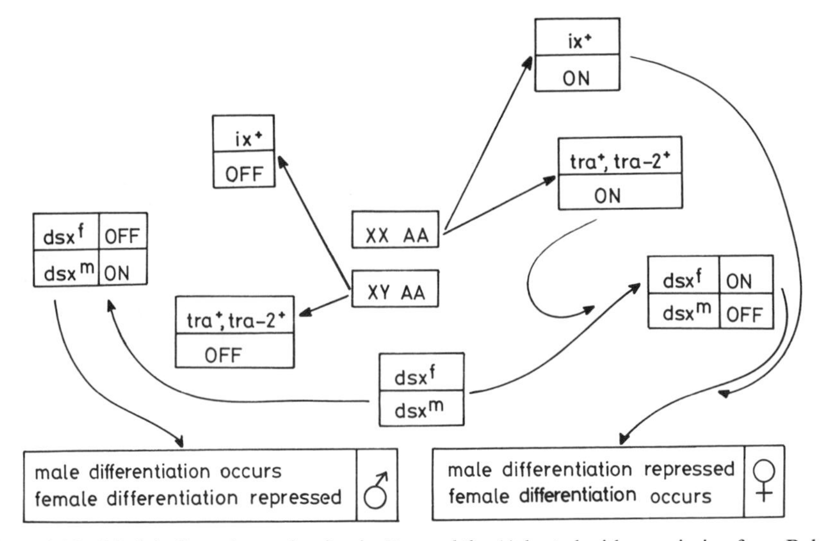

Figure 6.32 Model of sex determination in *Drosophila*. (Adapted with permission from Belote and Baker, 1982.)

shown in Figure 6.32. With a low X/A ratio (0.5 or less), *dsx*^{m} is expressed and *dsx*^{f} repressed, with the "neutral" tendency toward maleness following in consequence. With a high X/A ratio (1 or more), *dsx*^{f} is set for expression (and *dsx*^{m} is repressed), with *tra* and *tra-2* required to maintain this state. The role of *ix* and its position in the pathway have not been established, but it is depicted as acting after the *dsx/tra/tra-2* step to produce a further repression of male functions. If the pattern of epistasis reflects a pathway of determinative decisions, with later steps epistatic to earlier ones, then *ix* would act prior to the *tra/tra-2* step. This model suggests that its product might perform a preliminary stabilization of *dsx*^{f} activity.

Understanding the nature of the *tra/tra-2* interaction with *dsx* requires an elucidation of *dsx* itself. One possibility is that *dsx* specifies a diffusible product, either a protein or an RNA, possessing two separate gene-regulatory activities. Alternatively, there might be two *dsx* molecules with complementary activities. However, as Baker and Ridge note, a paradox develops from this interpretation, namely, that similar intersexual phenotypes arise in both *dsx*^{D}/*dsx*$^{+}$; X/X and *dsx/dsx*; X/X genotypes. In the first case, both *dsx* activities, *dsx*^{m} and *dsx*^{f}, should be expressed—*dsx*^{f} activity from *dsx*$^{+}$—in the presence of two X chromosomes. Intersexuality would then result from their simultaneous expression. In the second genotype, neither activity is expressed and intersexuality is produced. Naively, one would not expect the same phenotype to result from either the presence or simultaneous absence of the two products. One possibility is that the two products, when jointly expressed, are antagonistic to each other's activity. However, this paradox will probably be resolved only through a molecular characterization of the locus, its products, and their patterns of expression.

Dosage Compensation and Sex Determination

Dosage compensation is the mechanism for equalizing the effects of different X chromosome compositions between the two sexes. In *Drosophila,* it consists of a doubling of the rate of transcription of X chromosome genes in males relative to females; this doubling cancels the effect of the halving of the X chromosome gene dose in males relative to females. Compensation serves only to equalize X chromosome transcription between the sexes; within each sex, the amount of X gene product is proportional to the number of doses of each X-linked gene, so that a duplication will double the activity of a product in either sex.

Superficially, somatic sexual determination and dosage compensation appear to be divorced from one another; sexually transformed XX and XY flies retain a pattern of dosage compensation consistent with their chromosomal rather than their phenotypic sex. However, it now appears that while the presence or absence of the activities of individual sex-determining loci is unrelated to dosage compensation states, the initial setting of states of expression of these genes and of dosage compensation states is mediated by a single key genetic locus, the so-called *Sex lethal (Sxl)* gene of the X chromosome.

The body of observations, and correlative interpretations, that link sex determination and dosage compensation form a complex skein (see Baker and Belote, 1983). Nevertheless, four conclusions can be extracted from this web. Firstly, *Sxl* activity seems to be set early in development and differentially between the two sexes; it is apparently on in females and off in males. Secondly, its activity in females sets female-specific dosage compensation levels and turns on the expression of the sex-determining homeotic loci. Thirdly, the differential setting of *Sxl* between the two sexes is a consequence of the reading of the X/A ratio requiring the maternally stored products of several genes. Fourthly, the hyperactivity of the X chromosome in males requires the expression of several autosomal loci, collectively designated the "male-specific lethal" genes.

The importance of *Sxl* in dosage compensation was discovered fortuitously during an investigation of an autosomal sex-specific lethal mutant, *daughterless* (*da*; 2–41.5). The original *da* mutant is a temperature-sensitive maternal effect lethal; X/X embryos obtained from homozygous *da* mothers at 25°C die as larvae or pupae, while X/Y embryos are fully viable. At 18°C, a small proportion of daughters survive. However, a spontaneously arising mutant in the *da* stock was found to restore the viability of daughters nearly completely (Cline, 1978). The mutation suppressing the *da* maternal effect is on the X chromosome and was named *Sex-lethal, Male-specific* (Sxl^{Ml}). To rescue, Sxl^{Ml} must be present in the genome of the *da* daughters themselves. Most importantly, the mutation is lethal to male progeny. Sxl^{Ml} females are fully viable whether or not they come from *da* mothers.

The significance of the *Sxl;da* phenotypic interaction would be unclear, except that Sxl^{Ml} maps to virtually the same position (1–19.2) as a previously isolated female-specific lethal (formerly *Fl,* now designated Sxl^{Fl}). Cline (1978) has characterized these two mutations by the Mullerian test: Sxl^{Fl} is a hypomorph and Sxl^{Ml} is a hypermorph. Evidently, the mutants have opposite patterns of expression, with opposing consequences for viability in the two sexes. Female-lethal *Sxl* mutants, of which there are now several alleles, make effectively less Sxl^+ product, a condition that reduces or eliminates female viability; Sxl^{Ml} makes too much Sxl^+ product, a condition that is fatal to males but not to females. The viability effects do not depend on phenotypic sex but rather on X/A chromosome balance, since genetic sexual transformation of *da,* Sxl^{Fl}, or Sxl^{Ml} flies of lethal sex chromosome composition leaves lethality unaltered.

Cline (1979) has rationalized the findings as follows: maternally stored da^+ product switches on Sxl^+ activity when the X/A ratio is 1 but not when it is 0.5, the product being essential for female viability in some fashion connected with the X/A balance itself. In males, the product has the opposite effect, causing lethality. The process that requires *Sxl* action must be dosage compensation, since it is not sexual differentiation per se. Cline suggested that the *Sxl* product might be necessary to reduce X chromosome activity specifically in females. If this suggestion is correct, then absence of this gene product in X/X embryos should be accompanied by an elevated rate of X chromosome transcription. This prediction has been verified: Lucchesi and Skripsky (1981) found just such X chromosome hypertranscription in the polytene chromosomes of homozygous lethal Sxl^{F} larvae.

The normal male hypertranscription of the X chromosome in males is mediated by the products of at least four particular genes, the male-specific lethal (*msl*) loci. In an intensive search of mutagenized second and third chromosomes, Belote and Lucchesi (1980a) isolated four mutant alleles of three genes that show male-specific lethality. The genes are *msl-1* (2–53.3), *msl-2* (2–9.0), and *mle* (2–55.8); a fourth, *msl–3*, is on the third chromosome (at 3–26). These lethal effects are a function of the single-X condition, being independent of the presence of the Y chromosome or the expression of phenotypic maleness. In a second study, Belote and Lucchesi (1980b) showed that three X-linked enzyme activities were depressed to about 60% of normal in the lethal male larvae and that transcription on the X chromosome in homozygous mle^{ts} larvae was reduced to about 65% of normal at the restrictive temperature for this ts mutant. These male-essential genes must act specifically to boost X chromosome transcription. (While unnecessary for female survival, in otherwise wild-type backgrounds, they may nevertheless be expressed to some level in females, at least during oogenesis. The latter inference derives from the discovery that in the appropriate genetic context, the mutations have mild maternal effects on sex determination of progeny.)

The link between dosage compensation and phenotypic sex determination became apparent when small patches of mutant lethal *Sxl* tissue were examined in mosaics. These clones were found to be viable (unlike the whole organism homozygotes) but small. Most importantly, they were observed to be sexually transformed. In XX/XO gynandromorphs, where the single X in the XO tissue carries Sxl^{M}, sexually dimorphic regions are phenotypically female (Cline, 1979). Conversely, in X/X clones homozygous for a female-lethal *Sxl* allele, the transformation is to male tissue (Sanchez and Nothiger, 1982). The *Sxl* reversal of X chromosome transcription patterns must reverse the reading of the X/A ratio and, simultaneously, expression of the autosomal sex-determining genes (*tra, tra-2, dsx,* and *ix*). (The inviability of large patches of X/O tissue with female-level X transcription presumably reflects the insufficiency of X chromosome expression from a single X; the fact that clones of X/X tissue with male-level expression of the X are viable suggests that hyperexpression of the X is tolerable to a certain extent.)

The manner in which the X/A ratio is initially sensed to produce an on or off setting of *Sxl* remains mysterious. It presumably involves the *da* gene product and probably the maternally stored products of several other autosomal genes, including the male-specific lethal loci, as well (Baker and Belote, 1983, pp. 371–375). However, the autosomal and X-linked sites used by this collective set of autosomal gene products to measure the X/A ratio have remained elusive; there may well be multiple sites on both the X and the autosomes involved in the process (see Lucchesi, 1983).

Although the process by which *Sxl* activity is set by the X/A ratio is unknown, its timing has been ascertained. The X/A ratio is measured early, at or before the time of blastoderm formation, and then ceases to be registered for settings of either sex determination or dosage compensation. Employing an *X;3* translocation, in which the X chromosome portion can be lost through mitotic recombination, Sanchez and Nothiger (1983) generated X/O clones at successive intervals from blastoderm formation, which were either Sxl^- or Sxl^+. Sxl^- male clones, of both X/O and X/X composition, could be formed up to 48 hr in the genitalia and up to 72 hr in the sex comb region. However, male Sxl^+ clones could not be formed at any point after blastoderm. The inviability of X/O Sxl^+ clones is understandable if the X/A ratio sets *Sxl* activity at or before blastoderm; X/O cells set at the low female-level X transcription rate would not make enough X chromosome gene products to survive. If *Sxl* activity were continuously set in response to the X/A ratio, late-generated X/O clones should have male-level X transcription and should be fully viable.

A schematic depiction of the early events in the setting of the sexual phenotype is given in Figure 6.33. The actions of *Sxl* in setting dosage compensation and the expression capabilities of the sex-determining loci are viewed as independent of one another; it is possible that the latter effect is dependent on the former.

The major aspect of the sexual phenotype that remains outside this scheme is gametogenesis. Gamete precursor cells are seemingly oblivious to the effects of the autosomal sex-transforming genes and can function normally when homozygous for these mutations if placed within the appropriate somatic environment. As Schupbach (1982) has stressed, it is not known

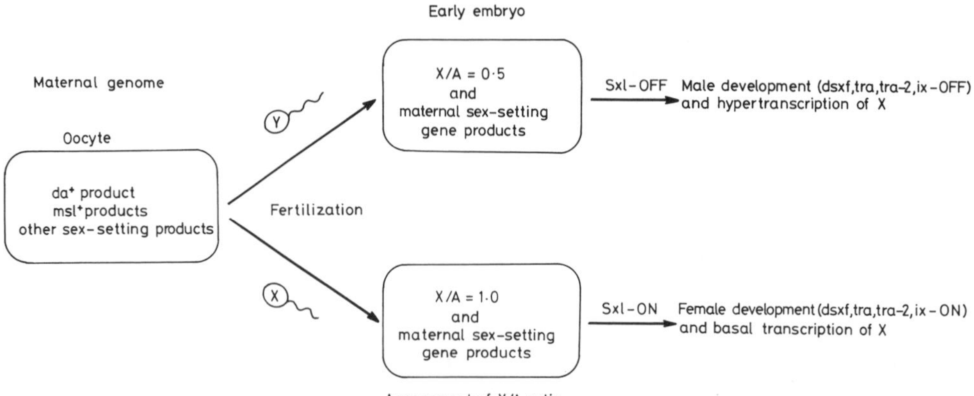

Figure 6.33 The first events in the setting of dosage compensation and sex determination; see text for discussion.

whether gamete precursor cells placed in the wrong somatic environment undergo no development at all or begin the sequence in response to the surrounding mesodermal cells but cannot complete it. It might be, for instance, that germ cells in *tra/tra;* X/X animals begin spermatogenesis but stop early in the sequence.

Gamete production evidently depends directly on the sex chromosome constitution of the germ cells. X/Y pole cells transplanted into X/X; *tra/tra* hosts proceed much further along the pathway of normal sperm development than the host germ cells, signifying that the transformed somatic environment in the *tra* hosts supports sperm production when the germ cells themselves have the appropriate sex chromosome composition (Marsh and Wieschaus, 1978). In contrast to somatic sex determination, the presence or absence of the Y may be crucial in setting gamete development on one pathway or the other; the fertility factors of the Y are certainly essential for spermatogenesis. However, the X chromosome number may also play a role; there is some evidence that possession of two Xs in gamete-producing cells is always inhibitory to sperm production, regardless of the presence or absence of the Y (Lifschytz and Lindsley, 1972). However, whether it is the direct physical difference between X chromosome numbers, or an indirect consequence such as *Sxl* activity, that is the crucial determinant, or even whether *Sxl* activity differs between male and female germ lines in a manner parallel to that of the somatic tissues are matters to be resolved.

SOME QUESTIONS

This chapter has described the major developmental changes in *Drosophila* that occur between the blastoderm stage and the period of pupal metamorphosis. The emphasis in this treatment has been on the establishment of differences between cells and the commitments that take place during the transformation of the early embryo into the adult. To provide an overview of the material, the major questions are summarized below.

The first set of questions involves segmentation and compartmentalization. These processes may be linked functionally; we will return to mechanisms of segment formation and a possible connection between intrasegmental compartmentalization and segment formation in Chapter 8. However, the formation of imaginal intrasegmental compartments raises several unanswered questions. How and when do the compartments arise? Are gradients involved, or is there some other form of geographical partitioning?

Above all, one would like to know the function of compartmentalization. An answer to this question might help to resolve the problem of the existence of compartmentalization in the internal tissues of the fly. The question of function is intimately related to that of universality. The primary function of a-p compartmentalization might be to facilitate segmental separation or to split a segment into growth control units in order to ensure that imaginal discs will have the correct size and shape when they are finally stitched together. If this is the case, then disc-type compartmentalization might be expected to occur only in those organisms with a comparable form of metamorphosis. On the other hand, cell equivalence groups in the nematode and the divergence of ICM and TE cell groups in the mouse embryo suggest that the setting aside of small groups of cells in defined regions for particular developmental pathways may well be a general mechanism. In both of the latter cases, however, a measure of differentiation precedes or accompanies the segregation process. To assess the comparability of these phenomena with *Drosophila* compartmentalization, we need to be able to examine compartmentalization events more directly than by examination of clones of tissue in the adult or even in the late discs. This task promises to be difficult, although the use of antigenic and other molecular markers may eventually provide the means.

Questions about cell behavior in compartmentalization inevitably lead to questions about genes and genetic control, and in particular to the bithorax and Antennapedia selector gene complexes that govern specific compartments and segments. What do their gene products do, and how do they do it? Specifically, are they true transcriptional regulators, comparable to the regulator genes of bacteria and bacteriophage, and do they directly regulate batteries of genes, as is usually assumed? Or, rather, is the direct action of each gene restricted to setting some aspect of the cells biology, which leads to a cascade of consequences resulting in distinctive segment- or compartmentwide properties? This question is connected to one of the most difficult issues in development: the relationship between a final body pattern of complex structural elements and the underlying cellular pattern of gene activities that gives rise to that pattern. Each gene function of BX-C may "control," individually or in combination with several others, the presence or absence of a particular feature (e.g., ventral pits, Keilin's organs, spiracular openings, the degree or width or shape of the denticle belts), but each of these is a complex property involving both a structural aspect and a distinct spatial placement, the problem of pattern formation within embryonic fields.

Beyond the puzzles of selector gene action, there is the problem of the maintenance of the segmental state and its genetic basis. Why are there so many external "regulator" loci, such as *Pc,* and how do they act? Is there, indeed, a clean separation between establishment and maintenance functions?

Finally, there is the problem with which the chapter began: the way that developmental diversity is related to overall patterns of structural (realizator) gene expression patterns. Do the differences between cell types stem from a relatively small number of qualitative differences in gene expression, or from quantitative changes in gene expression, or from a combination of the two kinds of differences? Might one even expect general principles of gene expression control to emerge? Or, in the end, will there only be a catalog of correlations? This subject will be taken up again in the final chapter.

SEVEN

MOUSE POSTIMPLANTATION DEVELOPMENT

DESCRIPTIVE EMBRYOLOGY

In appearance, a chicken egg and a mouse egg are very different. Most obviously, they differ vastly in size, roughly 5–6 cm average diameter in the former compared to 80 μm in the latter. Despite this difference, the basic embryology of the chick and the mouse have much in common. The difference in egg size reflects the different ways of provision for the nutrition of the developing embryo. The chick embryo has all of the nutrients that it needs prepackaged into the egg, in the form of yolk and albumin, while the mouse embryo must derive virtually all of the material and energy it needs for its growth and development directly from its mother. The mammalian arrangement requires that the embryo construct a much more elaborate set of extraembryonic membranes than the chick, membranes that will house it and connect it to its maternal parent.

Nevertheless, the relative simplicity of chick embryogenesis provides a suitable introduction to the mouse and will be briefly described here. The unfertilized chicken egg consists of a large mass of yolk capped by a small disc of cytoplasm containing the egg nucleus. Fertilization of the membrane-enclosed egg occurs as the egg passes down the oviduct, before the hard outer shell is formed, and is succeeded by rapid cleavage divisions, which generate a disc of approximately 80,000 cells by the time of laying. This cell layer or blastoderm consists of a relatively thin central area (the "area pellucida") and a thicker outer ring (the "area opaca") (Fig. 7.1). At one side of the area pellucida, a secondary thickening about five cells deep is apparent. This thickening is termed the "primitive streak" and constitutes the direct embryonic precursor region. By a series of cell movements, the

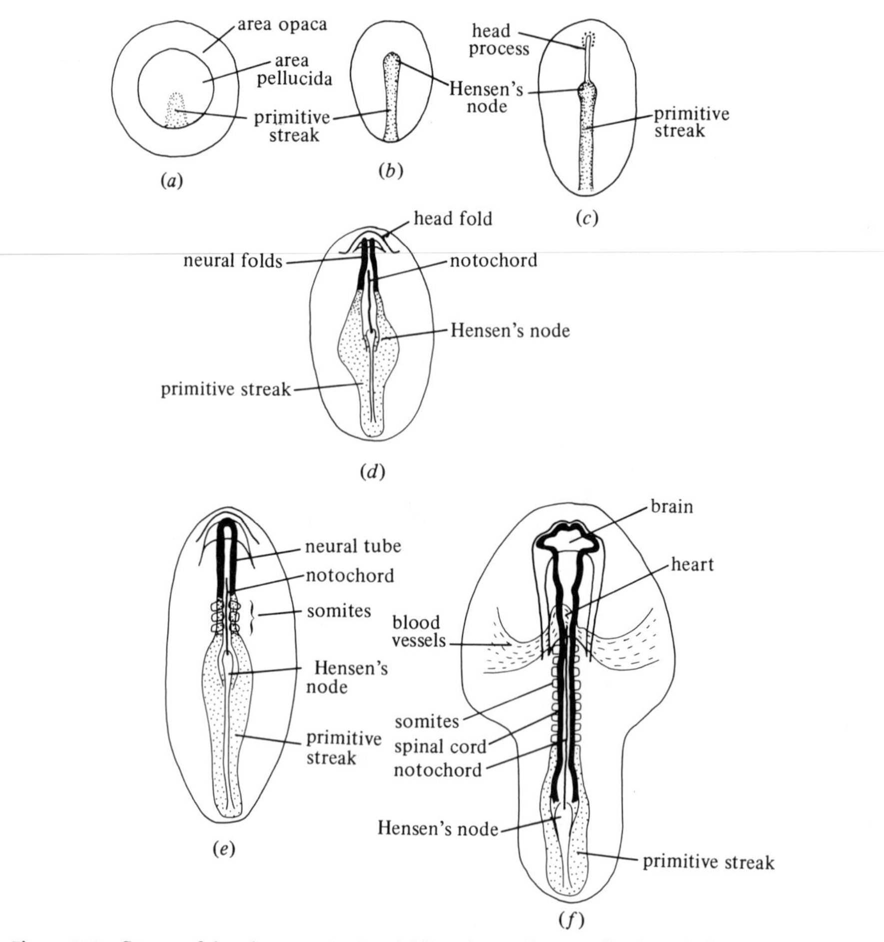

Figure 7.1 Stages of development in the chick embryo. See text for description. (Figure kindly provided by Dr. J. Slack; reproduced from *From Egg to Embryo,* Cambridge University Press, 1983.)

primitive streak soon comes to be composed of two cell layers. The upper, ectoderm-like layer is termed the "epiblast" and the lower, more endoderm-like layer the "hypoblast." The embryo itself develops exclusively from the epiblast, while the hypoblast gives rise to the major extraembryonic membranes of the yolk sac. These membranes separate the embryo from the yolk and serve as a conduit for nutrients from the yolk to the developing chick. The cell movements that generate and separate the epiblast from the hypoblast are complex and constitute the process of gastrulation; they involve migration of cells toward and through the primitive streak area.

The primitive streak defines the future a-p axis of the embryo. The next major phase of development involves the creation of the major longitudinal (axial) sequence of structures in the embryo. The primitive streak lengthens, and this lengthening is followed by the formation of a condensation at the anterior end, to form the structure known as "Hensen's node." Hensen's node is mesodermal and gives rise to the main mesodermal elements of the early embryo. A portion elongates in the anterior direction, underneath the outer ectodermal cells, and thereby lays down the precursor cells of the notochord, a transitory central element in vertebrates. The most anterior point of this migration marks the position of the "head process," the future cephalic region. The remaining portion of Hensen's node migrates posteriorly, under that portion of the ectoderm that gives rise to the neural plate, the precursor of the spinal cord. This migration is accompanied by the delimitation of the somites, which are paired mesodermal, segmental units. The somites are the ultimate source of the axial skeleton and ribs, the body musculature, and the major part of the dermis of the adult bird. As somite formation progresses, the neural plate closes to form the neural tube (Fig. 7.2). The ensuing developmental events in the chick embryo are so closely correlated with the stages of somitogenesis that chick embryos can be reliably staged by reference to the number of somite pairs that have formed at any given time point. By the end of somite formation, the body plan of all of the major organ systems in the chick is discernible.

The mouse embryo also develops from the primitive streak, but the preamble to its formation is considerably more complicated. At the time of implantation, which occurs in the mouse at about 4.5 days postconceptus

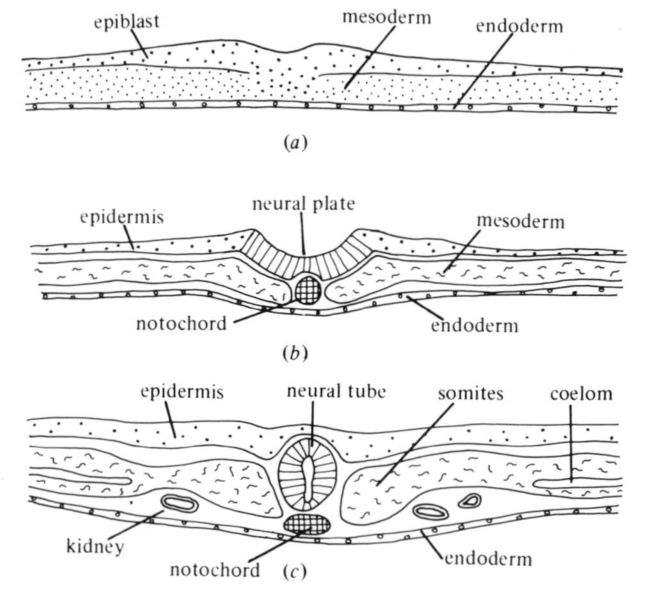

Figure 7.2 Development of the chick neural tube. (*a*) Primitive streak stage; (*b*) neural plate development; (*c*) neural tube stage. (Figure kindly provided by Dr. J. Slack; reproduced from *From Egg to Embryo,* Cambridge University Press, 1983.)

(p.c.), the blastocyst consists of just four cell types: the polar and mural TE cells and the primary ectoderm and primary endoderm. A full 2 days elapse during implantation before a detectable primitive streak arises. The complete sequence of events from implantation through the first stages of organ formation has been described by Snell and Stevens (1966), and is summarized here.

The sequence of changes is diagrammed in Figure 7.3. At the time of implantation, the internal endoderm covers the ectodermal ICM cells but has not spread around the inside of the blastocoele. During the next 1.5 days, while the embryo is embedding itself in the uterine wall, the primary endodermal layer comes to cover both the inside of the blastocoele and the inner ectodermal cell mass. The latter, covered with a stocking of endodermal cells, projects progressively farther into the blastocoelic cavity, and the whole structure is now referred to as the "egg cylinder"; the blastocoele at this point is designated the "yolk cavity" (which in the avian embryo is filled with yolk).

(This early invagination of embryo precursor material into the yolk cavity is peculiar to mice and rats, and does not take place in other placental mammals, even in more primitive rodents. The typical mammalian pattern is initially similar to the avian one; the ectoderm is a disc sitting atop the endodermal layer and invaginates only at a much later stage. The murine arrangement, in which ectoderm comes to be enclosed by endoderm, is referred to in classical embryological texts as the "inversion of the germ layers.")

The primary endoderm now consists of two distinct cell layers. The layer that covers the inside of the yolk cavity is referred to as the "distal" (or "parietal") endoderm, while that covering the egg cylinder is termed the "proximal" (or "visceral") endoderm. The proximal endoderm blankets both the embryonic ectoderm—so called because it is the precursor of the embryo proper—and the extraembryonic cells; the latter are derived from the polar TE cells, which retain their capacity for cell division. Capping the extraembryonic ectoderm in the 5.0-day embryo (Fig. 7.3*c*) is the ectoplacental cone, also derived from the polar TE and a major precursor of the placenta. At this stage, a small cavity within the embryonic ectoderm can be discerned, the proamniotic cavity, which is the precursor of the amniotic cavity, the fluid-filled sac containing the fetus.

The entire structure enclosed by the visceral endoderm is called the "yolk sac," and from this point on, the visceral endoderm can be equated with the hypoblast of the chick embryo. The yolk sac enlarges within the yolk cavity, and by 5.5 days (Fig. 7.3*d*) the embryonic ectoderm has begun to delaminate the second of the germ layers of the embryo, the mesoderm. Concurrently, the cavities within the embryonic and extraembryonic portions of the egg cylinder have enlarged and temporarily joined. Cavities have also begun to appear within the mesoderm; these cavities unite to form the beginnings of the exocoelom, which by 7.25 days (Fig. 7.3*e*) occupies the space between

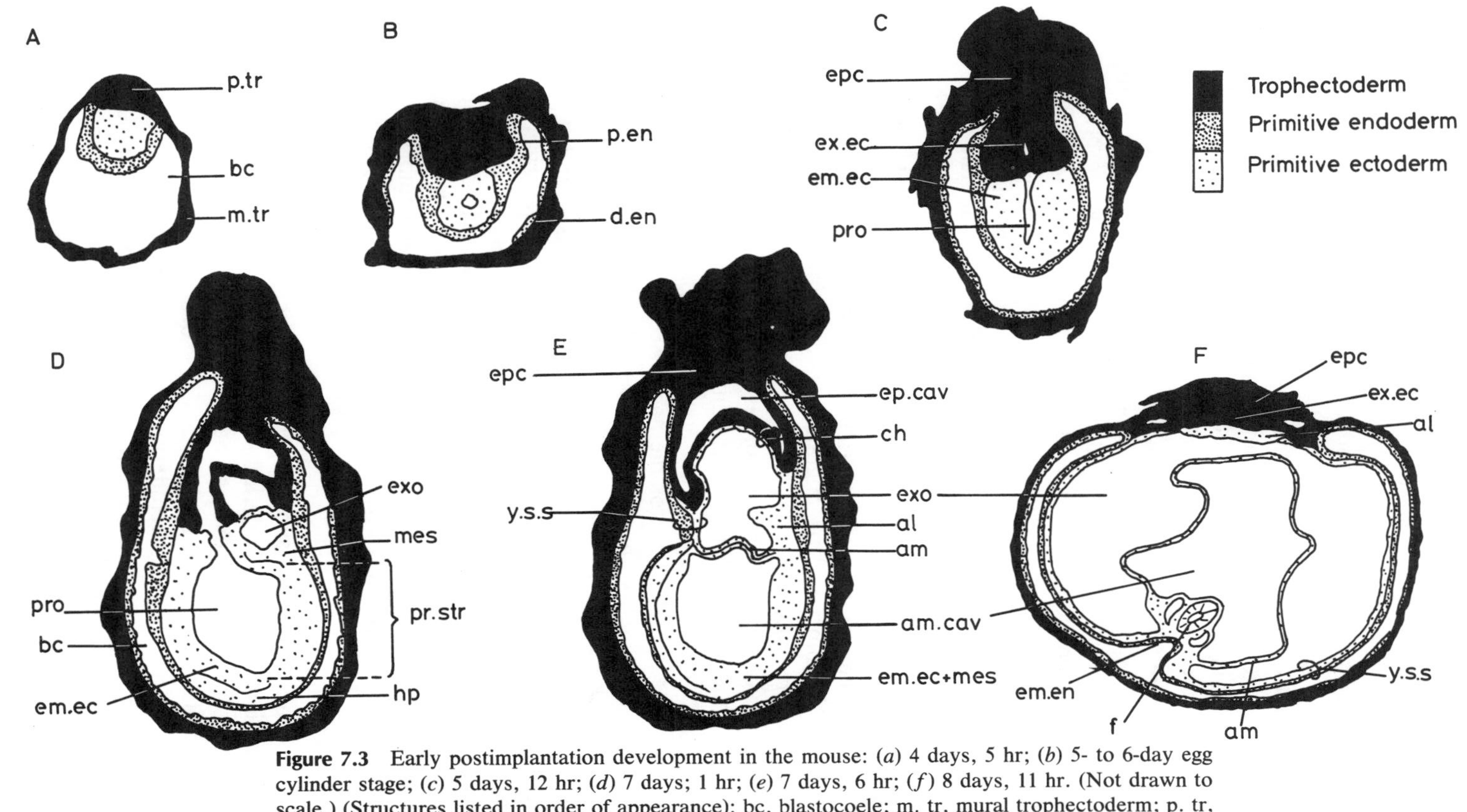

Figure 7.3 Early postimplantation development in the mouse: (*a*) 4 days, 5 hr; (*b*) 5- to 6-day egg cylinder stage; (*c*) 5 days, 12 hr; (*d*) 7 days; 1 hr; (*e*) 7 days, 6 hr; (*f*) 8 days, 11 hr. (Not drawn to scale.) (Structures listed in order of appearance): bc, blastocoele; m. tr, mural trophectoderm; p. tr, polar trophectoderm; p. en, proximal endoderm; d. en, distal endoderm; epc, ectoplacental cone; ex. ec, extraembryonic ectoderm; em. ec. embryonic ectoderm; pro, proamnion; exo, exocoelom; mes, mesoderm; pr. str, primitive streak; hp, head process; y.s.s., yolk sac splanchnopleure; ep. cav., ectoplacental cavity; ch., chorion; al, allantois; am, amnion; am. cav, amniotic cavity; em. en, embryonic ectoderm; f, fetus. (Adapted from Gardner, 1978.)

the ectoplacental and amniotic cavities. The exocoelom is separated from the amniotic cavity by the amnion, a membrane composed of a layer of mesoderm (on the exocoelomic side) and a layer of ectoderm. By 8.5 days, the exocoelom has surrounded the amniotic cavity, except for the region occupied by the fetus (Fig. 7.3 *f*). By this stage, the ectoplacental cone has fused with the chorion (the membrane bounding the exocoelom opposite the amnion) and the allantois, a structure derived from the mesoderm that serves as a connection between the maternal blood supply and the fetus. The chorioallantoic membrane is connected to the fetus through the richly vascularized membrane that bounds the exocoelom, the yolk sac splanchnopleure.

The fetus originates at 6.5 days, with the appearance of the primitive streak as a thickened region on one side of the embryonic ectoderm, near the developing allantois. Between 6.5 and 10 days, all of the major organ systems take shape within the primitive streak through a complex series of morphogenetic movements accompanied by extensive growth. The process beings with the production and proliferation of mesodermal progenitor cells at the proximal (allantoic) end of the primitive streak; this position marks the future caudal end of the fetus. As ectodermal cells migrate through the primitive streak, they move both laterally and distally toward the future cranial end of the embryo. This migration of cells into and through the primitive streak constitutes gastrulation in the mouse embryo. The a-p orientation of the embryo is thus established or revealed by the laying down of the mesoderm in a caudocranial direction. As this process continues, mesodermal cells progressively occupy the space between the visceral endoderm and the embryonic ectoderm, except for the narrow central section that forms the spinal region of the fetus; this early primitive streak stage is very similar to that of the chick.

At 7.0 days p.c., the head process has taken shape at the most distal end of the egg cylinder. As in the chick, the head process gives rise to the notochord; it also contributes to part of the endodermal lining of the gut. The head process grows both proximally, on the side of the egg cylinder opposite the primitive streak, and laterally. By 7.5 days, the growing head process and the mesodermal sheets have met; throughout most of the embryo, ectoderm is separated from primary endoderm by the mesodermal layer, except for the midsaggital strip of head process directly in contact with the ectoderm.

The early primitive streak stage is a period of exceptionally rapid growth and transformation. During this period, the longitudinal elements, namely, the digestive tract and the axial system, begin to form. The gut begins as two invaginations of the yolk sac on opposite sides, near the junctures of the embryonic and extraembryonic ectoderm. The foregut precursor pushes in at the cranial end, beginning at 7.0 days, and by 7.75 days has become a pronounced inpocketing; the hindgut has just begun to form at this point. The blind ends of the foregut and hindgut invaginations eventually break through to the outer surface of the embryo, giving rise to the mouth and anus, respectively. The invagination of the foregut also lays the foundation of two other key organs, the heart and the brain. The most dorsal region of ectoderm that is pushed in is the head fold structure, which develops into the brain. The heart arises from the mesoderm lying between the head fold and the outer endoderm.

The axial system comes into being from the lateral sheets of mesoderm that border the notochord, the so-called paraxial mesoderm. As in the chick embryo, the axial system arises in a strict, linearized sequence of somite formation that begins at the cranial end of the embryo, just posterior to the embryonic brain, and progresses to the caudal end. Somitogenesis therefore proceeds in the opposite direction to that of mesoderm formation. The delimitation of somites is shown in Figure 7.4. The first pair of somites arise at about 8 days. Initially, pairs of somites are laid down at the rate of one per hour; subsequently, the process slows to one pair every 2–3 hr (Tam, 1981). Somitogenesis is complete by about 13 days of development, when 65 pairs of somites have been formed; of these, 30 pairs will form the skeleton, musculature, and dermis of the body (neck, thorax, and lower back) and 35 pairs will form these elements in the tail. Each newly formed somite shows a trace of a boundary between the anterior and posterior halves, and each vertebra arises from cells of the posterior half of one somite and the anterior half of the next. Differentiation of the somites is accompanied by dispersal of the mesenchymal cells that compose them, leading to an obliteration of the intersomitic boundaries.

The stages of neural development, the precursor of the spinal cord and the other major component of the axial system, are closely correlated with the stages of somitogenesis. The neural tube develops from the closure of the neural plate area just above the notochord and between the sheets of paraxial mesoderm. The strict correlation of the somite stage with other events in normal development permits one to stage embryos according to their somite number, as in avian embryos.

By 10 days p.c., the major elements of the circulatory system, also formed from the paraxial mesoderm, have developed. The turning of the

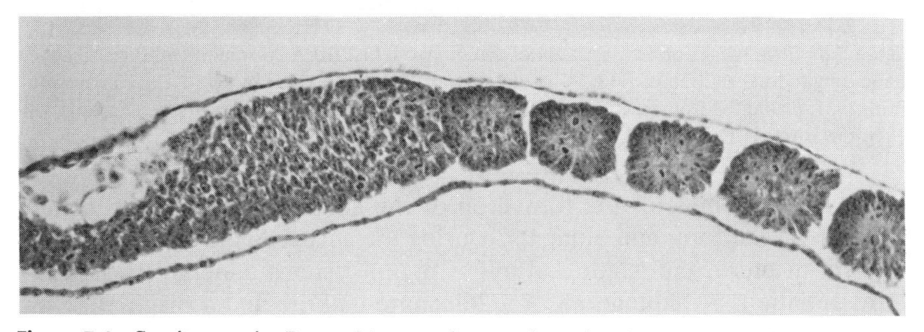

Figure 7.4 Somitogenesis. Presomitic mesoderm and somites in a seven-somite mouse embryo. (Figure kindly provided by Dr. P.P.L. Tam; from Tam, 1981.)

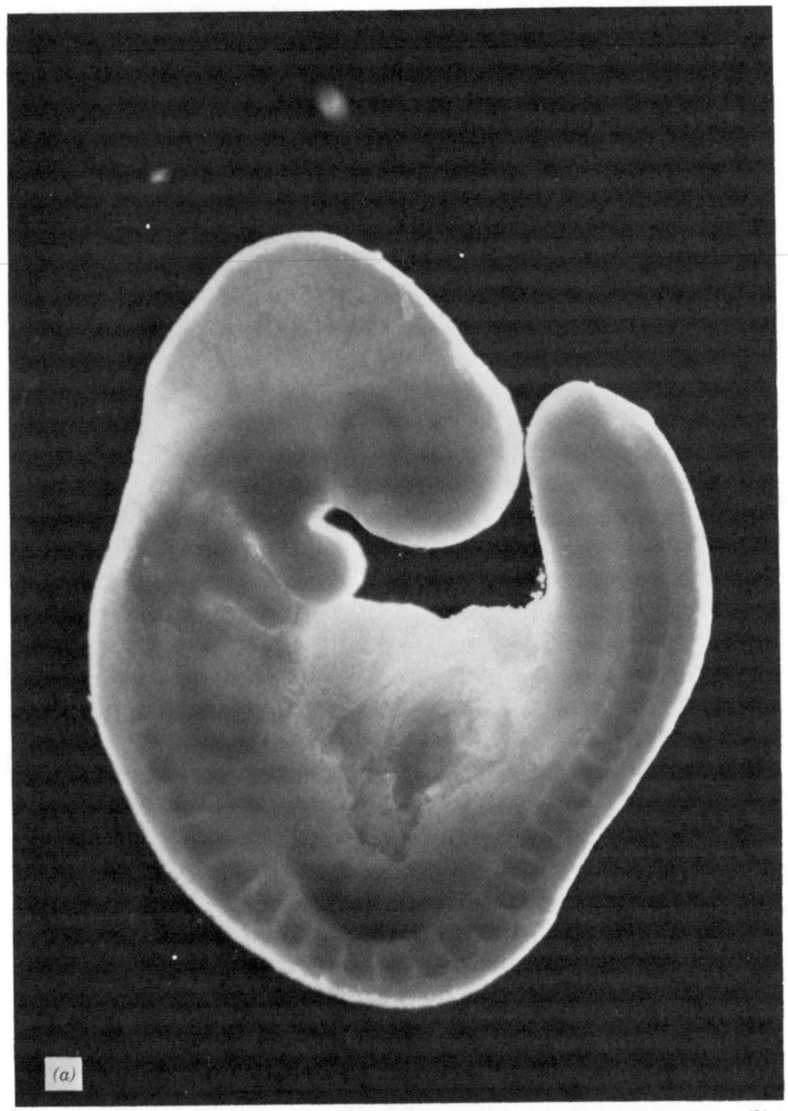

Figure 7.5 Two stages of development of the mouse fetus. (*a*) A 26-somite embryo; (*b*) schematic longitudinal section of a 13.5-day embryo. (*a* Kindly supplied by Dr. P.P.L. Tam, from Tam, 1981; *b* adapted from R. Rugh, *The Mouse: Its Reproduction and Development,* copyright 1968, the Burgess Publishing Co., Minneapolis, Minn.)

embryo also completes the formation of the gut by generating the midgut and joining the fore- and hindguts. During the succeeding 10 days of development in utero, the major and minor organ systems undergo progressive differentiation. A photograph of a 26-somite embryo and a diagram of the 13.5-day embryo, showing the major organ rudiments, are presented in Figure 7.5. A timetable of some of the major events in postimplantation development is given in Table 7.1.

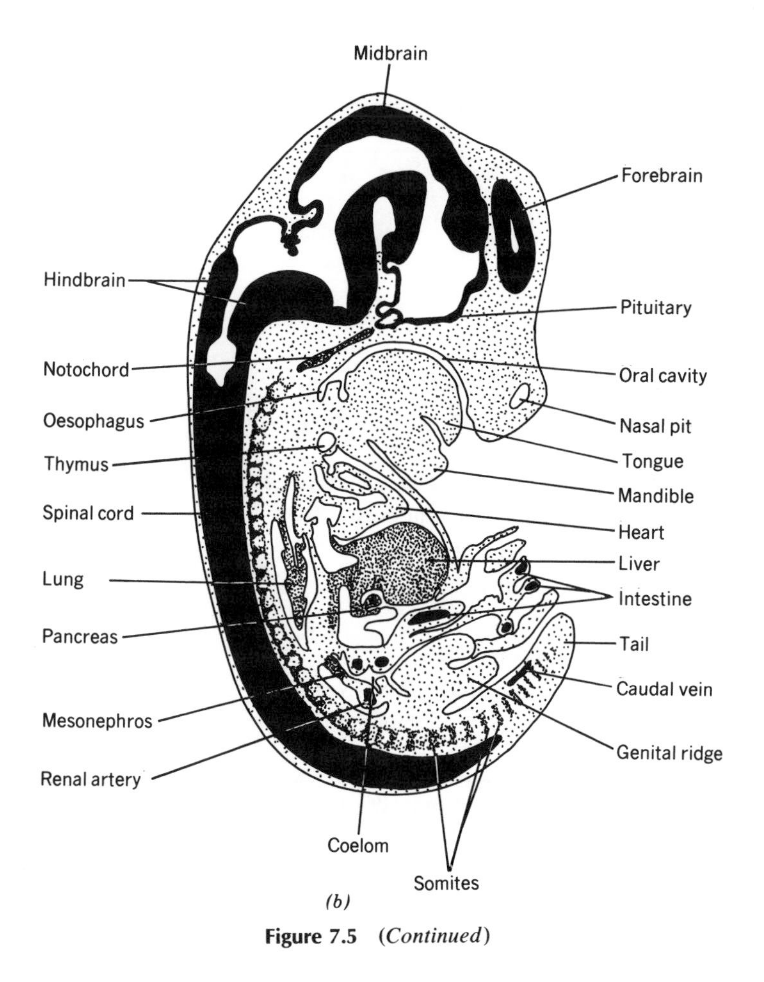

Figure 7.5 (*Continued*)

Growth, Size, and Morphogenesis

One of the most general and remarkable features of embryogenesis is the coordination of growth with the events of development. This coordination ensures that the newborn or newly hatched progeny of a given species or strain are of a relatively uniform size and that the internal and external structures are correctly proportioned.

The existence of mechanisms for coordinating development with growth becomes obvious when embryo size is altered experimentally (Slack, 1983). The most dramatic of such regulatory responses are those seen when isolated blastomeres from early embryos give rise to whole animals. For instance, if one separates the blastomeres of the four-cell sea urchin embryo, each cell develops into a perfectly shaped and normally functioning pluteus larva, although only one-quarter the normal size and containing one-quarter the normal number of cells. Similar regulative capacities are displayed by the

Table 7.1 Developmental Landmarks in Postimplantation Development[a]

Event	Days p.c.
Implantation	4.5
Proamniotic cavity appears	5.0–5.5
Primitive streak detectable	6.5
Foregut appears	7.0
Neural plate forms	7.75
Somite formation begins	8.0
Chorioallantoic placenta forms	9.0–10.0
Primordial germ cells appear at the base of the allantois	7.75–8.0
Primordial germ cells arrive in genital ridges	10.5–11.5
Melanoblasts leave the neural crest	8.5–9.5
Neural tube development completed	10.0
Forelimb buds appear	9.2
Hindlimb buds appear	10.0
Somite formation completed	13.0
Hematopoiesis begins in the fetal liver	9.5–10.0
Hematopoiesis begins in the fetal bone marrow	16–17
Birth	19–20

Source. Adapted from McLaren (1976a) and Snow et al. (1981).
[a] Times cited are approximate; inbred mouse stains frequently show a delayed timetable.

first blastomeres of frogs and starfish; in mammals, normal newborn rabbits have been obtained from isolated blastomeres of eight-cell embryos.

Major adjustments can also occur much later in development. Individual duck blastoderms, for instance, if operated on prior to the emergence of the primitive streak, can give rise to four independent embryos. In *Drosophila,* severe limitations of the food supply during the period of larval growth can lead to the production of miniature, although otherwise fully normal, fruit flies.

In the mouse, growth regulation can occur at both early and intermediate points of development. Chimeric blastocysts made from two or more embryos are invariably larger than normal blastocysts, yet develop into newborn mice of standard size. At some point, there must be a "downward" regulation of size to yield a normal-sized embryo. The point has been localized to between 5.5 and 6.0 days p.c.; it occurs, therefore, after implantation but before primitive streak formation. This response probably reflects a limitation of the nutrient supply to the larger blastocystic embryos, with a consequent slowing down of cell division rate.

"Upward" regulation of the growth of abnormally small embryos also occurs, although after primitive streak formation. Unlike the sea urchin embryo, viable mammalian embryos formed from single blastomeres of the two- and four-cell stage do not give rise to one-half or one-quarter size animals, but to newborns of normal size. The size adjustment apparently occurs between days 10.0 and 11.0 p.c., and takes place shortly after the maternal circulation is made fully available to the embryo, with complete development of the placenta. The nutrient supply may thus make possible the growth spurt, but the signal to do so must come from the embryo itself, which somehow "knows" that it is small.

The most striking instance of upward regulation of growth occurs following treatment with mitomycin C (MMC). This drug, when injected intraperitoneally into mothers at the beginning of primitive streak development, produces extensive random cell death in the embryo. At 7.5 days p.c., the epiblast of MMC-exposed embryos contains only 10–15% as many cells as that of control embryos, the dead cells being sloughed off (Snow and Tam, 1979). Despite this extraordinarily large reduction in cell number, the majority of embryos manage to catch up to control embryo size by 12.0–13.5 days, and many are born alive. The primary defect in the survivors is a partial or complete sterility, reflecting a permanent reduction in germ cell number effected by the MMC treatment.

Examination of MMC-treated embryos shows that the developmental time course of both ectodermal and mesodermal components is affected, with a visible retardation of neural tube development and of somitogenesis. However, the effects on the neural tube and the somites are partially uncoupled, with the neural structures experiencing the more severe delay. It follows that the stages of development of the two major components of the axial system are neither obligatorily coupled to, nor dependent on, one another during the period of recovery (nor, by inference, during normal development).

An intriguing aspect of somite development in MMC-treated embryos is that while the normal number of somites is produced in the catchup embryos, the individual somites are smaller and have fewer cells (Tam, 1981). The conservation of somite number at the expense of individual somite cell content suggests that there exists some form of global partitioning mechanism within the presomitic mesoderm. This mechanism operates prior to the first segmentation event and ensures that the total mass is divided into the correct number of somite pairs. The existence of such a pattern-setting mechanism is, at least superficially, reminiscent of the partitioning of the *Drosophila* blastoderm surface into segments, although somitogenesis appears to have a much greater regulative capacity (and operates with greater cell numbers).

GENE EXPRESSION IN POSTIMPLANTATION DEVELOPMENT

The events of implantation mark a point between the early but limited cell diversification of preimplantation development and a much more extensive set of cellular differentiations. Is this acceleration in cell diversification accompanied or driven by a corresponding set of qualitative changes in the gene expression pattern? The extensive evidence for *Drosophila* and the less complete data for *Caenorhabditis* indicate that successive developmental stages and different cell types exhibit a large set of conjointly expressed genes, with comparatively few qualitative differences between these gene sets. Is the mouse similar or different in this respect?

Answers to this question have been sought primarily by protein labeling and electrophoresis techniques and by RNA-DNA hybridization techniques. The results indicate that, as in the fruit fly and the nematode, there is a broad base of shared gene expression between different cell types and stages during mouse development. At the same time, they show that large-scale quantitative shifts in gene expression or gene product accumulation occur during development.

Labeled Polypeptide Patterns

The first developmental events in the postimplantation embryo involve the elaboration of the principal extraembryonic membranes and the development of the primitive streak. By 7.5 days of development (Fig. 7.3*e*), the polar TE has given rise to a distinct ectoplacental cone (EpC) region, a layer of giant cells (GCs) that surround the diploid core cells of the EpC, and a layer of extraembryonic ectoderm (EE), which includes the chorion. The future embryonic region is separated from the exocoelom by the amnion and consists of the region of the primitive streak and, opposite it, a region primarily composed of embryonic ectoderm (EmE).

Johnson and Rossant (1981) assayed the molecular correlates of this first stage of postimplantation change by comparing the polypeptide labeling patterns of the three polar TE lineages with each other and with embryonic ectoderm. The different regions were dissected, labeled in vitro, and the polypeptides then extracted and separated on two-dimensional gels. In a large number of such comparisons, 50 spots, from a total set of more than 500 detected, were found that distinguished the four cell types from one another. Of these, six were unique to EmE, one to EpC, and three to GCs. All of the remaining polypeptides in this set of 50 were shared between different subsets of the four tissues. Although the comparisons themselves provide no direct information about developmental relationships between the tissues, an in vitro culture experiment showed that cultured EE cells first develop an EpC-type profile that is then succeeded by a GC-type profile. By this criterion, the EE cells may be equivalent to the original polar TE cells and serve as a reservoir of EpC and GCs. The principal finding is that within the four ectodermal cell groups compared, which differ in cytology, function, and fate, there is a large degree of similarity in the detectable polypeptide synthesis pattern.

Later periods of development show much the same similarity of protein synthesis. A detailed study of polypeptide synthetic patterns in embryonic mouse organs, employing [^{35}S]methionine pulse labeling in vitro, was reported by van Blerkom et al. (1982). The organ systems tested included liver, brain, kidney, forelimb, hindlimb, yolk sac, lung, and fetal tail (essentially somites) from 10- to 13-day embryos. For all stages and all organs, a total of 850–1000 polypeptide spots were detected. For the various organs surveyed, an average of three organ-specific protein synthetic differences were found, or less than 1% of the total. In comparison, quantitative shifts specific to organs or stages were relatively more common. In the lung, for example, 5 qualitative changes in spot pattern and 15 quantitative ones were measured in the period from 10 days p.c. to early postnatal development. For the brain, from day 10 p.c. to day 10 postnatally, 1 qualitative change and 20 quantitative changes were detected. The very small number of observed changes in this survey was not a reflection of limitations in the labeling or electrophoretic procedures; labeling with ^{14}C amino acids or changing the pH range in the separation increased the total number of detected spots without altering the percentages of qualitative change in pattern.

The results of this and other studies not cited here are all in general agreement: there is little difference in polypeptide labeling patterns between cells that are markedly different in phenotype. The two-dimensional separation procedure, however, may only resolve the various housekeeping proteins, all of which should be present in comparable amounts between different cell types. However, not all housekeeping proteins are necessarily in the abundant or semiabundant classes. Galau et al. (1976) have presented arguments that physiologically significant metabolic functions can be carried out by proteins produced from the rare mRNA class. Conversely, some of the detectable polypeptide spots may be produced from relatively rare mRNAs if translation efficiencies differ between certain species. Nevertheless, whatever the cellular functions of the detected proteins or the relative abundances of their mRNAs, they must represent only a portion of the expressed protein-coding genes. In contrast, RNA-DNA hybridization methods provide a more inclusive measure of the complete set of the number of expressed genes.

mRNA Pool Studies

Several measurements of mRNA sequence diversity in mouse cells have been made. Sequence complexity was estimated in these studies from the kinetics of hybridization of poly A$^+$ mRNA to cDNA made from such RNAs; the advantages of this approach were described earlier (p. 249). The chief disadvantage is that only minimal estimates of sequence diversity are obtained. The true sequence diversity is underestimated in two ways. Firstly, any mRNA species that are transported to the cytoplasm without poly A tails will not be purified; hence, these sequences will be missing from the cDNA preparation. In *Drosophila,* at least, these molecular species apparently comprise a substantial portion of the total mRNA pool. Secondly, rare mRNAs that are present in some but not all cell classes will not be detected because of dilution in the total mRNA pool. Nevertheless, if the presence or absence of polyadenylation is characteristic of a particular mRNA sequence, the measurement of sequence complexities in different cellular poly A^+ mRNA pools will provide meaningful comparisons between different cell types. The three studies described below show that most mouse tissues have approximately 10,000 poly A^+ mRNA species, of which the great majority are shared between cells of very different types.

In a comparison of such mRNAs from adult liver and brain and from whole 14-day embryos, Young et al. (1976) found that the three mRNA pools are complementary to 0.7, 1.7, and 0.7% of the sequences of the mouse genome, respectively. If one takes 2000 b.p. as a typical mRNA length, these percentages correspond to approximately 12,000 different mRNA species for the liver, 25,000 for the brain, and 10,000 for the 14-day embryo. (Although the brain is well developed in 14-day embryos, the lower sequence complexity for whole embryos relative to the brain presumably reflects the dilution of rare brain-specific sequences in the former.) By performing cross-hybridizations to near saturation in order to detect sequence overlap, Young et al. found a large measure of qualitative similarity between the mRNA pools of adult liver and brain. However, the rates of cross-hybridization, which primarily reflect the more abundant species, were lower than those in the homologous hybridizations. The measurements therefore indicate that the most abundant species for each of the two tissues were less prevalent in the other.

A similar analysis by Hastie and Bishop (1976) of the mRNA populations of adult liver, brain, and kidney produced estimates of 13,000, 11,600, and 11,500 different structural gene sequences in the three messenger pools, respectively. (The distinctly higher sequence complexity of the brain mRNA pool reported in the previous study was not seen in this one; the reason for the discrepancy is unclear.) For all three tissues, three discrete abundance classes could be distinguished, consisting of the very abundant species (4–9 types), the intermediate (480–750), and the rare (11,000–12,200). Of the rare 11,000 poly A^+ mRNA sequences of the kidney, 9000–10,000 were estimated as shared by liver and brain. As in the study of Young et al., the abundant mRNA species for each tissue were also detected in the mRNA pools of the other two but were in the intermediate or rare class. Altogether, a total of 15,000–16,000 mRNA species are found within the three tissues. Each tissue sampled in the analysis represents one of the major germ layers of the fetus—the liver, endoderm; the kidney, mesoderm; the brain, embryonic ectoderm—yet the vast bulk of the structural genes expressed as poly A^+ messengers are shared between them. The main qualitative differences are in the rare mRNA classes, although quantitative differences of the abundant and intermediate classes are a marked feature.

Ideally, one would like to trace the entire pattern of transcript diversification that occurs from the beginning of fetal development, in the primitive streak stage, on. Unfortunately, this has not yet proven possible because of the difficulty in obtaining enough early embryonic material. An alternative approach is to use neoplastic cells whose characteristics mimic those of early embryonic cells. These cells are termed "embryonal carcinoma (EC) cells" and have been widely used as an analogue of the early embryonic cellular condition.

EC lines come in at least four grades of developmental capacity, ranging from those that give rise to all cell types (totipotent lines) to those with wide but incomplete capacities (pluripotent lines) to those incapable of giving rise to any differentiated cells (nullipotent lines). (Nullipotency is associated with long-term culture and accumulated chromosomal abnormalities.) A few lines have specialized differentiative capacities, such as a particular capacity to give rise to muscle cells under a differentiative stimulus. The different grades of developmental capacity may provide analogues of the various stages of embryonic cell specialization.

If toti- or pluripotent EC cells are truly similar to early embryonic cells, then a measurement of transcript diversity in EC cell lines should give a hint of the gene expression pattern in ICM or primitive ectoderm cells. Affara et al. (1977) measured EC mRNA transcript diversity in two pluripotent cell lines and two specialized, EC-derived myoblast (muscle precursor) cell lines. For the pluripotent EC lines, a transcript diversity of 7700 different poly A^+ mRNAs was estimated, and for the myoblast lines a transcript complexity equivalent to 13,200 poly A^+ mRNA species. By cross-hybridization tests, all of the EC mRNA sequences were found in the transcript pools of the myoblast cell lines, but the latter cells contained a substantial number of rare mRNA species not found in the EC cell line. As before, the abundant and intermediate abundant sequence classes that characterize one cell type are less abundant in other cell types.

Taking EC cells as an analogue of early embryonic cells, the comparison suggests that divergence of cells from the early embryonic cell state is accompanied by an expansion of gene expression, principally in the class of mRNAs present in just a few copies per cell. However, other factors, such as the different origins of the various cell lines, may have contributed to the differences between the EC and myoblast cell lines.

Another investigation showed that terminal differentiation is accompanied by relatively little change in overall transcript diversity. The cells of a Friend erytholeukemic line—a virally transformed cell line with the potentiality to differentiate into red blood cells—can be induced to differentiate into mature red blood cells by exposure to the permeabilizing agent dimethyl sulfoxide (DMSO). When these cells, whose sequences are all common to EC cells apart from a few in the most abundant class (Affara et al., 1977), are induced to differentiate, the principal changes occur in the most abundant species. These changes include a decline in the frequency of several sequence types and the appearance of new ones, including, most characteristically, the hemoglobin mRNAs (Affara and Dauras, 1979). Thus, neither commitment to a given pathway of differentiation nor the process of overt differentiation itself necessarily entails large-scale changes in overall gene transcript diversity relative to the (presumptive) early embryonic cell state. Whether the biologically significant changes are primarily those in the abundant classes or whether changes in all classes are equally important remains an open question. However, in all specialized cells whose physiological function depends on a certain protein or group of proteins, one change will always be an increase in the concentration of the abundant mRNA(s) that code for those protein(s).

Changes in cellular mRNA concentrations may occur either through regulation of transcription or through changing efficiencies of transcript processing. There is some evidence that control of processing plays some role in regulating mRNA abundances. When the nuclear RNA pools of different mouse cell types whose cytoplasmic mRNA compositions differ are examined, they are found to contain sequences for the same rare mRNA types (reviewed by Davidson and Britten, 1979). Thus, the nuclei of brain cells have copies of all of the sequences present as mRNAs in the cytoplasm of kidney cells and vice versa. However, the nuclear sequences that do not become converted to mRNAs are rare, being present at only one or a few copies per nucleus. These results indicate that there may be some transcription of all potential mRNA sequences in all cells.

On the other hand, the regulation of the more abundant species may be subject to specific transcriptional control. Derman et al. (1981) cloned several sequences from cDNA made from liver poly A^+ mRNA, and identified 11 cDNA clones corresponding to abundant or semiabundant mRNAs that are much rarer in two other cell types that were examined—hepatoma cells and mouse L cells. For all 11 of these "liver-specific" sequences, the rate of pulse labeling of these sequences in nuclei in vitro—which serves as a measure of the instantaneous transcription rate—was found to be significantly higher in liver nuclei than in brain nuclei. In contrast, the in vitro pulse rate labeling of mRNA complementary to five cDNA sequences shared by several cell types was found to be equivalent in liver and brain cell nuclei. The experiment indicates that the more abundant mRNA sequences in cells are more concentrated precisely because the transcription rate of their homologous gene sequences is specifically enhanced. Transcriptional regulation in eukaryotes may therefore involve primarily modulations in transcription rate that shift sequences between the more and less abundant categories, while the presence or absence of certain rare sequences as mRNAs may involve some degree of sequence-specific posttranscriptional processing.

Summary

The nature of the relationships between the changes in gene expression that take place during mouse development and the accompanying changes in cellular phenotype still remain obscure. However, the picture seems rather similar to that presented by *Drosophila:* development is accompanied by a mild increase in qualitative diversity (from perhaps around 7700 EC-type polyA$^+$ mRNA species to something over 10,000 in the fully differentiated cell types) and a large measure of shared expression. Quantitative modulations of synthesis affect both the mRNA and protein pools, and may constitute an important component of the processes of cellular phenotypic change.

POSTIMPLANTATION DEVELOPMENT: GENETIC ANALYSIS

Postimplantation development is a continuous sequence of events but may be divided arbitrarily into two phases. The first extends from implantation at 4.5 days to approximately 6.5 days p.c., during which the extraembryonic membranes and the primitive streak are formed. The second and longer phase, from 6.5 days to birth, is the period of organ rudiment formation, organogenesis, and fetal growth. Everything that precedes the primitive streak stage involves the elaboration of structures necessary for the housing and nutrition of the embryo proper that arise from the primitive streak.

A large part of contemporary murine developmental genetics has centered on the delineation of the various component cell lineages of the tissues and organs of the postimplantation embryo by means of genetic techniques. Because mitotic recombination is either rare or nonexistent in mammalian embryos, it cannot be used to reconstruct the detailed history of these lineages. Instead, principal reliance has had to be placed on the construction and analysis of chimeras. As in the study of preimplantation lineages, early embryo chimeras are first made either by morula aggregation or by injection of cells into blastocysts. The chimeric embryos are then implanted in pseudopregnant foster mothers and, following the requisite period of development, the animals are dissected and the tissue or structure of interest is analyzed in terms of the input marker phenotypes. From the pattern of distribution of the marked cells, deductions can then be formulated about the nature and/or number of the founding cells. This form of analysis has proven quite successful for preimplantation development, as we have seen, and for early postimplantation development. For post–primitive streak development, the method has been applied principally to estimating the number of founder cells for different organs or tissues; in these studies, the interpretations, while interesting, have been consistently controversial. A second form of clonal analysis is that based on the inactivation of single X chromosomes that occurs during the early development of two-X (female) embryos; it has also been employed to estimate founder cell numbers. However, these analyses also suffer from ambiguities that affect the interpretation of the results. The findings from these approaches, and the difficulties, are reviewed below.

Origins of the Extraembryonic Membranes and Fetal Germ Layers

At implantation, the TE lineage of the early blastocyst has split into two distinct cell groups. The mural TE cells, which occupy the sides of the blastocyst, have begun the transformation to primary GCs. These cells eventually achieve ploidy levels of several hundred; their function is to achieve the initial implantation of the blastocyst in the uterine wall. The other TE cell group, the polar TE, produces the EpC and EE cells. The presumptive cell derivation relationships in the polar TE lineage were described earlier; the EE cells may be the primary type and give rise to EpC cells, which in turn generate secondary GCs at the most proximal end of the embryo.

Although there has been little controversy about the TE lineages, the origins of at least two fetal tissues have been the subject of debate in the recent past. One of these tissues is the fetal endoderm. From the classic embryological description of the formation of the gut, it had been assumed that the proximal endodermal layer gave rise to the lining of the gut; without question, endodermal cells from this layer are carried along in the invaginations that give rise to the fore- and hindguts. However, this source is difficult to reconcile with the development of the avian embryo: in the latter, the entire embryo, including the gut, arises from the epiblast. The avian hypoblast, which appears similar in function and location to the proximal endoderm, does not contribute to the embryo proper. The experiments of Gardner and Rossant (1979; reviewed in Chapter 4) showed that embryonic endoderm similarly arises from the primitive streak. The initial EmE is, in fact, the source of the three germ layers of the fetus.

The origins of the germ line, the gamete-producing cells, have been the second contentious issue. Direct histological tracking of the primordial germ cells (PGCs) had placed their source, like that of the gut, in the proximal endoderm (Snell and Stevens, 1966). A number of observations have forced a revision of this view; it is now apparent that the germ line, like all the other components of the fetus, comes from the EmE (reviewed in McLaren, 1983a).

A feature of the PGCs that has made it possible to trace them in early development is their high level of alkaline phosphatase (AP) activity. By staining for AP activity, Ozdzenski (cited in McLaren, 1983a) was able to detect presumptive PGCs in the 8.0-day p.c. embryo near the caudal end of the primitive streak, within the base of the allantois. Twelve hours later, the main cluster of high AP cells was observed in the proximal endoderm and, 12–24 hr later, within the invaginated hindgut rudiment. From this position, the PGCs then move by an active, chemotactic process, along the dorsal mesentery within the developing body cavity, to the genital ridges, which they then colonize. (This pattern of movement occurs in embryos of both sexes; all of the differences in germ cell maturation between the sexes begin to be apparent after colonization of the genital ridges.) The process of migration is complete by about 11 days p.c.. From the point at which they first become distinguishable to the end of migration, the number of histochemically detectable PGCs increases from about 100 to over 1000 (Tam and Snow, 1981).

Two genetic approaches establish that the early epiblast or primitive steak is the source of the PGCs. The first involves the phenomenon of X chromosome inactivation in female embryos. Although female mammals, like female fruit flies, have two X chromosomes to the male's single X, mammals do not use the fruit fly's system of transcriptional dosage compensation. Rather, in each somatic cell of the female embryo, one X chromosome is totally inactivated through chromosome condensation (Lyon, 1961). The inactive state is then inherited by all of that somatic cell's progeny. This transcriptional inactivation happens after the blastocyst stage in all cells of the embryo. In the endodermal layer, inactivation occurs early, and it is always the paternal X chromosome that is inactivated. In the epiblast (the EmE and all its derivatives), inactivation is slightly later and is random, with half of the cells experiencing inactivation of their paternal chromosome and the other half that of their maternal X. If the germ cells were of endodermal origin, they should all show an active maternal X only. In fact, the PGCs within the genital ridges always show the epiblast pattern, a random inactivation of the maternal and paternal Xs (McMahon et al., 1981).

Chimera experiments supply an even more direct proof that the germ line comes from the primitive ectoderm. These experiments not only pinpoint the source of the germ line but also establish that it does not arise from a defined germinal plasm. This fact marks a significant difference from the eggs of many insects, nematodes, and anuran amphibians, among other animals. The first experiments indicating that the germ line arises from nonspecialized blastomeres were those of Kelly (1975), described in Chapter 4. She found that in each of two four-cell embryos dissociated to individual blastomeres, at least three of the donor cells gave rise to germ line chimeras. In all cases, the germ line chimeras also displayed mosaicism in their somatic tissues, showing the absence of segregation for germ line–forming ability by as late as the second cleavage division.

Even more conclusive are the experiments of Gardner (1977), who injected single ICM cells from 3.5-day p.c. blastocysts or single embryonic ectodermal cells from 4.5-day blastocysts and scored for subsequent somatic and germ line mosaicism. For the 3.5- and 4.5-day donors, there were, respectively, four out of six and two out of seven somatic chimeras that also showed germ line chimerism (as evidenced by progeny possessing the donor germ line marker). Although the number of chimeras in these experiments was not large, the frequencies were typical for single-cell injections. The experiments show that there is no specialized germ line determinant in the mammalian oocyte and that the primary ectoderm is the source of the germ line. In mammals, therefore, the continuity of the germ line is maintained through the soma rather than independently of it. The same is true of the avian embryo; reciprocal interspecies chimeric hypoblast–epiblast combinations of quail and chic reveal an epiblast origin for the germ line (Eyal-Giladi et al., 1981). A thorough discussion of the biology of the mammalian germ line can be found in McLaren (1981a).

The complete sequence of major tissue derivations in the mouse embryo,

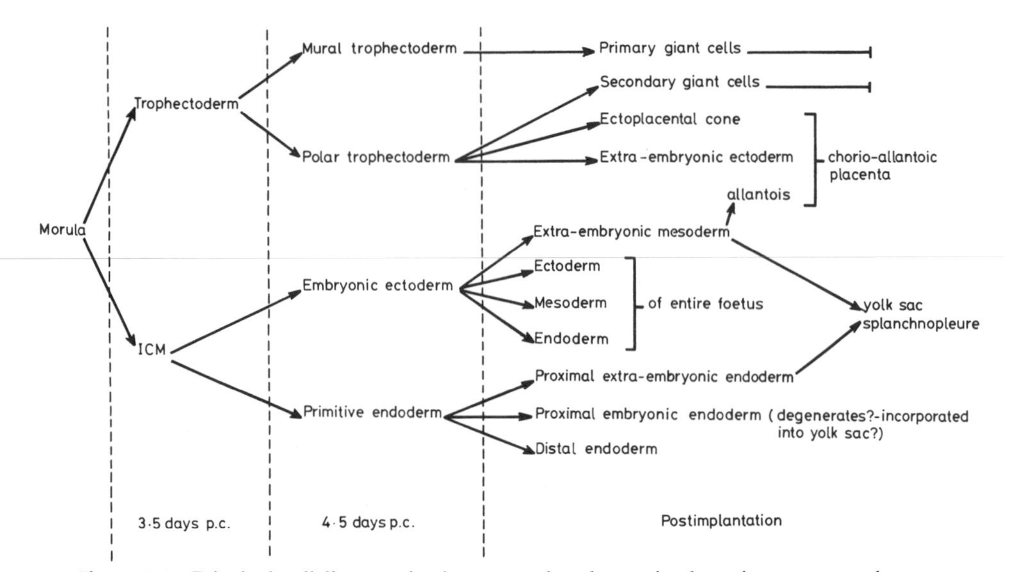

Figure 7.6 Principal cell lineages in the pre- and early postimplantation mouse embryo: a summary. (Reproduced with permission from Gardner, 1978.)

reconstructed primarily from chimera studies, is summarized in Figure 7.6. The chief conclusions are that all the cells of the fetus are derived from a portion of the EmE and that the placenta is composed of several discrete lineages of both the TE and the ICM.

Genetic Analysis of Tissue and Organ Foundation

The period of major tissue foundation in the mouse embryo occurs within 24 hr of the formation of the primitive streak. However, many of the details of histogenesis and organogenesis are obscure. The two aspects that have received most attention are the first cellular commitments in organogenesis—whether they constitute restrictions in developmental capacity (determination) or depend on cell movement—and the number of founder cells for each organ. One uncertainty that affects the analysis concerns the precise timing of the initial developmental assignments. Presumably, the first restrictions of developmental potential within the primitive streak occur between 6.5 days p.c. and the beginning of somitogenesis. By the late somite stages in avian embryos, distinct determinative states have been imposed; somites from the thoracic regions of the embryo generate vertebrae with ribs when transplanted to other regions of the embryo, but lumbar somites, those from the lower back region, never form ribbed vertebrae. If one assumes that cells of the early primitive streak mouse embryo are unrestricted, as they are in the avian embryo, a considerable degree of narrowing in developmental potential occurs between the early and late primitive streak stages.

Experimental embryology offers one set of approaches to establishing the nature and timing of the first developmental commitments in primitive streak embryos. One method is to cut out and culture pieces of these embryos at specific stages and to see what they form. The range of structures that can be produced by a given piece from a given stage serves as a measure of the autonomous developmental capacities of that piece; if specific embryo fragments always produce specific characteristic structures, the fragments may be presumed to have experienced some assignment of developmental role. Snow (1981) has reported the results of such a study. Isolated pieces of early (7-day p.c.) to late (8.5-day p.c.) primitive streak embryos and the complementary deficiency embryos were cultured in isolation for 24 hr in vitro and then examined for the structures formed. In these tests, examination of the deficiency embryo in conjunction with the fragment allows one to determine whether duplication or regeneration of structures occurs. A high frequency of differentiation (>70%) of the cut pieces and an even higher frequency for the deficiency embryos (>90%) was observed.

These results can be related to the prospective fates of the different regions assayed. In all cases, the autonomous development of those tested pieces that produced recognizable structures corresponded to the prospective fate of that region in the unoperated primitive streak embryo. Thus, the most anterior third of the 7.5-day embryo, which is destined to give rise to the head fold, brain, and foregut, produced those structures in vitro. The middle third, which gives rise in vivo to the major mesodermal elements, was found to produce somites, neural tube, and some heart structures in vitro. Finally, the most posterior section, whose fate is to produce tailbud, hindgut, and PGCs, produced those structures plus allantois, and no others. A further trisection of the posterior region permitted a mapping of the allantois to the most posterior section of the primitive streak, and the PGC-forming capacity to the section immediately anterior to this one.

What developmental property is mapped in this experiment? Snow (1981) has described the result as being that of an "allocation map," to indicate a preliminary regional localization of capabilities (without necessarily being fixed, determinative states). On the other hand, Beddington (1982) has argued that such an inference is invalid because the tested pieces are large and consist of more than one tissue type. In principle, the presence of a combination of cell types might establish or stabilize a range of specific inductions, each of which triggers a differentiative pathway whose end product is scored at the termination of the culture period. On the basis of the ability of late primitive streak ectodermal cells, labeled with [^{3}H]thymidine, to colonize different regions and germ layer derivatives, Beddington has suggested that the ectodermal cells at least are undetermined in late primitive streak embryos. A difficulty with this view is that colonization and replication of cells in a heterotopic site do not necessarily signify that the cells have taken on the developmental state of the cells with which they cohabit.

To date, therefore, the techniques of classical embryology have left unsolved the problem of when determination in the primitive streak embryo takes place. It may be, however, that before states of determination can be assessed, a fuller description of cell allocation in the early embryo is necessary. Allocation presumably precedes or accompanies determination, and characterization of the timing and number of founder cells allocated to various tissues or organ primordia should provide a basis for the subsequent analysis of determination.

In principle, clonal analysis should be able to furnish information on the number of founder cells present at allocation from the estimates of numbers of descendant clones produced within the organ or tissue at the time of allocation. In the fruit fly, one first determines the approximate time of allocation by finding that point at which mitotically generated recombinant clones cease to mark neighboring primordia (whose proximity can be judged from the gynandromorph fate map). Mitotic recombination is then induced at that stage, and from the fraction of tissue or organ occupied by the marked cells, the number of founder cells can be calculated.

The mouse embryo demands a different strategy because marked cells cannot be induced by mitotic recombination in mammals; because one cannot generate marked clones at will during development, one cannot directly estimate the time of allocation in mammalian embryos. Rather, one must introduce the marked cell(s) at arbitrarily early times, either by direct physical placement to create chimeras or by the use of X chromosome inactivation in female embryos. If a female fetus is heterozygous for an X-linked gene—for instance, for a gene encoding electrophoretically distinct forms of an enzyme—half of its cells will express one form, the other half the other form. X inactivation therefore provides a natural form of genetic mosaicism within heterozygous female embryos and thereby furnishes a second potential means of clonal analysis in the mouse embryo. By measuring the proportion of cells within a tissue that express one allele or the other, the experimenter can arrive at estimates of founder cell number. Unfortunately, as in chimera construction, the genetic heterogeneity is not under experimental control and occurs before primitive streak formation.

With either chimera construction or X inactivation, therefore, the genetic marking of cells occurs before the presumptive times of fetal tissue and organ allocation. As discussed earlier, the consequence is that marked structures tend to consist of disproportionately excessive amounts of marked tissue, producing systematic underestimation of founder cell number (Fig. 6.12).

There is a further aspect of mammalian cell development, typical of vertebrate development in general, that can complicate the analysis. In *Drosophila,* the descendants of single cells tend to remain contiguous; hence, most visibly coherent clones are true individual descendant clones. In the mouse, this situation is much less true. Extensive cell mingling and cell migration take place, particularly in later development. The result is that a given early clone will usually fragment into separate daughter subclones. If one mistakenly equates patches of marked cells with individual descendant clones, the

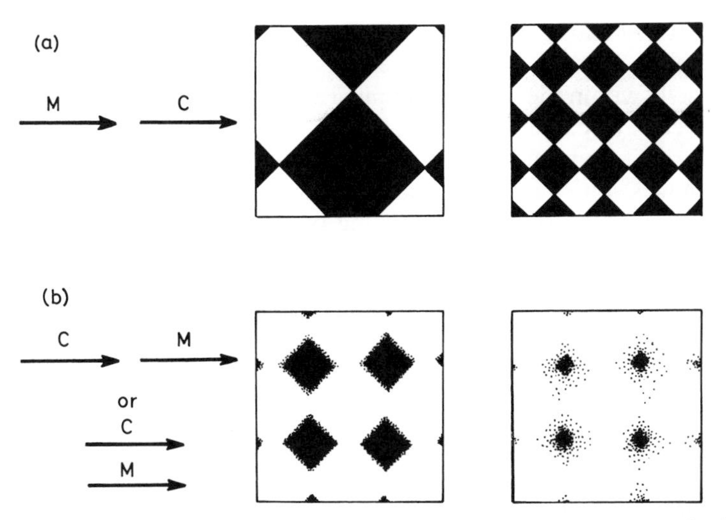

Figure 7.7 Patch pattern as a function of the timing of coherent clonal growth and cell migration. (*a*) Change in a pattern of large patches (left) produced by a period of migration (M) followed by coherent daughter clone growth (C) without subsequent migration; (*b*) patterns produced by clonal growth followed by migration or by a balance of clonal growth and migration. (Adapted from A. McLaren, *Mammalian Chimaeras,* Cambridge University Press, 1976.)

estimate of primordial cell numbers will be inflated to a degree corresponding to the fragmentation of descendant clones (Nesbitt, 1974). Furthermore, the estimate can be affected depending on whether cell migration occurs before or after tissue allocation. Some of the effects of different patterns of migration and coherent clonal growth on final patch sizes and discreteness are shown in Figure 7.7. When mingling is very extensive, individual patches may consist of two or more subclones that happen to be contiguous.

The consequences and complications of all of these factors in estimating founder cell numbers in the mouse have been reviewed by McLaren (1976a) and West (1978). The ultimate result is that estimates from clonal analysis in the mouse are considerably less certain than the corresponding estimates in the fruit fly. Nevertheless, the mammalian studies have yielded minimum estimates of founder cell number and have illuminated the detailed patterns of cell growth in mammalian development. The chimera studies will be discussed first, and then the X chromosome mosaicism studies will be reviewed.

Chimera Studies

Chimeras have been analyzed in two distinct ways to estimate founder cell numbers. One approach employs indirect cell markers—such as enzyme electrophoretic variants, which can be scored only in bulk preparations—and involve statistical estimates of founder cell number from the proportions of the component genotypes in the chimeras. The second set of methods utilizes direct cell markers, which can be scored in situ by direct inspection or histochemical means. In these methods, the physical disposition of the marked cells within the total cell population provides clues to the number of founding cells or clones and to their growth patterns within the tissue.

The earliest approach to estimating founder numbers with the chimera technique was statistical. It was based on the assumption that formation of a tissue or organ primordium, consisting of a fixed number of cells, can be likened to a lottery drawing in which cells are picked at random from a common precursor pool until the correct number has been drawn. In any large set of two embryo-aggregation chimeras, this method of primordium formation should generate a range of binomially distributed genotype compositions; the shape of the distribution is a function of the number of founding cells.

Instances involving very small numbers of founder cells will illustrate the point. If a given tissue primordium is always started by one cell, then none of the samples of that tissue from a group of chimeras will be of mixed composition; all will be genetically pure. If a primordium is always composed of two founder cells, then the fraction of chimeras that will have tissue of pure genotype a will be one-fourth (= ½ × ½), that of pure genotype b will also be ¼, and the fraction of chimeras with mixed tissue will be one-half (= 2 × ½ × ½). By the same reasoning, if the cell number is three, then one-fourth of the chimeras will be genetically pure (2 × ⅛ for each genotype) and three-fourths will be mixed. Since, in practice, it is often easiest to score the fraction of embryos with nonchimeric tissues, some of the early estimates of founder cell number were based solely on the fraction of mixed embryos that were judged nonchimeric for the structure in question.

A variant of this approach is to use relative similarities in composition between different tissues as an indicator of possible common tissue origins. An example is that of the analysis by Wegman and Gilman (1970) of two blood cell tissues in comparison to the skin of the mouse. Mixing two input genotypes that differed for a white blood cell marker (an immunoglobulin), a red blood cell marker (hemoglobin), and a coat color marker, they observed a correlation in nine chimeras between the composition of the immunoglobulin and hemoglobin markers, with a much smaller degree of correlation between these blood cell markers and that of coat color. The correlation between white and red blood cell genotypes was inferred to indicate a common cellular origin of the white and red blood cell lineages. From the observed frequency of chimerism in each of the markers, primordial cell numbers for both hemopoietic and surface ectoderm were calculated as being no fewer than five cells each.

There are two principal difficulties with all such statistical estimates. The first is the most important: the calculations are based on the implicit assumption that there is *complete and random cell mixing until the moment of tissue foundation*. The consequences of this assumption were first described by McLaren (1972). If there is a period of regional coherent clonal growth preceding the moment of tissue allocation, then the estimate of founder cell number will in reality be an estimate of the number of marked coherent *clones* from which the primordium originates. The greater the extent of coherent clonal growth preceding tissue allocation, the lower the numerical estimate of cells (= clones) will be. Indeed, all such estimates for fetal tissues probably reduce to the number of *original fetal progenitor cells present at the introduction of genetic heterogeneity*. For chimeras made by morula aggregation, the estimates of presumptive founder cell number for a given fetal structure therefore, in all probability, refer to the number of cells present at aggregation that contribute descendants to the structure in question, or about five to eight cells.

The estimate of five to eight cells derives from the following considerations. From single cell injections into blastocysts, it is apparent that single ICM cells can contribute to most or all fetal tissues and hence organs (Gardner and Rossant, 1979). Therefore, most or all ICM cells in the late blastocyst probably contribute to all fetal tissues. (Such widespread decendancy of single ICM cells presumably reflects a thorough mixing of cells prior to or during primitive streak formation.) Tracking of prelabeled cells in morula aggregates, however, shows little cell mixing until blastocyst formation (Garner and McLaren, 1974), with the consequence that the ICM derives primarily from the five to eight cells that are present internally in the aggregate. Indeed, most binomially derived estimates of fetal tissue founder cell number are in the range of two to five or slightly more cells (see Table 5 of West, 1978).

There is a second problem with the statistical approach: the influence of selection during cell growth within chimeras. When the cells of the two input strains differ in background genotype, they often contribute differentially to different tissues or organs (Mintz, 1974; McLaren, 1976a). A consistent underrepresentation of one genotype in a tissue relative to its frequency in other tissues is a certain indication of relative tissue-specific differences in growth or competitive ability between cells of the two genotypes. Such selective differences can, of course, influence the relative contributions to several tissues, undermining the presumptive significance of correlative frequencies for inferences about similar or disjoint cellular origins.

The analysis of chimeric tissue in situ, using direct cell markers, eliminates some of the ambiguities inherent in the statistical approaches. By permitting the direct visualization of marked cells within the analyzed tissues, the in situ screening techniques directly yield information on clonal growth patterns. This information, in turn, can help to furnish some minimal estimates of founder cell numbers for the structures or tissues in question. When the screening is carried out only on adult or newborn mice, the inferences remain indirect and are affected by the uncertainty as to whether individual founder cells or clones of adjacent founder cells are being estimated. However, if the marker or markers of choice are expressed throughout development, and if screening is carried out throughout this period, the method can yield clear results. In practice, few such studies have yet been carried out, but the use of cell autonomous antigenic markers may make this approach feasible in the near future.

The first example of an in situ analysis of chimeric patterns concerned the origins of the melanocytes, the cells responsible for pigmentation of the mouse coat. The melanocytes are the daughter cells of the migratory mesenchymal cells termed "melanoblasts" that originate in the neural crest, the region of epithelium that marks the site of dorsal closure of the neural tube. The neural crest gives rise to two major populations of migratory cells, those that migrate deep into the body between the neural tube and somites to give rise to neurons of the autonomic nervous system, and a second population that migrates just underneath the surface of the ectoderm (Fig. 7.8). It is the latter group that includes the melanoblasts.

By making chimeras between strains that differ in melanocyte pigment-forming capacities, it is possible to determine something about the clonal histories of the melanoblast cell populations. Such experiments were first performed by Mintz (1967). Chimeras were made between albino (*c/c*) and wild-type (*C/C*) strains and between strains differing in melanocyte genotype for color (e.g., black *B/B* and brown *b/b*). In all of the experiments, the striking outcome was the production of striped chimeric mice. An example is shown in Figure 7.9. Initially, Mintz reported that the patterns showed a regular alternation of the different colors, but larger samples have shown that there can be considerable variability in the width of the stripes.

Despite this variability, an underlying or "archetypal" striping pattern can be discerned when all the chimeras are compared. The wider bands observed in the animals simply consist of multiple unit widths, where the unit width

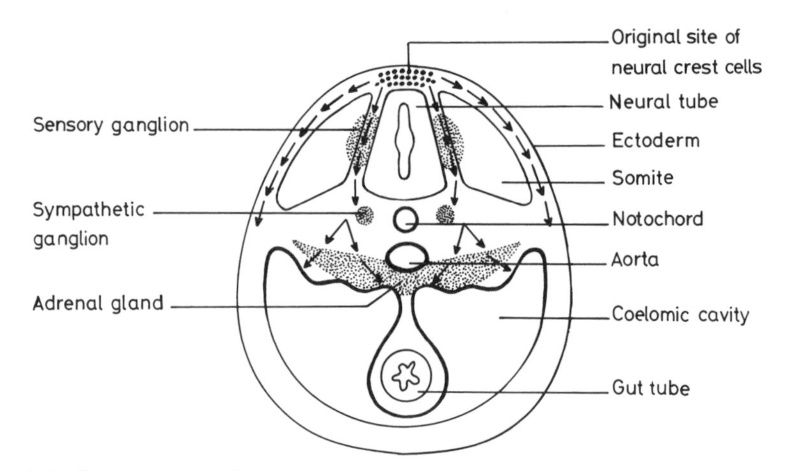

Figure 7.8 Two pathways of neural crest migration. The outer arrows indicate cells that give rise to melanocytes; the internal arrows denote neural crest cells that give rise to parts of the neural system and the adrenal gland; stippled areas indicate neural crest—derived areas. (From Alberts et al., *Molecular Biology of the Cell.* New York: Garland Publishing Inc., 1983.)

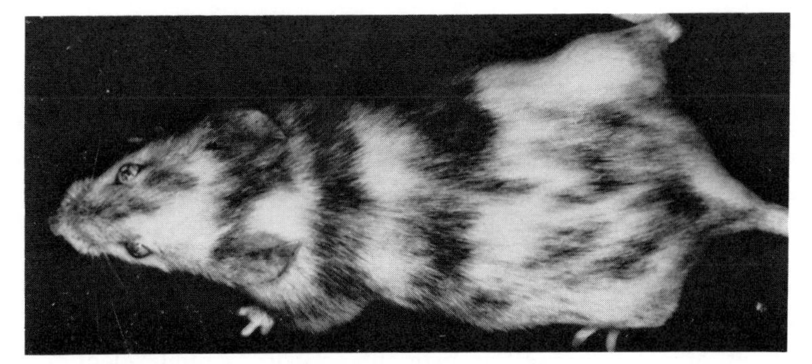

Figure 7.9 Adult chimeric mouse consisting of black (*C/C*) and albino (*c/c*). Note the striping pattern in this melanocyte chimera. (Figure kindly provided by Dr. B. Mintz.)

corresponds to the narrowest single stripe seen. The striping pattern itself is explicable if coloration in the coat is produced by small numbers of melanoblast clones arranged in an initial linear series along the longitudinal axis in which the clones that neighbor any particular clone may either be of the same genotype (giving a broader band) or of the alternative genotype (giving a central band of unit width). Comparison of the banding patterns in many chimeric mice suggests that the head region consists of three unit stripe widths, the body six, and the tail eight. Because the left and right sides at any position along the longitudinal axis can be of different color, the clonal origins of the left and right sides must be independent. Equating stripe widths with founder clones, one obtains 17 founder clones per side, or a total of 34 melanoblast founder clones. The independence of the left and right sides suggests that the cells that give rise to melanocytes are commtted to do so before dorsal closure of the neural tube is complete and also that the founder cells/clones begin their lateral migration before closure takes place (Mintz, 1974). Were it otherwise, cross-contamination would produce a high frequency of mixed unit stripes. With dorsal closure beginning at 8.5–9.0 days p.c., the migration of melanoblasts must take place before this period.

Does the estimate of 34 melanoblast founder clones mean that there are only 34 founder cells of the melanoblast lineages? Not necessarily. From the fact that the clarity of the striping is sharper in chimeras composed of cells from more distantly related strains than from more closely related ones, it appears that there is a partial sorting out according to genotype (McLaren, 1976a). If a certain amount of self-segregation is occurring early in the melanoblast lineage, the founder clones might consist of two or more cells of like genotype. However, the number is unlikely to be more than two or three, since the higher the number the more likely would be the occurrence of "split" unit stripes, and these are not seen.

The technique of chimera analysis has also been used to explore the origins of mesodermal structures derived from the somites. It has been possible to analyze the clonal origins of vertebrae, for instance, because certain common laboratory strains differ in their details of vertebral morphology. Utilizing these morphological differences as indices of genotype and making chimeric mice from such strains, Moore and Mintz (1972) concluded that both the left and right halves and the anterior and posterior halves of each vertebra have separate and distinct clonal origins; these findings indicate that each vertebra is derived from a minimum of four clones. This result fits the known facts of somite development well: each vertebra is formed by the fusion of the caudal (posterior) and cranial (anterior) halves of neighboring somites and by those of the left and right somite pairs. The estimate of four clones is a minimum estimate of the number of founder cells because of the possibilities of cell type self-assortment and differential selection within founding cell groups.

A different approach, using the indirect cell marker glucose phosphate isomerase (GPI), was used to examine the origins of skeletal muscle: do these muscles arise by cell fusion or by repeated nuclear division without cytokinesis? The approach relies on the fact that many enzymes are made up of two or more polypeptide subunits. For such polymeric enzymes, the existence of two alleles specifying different electrophoretic variants within the same cell results in the production of "hybrid" enzyme molecules; these molecules are distinguishable from the parental enzyme forms by their intermediate position on gels after electrophoretic separation. If skeletal muscles arise by fusion of cells, then chimeric muscles from strains that differ in GPI alloenzymes should make hybrid GPI. On the other hand, if each syncytial muscle arises by repeated nuclear divisions from a single cell, then the muscles of chimeric mice should make only the two pure forms of the enzyme. The analysis of skeletal muscle from such chimeras revealed the presence of substantial amounts of hybrid enzyme in addition to the two parental alloenzyme forms. Muscle syncytia must therefore arise through fusion of cells rather than as clones of individual nuclei (Mintz and Baker, 1967).

In a later study, Gearhart and Mintz (1972) analyzed the origins of both somites and muscles, using GPI as the marker. They found that the great majority of individual somites from 8- to 9-day embryos contained both forms of GPI. Individual somites are therefore not clones, but originate from at least two mesodermal cells. In the analysis of certain individual eye muscles, each of which was formed from a single myotome of a single somite, 22 out of 33 muscles assayed were found to contain both GPI alloenzymes and hybrid enzyme. The results show that each somite-derived myotome is also formed from two or more myoblasts. It was also observed that neighboring somites were often correlated in composition, which suggests that adjacent somites may share some cellular ancestry.

All of the results discussed above have involved retrospective analyses of events: inferences about the earliest events are made from the tissue composition of the adults. Relatively few studies have attempted to determine the history of a tissue by tracing the development of marked clones in chimeric tissue. One example of such an analysis is that of West (1976) on the growth patterns of cells in the retinal pigment epithelium (RPE) of the eye. This monolayer of epithelial cells lies between the neural retina and the enclosing outer structure of the eyeball, the choroid. By making wild-type/albino chimeras and examining cross sections of the RPE, West showed that cell mixing in the RPE is extensive until about 12 days p.c. At that point, each clone is little more than one cell on average. This period of extensive cell mingling is then succeeded by a period of more coherent cell growth, with some cell migration taking place during continued growth of the monolayer. Final clone sizes of marked tissue seen in section consist of five to six cells.

Such one-dimensional sections through a monolayer give a fine-grained picture of clonal growth. Examination of the total area of the RPE in chimeras provides a more complete picture of its history. When the two-dimensional surface of the RPE is analyzed in wild-type/albino chimeras in which the pigmented cell population in the RPE is a minority component, distinct radial patterns of growth can be discerned (Sanyal and Zeilmaker, 1977). From the size of the sectors, and on the assumption that each sector corresponds to a founding clone, the results indicated that the RPE is initiated by a relatively small number of founding clones, probably more than 10 but fewer than 250 (West, 1978). Synthesizing the cross-sectional and whole surface observations, it appears that the RPE is founded by a relatively small group of founder cells that generate descendants in a radial pattern of outward growth, initially with much local cell mixing and subsequently with greater coherent clonal growth.

X Chromosome Inactivation Analyses

The naturally occurring phenomenon of X chromosome inactivation in female mammalian embryos furnishes the second major form of clonal analysis in the mouse. Although X inactivation, like chimera production, produces two phenotypically different populations of cells, the difference is generated intracellularly within individual heterozygous embryos rather than by the forced mixture of cells from two genetically distinct embryos at the outset. This basic difference is diagrammed in Figure 7.10. An additional difference between the methods is that X inactivation analyses usually require the use of X chromosome markers, while chimera analysis can employ any chromosomal markers, whether autosomal or sex-linked. The few autosomal genes that can be utilized in X chromosome mosaicism studies are those located on pieces of autosomes that have been translocated to the X. These translocated fragments experience the X-inactivation process, and if the translocated piece and the normal homologue differ for a scorable marker, a mosaic pattern of expression is produced. The most useful of these translocations is Cattanach's translocation, which is a transposition of part of chromosome 7, containing the wild-type allele for the albino locus (*C*), into the X. When the standard seventh chromosome homologue contains the recessive (*c*) allele,

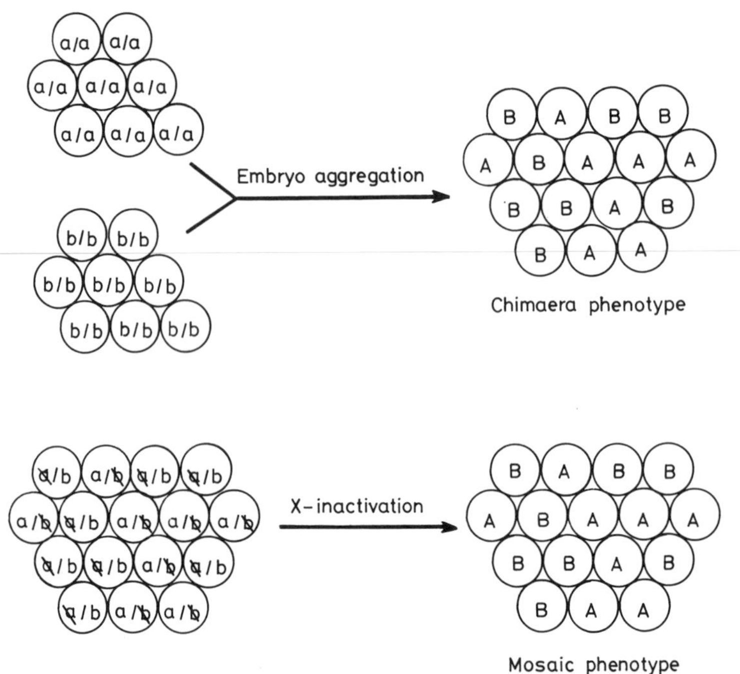

Figure 7.10 Two ways to produce mice with a dual-composition cell population. Top, chimera construction; bottom, X chromosome inactivation in an X chromsome heterozygote. (Redrawn from A. McLaren, *Mammalian Chimaeras,* Cambridge University Press, 1976.)

inactivation of the translocated chromosome produces an albino clone in the descendants of that cell.

As drawn schematically, the phenotypic outcomes of chimera construction and X inactivation look identical. In reality, the patterns of spatial heterogeneity in chimeras and X inactivation mosaics often look somewhat different even when the same markers are employed. In general, the chimeras show more variable patchiness and larger patches. Indeed, the two properties are directly related. Variable tissue composition between cells of two genotypes tends to produce larger patches, with any shift from a 1 : 1 ratio increasing the probability that cells of like genotype will find themselves as neighbors. However, when cell mixing is extensive, as in the early period of development of the RPE, the differences between chimeras and X chromosome mosaics are reduced (West, 1976). Only when there are large differences in background genotype between the component strains of chimeras does the chimeric pattern begin to look different from the X mosaicism pattern because of selective self-assortment.

The principal methods for estimating founder cell number in X chromosome mosaics are statistical. The central premise of these methods is that if sampling for tissue foundation is from a randomly mixed cell pool, the variance in composition for a given tissue between different mosaics in the same series will be an inverse function of founder cell number (Nesbitt and Gartler, 1971). When the founder cell number (*n*) is very small, a large variance between like samples will be observed; when *n* is large, the variance should be small. If tissues or organs show similar compositions and similar variances, they may stem from a common precursor pool; the residual variance between different samples would then reflect differences in the founder cell numbers of the individual primordia.

The first study of this kind was reported by Nesbitt in 1971. The cell marker she employed was Cattanach's translocation, whose frequency of inactivation can be scored in chromosome spreads from mitotic cells labeled with [^{3}H]thymidine early in S phase and scored autoradiographically. (Because the inactive, heterochromatinized X replicates late in S phase, the portion of the cell cycle devoted to chromosome replication, it appears as the nonlabeled X in these conditions.) Samples were prepared from several different tissues—spleen, lung, liver, and coat melanocytes—of newborn mice, cultured for 4–5 days to amplify the fraction of dividing cells, and then analyzed by the mitotic spread procedure. The analysis indicated that there are about 20 epiblast cells that act as the precursor of the fetus at the time of X chromosome inactivation and that 20–50 founder cells are allocated within the primitive streak for each of the organ systems analyzed.

A more recent approach to the same problem has been reported by McMahon et al. (1983). They used an indirect cell marker, the X-linked enzyme phosphoglycerate kinase (PGK), which exists in two electrophoretically distinguishable forms, and a sensitive assay, suitable for small amounts of electrophoresed enzyme from tissue samples. The samples were prepared from 12.5-day female embryos, heterozygous for the two alleles of PGK, and analyzed for PGK electromorph composition. The tissues examined were representatives of the three primary germ layers of the fetus—neural ectoderm, heart mesoderm, and liver endoderm—and the germ cells. The analysis indicated that there are 47 fetal precursor cells and about 190 precursor cells for each of the four tissues examined (at the time of germ layer delimitation within the primitive streak). Although the last number is an average, a high correlation of alloenzyme composition between different tissues from each embryo was observed, suggesting that all three primary germ layers and the germ cells arise from similar-sized precursor cell pools. The higher estimates of founder cell number by McMahon et al. (1983), compared to those of Nesbitt, (1971), reflect the lower variances observed in their study.

The estimate of 190 germ line precursor cells (PGCs) is of special interest for two reasons. Firstly, this number is considerably greater than the number of PGCs detectable by AP staining at 8.5 days, the first point at which the cells can be observed directly. If the statistical estimate is correct, then high AP activity may identify only a fraction of the true number of PGCs. Secondly, the similarity of this number to that for the other three germ layers raises the possibility that PGCs are not actively instructed to be such but may simply be the residuum of unassigned cells (Snow and Monk, 1983). In this view, the retention of totipotency by PGCs is a consequence of their failure to be restricted in fate. It may even be that the caudal end of the primitive streak, from which the PGCs arise, is a region that protects against restriction of totipotency. As discussed earlier in connection with germ line determination in *Drosophila,* the posterior polar plasm from which the germ line of the fruit fly originates may exert its determinant function by providing comparable protection.

As in the chimera studies, the estimates of founder cell number from X mosaicism studies are probably estimates of the number of coherent clones contributing to the tissue rather than the number of founder cells per se. When more reliable information from other techniques on the times of tissue allocation becomes available, both forms of clonal analysis should contribute more precise estimates of founder cell number than are presently obtainable.

DEFECTS IN ORGANOGENESIS: GENETIC ANALYSIS

A very large number of mouse mutants are affected in organogenesis. Many of these mutants were first detected on the basis of their dominant heterozygous phenotypes—for instance, a degree of runting or a change in coat color from the parental stock. In homozygotes, the mutant defects often result in embryonic or perinatal mortality. Another class of mutants affected in the development of organ systems are those initially identified on the basis of their behavioral phenotypes; such mutants are usually defective in the nervous system. The *shaker* (*sh*) and *waltzer* (*w*) mutants, whose phenotypes are conveyed by their names, are examples of neural defectives.

Until the advent of chimera and mosaic studies, the developmental genetics of the mouse consisted almost entirely of the study of the developmental defects associated with particular mutants. The procedure is the classical one of developmental genetics. One first identifies the set of mutant phenes and then works backward in development to the first observable aberrancy. From these observations, one then attempts to reconstruct the cellular site, and if possible the biochemical basis, of the initial developmental derangement. As discussed in Chapter 1, the method is replete with difficulties and is only occasionally successful. In a number of instances, however, the etiology is similar to that produced by known specific enzymatic deficiencies in humans and the mouse syndrome proves to have an identical basis (Bulfield, 1981).

For the most part, however, the mutant defects that produce severe aberrations of early or middevelopment have not been characterized at the biochemical level. Nevertheless, these mutants represent a valuable store of genetic source material for future investigations. As the techniques for biochemical and cytological characterization advance, the precise biochemical basis of some of these mutant defects will become clear and the characterization of the genes themselves, isolated via recombinant DNA procedures, will elucidate the genetic basis of individual defects in an increasing number of instances.

Perhaps the single most valuable tool in characterizing the cellular basis of mutant defects has been the electron microscope. With electron micrographs, it is often feasible to determine which cells are defective and to reduce the range of possible explanations of the mutant phenotype (although ambiguities usually still remain). An example of such an analysis will illustrate the usefulness of the approach.

One of the first morphological mutants of the mouse to be characterized was the dominant short-tailed *T* or *Brachyury* mutant, reported by N. Dobrovolskaia-Zavadskai in 1927. When short-tailed (*T*/+) mice are crossed, one obtains a 2 : 1 ratio of short-tailed to normal mice instead of the expected 3 : 1 dominant to recessive ratio (1 *T*/*T* : 2 $T/^{+}$: 1 +/+). This observation suggested that, like A^y, *T* is lethal when homozygous. This suggestion was confirmed by Chesley (1935), who performed the first detailed examination of the embryos. Using the standard light microscopic techniques available at the time, Chesley observed that the first abnormalities in *T*/*T* embryos appeared in early somite, 8.5-day p.c. embryos. In the posterior third of the embryo, the notochord is absent, the somites are disorganized, and the neural tube shows aberrant morphology. These early defects are succeeded by progressive disorganization of the entire notochordal and neural tube regions and a failure of hindlimb development. Heterozygous (*T*/+) litter mates, in contrast, do not show any defects until about 11 days p.c. At this point, they exhibit a diminished neural tube and notochord in the tail region and minor irregularities of the notochord anteriorly. The heterozygous syndrome is thus a milder version, with a later onset, of the homozygous developmental defect. The cause–effect relationships between the various abnormalities could not be determined, but Chesley speculated that the notochordal defect was primary and the neural tube defects secondary.

In a detailed electron microscopic study, Spiegelman (1976) confirmed and extended the earlier findings. By 8 days p.c., the *T*/*T* embryo has a reduced trunk and a correspondingly enlarged primitive streak. In these embryos, the conversion of embryonic ectoderm to mesoderm is blocked; in consequence, instead of forming notochord and somites in the posterior region, the blocked mesodermal cells form excess neural tube material (Fig. 7.11). In effect, the block in mesodermal differentiation from ectoderm leads to a temporary hyperplasia or overgrowth of neural tissue; these epithelial tubes subsequently undergo degeneration, concomitantly with the disintegration of the anterior mesodermal axial structures. (This syndrome of neural overgrowth, following a block in ectodermal differentiation, is reminiscent of the *Drosophila Notch* mutant syndrome discussed in Chapter 6.)

The specific cellular defect producing the *T*-lethal syndrome is revealed by the electron microscope. Unlike the wild-type embryo, in which the neural tube and notochord are separated by a discrete basal lamina, these two cell types meet in the lethal *T* embryos at the ventral margin of the

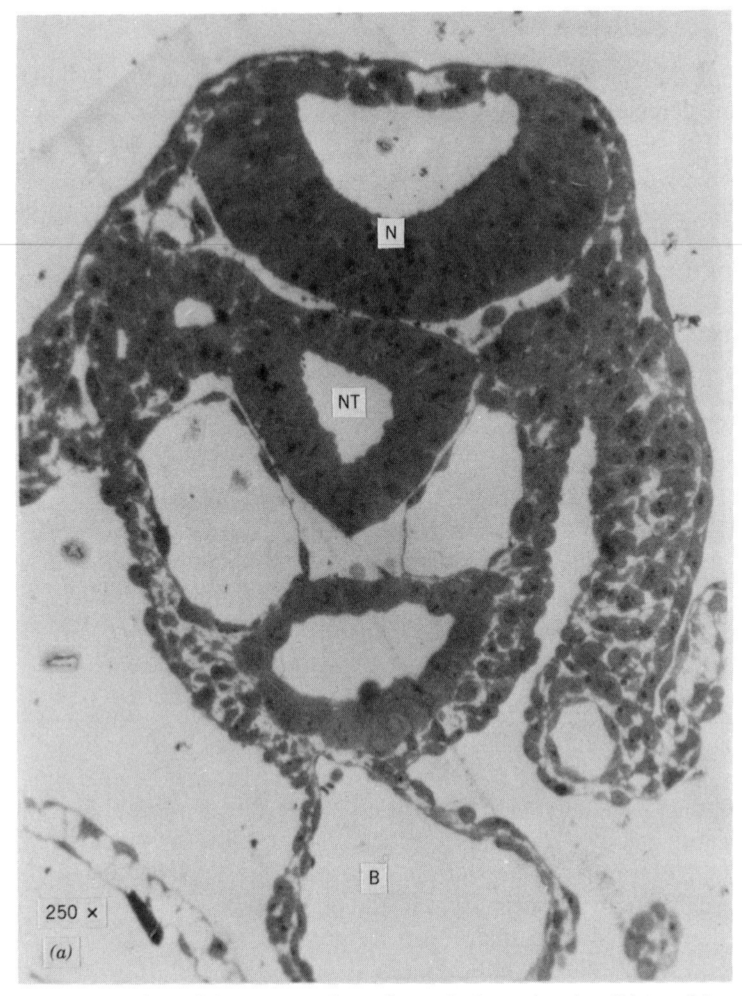

Figure 7.11 Comparison of the cross sections of neural tube regions in wild-type (*a*) and *T*/*T* (b) embryos. Magnification: 250X. N, neural tube; NT, neural tube hyperplasia; C, notochord. (Photographs kindly provided by Dr. M. Spiegelman; from Spiegelman, 1976.)

neural tube (Fig. 7.12). There are no basal laminae in the region of contact, and the neural tube and notochordal cells establish junctions with one another. This failure of normal cell recognition, with subsequent failure to establish a normal barrier between the notochord and the neural tube, must play a role in the ensuing multiple failures of mesodermal organization and differentiation.

The difference in notochordal cell surface behavior in the mutant could be a direct reflection of a cell surface property that directly impedes normal primitive streak differentiation. It is equally possible that the cell surface

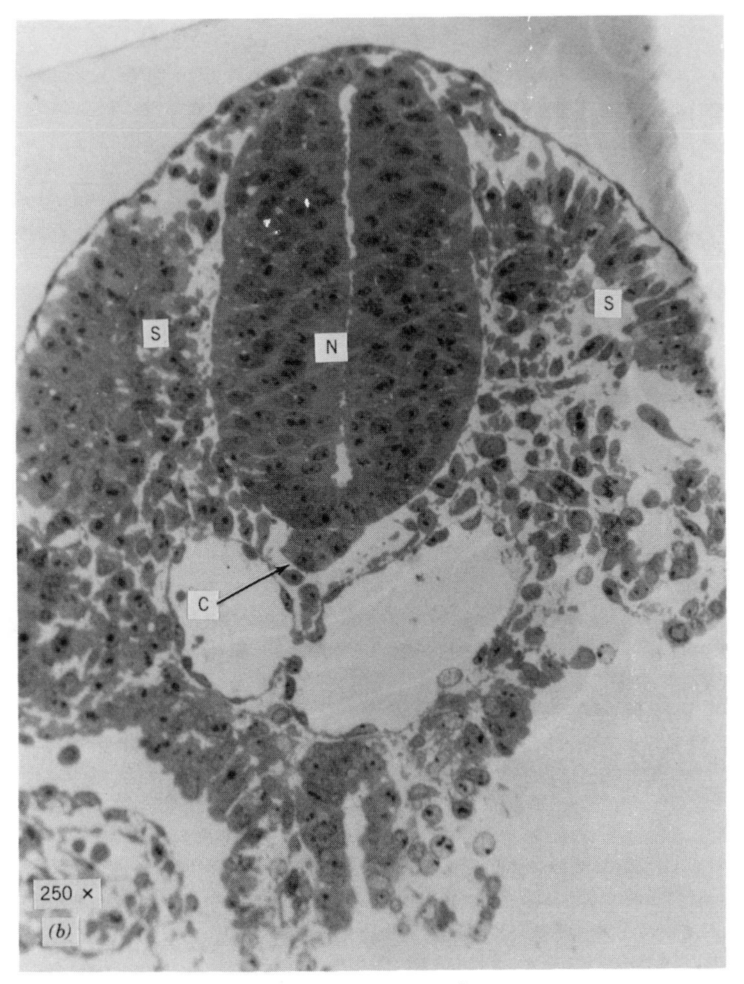

Figure 7.11 (*Continued*)

change is a secondary event and perhaps only one element of the failure of mesodermal differentiation. The pictures themselves do not settle this point, but only raise it.

To characterize the role of a wild-type gene in development, one requires a complete description of both the cellular and developmental phenotype and the genetic character of the mutation itself. With *Drosophila,* one can usually perform the Mullerian dosage test in order to classify the expression property of the mutation. In the absence of the necessary duplications and deficiencies, one can resort to isolating additional mutant alleles of the same gene. The various mutants can then be compared, and the phenotype produced by a complete or nearly complete inactivation of the gene can be inferred.

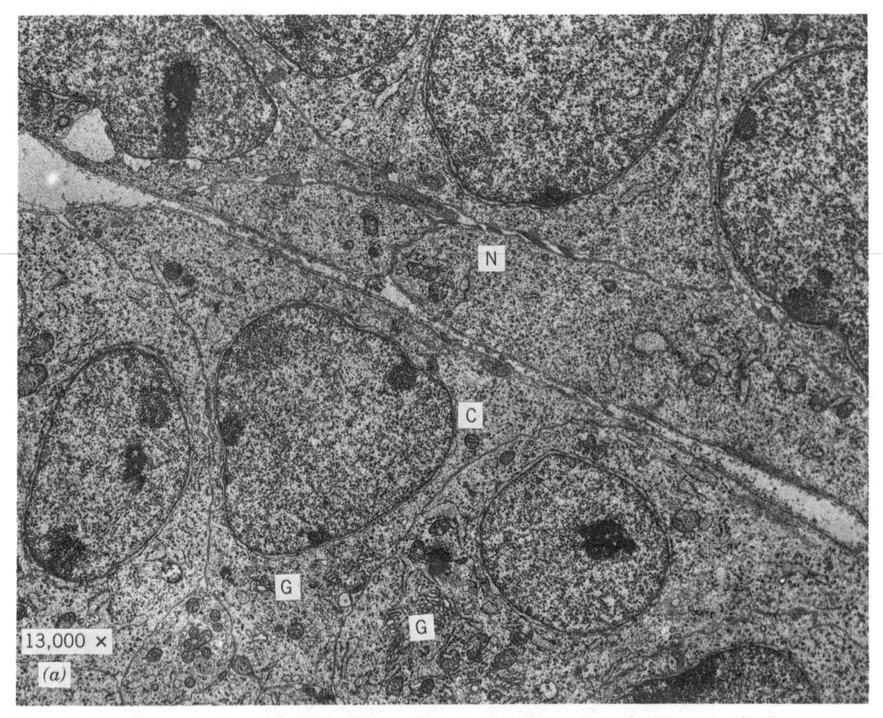

Figure 7.12 The notochord-neural tube in -day mouse embryos. (*a*) wild-type, 13,000X; (*b*) *T*/*T* embryos, 13,000X. N, neural tube cells; C, notochord cells; G, golgi apparatus. (Photographs kindly provided by Dr. M. Spiegelman; from Spiegelman, 1976.)

In the mouse, these genetic stratagems are usually impossible. For the most part, the necessary duplications and deficiencies are not available for the dosage test, and large-scale searches for noncomplementing lethal mutants are impractical.

Only when a mutation creates a distinctive visible phenotype without attendant lethality can one hope to isolate more mutants of the same gene by direct screening, permitting comparative studies. To isolate new mutants, one screens for noncomplementer chromosomes, as in *Drosophila*. Heterozygotes carrying the initial mutation ($+/m$) are mated to mutagenized wild-type animals, and the progeny are examined. Any individual in the F_1 showing the mutant phenotype must carry the original mutation, *m,* and a new mutant allele, m^*. These individuals are then outcrossed to isolate the new mutant, which can be distinguished from the original if the original mutation is linked to scorable markers.

Because this method is limited to genes in the mouse that give a visible but fully viable phenotype, it is inapplicable to many genes of developmental interest. Where it can be applied, it has been informative. The procedure is illustrated by the search for mutants of the *albino* (*c*) locus on chromosome 7. This gene is probably the structural gene for tyrosinase, the enzyme that

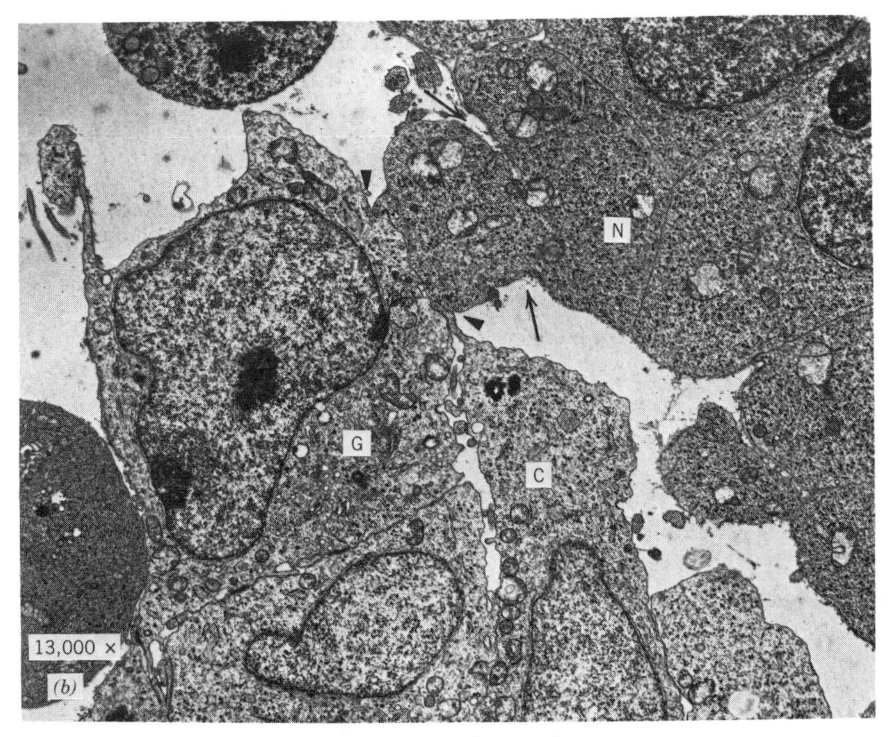

Figure 7.12 (*Continued*)

synthesizes the major melanic pigments. To collect new albino mutants, female heterozygotes (*c*/*C*) The *C* allele being the dominant, wild-type allele) were mated with irradiated wild-type males, and all albino progeny were outcrossed to isolate the new *c* alleles. A number of these new mutants are lethal when homozygous. The reason for the lethality is not the absence of *c* activity but the deletion of the mutant chromosomes for adjacent essential loci.

Gluecksohn-Waelsch (1979) has described a set of six of these lethal deficiency *albino* chromosomes. One mutant homozygote dies in early cleavage, one dies in the early egg cylinder stage, and the remaining four exhibit perinatal mortality; these last four exhibit a specific deficiency for six particular liver enzyme activities (while having normal activities for several other liver-specific enzymes). By making heterozygotes of pairs of the different lethal deficiency chromosomes and scoring for the presence or absence of particular mutant phenotypes, one can position the wild-type genes whose activity prevents particular mutant phenes. Such a gene localization procedure is termed "complementation mapping." The principle is illustrated in Figure 7.13*a*. The results of this deficiency complementation mapping are given in Figure 7.13*b*. The liver enzyme defect is between *c* and *Mod-2* (the gene for mitochondrial malic enzyme-2), and the essential early embryonic

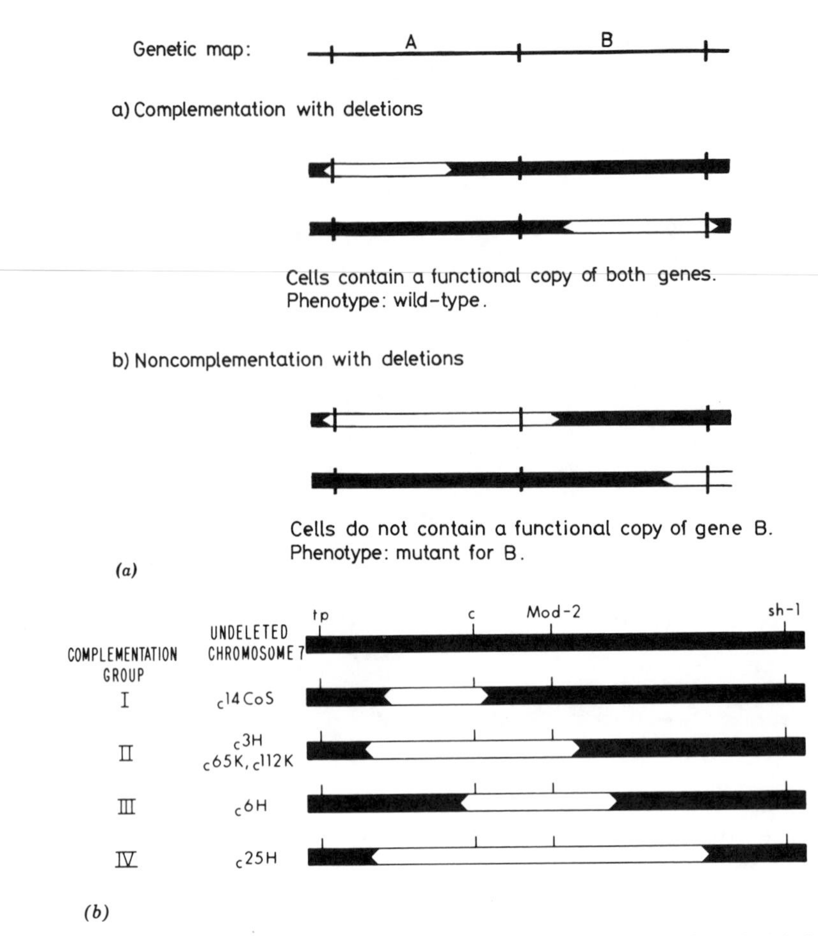

Figure 7.13 Deletion mapping of functions. (*a*) Principle of deletion mapping; (*b*) deletion map of the c region, from Gluecksohn-Waelsch (1979). See text for discussion. (The figure in *b* kindly supplied by Dr. S. Gluecksohn-Waelsch; from Gluecksohn-Waelsch, 1979, copyright MIT Press.)

function(s) lie between *Mod-2* and *sh-1* (one of the genes that gives the shaker phenotype). The results show that isolating new mutants of a locus can allow one, if some of the mutations are deletions, to work "outward" along the chromosome from that locus, identifying new essential gene functions and their linear sequence. In effect, the locus provides a bridgehead for exploring the genetic organization of a chromosomal region. (An analogous molecular approach to the analysis of chromosome genetic organization will be described in Chapter 9.)

The defect in the four perinatal mutants may be of special interest with respect to genetic regulation. In these mutants, the cause of death appears to be a combination of defects in the liver, kidney, and thymus, with the liver defects being the most serious. The enzymatic deficiencies of the mutants may reflect the loss of a specific regulatory locus essential for their activity. When mouse/rat hybrid cells are constructed by fusing mouse cells homozygous for one of the deletions with rat cells, the hybrid cells are found to express the mouse G6PD activity (one of the deficient activities of the mouse cell); evidently, the rat genome can supply the missing regulatory factor (Cori et al., 1983). The immediate cause of the inability to synthesize the affected enzyme activities may be a reduction in receptors for glucocorticoid hormones found in the mutant cells. However, the presence of some receptors in the cells of the deletion homozygotes suggests that the structural gene for the receptors is not missing (Gluecksohn-Waelsch, 1983).

The mortality effects associated with the late-acting mutants may involve certain aberrancies of cellular organelle development. Electron microscopic inspection of the affected organs in the four perinatal lethal homozygotes shows that the membranes of the rough endoplasmic reticulum, the Golgi apparatus, and the nucleus are all abnormal in the liver and the kidney. However, these membrane organelles are normal in appearance in the thymus of the affected embryos, although the thymus itself shows morphological abnormalities. These organelle effects reveal tissue-specific differentiative aspects of certain commonly shared organelles. The existence of such subtle differences had not been anticipated on the basis of conventional cytological studies and adds a further dimension to the known complexity of tissue differentiation. However, some caution is required in accepting this interpretation: the liver organelle defects may be particular secondary consquences of the enzyme or other deficiencies that do not pertain to the thymus.

THE *t*-COMPLEX: MAMMALIAN SWITCH GENES OR DEVELOPMENTAL ANOMALY?

One of the fundamental questions in animal developmental genetics is whether development is governed by major regulatory "switch" genes. We have seen instances of apparent switch genes in the two invertebrate systems discussed earlier. In the nematode, replacement of various cellular lineages can be produced by mutation in a few specific genes, with consequent major alterations in the developmental pathway. And in *Drosophila,* the homeotic mutants, particularly those of the *bithorax* and *Antennapedia* gene complexes, show that certain genes play key roles in the assignment of segmental phenotype. Are there comparable loci in mammalian development?

During the 1970s, several investigators proposed that the so-called *t*-complex on chromosome 17 of the mouse played just such a key regulatory role in early mouse development. Several results have rendered this hypothesis increasingly unlikely, but the developmental genetics of the *t*-complex have produced much information of interest.

The first *t* mutations bearing chromosomes were picked up fortuitously because of an unusual reaction they display with *T*. As described earlier, *T*/+ heterozygotes are short-tailed mice and *T*/*T* homozygotes are embryonic lethals. When certain mice from wild populations were crossed with *T*, a new class of viable progeny appeared: completely tailless mice. Surprisingly, the condition of taillessness was found to breed true in brother–sister matings from each cross; that is, only tailless progeny were produced in each ensuing generation. Outcrossing revealed that each tailless mouse carried a new recessive lethal, a *t* mutant, in addition to *T*. Crude mapping placed the new mutants near *T* on the chromosome now designated number 17. Each tailless strain therefore carries two heterozygous lethals, *t* and *T*. Such strains, in which each member of a homologous chromosome pair carries a lethal mutation and in which only the heterozygous combination is viable, is said to be a balanced lethal strain. The explanation for the perpetuation of the tailless phenotype is that the *T*/*T* and *t*/*t* embryos are lethal and only tailless heterozygotes (*T*/*t*) survive.

The genetic complexity of the *T*/*t* locus, as it was first designated, emerged only when the different tailless strains were crossed with one another. If all of the *t* mutations are in the same functional genetic unit, then the intercrosses should yield only tailless mice. Thus, if two different *t* mutants are designated t^a and t^b, the intercross ($T/t^a \times T/t^b$) should give only parental genotype, tailless progeny. In fact, some crosses gave normal-tailed mice in as many as one-third of the surviving progeny. These mice were subsequently shown to be of genotype t^a/t^b. In other words, some of the *t* mutants were found to complement each other, permitting normal development. As more *t* mutants were isolated, it became apparent that the different lethal alleles could be placed into six different complementation groups.

In addition to *t* mutants from wild populations, new *t* mutants were isolated in the laboratory from the true-breeding tailless (*T*/*t*) lines, at a rate of about 1 in every 500 or 1000 progeny. These new mutants were isolated as exceptional, normal-tailed mice within the stocks, and were shown by the complementation test to carry new *t* alleles, derived by some form of unequal crossing over within the *T*/*t* complex. Many of these laboratory-derived strains are semi- and fully viable *t* mutants, each of the new mutants being identified as a *t* mutant by the *T* interaction test.

However, direct genetic mapping of the mutant sites could not be carried out because all of the *t* mutants exhibited a region of depressed recombination within the area of the *t*-complex itself. Furthermore, many lethal *t* mutants show unusual inheritance patterns, in particular, a tendency for the mutant allele to be preferentially transmitted from wild-type heterozygotes (*t*/+), but only in the male line. In contrast to this behavior of the lethals, some of the viable *t*'s are present in less than 50% of the sperm of heterozygotes. An additional complexity is that combinations of high- and low-transmission complementing *t* alleles do not always behave predictably, but can show partially reversed relative transmissibility in each other's presence.

The causes of these distortions in transmission ratio are obscure but reflect both early and late developmental effects on spermatogenesis. The preferential transmission effects from *t*/+ heterozygotes apparenly reflect direct alterations in the fertilizing ability of the haploid mature sperm. These changes in sperm behavior are correlated with specific antigenic changes in the sperm head. Each lethal complementation *t* group is associated with a particular set of sperm antigenic determinants; some antigenic determinants are shared between different groups, while others are unique to a particular complementation group.

The potential developmental significance of the *t*-complex became apparent only when the specific lethal syndromes of homozygous inviable embryos were closely examined. It was found that members of each complementation group show arrest of embryonic development at characteristic stages, ranging from early morula to early somitogenesis. The developmental blocks associated with each lethal complementation group are summarized in Table 7.2. Within each complementation group, different mutants often show some differences in severity or terminal stage, but the members of each group resemble each other more closely than do members of different groups.

In a review of the genetics and biology of the *t* mutants, Bennett (1975) proposed that the developmental effects reflect a major governing role of the presumed wild-type *t*-complex in early mouse development. This hypothesis amplified an earlier suggestion made by Gluecksohn-Waelsch and Erickson (1970). Based on the known effects of the *t* mutants on sperm antigens (and some immunological work on teratocarcinomas to be described below), Bennett proposed that normal development involves a fixed sequence of chang-

Table 7.2 Developmental Blocks in Lethal *t* Mutants

t Complementation Group	Stage of First Abnormality	Stage of Developmental Arrest	Cell Types Affected
t^{12}	Morula	Morula	All blastomeres
t^{w73}	Late blastocyst	Early egg cylinder	TE
t^{0}	Early egg cylinder	Late egg cylinder	TE, EmE, endoderm
t^{w5}	Late egg cylinder	Preprimitive streak	Mainly EmE
t^{9}	Early primitive streak	Late primitive streak	Mesoderm and EmE
t^{wl}	Late primitive streak	Early organogenesis	Ventral neural tube, brain

Source. Observations summarized from Bennett (1975) and Paigen (1980).

ing cell surface properties, each essential for the next developmental transition. The role of the *t*-complex would be to determine this programmed sequence of changing cell surface properties. Because the earlier embryological studies pinpointed ectodermal cells and derivatives as the chief foci for *t* mutant effects, the action of the *t*-complex was provisionally characterized as the mediation of a series of switches within the ectodermal lineages. In terms of this hypothesis, the developmental effects of *T* were interpreted as stemming from the cell surface defect observed in the electron micrographs (Fig. 7.12).

The most telling argument against the developmental switch hypothesis comes from an analysis of the expression of *t* antigens during development. If the developmental transitions in the early embryo are mediated by the appearance of specific cell surface molecules on particular cell types at specific developmental stages, then one would predict that the sequence of *t* mutant syndromes, from morula to late neural tube stages, would be reflected in a sequence of *t* antigen appearances on the cell surface(s). Kemler et al. (1976) tested this hypothesis by preparing antisera specific to the various *t* complementation groups and monitoring the time course of appearance of the *t* antigens on embryos. (The antisera were prepared by immunizing mice with spermatozoa from heterozygotes of the various *t* lethals and then absorbing the antisera first with wild-type sperm and then with lymphocytes, the latter step to remove other nonspecific antigenic determinants.) The time course of appearance of the different antigens on heterozygous (*t*/+) embryos was then monitored by treating the embryos first with the specific sera and then with a fluorescent anti-mouse Ig antibody. Under these conditions, only embryos with the characteristic *t* antigen on their surface fluoresce.

For four mutants representing four different *t* lethal complementation groups (the t^{12}, t^{0}, t^{w5}, and t^{9} groups), t-specific antigens were detected on embryos as early as the eight-cell stage in all cases. By this test, the *t* mutant cell surface changes all appear early and without temporal relationship to the time of onset of the various developmental aberrancies.

The developmental switch hypothesis also requires the existence of wild-type alleles of the *t* "mutations." Much evidence now indicates that the *t* mutants differ from the wild-type in large blocks of DNA within the proximal region of chromosome 17, rather than in single-site mutations (Lyon et al., 1979). The absence of recombination between wild-type chromosomes 17 and *t* mutant chromosomes in this region reflects their absence of fine-structure homology. Instead of mutant alleles, therefore, particular *t* mutations are said to be "*t* haplotypes." Furthermore, the different developmental aspects of the *t* mutant phenotype—interaction with the *T* locus, embryonic lethality effects, enhanced transmission ratio in *t*/+ heterozygotes, male sterility in many double heterozygotes—are caused by different loci within the block of *t*-chromatin. It has been possible to dissect these different loci genetically because of the occurrence of rare crossovers in the proximal region between *t* haplotypes and marked wild-type chromosomes; the re-

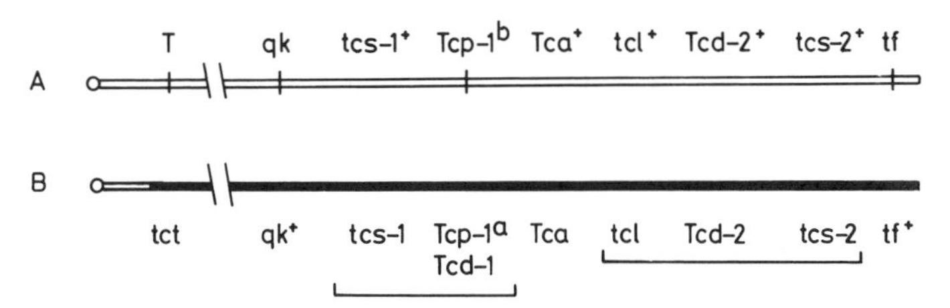

Figure 7.14 Map of the *t*-complex region in wild-type and complete *t* haplotype. The solid line represents *t* chromatin. *T, short-tailed*; *tct,* tail interaction factor; *qk,* quaking; *tcs-1* and *tsc-2,* proximal and distal sterility factors, respectively; *Tcp-1,* testicular *t* complex protein; *Tca* and *Tcd-2*, proximal and distal segregation distortion factors respectively; *tcl, t*-lethal factor(s); *tf,* tufted. (Reproduced with permission from *Genetic Research* (*Cambridge*), vol. 38, p. 116, Cambridge University Press; Silver, 1981)

combinant chromosomes bearing portions of the *t*-complex experience recombination with other *t* chromosomes at normal rates, because of their shared homology, and can be further analyzed (Silver and Artzt, 1981). A genetic map of the *t* region, and its wild-type analogue has been constructed by Siler (1981) and is shown in Figure 7.14. The results explain why no *t* mutants have ever been isolated from non-*t*-carrying populations in the laboratory. In contrast, several x-ray–induced mutants of *T*, which was once thought to be part of the *t*-complex, are known.

The developmental differences between *t* haplotypes reflect differences in the amount and sequence composition of *t* chromatin they contain. Indeed, the observations suggest an unusual origin of *t* chromosomal material, originally proposed by Dunn and Bennett in 1971, namely, that all *t* mutants share a common ancestral chromosome, introduced from a distinct and partially diverged mouse population. Because *t* mutants are found in all wild *Mus musculus* populations, the hypothesis demands that the progenitor chromosome has spread throughout the world via these populations. Such diffusion is explicable in terms of the genetic properties of the complex itself: recombination suppression would tend to keep the transmission-enhancing factors together with the lethality and sterility factors, with the former acting to keep the *t* chromosomes at high frequency within each population. Only the deleterious effects associated with the *t* chromosomes prevent them from driving the non-*t* chromosome out of the genetic pool.

The direct evidence that all *t* chromosomes derive from a single source is a major testicular protein encoded by a gene (*Tcp-1*) in the region of the *t* complex. All non-*t*-bearing mouse strains, including *T*/+ strains, produce a slightly basic form of the protein, while all *t* mutants possessing the proximal region of the complex produce an acidic form of this testicular protein (Silver et al., 1979).

The *t*-complex is therefore extremely interesting from the dual perspectives of population genetics and genome evolution. The basis of the developmental effects, however, remains elusive. The *t*-complex contains analogues of several genes within the comparable region of the wild-type chromosome (Paigen, 1980), yet homozygosity for particular sets of these *t* gene products produces distinct developmental aberrancies. In some sense, the gene activities of the *t*-complex are mismatched with the developmental activities that arise from the rest of the genome of *M. musculus,* in effect a form of hybrid lethality. Such lethality is usually believed to reflect a failure of the transcriptional regulatory apparatus to cope with gene products or regulatory sites that are slightly different (Wilson et al., 1974).

Many questions about the *t*-complex and *t* mutants remain. Solutions to the puzzle of how particular *t* lethals affect critical early transitions in the mouse embryo may prove highly informative about these transitions, but the concept that the complex consists of a set of switch genes that operate in normal development is probably incorrect.

TERATOCARCINOMAS: MODEL SYSTEMS FOR DIFFERENTIATION

Some of the most interesting events in postimplantation development are those of tissue allocation and determination and the beginning of tissue differentiation. Unfortunately, these are also among the events most inaccessible to the experimenter. The methods of clonal analysis furnish some information about tissue foundation, but the conclusions are necessarily indirect and tentative and, by their very nature, cannot elucidate the underlying mechanisms in these early cell assignments. On the other hand, direct biochemical analyses of the first commitments and tissue differentiations in primitive streak embryos are rendered difficult or impossible because of the limited amounts of material available.

In recent years, an alternative approach to early development has been made possible by the realization that EC cells, a class of neoplastic stem cell, may provide an analogue of the early totipotent cells of the mouse embryo (Martin, 1980). EC cells were initially identified as the stem cells of certain tumorous growths called "teratocarcinomas." As the name implies, these growths are both tumorous and "monstrous." First discovered in humans, they appear to be highly disorganized embryos consisting of a nearly random mixture of tissues and structures such as teeth and hairs. The first class of teratocarcinomas discovered in mice was reported by L. C. Stevens and his colleagues in the 1950s. These were testicular tumors arising at a high frequency in the inbred mouse line 129. Analysis of these growths revealed that they contain two classes of cells, the various differentiated cell types and a core of cancerous stem cell, the EC cells. Culturing of the tumors showed that the differentiated cells are benign (noncancerous), while the EC cells are neoplastic. In general, serial culture of teratocarcinomas in vivo leads to a loss of virulence expressed in the production of benign growths called "teratomas." However, the EC cells can be maintained as cell lines by serial culture in the intraperitoneal space. Such in vivo ascitic cultures maintain their malignancy and differentiative capacities indefinitely.

The first teratocarcinomas to be described were testicular tumors. However, EC cells can be derived from any mouse strain by transplantation of early (3- to 7-day) embryos to nonuterine sites, such as the kidney or testis capsule of the male adult. It is also possible now to derive EC lines directly from early embryos using in vitro culture (Evans and Kaufman, 1981). The frequency with which embryos give rise to teratocarcinomas is a function of their own genotype—different inbred lines giving different results—and of their maternal source, suggesting that there is a maternal effect in the potentiation to form EC cells (Damjanov and Solter, 1982). Furthermore, spontaneous ovarian teratocarcinomas occur with measurable frequency in one strain; these apparently arise from early oocytes. Teratocarcinomas can therefore develop from either male or female germ line cells, although always from diploid germ line precursors, or from early embryos. When the source is an early embryo, the neoplastic stem cells presumably originate either from the normal totipotent cells of the ICM or from the primary EmE. Regardless of source, all EC lines can be cultured in vitro (Martin and Evans, 1975). Depending upon its history and a variety of little understood factors, each line comes to be characterized by a very broad, intermediate, or narrow range of differentiative capabilities.

When cultured intraperitoneally, EC cells typically give rise to two kinds of embryonic bodies. The first, simple embryoid bodies, consist of a core of EC cells surrounded by a layer of endoderm (Fig. 7.15); superficially, they resemble the ball of ICM cells covered by endoderm, typical of the early blastocyst stage. The second class, cystic embryoid bodies, arise after more prolonged culture of EC cells and probably develop from simple embryoid bodies. They resemble the embryonic region of late egg cylinder embryos, but are substantially more disorganized and exhibit a mixture of differentiated tissue regions. Certain in vitro EC lines also differentiate in suspension, under the appropriate conditions, and produce both kinds of embryoid bodies. Other EC lines typically show differentiation in vitro in monolayers on the inner surface of the culture vessel.

Although EC cells resemble early embryonic cells and can be derived from them, there has been some debate as to which specific embryonic cell type they most resemble. From their patterns of synthesized polypeptides, they appear closer to primary EmE in developmental state than to ICM (Martin, 1980).

The most productive approaches to the characterization of early embryonic differentiation through EC cells have been immunological. In one procedure, rabbits are injected with EC cells to produce an anti-EC serum, and the purified antiserum is then applied to staged early embryos or to other EC lines. The presence or absence of the target antigen on the embryos or cells being screened is determined either by the detection of a biological effect or

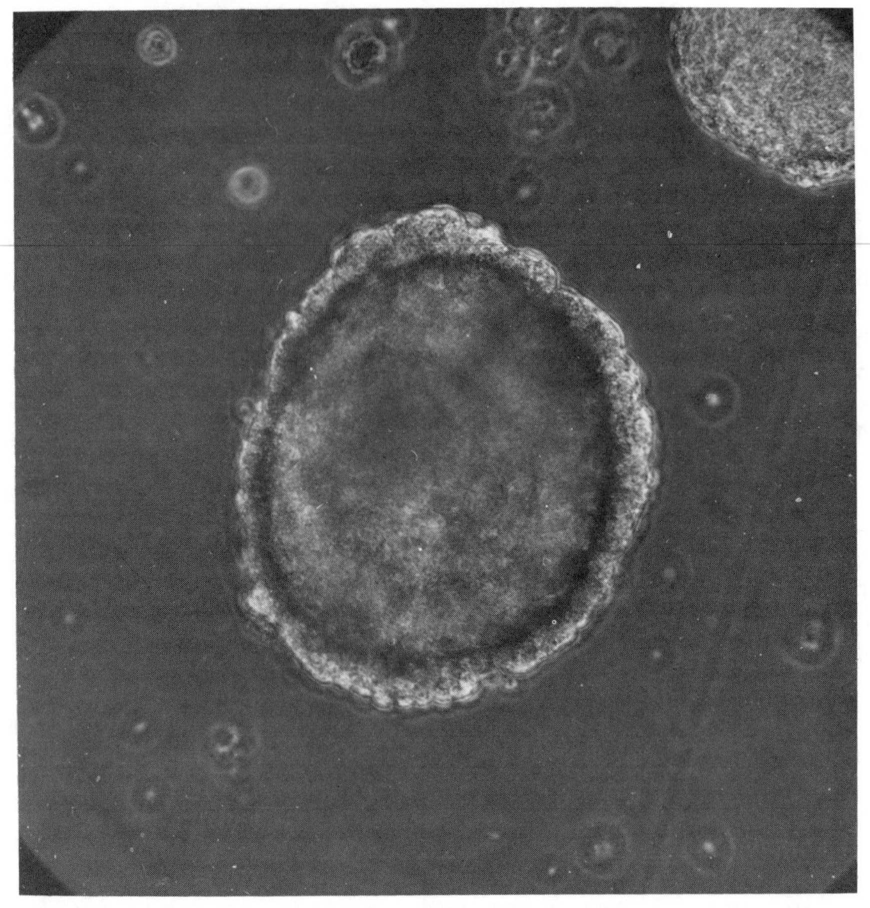

Figure 7.15 A simple embryoid body derived from EC cells. (Photograph courtesy of Dr. E. Robertson.)

by using fluorescein-conjugated reagents that recognize and bind to the first antibody. [The latter approach was used by Kemler et al. (1976) to detect *t*-specific antigens.] A different strategy is to prepare antibodies to purified components found only in certain differentiated cells and then to monitor the time course of appearance of these antigens, usually by means of the fluorescence assay, in differentiating cell lines.

An example of the immunological strategy for identifying events in early development using EC cells involved the production of an antiserum to one nullipotential line of cells, cell line F9; the F9 antiserum has served to identify a cell component required for compaction of the early embryo. This antiserum is complex, but its specific gamma globulin component detects a cell surface molecule present throughout early development. It is present on the surface of all cells of premorula embryos and in early and late blastocyst stages, becomes localized subsquently to the EmE, and finally disappears on the ninth day of development. In adults, the F9 antigen is found only on cells of the adult male germ line, including sperm cells. When early embryos are treated with the antigen-binding fragment of the gamma globulin fraction, compaction of the eight-cell embryo is prevented. Both the anti-F9 serum and several monoclonal antibodies that prevent compaction bind to a specific glycoprotein termed "uvomorulin" (Hyafil et al., 1981). Uvomorulin undergoes a calcium ion–mediated conformational change that is essential for the process of compaction; the antibodies that block compaction evidently prevent this conformational change. Although the structural gene for uvomorulin has not been identiied, two lethal *t* complementation groups block the appearance of F9 antigen (Kemler et al., 1976). These *t* mutations presumably interfere with either the synthesis or surface localization of uvomorulin.

From the standpoint of genetics, the interest of EC cells lies in their potential use as vehicles for genetic alteration of mouse development. This possibility exists because the cells of certain EC lines can participate in normal embryonic development when injected into blastocysts, contributing descendant cells to all tissues and organs. This ability also shows that the neoplastic state of the EC cell is reversible when the cell is placed in a normal embryonic environment.

If normal EC cells can be reinserted into the developmental program, EC cells that are mutant but viable in culture may be similarly capable of reintegration. If so, the way is open for the construction of new lines of mice bearing mutations induced in vitro. The only conditions for success are that the introduced mutant cells be capable of participating in somatic development and giving rise to germ line tissue in the viable chimeras. This procedure, if successful, would be considerably more efficient than standard mutant screening of whole animals because vastly more cells can be examined in culture for mutant clones than whole animals screened for mutant phenotypes.

The first demonstration that EC cells can participate in whole animal development was reported by Brinster (1974). His experiment involved the injection of *agouti* cells into blastocysts of an albino strain. The source of EC cells that Brinster used was the ascitic fluid of a tumor line derived from strain 129, serially propagated over many years by intraperitoneal injection. Of 60 surviving animals, each derived from a blastocyst injected with two to four EC cells, one animal showed a coat with partial gray striping. The source of this agouti color must have been the donor EC cells.

This particular animal did not sire agouti progeny when mated with an albino female—the dominant *agouti* allele would have been detected in any progeny from agouti germ line cells—and was therefore probably not a germ line chimera. However, a germ line chimera was subsequently obtained by Mintz and Illmensee (1975), who used the same ascites tumor as donor. They obtained three chimeric animals, detected by patches of donor coat color, which were chimeric in their red and white blood cell–forming tissues and in their livers (as shown by the presence of donor cell GPI). One of these mice, a male nicknamed "Terry Tom," sired progeny bearing tumor cell line markers and was therefore a germ line chimera.

The 129 cell line is an in vivo line of EC cells. Using EC cells to transmit new mutations requires the utilization of EC cells from individual cell clones, and such clones can be obtained only from in vitro EC cell lines. Somewhat surprisingly, the great majority of such cell lines, when tested, including those with large pluripotential somatic developmental capacities and lacking any detectable chromosomal abnormalities, have failed to give germ line chimeras. Either these cell lines are genetically deficient in one or more ways or the PGCs derived from them fail to migrate to the gonadal ridges in sufficient number.

However, successful transmission of an in vitro cultured cell line, METT-1 (for "mouse euploid totipotent teratocarcinoma-1") has been reported by Mintz and Stewart (1981). The strategy is diagrammed in Figure 7.16. The cell line, also derived from strain 129, shows a normal karyotype and is chromosomally female; it was repeatedly passaged in vitro before being injected into blastocysts of a host nonagouti (black) strain. Measured either in terms of the percentage of visible chimeras (as scored by the frequency of agouti contribution to the coats of the surviving animals) or by the extent of the contribution of donor genotype cells to internal tissues (as shown by the GPI marker), this cell line is significantly more normal than other previously tested in vitro lines. Of nine coat color female chimeras, one proved to have strain 129 oocytes, giving rise to three agouti F_1 progeny. These three animals in turn gave rise to agouti F_2 progeny.

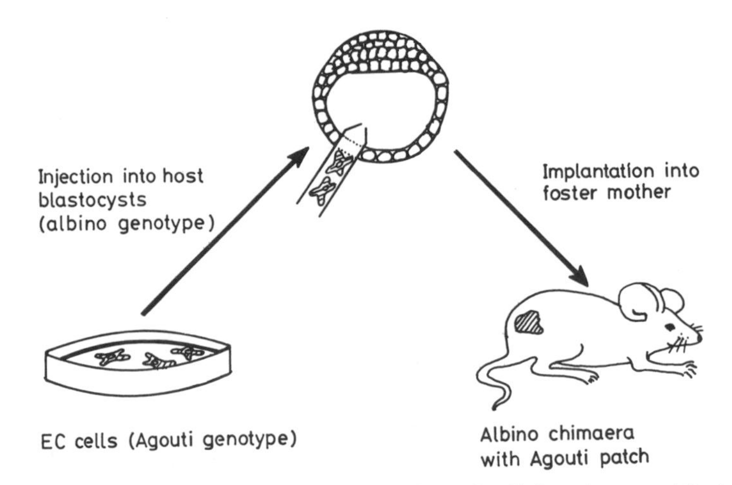

Figure 7.16 Transmission of *agouti*-genotype in vitro EC cells into the normal fetal development pathway.

A similar approach with embryo-derived pluripotential (EK) cells may be even more efficient. Bradley et al. (1984) report that over 50% of surviving blastocysts injected with EK cells are visibly chimeric. Of these, approximately one-third proved to be germ line chimeras. Induction of new mutations in EK cells for passage into the germ line promises to be superior to the use of EC cells in this respect.

It may soon be possible to transmit mutations, induced in EC or EK cells, into new strains of mice. The principal challenge may be in devising selection procedures for isolating mutant cell lines in vitro that show interesting developmental phenotypes when integrated into developing embryos. However, the development of direct modification of the germ line through transformation of the early embryo, including the ability to induce mutations with the inserted genes (Chapter 9), provides a simpler method of producing genetically modified offspring than the EC cell route.

COAT COLOR: DEVELOPMENTAL BIOLOGY AND GENETICS

The hereditary basis of coat color is the oldest problem in mammalian genetics. Genetic investigtion of coat color predates the rediscovery of Mendelian genetics in 1900, but the modern analysis of coat color genetics begins with the work of Cuénot, Castle, and Wright in the first two decades of this century. The work of Wright on the genetics of guinea pig coat color was particularly important; it was the first attempt to use genetic differences to probe the biochemistry of development. The reason for this early attention to the coat color of rodents is its ready accessibility to genetic analysis. Although the mammalian coat still conceals many of its secrets, the broad outlines of its biology are now visible. Excellent, detailed treatments of the field can be found in Searle (1968) and Silvers (1979).

The mammalian coat serves two functions. The primary one is insulation. By constituting an efficient air-trapping layer next to the skin, the packed hairs of the coat provide substantial protection against thermal loss. (The feathers of birds furnish similar protection.) The coat typically consists of two classes of hair: the shorter and thinner underhairs—about 80% of the total hair on a mouse—and the longer, thicker overhairs. Within each category, several different hair types exist. The underhairs collectively provide most of the direct insulating capacity of the coat, while the chief function of the overhair layer is to protect the underhair coat. In addition, some overhairs have a specialized sensory role, as do the vibrissae ("whiskers") of the mouse.

The second major function of the coat is to serve as a visual signaling device; in this role, color becomes crucial. The color pattern serves as a recognition device for other members of the species, a function that is particularly important in mate selection. In some species, for example, the skunk, it serves additionally as a warning signal to other animals. In some mamma-

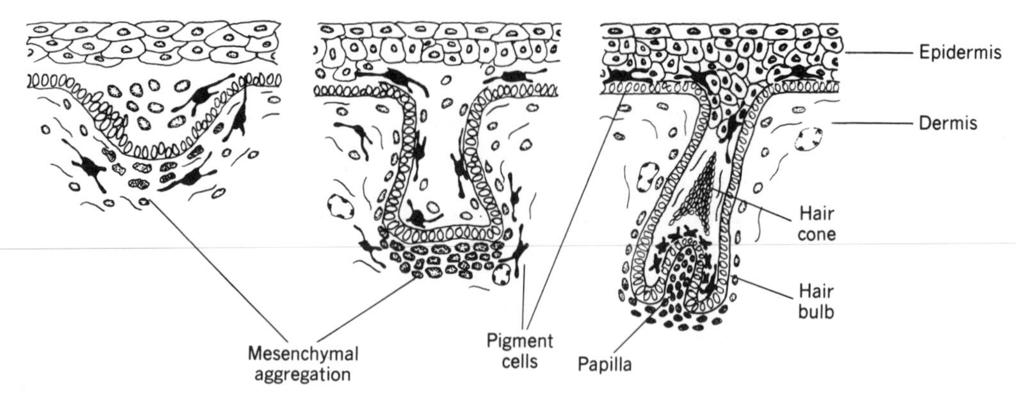

Figure 7.17 Sequence of hair follicle development in the mouse coat. (Figure kindly supplied by Dr. W.K. Silvers; from Silvers, 1979.)

lian species, color serves the function of concealment through camouflage, as in the snowshoe rabbit.

All hair shafts develop from multicellular structures termed "follicles," which arise late in fetal development as inpocketings of the epidermis into the dermis. The process of hair follicle development is diagrammed in Figure 7.17. As the epidermal cells push down to form an inverted blunt-ended cylinder, mesenchymal cells of the dermis aggregate at the base or hair bulb. These mesenchymal cells form a thickened papilla at the end of the follicle that is partially enclosed by the epidermal cells of the hair bulb. It is from the hair bulb that the shaft originates. Each hair consists of an outer cylinder of thin, compressed cells, the cortex, and an internal set of cells, the medullary cells, separated by spaces within the shaft of the hair (Fig. 7.18). Both cortex and medullary cells, but especially cortex cells, are rich in the structural protein keratin. Both cell types are also repositories of pigment. These melanic pigments are secreted directly into the cells of the hair shaft by melano-

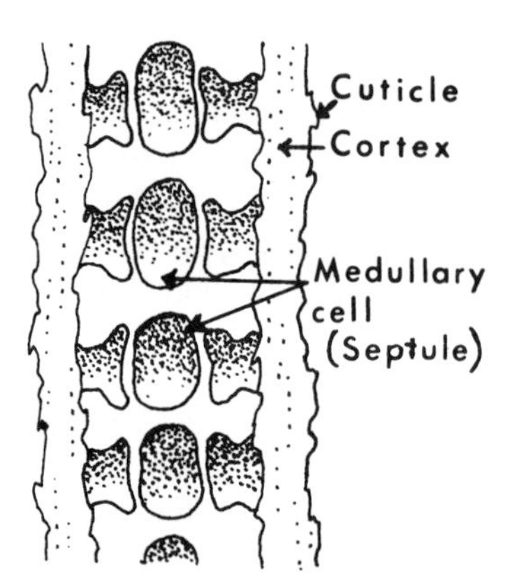

Figure 7.18 Internal structure of a hair shaft in the mouse coat. (Drawing courtesy of Dr. W.K. Silvers; reproduced from Silvers, 1979.)

cytes located in the region of the hair bulb. Pigment is transferred in the form of small pigmentary granules termed "melanosomes" to the basal cells of the shaft throughout the period of hair growth.

The developmental biology of melanocytes is complex. As discussed earlier, the melanocytes of the skin are derived from the migratory melanoblasts of the neural crest. They arrive at the dermis–epidermis interface at about day 12 of development and form a network of interconnected cells throughout the epidermis. Each active melanocyte eventually transfers its melanosomes either into cells of a developing hair shaft or into neighboring epidermal cells that are not part of a follicle. The transferred melanosomes are complex entities in their own right; each consists of a fibrous protein granule attached to a melanic pigment. The pigments are of either of two kinds: eumelanin and pheomelanin. Eumelanin is the prototypical black melanic pigment, while pheomelanin is yellow. Both are synthesized from tyrosine in a series of sequential oxidations carried out by tyrosinase. Pheomelanin differs chiefly from eumelanin in having attached cysteinyl groups. Maturation of the protein moiety of the melanosome takes place gradually, and only mature melanosomes are transferred to epidermal cells. The final color of a hair or hair subregion is a function of which pigment is deposited, the number and spatial arrangement of the melanosomes within the cells of the hair shaft, and the shape and size of the melanosomes.

Given the developmental and biochemical complexity of hair construction, it is not surprising that many genes directly or indirectly exert effects on the color, color pattern, or structure of the coat. To date, more than 50 loci have been identified on the basis of their mutant effects on coat color; a large proportion of these loci have pleiotropic effects and for these, the relationship between color and other mutant phenes is unknown.

Only a minority of the loci identified on the basis of their coat color mutant phenotype are directly involved in the synthesis of the melanic pigments. Of the few genes in this category, the albino or *c* locus is the most important. The mutants of the *c* locus form a phenotypic series whose effects range from only a mild diminution of the pigments to a complete absence of melanin. This last is the original *c* mutant, characterized by a completely white coat and pink eyes. In standard genetic backgrounds, the wild-type *C* allele is dominant to all other members of the series. However, in the presence of certain nonallelic genetic differences that affect coat color, *C*/*c* animals can be observed to make less pigment than *C*/*C* homozygotes.

The *albino* locus is almost certainly the structural gene for tyrosinase, the key enzyme in melanin synthesis. Not only is there a correspondence between the severity of the phenotypic defect and the amount of tyrosinase in the allele series, but a particular *c* mutant, the Himalayan variety (c^h), which is distinguished by a phenotypic temperature sensitivity in pigmentation—such that the extremities of the animal, which are always cooler, are darker—possesses a correspondingly temperature-sensitive tyrosinase activity (Coleman, 1962). (Siamese cats, no near relation of the Himalayan mouse, have a similar problem.) Unlike the mutants of several other genes whose hair follicles lack melanocytes, the *c* mutant has normal numbers of melanocytes within the follicles, but they are devoid of melanosomes.

A second locus, *brown* (*b*), is also involved in the biosynthesis of melanosomes. Mutants of this gene cause the replacement of the black granules with brown melanosomes. The effect of mutations in *b* appears to be on the proteinaceous granules that bind the melanic pigments. When examined microscopically, the melanosomes of *b* mutants are seen to be smaller than those of the wild-type; the visible alteration in color is a direct consequence of this change in melanosome size.

However, the great majority of coat color mutants produce their effects on color indirectly, either through interference with melanocyte development or morphology or through subtle influences on the interactions of melanocytes with the other cell types in the hair follicle. Examples of two genes that affect color through an effect on melanocyte morphology are the loci *dilute* (*d*) and *leaden* (*ln*). Mutants of both genes exhibit an apparent diminution of pigment intensity, the reduction depending upon the allelic composition of the other pigment genes. Thus, homozygosity for *d* causes mice that would otherwise be black to be Maltese blue and those that would be dark brown to be light brown. The dilution effects for mutants of both genes stem from a similar change in melanocyte morphology, although the precise biochemical basis of the change probably differs between *d* and *ln* mutants. Normal melanocytes have a highly branched, or dendritic, morphology that facilitates the transfer of melanosomes. In the mutants, the melanocytes are much more compact. In consequence, melanosome transfer takes place in an irregular, bottlenecked pattern of release. The individual hairs of both mutants show large aggregations of pigment that are unevenly spaced within the hair shafts. Although the macroscopic impression is one of a reduction in pigment, there is as much pigment present as in the wild-type.

A second group of mutants that are indirectly affected in pigmentation are the various white-spotting mutants. Some are dominants, producing the spotting phenotype in heterozygotes; others are recessive. Two of the white-spotting mutants show severe anemias and partial sterility. Although white spotting looks like regional albinism, the cause is fundamentally different: hair follicles in the white areas lack melanocytes. In some mutants, the fundamental defect may be in the neural crest, involving a failure of formation or of migration of melanoblasts. In others, the melanoblasts migrate but die en route. The latter defect has been studied in chimeras and found to occur in the melanoblasts of the mutants, suggesting that the mutation causes autonomous cell death (Mintz, 1974).

Perhaps the most interesting locus affecting coat color is the *agouti* gene because of the range and patterns of color associated with its various alleles. Seventeen alleles of the locus have been identified, ranging from two causing dominant yellow expression (one, the classic lethal, the other a homozygous viable) to various black or dark brown recessive alleles. Thus, in an otherwise wild-type background, A^{y} and A^{vy} produce a yellow coat; the A, agouti, standard allele, produces a gray coat; and a^{e}, extreme nonagouti, produces a completely black coat in homozygotes. The *agouti* locus evidently governs, in some fashion, the balance between pheomelanin and eumelanin synthesis. A diagram of several of the *agouti* allele hair color phenotypes is shown in Figure 7.19. A general feature of these phenotypes, except for the extreme black-pigmented form, is that hairs on the ventral side are always lighter than those on the dorsal side.

The puzzle of the *agouti* locus action is revealed by considering the color pattern of individual hairs in the standard A strain. These hair shafts have black tips, yellow shafts, and black bases. The apparent grayness of the coat in the A strain results from this internal distribution of the two pigments. The transitions between black and yellow in the hair are not abrupt but take place over the length of three to four medullary cells. Indeed, some medullary cells in the transition zone contain both pheomelanin and eumelanin granules. This finding suggests that the gene somehow affects the synthetic behavior of the melanocyte rather than its intrinsic synthetic capacity for the two melanic pigments.

In fact, the *agouti* locus acts in the epidermal or dermal cells of the follicle rather than in the melanocytes, as shown by the creation of skin chimeras in grafting experiments. When embryonic or neonatal skin is grafted between strains, the developing follicles are populated by host melanoblasts. In all

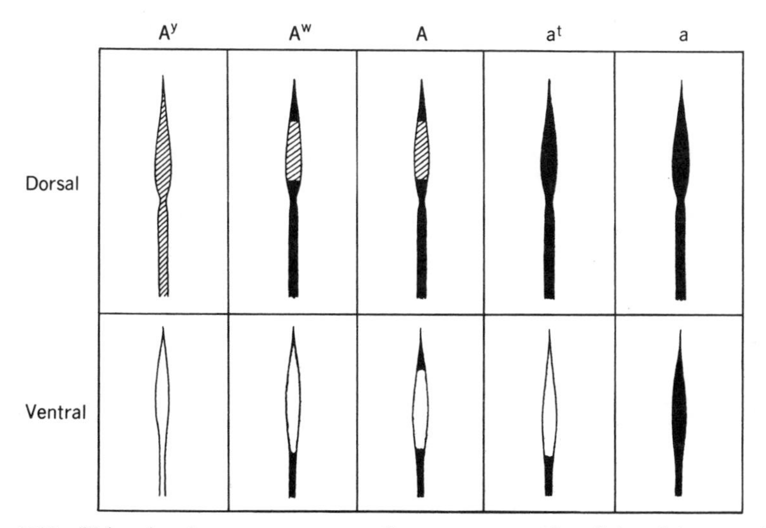

Figure 7.19 Hair color phenotypes corresponding to representative alleles of the *agouti* locus. (Figure kindly provided by Dr. A. Searle; from Searle, 1968.)

cases, the phenotype of the developing hairs, with respect to *agouti*, is characteristic of the genotype of the grafted skin and not of the host melanoblasts.

Thus, if one grafts skin of nonagouti, *a*/*a* (black) to A^y (yellow) hosts, the developing skin graft produces a patch of black hair, although the hair follicles of the graft are populated with A^y-genotype melanoblasts (Silvers and Russell, 1955). The results do not reflect the action of an indigenous population of melanoblasts carried along in the skin grafts because the same result is obtained when the skin graft is taken from white patches of white-spotting mutants (which lack melanoblasts); the color phenotype of the developing hair on the graft still follows that of the *agouti* allele of the graft.

Is the instruction affecting pigment synthesis given by the epidermal or dermal cells of the follicle? One can answer this question by making reciprocal dermis-epidermis hybrid grafts using embryonic skin. In such experiments, skin is taken from 13- to 14-day embryos of both *agouti* genotypes and separated into its two components by light trypsinization. The reciprocal hybrid grafts are then made by recombining the layers between the two genotypes, allowing a day for the grafts to heal, and then transplanting them individually to the testis of newborn males (a particularly favorable location for growth of the new grafts). If the hair phenotype follows the dermal genotype, one concludes that the gene acts in the dermis. Alternatively, if the phenotype follows the genotype of the epidermis, the latter must be the site of gene action.

The observations have been surprisingly complex. For some *agouti* allele combinations, the dermis seems to be crucial, but for two others, which produce a "mottling" phenotype (A^{vy} and a^m), the hair phenotype follows the genotype of the epidermis. Poole (1980) has suggested that the immediate instructions to the melanocytes are given by the epidermal cells of the follicle but that the epidermal cells receive their orders from the dermis. In those genotypic combinations in which the epidermal cells govern the hair phenotype, the epidermis may have "set," in other words, lost its competence to respond to a new dermal signal.

The complete set of results shows that the final pigmentation pattern is the product of a subtle interaction between two very different cell types, dermal cells and melanocytes. As we have seen, these cell types have distinctly different developmental origins. In fact, chimera patterns can reveal such differences in origin and sites of gene action. The melanocytes are ectodermal and derive from the melanoblasts of the neural crest; the major part of the body dermis is mesodermal and derives from the somites. Because the melanocytes and somites have different tissue origins, genes expressed in one of these hair follicle components but not the other show very different spatial patterns of expression from one another in chimeras. All the melanocytes of skin, as we have seen, may ultimately trace their origins to 34 melanoblast clones in the body; what is certain is that the coat patterns of melanoblast chimeras can be reduced to a basic archetypal pattern of 17 stripes on each side. In contrast, agouti-nonagouti (gray-black) chimeras show a different and much more finely striped pattern (McLaren, 1976a). Examination of large numbers of these animals suggest that there may be as many as 85 basic stripe widths per side, including about 18 in the head region, altogether a much larger number than seen in the comparable melanoblast-generated patterns.

Two interpretations of the agouti chimera pattern have been offered. The first (Mintz, 1974) is that each unit stripe derives from a hair follicle progenitor clone. Apart from the 18 or so in the head, there are about 65 per side for the body and tail, a number that matches the somite number for these regions. In effect, each somite may contain one hair follicle progenitor clone. (In this view, the source of the stripes on the head raises the interesting possibility that there are rudimentary "invisible" head somites, a subject that has been debated by mammalian developmental biologists.) The other interpretation of the fine-grained pattern is based on the observation that there is much greater regularity of striping in these hair follicle chimeras than in melanocyte chimeras. McLaren (1976a) suggested that this regularity signifies a systemic patterning of some kind that is superimposed on the chimeric skin rather than an alternation of genetically determined clones. On the basis of random clonal placement, such regularity is unexpected. That there are *some* systemic influences that regulate *agouti* locus expression is apparent from the consistent dorsoventral differences in pigmentation within single genotypes (Fig. 7.19).

Solution of the various puzzles associated with the *agouti* locus ultimately requires determination of what the gene product does. Earlier results were interpreted to mean that the gene regulates the pheomelanin-eumelanin decision through control of the synthesis of sulfhydryl (SH)-containing compounds, with higher levels of SH compounds favoring eumelanin synthesis (Cleffman, 1963). However, this hypothesis has received little subsequent support (see Silvers, 1979, pp. 22–23). The action of the *agouti* gene product remains a puzzle.

Apart from the systemic effects on the expression of *agouti,* the idea that there are general systemic influences on the development of coat color pattern is inescapable from the fact that many mammals show a species-specific coat pattern of spots or stripes, or sometimes both. Some common examples are zebras, giraffes, and many members of the cat and squirrel families. In contrast to the color patterns of genetic chimeras, constructed in the laboratory, or those of X inactivation mosaics, each of these species-specific coat patterns is produced in animals whose cells are essentially identical in genotype. The question of the origins of such patterns is therefore one of developmental physiology rather than developmental genetics (although the comparative genetics of the different species' patterns is an interesting matter). In formal terms, these patterning processes can be thought of as morphogen systems acting throughout the presumptive field of pigmentation in the embryo or fetus. The morphogens may either suppress or enhance pigment formation. A two-morphogen model that produces a number of the typical spotting and striping patterns, and a general discussion of the problem, can be found in Bard (1981).

SEX DETERMINATION

Sex Chromosomes and Gonadal Differentiation

The ultimate source of the differences between male and female mammals is the sex chromosomes. As in the fruit fly, females are the homogametic sex, being XX, and males are the heterogametic sex, being XY. Unlike *Drosophila,* the crucial difference is the possession or absence of the Y rather than the number of Xs. Thus, female XO mice are phenotypically female and fertile (in humans, the XO condition produces sterile females) and XXY animals are phenotypically male, the reverse of the fruit fly situation. Evidently, in mammals, the Y carries one or more sex-determining factors, and maleness is the dominant sexual phenotype (in the biological sense); in the absence of the Y, femaleness results.

There is a further difference in the sex determination mechanisms. In *Drosophila,* sexual phenotype is determined autonomously, cell by cell, whereas in the mouse, gonadal sexual determination is primary, with all other aspects of sexual phenotype following from it. Secretion of the hormone testosterone by the male gonad triggers male genital development; in the absence of testosterone, the gonad secretes estrogens and the secondary sex characteristics that develop are female. This mechanism is general in mammals.

The first differences in gonadal differentiation become apparent at about 12.5 days p.c., 2 days after the arrival of the first PGCs. The male gonad begins to organize itself into testis cords, consisting of solid strings of spermatogonial cells encased in mesodermal somatic cells. The early female gonad retains a compartmented appearance, with groups of oogonial cells surrounded by a matrix of mesodermal cells. The urogenital system undergoes an accompanying differentiation. Initially, both sexes possess both kinds of urogenital structures, the female Mullerian ducts and the male Wolffian ducts. In each sex, one duct system develops while the other degenerates.

Although the primacy of gonadal differentiation in setting sexual phenotype is clear, it is not known with certainty whether the initial events occur in the gonadal soma or the germ line. In the chick, it has been possible to show that the soma of the gonad plays the crucial role: when PGCs are prevented from colonizing the genital ridge, the gonad nevertheless develops according to the sex chromosomal composition of the (somatic) ridge. In mammals, this kind of unequivocal demonstration has not yet been possible (McLaren, 1981a). What can be done is to construct chimeras, and analyze for correlations between phenotypic sex and the respective chromosomal composition of the somatic and germ line tissues of the gonad. Because this procedure almost never produces an animal that has reverse compositions for the two components, the interpretations are always uncertain.

However, the chimera results show two things. The first is that most of the XY/XX chimeras (which comprise about half of the experimental animals) are male, a consequence of the dominance of maleness. Above a certain threshold of male (XY) cells, perhaps as little as 30% (McLaren and Monk, 1982), a chimeric gonad tends to develop as a testis, while around that threshold, an ovotestis (a mixture of testicular and ovarian material) forms. Development of a testis then brings in train the development of secondary male sex characteristics.

The other finding of interest is that in XY/XX chimeras, regardless of the phenotypic sex of the gonad, there is little or no conversion of germ line cells of the "wrong" composition into functional gametes. For XX germ line cells developing within a testis, the prohibition seems to be absolute and may reflect an absolute block of spermatogonial development in cells with two X chromosomes (Lifschytz and Lindsley, 1972). For XY cells, transfer of germ line cells out of the somatic gonad before these cells encounter the normal meiotic block typical of the developing male gonad can sometimes permit oocyte development. Possession of the Y chromosome, therefore, is not an intrinsic block to oocyte development. [The respective roles of the sex chromosomes in gamete differentiation have been most recently reviewed in McLaren (1983b).]

The probable sequence of events therefore is an initial commitment in the somatic portion of the gonad created by the sex chromosome composition of that gonad, followed by a distinctive course of gamete maturation, and finally by the entrained development of secondary sex characteristics by the secreted male or female sex hormones. The chimera studies are discussed in McLaren (1976a, chap. 6) and the broader developmental biology in McLaren (1981a).

Mutants and the Role of the Y Chromosome

Sexual development in animal systems with sex chromosomes involves two stages, an initial phase of sex determinative decisions and an ensuing period of sexual differentiation. Both sets of processes are subject to genetic derangement. Because of the temporal separation between the initial event (gonadal determination) and the subsequent differentiation events, it is possible to make informed guesses as to which process has been genetically altered.

In mammals, the differentiation period is dependent on the continued presence of the secreted sex hormones: if androgen production is stopped by removal of the testes in male fetuses, the further course of somatic sexual development becomes shifted toward the production of female characteristics. Similarly, mutations that interfere with either gonadal sex hormone production or hormone binding would affect the later stages of sexual development.

Although mutants deficient in hormone production have not been reported, a sex differentiation mutant, characterized by insensitivity to testosterone, has been described. It is the X chromosomal testicular feminization (*Tfm*) mutant, and its phenotype is that of partially feminized males. (The mutation has no effect in females.) XY mice that carry *Tfm* develop testes, but these are small and underdeveloped; externally, the animals resemble females, although not perfectly (Lyon and Hawkes, 1970). The animals have normal testosterone levels, and administration of neither testosterone nor dihydrotestosterone rescues the male phenotype. Evidently, the defect is not in testosterone production but in the response to testosterone. Similar genetic syndromes have also been reported in rats and humans. In all cases, the specific defect appears to be a deficiency of the androgen receptor protein, which is normally distributed in all or most cell types (Ohno, 1976).

The most dramatic effects of *Tfm* are on secondary sexual characteristics, but there is also a block to spermatogenesis. In *Tfm*/Y individuals, spermatogonia are prevented from giving rise to spermatocytes, the immediate precursor cells of the sperm. To determine whether the effect is a direct one in the germ line or a secondary effect of an inappropriate gonadal soma, Lyon et al. (1975) constructed chimeras of *Tfm*/Y and +/Y embryos and then mated the mature chimeric males with *Tfm*/+ females. Some of the progeny obtained were found to be *Tfm*/*Tfm* embryos, showing transmission of some *Tfm*-carrying sperm from the male chimeras. The result indicates that the presence of some *Tfm*$^+$ in the gonadal soma can rescue spermatogenesis in some of the *Tfm* spermatogonia, and that the effect of the mutation on spermatogenesis is in the soma rather than the germ line.

Sex determination mutants should differ from sex differentiation mutants in having transformed gonads as well as transformed secondary sex characteristics. Although hunts for sex determination mutants in mammals are infeasible, because large numbers of animals cannot be screened, a number of putative autosomal and X chromosomal sex determination mutants have been identified in mice, goats, wood lemmings, and humans. In mice, several genetic conditions that affect sex determination have been identified (Eicher, 1983; Washburn and Eicher, 1983). However, the precise genetic basis of most of these conditions remains ill-defined, including the number of loci involved. For several, the effect involves an aberrant interaction between the Y and the autosomes, where the Y has been introduced from other subspecies or lines, resulting in a partial feminization of XY individuals. The genetic basis of these syndromes is obscure, but their existence shows that simple possession of a Y is insufficient to guarantee normal male development—that Y–autosome interaction in some manner is essential for this development.

Of the several murine genetic conditions that alter sex determination, that of *Sex reversal* (*Sxr*) is the best understood. It was first described by Cattanach et al. (1971) as a dominant mutation that transforms XX animals into sterile males. Because of this sterility, the initial mapping had to be carried out in crosses between normal females and XY *Sxr* carrier males; the results indicated that the condition was inherited independently of both the X and Y chromosomes. In consequence, the authors concluded that the trait was autosomal. However, all subsequent attempts to localize it to a particular autosome met with failure.

The solution of the *Sxr* puzzle was made possible by the use of a molecular marker specific to the Y chromosome. This marker is a particular satellite DNA that is unique to the Y in mammals but was first isolated not from mice or humans but from a snake, the banded krait. More surprisingly, it was isolated from the heterochromatic *female*-specific W sex chromosome of snakes. In the banded krait, as in most snakes, sex determination is carried out by the ZW-ZZ sex chromosome system. In this system, males are the homogametic sex, with two euchromatic Z chromosomes, and females are the heterogametic sex, with a ZW constitution. By looking for a satellite DNA species unique to the heterochromatic W chromosome, Singh and Jones (1982) isolated such a satellite, which they designated Bkm (for "banded krait, minor satellite").

The Bkm satellite is found throughout eukaryotes, from yeast to humans, although in varying amounts and different locations. The fact that it is found on the female-specific chromosome of snakes and on the male-specific chromosome of mammals implies that it does not determine femaleness or maleness per se but that it is a marker of sex chromosomes. When radioactively labeled Bkm satellite is hybridized in situ to metaphase spreads of chromosomes from XX *Sxr* animals, it is found to be located at the distal tip of one of the X chromosomes (Singh and Jones, 1982). The result indicates that the *Sxr* condition derives from the translocation of Y chromosomal material to the distal tip of the X, presumably by recombination during meiosis. The transfer of Y chromosomal material to the X by a high-frequency recombination event explains the absence of genetic linkage between the Y and the *Sxr* condition found in the earlier studies.

Eicher (1983) has proposed a hypothesis for the event. The idea is that the *Sxr* condition is characterized by a Y chromosome carrying two sex-determining, Bkm-containing regions, and that transfer of the more distal region from one chromatid to the X invariably occurs during meiotic pairing in the XY *Sxr* animals. The model is diagrammed in Figure 7.20. Note that the process, as diagrammed, produces equal numbers of X^{Sxr} and normal Y chromosomes. Indeed, XY *Sxr* carriers give rise to normal males in addition to transformed XX females and carrier XY *Sxr* sons (who bear the nonrecombinant, duplication-bearing chromatid). A similar explanation has been proposed by Burgoyne (1982).

The central question about mammalian sex determination concerns the manner in which the Y chromosome confers maleness. The general belief is

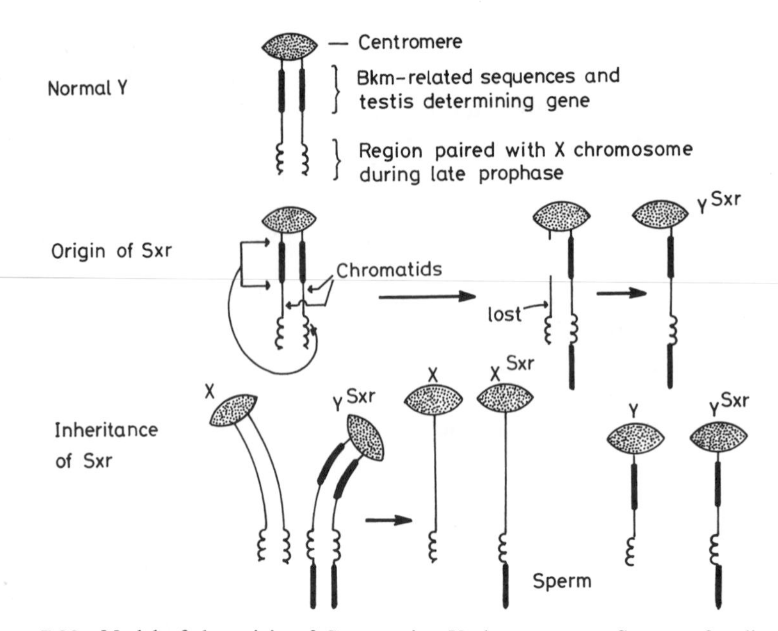

Figure 7.20 Model of the origin of *Sxr*-carrying X chromosomes. See text for discussion. (Reproduced with permission from Eicher, 1983; copyright Animal Reproduction Laboratory, Colorado State University.)

that the Y carries one or more genes that actively promote testicular development. Much speculation has focused on the possibility that one of the known male-specific antigens might play this role and be encoded by genes on the Y. One of these is detectable as a transplantation antigen, causing the rejection of male skin in females of the same inbred strain. This antigen has been dubbed the "H-Y antigen" (for the Y histocompatibility antigen) and is both highly conserved and male specific throughout the vertebrates. The possibility that the H-Y antigen is encoded by a gene on the Y and is the male-determining factor was originally proposed by Ohno (1976). A second antigen that is male specific was subsequently discovered in sera and designated SDM. For a long time, it was believed that SDM and H-Y were the same antigen. It is now apparent that they are not. More importantly, the current evidence shows that neither H-Y nor SDM is invariably associated with phenotypic maleness, nor is their absence invariably correlated with femaleness (Silvers et al., 1982). These antigens are therefore probably consequences rather than causes of male determination. The nature of the relationship of the putative male-determining factors to the Bkm satellite also remains to be elucidated. The Bkm sequence may contain protein coding functions but these are presumably not directly male-determining because Bkm is not invariably associated with maleness in vertebrates. If there are male-determining, protein-coding functions on the Y, they are presumably closely linked to the Bkm satellite.

A different view of the role of the Y in setting maleness has been proposed by Chandra (1984), who suggests that it is purely passive: to absorb repressor molecules, thereby permitting the function of an X chromosomal gene essential for testicular determination, designated *Tdm*. In this view, the Bkm satellite might be the male determinant itself by functioning as the repressor binding sequence. Chandra's specific proposition is that an autosomal locus encodes a repressor that binds with high affinity to the repeated Y chromosome sequence and with lower affinity to *Tdm*. In animals possessing one or more Y chromosomes, the repressor is bound by these chromosomes and *Tdm* expression is turned on; in XX animals, there is no sink for repressor and *Tdm* is repressed. Although this model is at best an approximation, it fits most of the known facts and provides a provocative alternative to the conventional view of the Y chromosome's role in sex determination.

Superficially, sex determination in mammals appears to operate by very different rules than in fruit flies or nematodes. The incontrovertible difference is that in fruit flies, the decision is made autonomously in all cells capable of a sexually dimorphic response, while in mammals the key sex determination events occur in one organ, the gonad, with all of the remaining sexual dimorphism following from this first step. There may, however, be certain underlying unities. In both organisms, the decision is a binary one in the cells that make it—either a "yes" or "no" to development along the pathway of the dominant sex. In addition, the key sex chromosome (the X in fruit flies, the Y in mammals) interacts with one or more autosomal factors in making the sex determination decision. These autosomal genes have been identified in *Drosophila,* while those of mammals are still obscure. However, the general logic of such an interaction, if not the precise molecular details, may prove to be surprisingly similar. The future of this problem is certain to be interesting.

PROSPECTS

The development genetics of the mouse can be divided into three phases. In the first, which lasted for approximately 60 years, the primary emphasis was on studying the developmental effects of particular mutations—the discipline of classical developmental genetics. The second phase occupied much of the 1960s and 1970s, and emphasized chimera studies as probes of cellular fates and capacities in development. The third phase is just beginning and promises to be the most informative: the detailed genetic and molecular analysis of genetic changes in the mouse and their developmental consequences.

Our knowledge of genetic control of development in mammals is rudimentary at present. In consequence, one cannot even begin to frame the kinds of questions about control that one can ask about *Drosophila* or *Caenorhabditis*. One obvious difference in mutant effects between these two invertebrates and the mouse is the absence of homeotic mutants in the latter. Of the 500 or so mouse mutants that are known, none gives an obvious homeotic effect. In any randomly chosen collection of 500 fruit fly mutants, in contrast, the number producing some homeotic effect would be significant. One can posit at least three possible reasons for this difference. The first is that fruit fly homeotics stand a much greater chance of surviving until adulthood because their effects are often principally in imaginal structures, unneeded until the pupal period, while mouse homeotics are early embryonic lethals, for the most part, and do not survive long enough to permit detection of their homeotic effect. The second explanation is that many of the homeotic effects in *Drosophila* reflect some form of regulative replacement; the mouse may lack such replacement mechanisms. The third explanation is that there is a genuine difference in the underlying genetic regulation between the two organisms, with blocks of recognizable morphology being under single-gene control in the fruit fly but under multiple regulatory elements in the mouse. There is not enough information at present to allow us to decide between these possibilities.

In fact, future progress will require both more information on existing mutants and more genetic source material. In the first place, there is a strong need for a thorough characterization of the existing mutants by the complete armory of observational and molecular techniques. The analysis of the *albino* deletion mutants provides a model for this kind of study. The other side of this analysis is that of the genes, and the mutational lesions themselves, by the methods of recombinant DNA analysis. To understand a mutant effect, it is almost never enough to know the developmental phenomenology; the character of the mutation itself must be understood.

At the same time, new mutants must be obtained to expand the available genetic source material. These mutants will become available through more efficient methods of chemical mutagenesis and DNA insertional mutagenesis. The latter method, which will be described in Chapter 9, not only generates new mutants but allows the subsequent purification of the altered gene. The mutant gene can then be used to isolate the normal gene from the DNA pool, permitting a molecular characterization of both the wild-type and mutant forms.

The new approaches promise a dramatic expansion of opportunities for investigating the genetic basis of mammalian development. The developmental genetics of the mouse is, in a certain sense, only just beginning.

EIGHT

PATTERN FORMATION

The way in which the activities of the cells of an embryo are coordinated in space and time is one of the great remaining enigmas of biology, and it is a problem which has not, so far, benefited significantly from recent advances in the methodology of molecular biology. Instead, our efforts at understanding the formation of spatial pattern in developing animals are still comparable to the pre-Mendelian stage of genetics; we are still searching for the formal "rules" by which we can predict the behavior of embryos under various experimental treatments.

S. Bryant, V. French, and P. Bryant (1981)

If one were to dissociate different mammals into their constituent cells and then attempt to reclassify them on the basis of their cell types, it would be a difficult task. Apart from size, the major differences between a shrew and a grizzly bear reside in their relative proportions—ultimately, in the spatial arrangements of their cells. The processes that generate these characteristic differences are termed those of "pattern formation."

Pattern formation is one of the most important phenomena of animal development and, at the same time, one of the most elusive. Like such blanket terms as "commitment" or "competence," the phrase itself undoubtedly covers a variety of mechanisms. For instance, the body of the nematode comprises a recognizable morphological pattern, whose formation involves a rigidly fixed set of cell division programs. On the other hand, the development of a fruit fly or a mouse relies much less on fixed schedules of cell lineage than on cell–cell interactions and coordinating, systemic effects.

The fact that recognizable animal morphologies arise during development does not in itself prove that systemic or general mechanisms of pattern formation exist. In principle, the final, complex end result might be produced by a summed series of piecemeal events, as in the nematode. The evidence for the existence of general mechanisms of pattern generation comes almost entirely from observations on the disruption of early development. Such disruptions are nearly always followed by growth that regenerates either the original pattern or large sections of it. It is the phenomena of pattern *re*-formation that provide grounds for believing in the existence of coordinating mechanisms of pattern formation in normal development.

This chapter looks briefly at pattern formation and its genetic basis. We will begin by considering some of the general ideas that have been offered to explain pattern formation and then examine a few kinds of patterns that may represent a series of increasing complexity. Because the emphasis will be on the genetic basis and genetic analysis of patterns, the treatment will be largely confined to *Drosophila,* which remains the only organism susceptible to a genetic analysis of pattern formation mechanisms. For fuller treatments of pattern formation, the reader should consult Bryant (1978) and Meinhardt (1982).

CONCEPTS AND PROBLEMS

The first hypothesis of pattern formation was the "prepattern hypothesis" of Curt Stern (1954). Stern's formulation was designed to explain the origins of

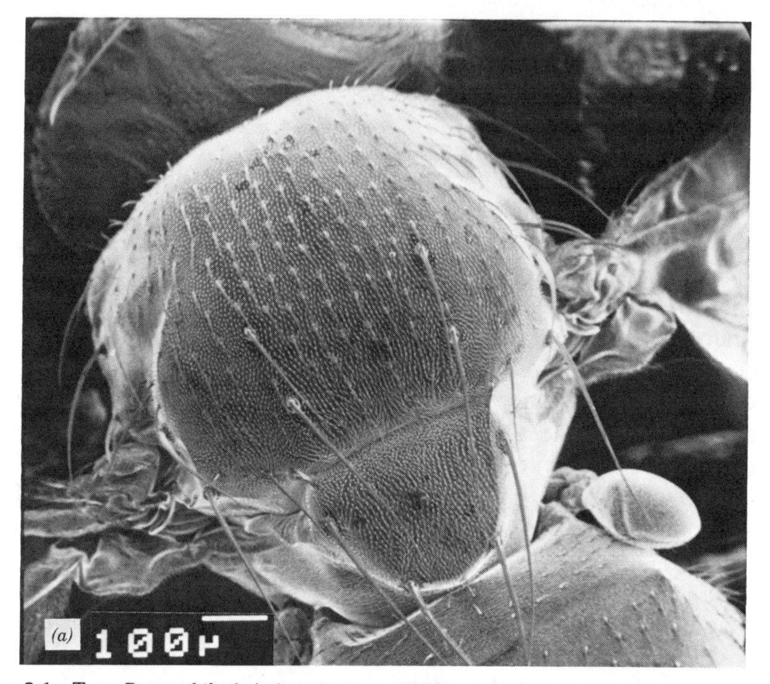

Figure 8.1 Two *Drosophila* bristle patterns. (*a*) The macrochaetes of the mesothorax; (*b*) bristle rows on the abdomen. The scale bar marks 100 μm. (Kindly supplied by Dr. A. Ghysen; reproduced with permission from Richelle and Ghysen, 1979.)

certain two-dimensional patterns, those of the bristle arrays on the surface of the fruit fly. A striking feature of these bristle patterns is that each external region has its own characteristic arrangement; two different patterns are shown in Figure 8.1. The surface of the mesothorax possesses both standard-sized bristles (microchaetes) and large bristles (macrochaetes). On each side of the body, the macrochaetes occupy fixed positions but are not uniformly spaced; however, they are mirror symmetric with respect to the midline, reflecting the origins of the mesothorax from two half-thoraces. In contrast, the abdomen always expresses rows of bristles of uniform size and spacing. In both of these regions, each bristle consists of an organule of four cells that arise from an apparently homogeneous sheet of epidermal cells.

Stern's hypothesis was that the distribution of a morphogen determines where bristles will appear. The prepattern consists of the spatial distribution of the morphogen, its peaks and valleys. Whether or not a bristle appears at one of the singularities depends on the cells' ability to respond to the prepattern signal. In wild-type strains, each set of cells that should respond does so, and the final pattern mirrors the high points of the prepattern (Fig. 8.2*a*). This hypothesis explains the observed differences in bristle pattern between different regions as reflecting different underlying prepatterns. In principle, there might be as many different prepatterns as there are wild-type bristle patterns.

A variant of this idea, the "pattern field" hypothesis, proposed by Waddington (1973), allows for the possibility of a smaller number of pattern-

Figure 8.1 (*Continued*)

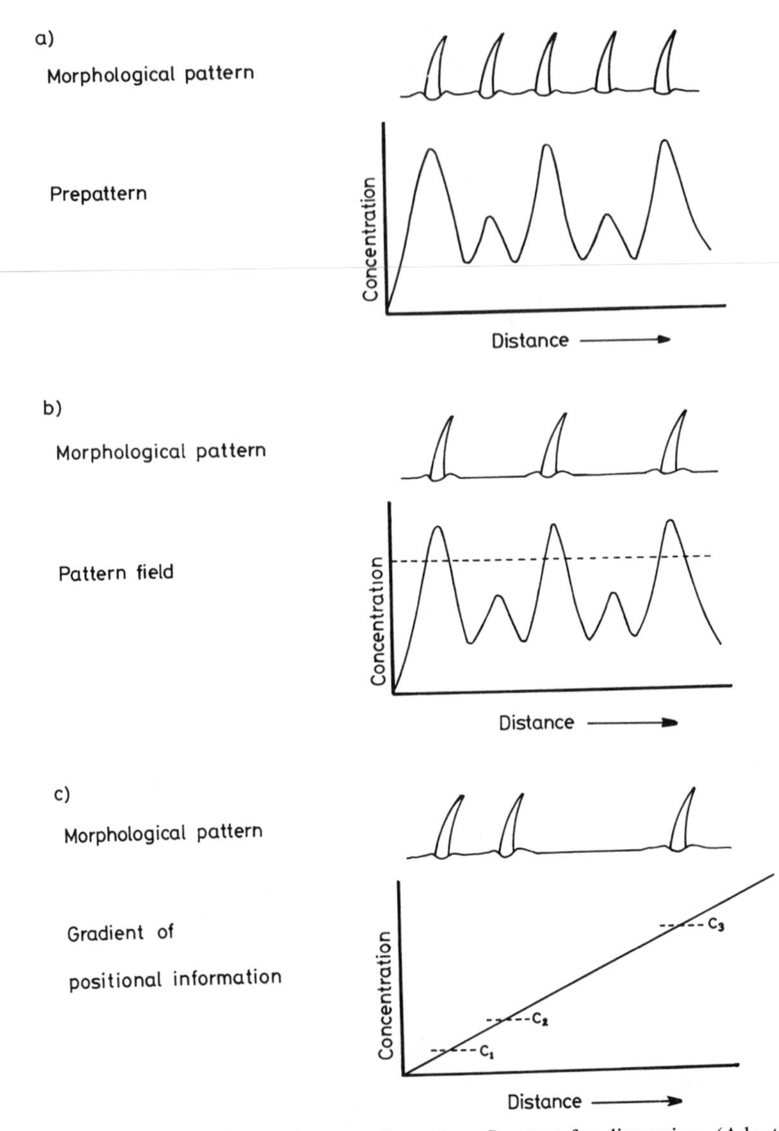

Figure 8.2 Three mechanisms of pattern formation. See text for discussion. (Adapted with permission from Wolpert, 1971.)

evoking signal distributions than of observed patterns. The central tenet is that different patterns may be produced by differences in the threshold of the cellular response to the signal rather than by the distribution of the signal itself. In effect, a smaller or greater number of singularities may evoke different numbers of bristles, depending upon the threshold for bristle induction (Fig. 8.2*b*). Although the pattern field idea allows a smaller number of prepatterns for a given number of patterns, both hypotheses entail fairly direct relationships between the singularities of the underlying pattern-evoking signal distribution and the final observed pattern.

There is a third idea, however, that completely uncouples the shape of the final pattern from the underlying evocative signal. In the hypothesis of "positional information," proposed by Wolpert (1969, 1971), there need be no obvious relationship between the distribution of the signal and the form of the developed pattern. The hypothesis posits the existence of a common, perhaps universal, distribution of a pattern-evoking chemical signal—the positional information—within an embryonic field; at each point in the field, a cell reads and interprets its position to acquire a position-specific state, the "positional value." The differences between different observed patterns, in this scheme, arise entirely from differences in cellular interpretation of the positional information; these differing interpretations, in turn, reflect differences in states of gene expression. The positional information itself may be a simple monotonic concentration gradient of a morphogen across the field or may be produced by a more complicated signal-generating scheme; positional information-by-gradient is diagrammed in Figure 8.2*c*.

Of these three hypotheses, that of positional information is most widely accepted. The evidence for some form of general positional information comes from several sources. One involves limb bud transplantations between equivalent or nonequivalent positions in the limb bud field. For instance, if a distal piece of early (proximal) wing bud is transplanted to a later-stage (distal) leg bud in an avian embryo, the transplanted tissue will produce wing structures, but those characteristics of distal wing: it has somehow interpreted its position with respect to the stump and responded accordingly. Other examples come from *Drosophila*. In gynandromorphs, for example, if the XX/XO boundary runs through the sex comb area, the bristles will respond solely according to their X chromosome constitution, with little distortion of the area or evidence of interactive effects; the result seems to indicate a common field of information that can produce either of two disparate responses according to the genetic composition of the cells in that field.

Perhaps the most striking demonstration of some form of common positional information in different structures is provided by the *Antp* leg. Although the extent of replacement of antennal regions by leg regions is variable, there is always a specific spatial pattern of corresponding replacements between the two appendages (Postlethwait and Schneiderman, 1971) (Fig. 8.3). Thus, the region that normally gives rise to the most distal segment of the antenna, the arista, is always converted to the most distal tarsal segment; the next most distal antennal segment, AIII, is replaced by the tibia or femur; and the most proximal leg segment, AI, is replaced by the most proximal leg segment, the coxa. In some sense, cells know their position and can decode that information to produce the appropriate structure for the leg or antenna.

The two strengths of the positional information concept are that (1) it

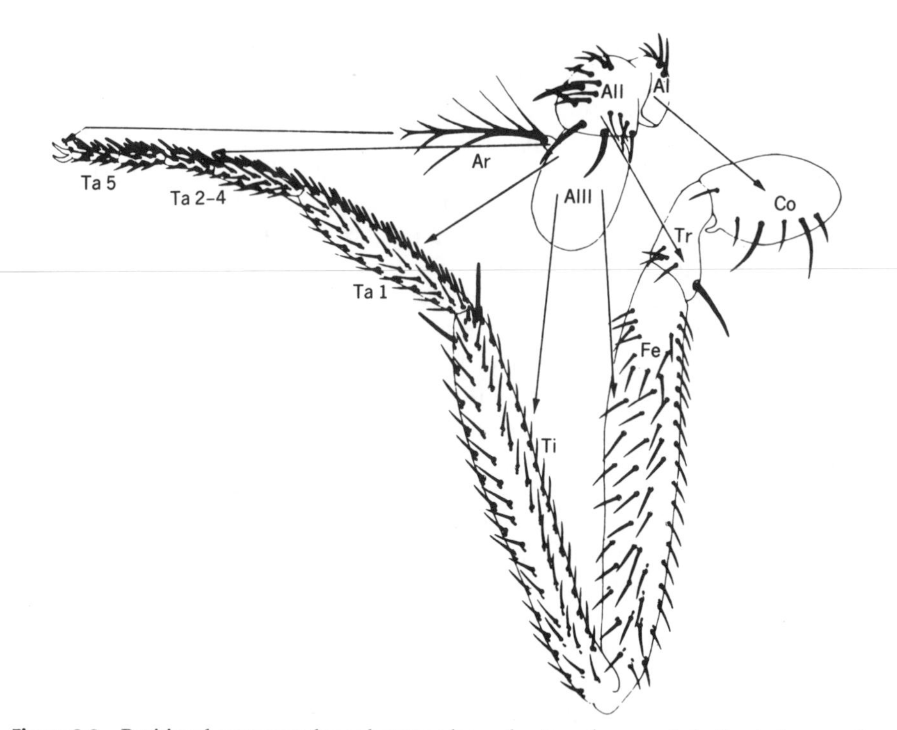

Figure 8.3 Positional correspondence between leg and antennal segments in the *Antennapedia* transformation. (Reproduced from Postlethwait and Schneiderman, 1971, with permission.)

provides a framework for understanding homologous replacements of the kinds described above and (2) it liberates the observed pattern from the shape of hypothetical underlying morphogen gradients. In this second respect, therefore, the idea constitutes an important conceptual simplification. The possibility that there are few, or perhaps only a single, universal form of positional information is appealing.

However, there are associated difficulties with this hypothesis as well. The principal problem is the obverse of its strength: it is so flexible that it can explain *any* pattern or pattern change as an alteration in the interpretive response of the cells. In effect, the idea is not subject to disproof. A second weakness is that it may serve to conflate certain position-associated properties of cell state with position per se and thereby obscure the basis of differential cell responses. The growing limb bud of the chick provides a possible example. In the limb bud, as noted above, there are developmental states associated with position along the proximodistal (p-d) axis, as we have noted, but these positions are also associated with an average number of cell divisions experienced from the time of limb bud emergence (Lewis, 1975). Positional value in this case may relate less strictly to geographical position along the p-d axis than to the number of cell divisions experienced.

While there are many uncertainties in grappling with the molecular and cellular bases of pattern formation, the problem of analyzing its genetic basis is even greater. Every mutation that affects visible development alters the final morphological pattern, yet there are few reliable criteria for discriminating between those genetic changes that affect the initial establishment of the pattern and those that alter secondary events in the ultimate development of the realized pattern. However, it is probable that the great majority of pattern-disrupting mutants are affected in secondary processes.

If positional information consists of morphogen gradients, identifying the morphogen molecular species might provide an essential clue to understanding the mechanism. However, the molecule(s), even if unambiguously identified, might be completely uninformative as to the mechanism. The molecular species could be a simple nucleotide or even an amino acid. It is known, for instance, that different slime mold species employ different small molecules to mediate very similar cell aggregation behaviors. In other words, all of the informational specificity might lie in the interpretive process, with the positional information itself being a simple trigger.

The class of mutants that immediately recommend themselves as pattern-altering mutants are the homeotics of *Drosophila,* which transform the pattern typical of one disc or disc region into that of another. However, the corresponding wild-type genes, while necessary for the normal patterns, cannot be solely responsible for them. The particular gene in any given case may be expressed throughout the compartment or segment, yet characteristic structures appear only at particular locations. Even if expression is not uniform and the pattern reflects these nonuniformities, the spatial modulations of expression will require explanation. To put the matter in slightly different terms, the homeotic genes whose wild-type gene products are active in particular regions may be necessary for the patterns of those regions, but the fact of their activity is insufficient to explain the details of those patterns (Gubb, 1985).

There are, in sum, no simple rules for obtaining answers about the genetic basis of pattern formation. Each kind of pattern poses its own problems and requires its own form of analysis; from that analysis, clues to the underlying mechanisms may become apparent.

ONE-DIMENSIONAL PATTERNS: REPETITIVE PATTERNS ALONG ONE AXIS

An example of a one-dimensional pattern is the linear array of segments in the insect body. This pattern may be regarded as two superimposed patterns: a periodic, repetitive one and a sequential array of slightly different units (Meinhardt, 1982, chap. 14). In all probability, the repetitive, periodic spacing of segments is the primary and initial pattern: the formation of evenly spaced segments is one of the earliest events in embryogenesis, with the individual differences between segments becoming apparent only later. Unlike some of the more complex imaginal patterns that arise late in development, there is a relatively unambiguous source of genetic material for the analysis of segmental patterns. It consists of the maternal coordinate mutants that affect segmentation along the a-p axis (Table 3.6) and the zygotic genome functions that alter the basic repetitive segment pattern (Table 6.2).

The simplest explanation of the repetitive sequence of segments is that it is produced by a repetitive chemical motif in the early embryo. One form of such a repetitive signal is that of a repeated concentration gradient recurring every segmental unit. In such a repetitive gradient, a given morphogen concentration would correspond to and produce a particular pattern element; the maxima or minima could be responsible for the singular discontinuities in pattern. The obvious singularities in the segmental pattern are the segmental boundaries themselves. A sine wave, in which the amplitude at any point represents a particular concentration of morphogen, could fulfill these requirements. The segmental boundaries can be visualized as arising from either the maxima or minima in such a chemical wave, or perhaps from both (Fig. 8.4*a*).

Viewing the segmental gradient as a sine wave permits one simple interpretation of the pair-rule mutants, those zygotic mutants that produce double-segment width patterns. As noted by Sander et al. (1980), if segmental boundaries arise at both maxima and minima, then the pair-rule mutants may be strains whose cells can correctly sense only the maxima or the minima. Mutants of this kind would show only half the number of segment boundaries, as is indeed the case in the pair-rule mutants.

Attractive as this formulation is, it probably oversimplifies the situation, since in none of the mutants does the anterior margin of the missing region correspond neatly to a segmental boundary (Nusslein-Volhard and Wieschaus, 1980). It may be rather that the pair-rule mutants experience a shift in the underlying gradient itself rather than in their reading mechanism; this "floating grid" hypothesis is discussed by Sander et al. (1980).

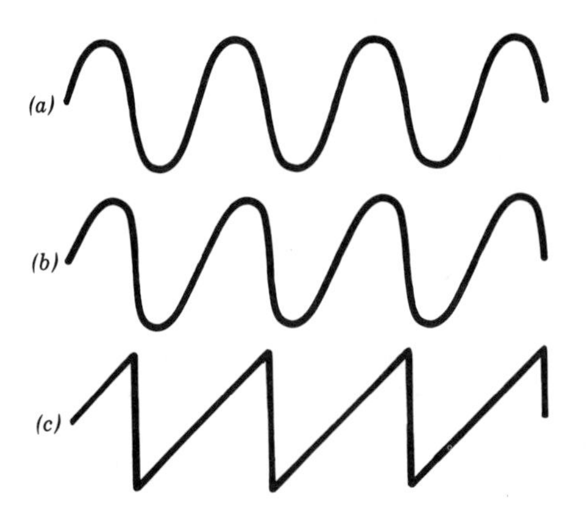

Figure 8.4 Three kinds of repetitive segmental gradients. Top: a morphogen gradient with a sine wave distribution; middle: a skewed sine wave gradient; bottom: a sawtooth distribution.

Another possibility is that the pair-rule mutants reveal an important but transient organizing step in the creation of a segmental pattern. From the fate-mapping experiments, it appears that segmental boundaries are laid down at distances no greater than three or four cell widths. Intuitively, this seems to be an uncomfortably small period for a chemical sine wave. Perhaps the initial organization is a sine wave with a period twice as large, succeeded by some partitioning device that divides each double segmental unit into two. Pair-rule mutants might be defective in either "fixing" the first step or activating the second step. Molecular evidence that some of the wild-type pair-rule genes are expressed at *double-segment* intervals has recently been obtained (Hafen et al., 1984), supporting the idea that the initial division is indeed in double-segment widths.

Segment spacing is the most obvious repetitive morphological motif in the segmented embryo, but there is, in addition, a finer-grained repetition, that of polarity of cuticular processes within each segment. One possibility is that this intrasegmental polarity is also a product of morphogen concentration in which a particular value assigns a particular degree of anteriorness or posteriorness within each segment. If this is the case, then a symmetrical sine wave, with identical values on either side of a maximum or minimum, may be an inappropriate representation of the presumptive repetitive gradient. A skewed sine wave (Fig 8.4*b*) or even a sawtooth curve (Fig. 8.4*c*) may be better representations. For the former, a given degree of anteroposterior value can be given by the slope of the gradient and, for the latter, by the numerical value of the concentration. (Asymmetric gradient shapes, however do not lend themselves as readily to explanations of abolition of alternative segments seen in pair-rule mutants as does the simple sine wave.)

There is some experimental evidence to support the idea of distinct positional values within each segment, although it is derived primarily from surgical experiments on *Oncopeltus*. Like *Drosophila,* the larval segments of *Oncopeltus* are characterized by a visible polarity. The anterior region of each segment is distinguished by a transverse elongation of the epidermal cells and by a dark orange color that grades into the paler color of the posterior half. Each segment border consists of a shallow groove overlying an internal ridge of endocuticle, and both sides of the border are free of bristles. The bristles that are adjacent to this border zone on both sides of the boundary all point posteriorly. This intrasegmental patterning permits an unambiguous assignment of position within each segment.

If parts of the larval integument of *Oncopeltus* are surgically removed or destroyed by microcautery in early larval instars, the integument can be regenerated during successive instars. (Unlike the fruit fly, the milk weed bug possesses five larval instars in all, and its larval cells are capable of cell division.) During regeneration, the segmental pattern is renewed. However, the new pattern is a function of both the *position* of the original lesion and the *amount* of material that is removed. The specific patterns of repair observed testify to an underlying cellular/biochemical distinctiveness asso-

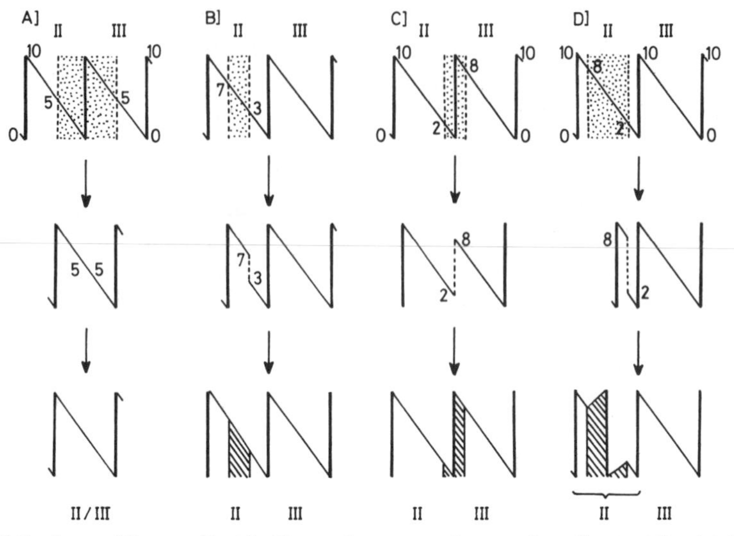

Figure 8.5 A repetitive gradient in *Oncopeltus* segment generation. See text for details. (Redrawn and reproduced with permission from Lawrence, 1981; copyright MIT Press.)

ciated with different positions within the polarized segmental unit. The results are schematized in Figure 8.5 and can be rationalized within a simple set of rules (Wright and Lawrence, 1981a; Lawrence, 1981).

Let the positional values within each segment be represented by numbers 10 through 0, with 10 corresponding to the most anterior value and 0 to the most posterior one. The first result is the simplest. If a whole segment width is excised or destroyed, such that identical values from two adjoining segments are brought together, there is healing without growth (Fig. 8.5*a*). If a small intrasegmental piece is deleted—for instance, one between positions 7 and 3—regenerative growth occurs, restoring the missing piece (Fig. 8.5*b*). Small pieces that remove boundaries can also be re-formed, generating a new boundary. Thus, if material between positions 2 and 8 is deleted, removing the most posterior part of one segment and the most anterior part of the next, along with the intervening boundary, the missing positions are intercalated along with a new boundary (Fig. 8.5*c*). Even though boundaries are singular morphological features, they may represent only a particular set of positional values within a continuum of values. (However, some evidence discussed below suggests that segmental boundaries in insect legs have special properties in the way they influence regenerative behavior.)

The most interesting result is produced by bringing together positions that are *more than half a segment apart* (Fig. 8.5*d*). When positions 8 through 2 are deleted, as diagrammed, an extra boundary is formed; material from 8 to 10 is regenerated in mirror-image orientation, along with a new boundary. New material from 0 to 2 is also formed, yielding a small additional segment between the new ectopic boundary and the old one.

This result is explicable if regeneration always takes the *shortest numerical route* between positional values. The shortest distance between 8 and 2, assuming equality of positional distance between numerical values, is 8–9–10–0–1–2, rather than 8–7–6–5–4–3–2. Only four values instead of five need to be intercalated. The rule can be seen to apply to all of the other cases as well, in which only the normal pattern is regenerated. The difference between case *d* and the others is that the regenerative route produces a structure that, in this instance (the boundary), is not present within a segment between these positions. In more general terms, when the shortest numerical route is in the *opposite direction* from the normal pattern sequence, the consequence is the *duplication of pattern elements*. (More will be said about this mechanism below, concerning appendage regeneration.)

The *Oncopeltus* regeneration results bear at least a superficial resemblance to some of the segmental pattern mutants in *Drosophila*. The phenotype of the segmental polarity mutant *pat,* with its mirror-image partial duplication of anterior segment regions and double boundaries (Fig. 6.5), is similar to that produced by intrasegmental deletion of more than half of a segment. Furthermore, the production of composite segments, as in some of the pair-rule mutants, is the kind of result anticipated from deletion of slightly more than a segment width beginning in internal positions and skipping every other segment. The *Oncopeltus* results do not explain the alternate segment effect in the pair-rule mutants, or the complete elimination of segmental boundaries in certain of the polarity mutants (*wg* and *hh*), or the neat intrasegmental duplications in *fu* and *gsb*. Nevertheless, they suggest that the deletion of cells or an equivalent process could produce some of the main features of the pattern mutants. One form of possible cell elimination would be extensive cell death, but examination of *prd* embryos shows no excess cell death in this mutant (Sander et al., 1980). However, an equivalent loss might be produced by the failure of cells to respond to particular morphogen concentrations.

It is important to remember that the idea of a repetitive segmental gradient is simply a formal construction. There seem to be graded properties within segments that repeat in a characteristic fashion, but there is no independent evidence for a molecular concentration gradient that repeats itself at segmental intervals. The cells within larval segments may only gradually acquire their graded properties in a succession of events that follow the initial laying down of singularities (prospective boundaries) at regular intervals.

Indeed, the presumptive maternal gradients involved in global embryonic determination (Chapter 3) may be a sufficient set of instructions for the spacing and polarity of segments. There is some indirect evidence to support this idea. When cleavage-stage embryos are ligated, an increase in the width of the denticle band coupled with a change in the polarity of the denticle rows occasionally occurs. These changes can appear as much as two segment widths distal to the site of ligation (Vogel, 1977). This finding suggests that polarity within segments is governed by the same maternal instructions that place segment boundaries.

The idea that one or more global determinative gradients impose the repetitive segmental pattern and lay the basis for the differences in segment character demands less biological complexity than the provision of both global monotonic gradients and superimposed repetitive segmental gradients. However, if the only gradient system is that of global gradients, the question arises as to how monotonic gradients can create a periodically repeated structure; it is not immediately obvious how a steadily increasing concentration of a substance can produce a periodic structure.

Meinhardt (1982, chap. 14) has proposed a set of postulates that would allow cell–cell interactions to break a monotonic gradient into a repetitive segmental pattern. These are that (1) cells exist in either of two states, labeled A and P, and tend to oscillate between them; (2) A cells stabilize P cells in their immediate vicinity and vice versa, creating stable borders where groups of A cells meet groups of P cells; (3) increasing concentrations of morphogen cause P cells to flip into the A state; (4) the higher the morphogen concentration, the greater the number of P-A transitions experienced, with segmental state being a function of the number of transitions. In effect, the creation of successive segmental states resembles the upward passage of a ship through a series of locks: each transition is similar to the previous one, but effects a new incremental state.

The hypothesis accommodates the creation of a superimposed repetitive and sequential segmental character and at the same time—if one equates the A and P states with those respective compartmental characters—the creation of a stripe-like compartmental organization. The function of A-P compartmentalization, in this view, is to participate in the creation of segmental divisions. Compartmentalization becomes part of the gating mechanism in the transition from one segmental state to the next. In this scheme, the absence of clear-cut segmentation in the cephalic region of the *Drosophila* embryo may be causally related to the absence of early A-P compartmentalization in this region. However, extrapolating from the formalism of the Meinhardt model to the facts requires caution: while division of the early embryonic segments into domains is an attractive hypothesis, the existence of A and P cellular states in the early embryo is unproven, and whether the postulated embryonic domains necessarily bear a fixed spatial relationship to the A/P compartment boundaries observed in adult structures is not at all clear.

TWO-DIMENSIONAL PATTERNS: BRISTLE ARRAYS

There are many recognizable two-dimensional patterns of elements in the animal world. They include arrangements of sensillae, hairs, bristles, and various kinds of scales. Among these patterns, insect bristle arrays have been subject to the most intensive analysis, particularly by genetic means.

In *Drosophila,* each external body region that carries bristles exhibits its own characteristic bristle pattern (Fig. 8.1). The fundamental assumption in all thinking about bristle pattern is that the number of epidermal cells that are *potentially* capable of generating bristles is much greater than the number of cells that do so. Various wounding and regeneration experiments substantiate this assumption; after wounding, new bristles are produced by cells that normally do not form them. Furthermore, manipulation of the genetic background of certain known mutants can substantially increase the number of bristles formed. Evidently, the "decision" of a particular epidermal mother cell on whether or not to form a bristle involves a regulated response to its neighbors.

The simplest form of bristle pattern, seen in many primitive insects, is that of equidistant spacing. The first mechanism to account for such spacing was proposed by Wigglesworth in 1940. He suggested that bristle formation is induced by a specific morphogen, a "chaetogen," produced initially by every epidermal cell. Induction first occurs randomly and is immediately followed either by reduction in chaetogen in the immediate vicinity or by production of an inhibitor by the induced cell. The consequence of bristle induction, therefore, is an inhibition of further bristle formation in its immediate vicinity. This model demands a certain degree of randomness in the positioning of bristles. However, all bristle patterns of this class do show a degree of randomness. The Wigglesworth model for equidistant bristle patterns has held up well (reviewed in Lawrence, 1973a).

A second bristle pattern consists of ordered rows of bristles, as seen on the abdomen (Fig. 8.1*b*) or on the segments of the legs. Held (1979) analyzed the factors that govern the number and spacing of the bristles within the eight bristle rows of the basitarsus (the most proximal tarsal segment) by employing environmental and genetic factors that shorten or lengthen the basitarsus. His conclusion, reinforced by a study of the basitarsus in 21 other species of *Drosophila,* is that the bristles in each row are always situated a fixed number of cells apart. As he points out, this conclusion is not readily compatible with segmental gradient models, in which bristle positions are specified by diffusible morphogens within each leg segment. The results support either a mechanism based on direct cell counting or a modification of Wigglesworth's model in which each cell produces a quantum of morphogen.

The most difficult bristle pattern to explain is nonequidistant spacing—for instance, the array of macrochaetes in the mesothorax (Fig. 8.1*a*). Within each half-mesothorax, there is a characteristic placement of nine macrochaetes on the main portion (the scutum) and of two on the posterior portion (the scutellum). The positioning of these bristles demands a more precise mechanism than the Wigglesworth model can provide. Stern's prepattern conception was an explicit attempt to solve this problem, with the positions of the bristles postulated as the high points of the prepattern.

Tests of the prepattern hypothesis are possible because of the existence of mutants that delete specific mesothoracic macrochaetes. Under the hypothesis, these mutants must either have an altered prepattern or be defective in

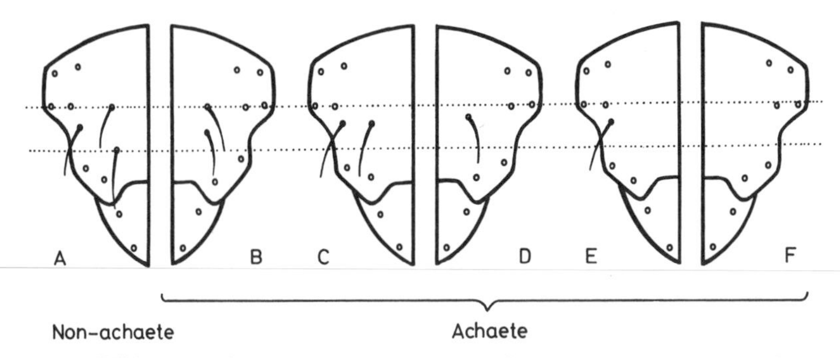

Figure 8.6 Wild-type and *ac* macrochaete patterns of the mesothorax; see text for details. (Adapted with permission from Stern, 1954.)

the specific regional competences of the mesodermal cells to respond to the prepattern. If the prepattern itself is altered, then large areas of mutant tissue should have the effect of disrupting the pattern *outside the mutant area*. If the mutants are affected only in specific regional competences to respond to the prepattern, then wild-type tissue should always produce bristles when it covers one of the mutant-affected sites, and mutant tissue in those sites should always lack bristles; if competence is affected, the mutants should be strictly autonomous in their effects.

The phenotype of the *ac* mutation is illustrated in Figure 8.6. It consists of the loss of one or both dorsocentral macrochaetes and the laterally located supra-alar bristle. The supra-alar may or may not be present independently of the dorso-centrals (d.c.s), but the two d.c.s seem to comprise an interacting unit: the anterior d.c. can be present without the posterior d.c., but if the posterior d.c. is present, the anterior one always develops. Furthermore, the absence of one of the bristles is usually accompanied by a slight shift in the position of one of the others. These mild interactive effects would appear to signify that the mutation alters some bristle-field property.

As *ac* maps to the X, Stern was able to make gynandromorph mosaics in which the *ac* tissue was also marked with *y* and *sn*. His initial results were generally consistent with total autonomy of the *ac* effect: sites covered by mutant tissue were always bristleless and, when genetically wild-type, made bristles (Stern, 1954). Stern interpreted his results to mean that *ac* is not a prepattern mutant but is deficient in local competences to respond to the prepattern; the slight interactive effects associated with the presence or absence of bristles in mutant mesothoraces were attributed to secondary causes. However, in a larger collection of mosaics, Roberts (1961) reported a significant number of cases of nonautonomy. In 19 out of the 110 mosaics in which one or more of the critical sites was occupied by *ac* tissue, bristles developed in genetically mutant tissue. In all of these cases, ac^+ tissue was contiguous to the bristle-producing site. However, the distance between the mutant and nonmutant tissue was not the only factor; only certain near-enveloping configurations of wild-type tissue sufficed to produce a bristle in mutant tissue.

The combined results indicate that *ac* is nonautonomous in expression over short distances and autonomous over long distances. These findings suggest that the wild-type locus specifies a product that is diffusible over short distances and plays a part in bristle induction. In this light, the gene function may be viewed as specifying a chaetogen. Some suggestive evidence along these lines is supplied by the behavior of the *Hairy wing* (*Hw*) mutant, which maps at the same position as *ac*. This dominant mutant is characterized by the production of extra macrochaetes in the vicinity of the standard macrochaetes. Like *ac, Hw* shows short-range nonautonomy in mosaics. One interpretation of these results is that while *ac* is hypomorphic for production of the putative chaetogen, *Hw* is hypermorphic, being an excess producer. The finding that extra doses of *Hw* produce even more macrochaetes throughout the mesothorax is consistent with this possibility (Ghysen and Richelle, 1979).

A general model of bristle formation that provides a context for these results has been formulated by Richelle and Ghysen (1979). It incorporates elements of the prepattern and Wigglesworth hypotheses and has four elements:

1. There is a probability distribution within each bristle-producing field for chaetogen synthesis by each epidermal cell.
2. Chaetogen diffuses freely between cells.
3. An epidermal mother cell is induced to form a bristle when the local chaetogen concentration exceeds a certain threshold value.
4. Each induced cell secretes a diffusible inhibitor, whose action is to prevent neighboring cells from making bristles.

In this model, the prepattern for bristle development is the probability distribution for synthesis rather than the spatial distribution of the morphogen itself. In turn, the probability of chaetogen synthesis is a function of cell position within the field. With the appropriate adjustment of the various parameters, the model can account for both the distinctive placement of the macrochaetes and the production of bristle rows. Furthermore, it permits a fairly precise placement of bristles within a probabilistic framework: within each region, cell responses average out to give precisely positioned bristles. The model therefore demands neither the precision of preformed morphogen distribution, as in the prepattern hypothesis, nor the precise interpretation of position required by cells in the Wolpert hypothesis.

The genetic results fit the model if the *ac-sc-Hw* locus is responsible for chaetogen synthesis. (*sc* is an allele of the locus that removes a different set of macrochaetes from the mesothorax.) Because the model is abstract, it provides no clues to the biochemical identity of the postulated chaetogen. In addition, the proposition that the *ac-sc-Hw* locus specifies the chaetogen does not obviously tally with the finding that the locus is essential for both viability and the development of the CNS (Garcia-Bellido and Santamaria, 1978). It is, of course, possible that this genetically complex locus specifies different proteins for the different functions.

In any event, none of the *ac* results are consistent with the simple notion that the gene specifies a prepattern in Stern's original sense. Furthermore, when the test for nonautonomy of effect is carried out on any of the other mutants that affect bristle patterns, none show long-range nonautonomy; all fail to meet this minimal criterion of a prepattern mutant. In consequence, the last 15 years have seen the concept of bristle prepatterns give way either to specialized variants such as the Richelle–Ghysen model or to more general ideas of positional information.

Topographical Polarity of Bristles and Hairs

There is an additional aspect of bristle pattern that requires mention. This is the feature of topographical polarity—the direction in which cuticular processes point with respect to body axes—and it is shown by both bristles and hairs on the surface of the fruit fly body. These orientations are regular and characteristic for the different regions of the body surface. Each set of bristles or hairs effectively constitutes an array of vectors, all with a similar orientation.

Polarity of these processes is expressed in two ways. The primary aspect is the direction of their slant. For most appendages, this slant is proximodistal (p-d), with the bristles/hairs pointing distally (away from the main body axis). The other manifestation of polarity is in the small, adventitious structures termed "bracts" that are associated with bristles in certain regions. For any given region of the body surface, these bracts are always found on one side of the bristle. In the legs, for instance, the bract-bearing bristles always have these projections on the proximal side.

The fact of bristle polarity was mentioned earlier in connection with the abdominal segments of *Oncopeltus*. Following surgical removal, regeneration can produce reversal of polarity in short segments and/or duplication of elements. As discussed earlier, these findings can be accommodated within the framework of a segmentally repeating gradient; in this scheme, the anterior boundary is the high point of the gradient (Fig. 8.5). Comparable reversals of bristle (and hair) polarity and duplication of elements can also be seen in the legs of certain *Drosophila* mutants (Poodry and Schneiderman, 1976). These abnormalities were found to occur only near the intersegmental membranes and only when the structure of the intersegmental boundary was disrupted. These results suggest that there is a singularity of some kind at the intersegmental boundary for the setting of segmental polarity. Poodry and Schneiderman interpreted their results within the framework of the segmental gradient hypothesis, postulating that the effects reflect the loss of certain "positional values," either from cell death or from other causes. It is interesting, however, that the *number* of bristles in the legs is (as also noted previously) not readily accounted for by the segmental gradient hypothesis (Held, 1979).

Other instances of genetically caused polarity alteration of bristles and hairs take place wholly within segments and without damage or measurable reference to the intersegmental boundary. Five mutant phenotypes in this category have been described by Gubb and Garcia-Bellido (1982). The mutant genes responsible for these effects are *prickle* (*pk*; 2–55.3), *spiny legs* (*sple*; 2–56), *frizzled* (*fz*; 3–41.7), *inturned* (*in*; 3–47), and *multiple wing hair* (*mwh*; 3–0.0). Amorphic mutants of *pk, sple, fz,* and *in* were isolated by standard genetic procedures.

The mutants are characterized by a change in vectorial properties of hairs and/or bristles, without accompanying changes in either the shape of the appendage or the number of cells present within it. Except for *sple,* whose polarity disruptions are in the abdominal tergites and leg bristles, the mutants all show an altered polarity pattern of hairs in the wing (and polarity effects in variable numbers of other regions). The wing effects are shown in Figure 8.7. The effects are all autonomous in large clones. Furthermore, various manipulations that alter wing size, wing shape, cell size, or wing venation pattern all fail to alter the polarity property diagnostic of the mutants. Furthermore, clones of marker mutations within homozygous mutant wings have been found to be wild-type in shape; clearly, polarity does not reflect patterns of clonal growth but is independently controlled.

The only clue to the nature of the defects is provided by *mwh,* which is a polarity mutant. This mutant, used commonly as a marker in clonal analysis, produces several hairs per cell instead of the usual single hair. This mutant phenotype can be phenocopied by heat shock at very specific times in development; the precise period in which a particular region can develop the *mwh* phenocopy is a characteristic of that region (Mitchell and Lipps, 1978). As noted earlier, heat shock stops ongoing protein synthesis and stimulates the synthesis of a special group of proteins, the so-called heat shock proteins. The basis of the *mwh* phenocopy has now been elucidated. In hair development, the growing hair is covered with a substance, cuticulin, that serves as a sheath. Following a phenocopy-producing heat shock, the cell extrudes material through the cuticulin shield, evidently because the synthesis of the shield has been interrupted; these extrusions produce either distinct hairs or a bundle of hairs united at their base (Mitchell and Petersen, 1981; Petersen and Mitchell, 1982). If one assumes that the polarity defect in the mutant results from a comparable defect in protein synthesis, then the other polarity mutants may also indirectly reflect such basic defects in hair synthesis. And, in some manner, the cells may thereby become slightly "skewed" with respect to one another, the result being an autonomous clonal defect. A proper understanding of these mutants awaits a characterization of the molecular and cellular biology of the defects.

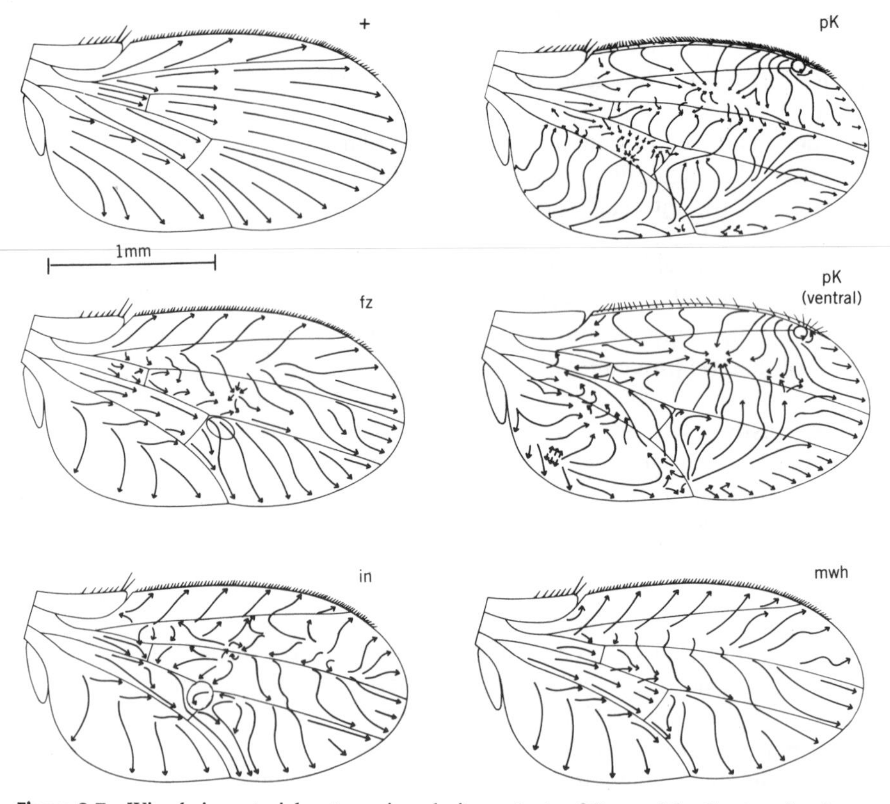

Figure 8.7 Wing hair vectorial patterns in polarity mutants of *Drosophila*. See text for discussion. (Figure kindly provided by Dr. D. Gubb; reproduced with permission from Gubb and Garcia-Bellido, 1982.)

TWO-DIMENSIONAL PATTERNS: APPENDAGES AND THE POLAR COORDINATE MODEL

The set of elements that comprise the patterns of appendages are considerably more complicated than the bristle patterns we have been discussing. This complexity becomes obvious when one considers the several features that collectively describe the pattern (Gubb, 1985). These include the overall topography of the region (its gross three-dimensional outlines), the distribution of the specialized cell types within the region, the standard array of cell sizes within each of these specialized cell types, and (where cuticular processes exist) the fact of polarity mentioned above.

However, in recent years, evidence has accumulated to indicate that the construction of appendages as different as cockroach legs, fruit fly wings, and salamander legs is based on a common pattern-forming mechanism of relative simplicity. In standard embryological terminology, all appendages constitute "secondary embryonic fields," which arise during later embryonic development. (The primary field is the early embryo itself, with its regulative properties.) The evidence that secondary embryonic fields share a common pattern-forming mechanism comes from the ways that these fields reconstitute their pattern elements following initial damage and subequent growth.

The conception that describes this mechanism is the polar coordinate hypothesis, formulated by Vernon French and Peter and Susan Bryant. The discussion that follows is focused on the derivation and predictions of this model from the regeneration behavior of imaginal discs in *Drosophila*. However, the central tenets of the hypothesis apply equally to the other secondary fields whose regeneration behavior has been tested (French et al., 1976; Bryant et al., 1981).

When an imaginal disc everts, it changes shape from a flattened disc to a tubular structure (although in the case of the wing, the top and bottom surfaces are apposed to form a flat sheet). It is clear that the patterning of a disc can be represented as a two-dimensional fate map. This remains true of the adult structure in the sense that the flattened pattern has now been mapped onto the surface of a three-dimensional structure. Thus, the phenomena of pattern formation in *Drosophila* involve primarily if not exclusively two-dimensional sheets of cells. (This pattern may well include the process of segmental division, even though this can be analyzed as a problem in one dimension.)

We have seen that the primary processes in embryogenesis also occur in two-dimensional epithelia (Chapter 3), although they later come to shape three-dimensional structures. This appears to be a general principle in pattern formation. Even when it occurs in solid blocks of tissue, as in the amphibian limb bud, a two-dimensional zone of proliferating cells is used. It seems that the setting up of patterns in two dimensions is complicated enough; organisms may generally avoid pattern formation in three dimensions.

Within each disc, there is little cellular difference between the different regions, but this seeming uniformity is deceptive. When specific fragments are cut from particular locations of each disc and implanted in the abdomens of mature larvae, each metamorphosed implant develops into a specific part of the adult appendage. Indeed, each mature disc is a highly structured mosaic of regional fates, allowing very precise fate maps to be constructed.

If, however, one first allows each implant to grow before metamorphosis, additional structures will be found in the fully developed piece. Typically, the fragment either duplicates its pattern elements or regenerates, the specific outcome depending on the fragment's initial position and size. The either/or duplication/regeneration choice is fundamental.

Both duplication and regeneration involve the formation of structures that were not originally specified by the fragment. Hence, in a general sense, both involve regenerative processes. Early in this century, T.H. Morgan distinguished between two kinds of regeneration. The first, requiring growth to form the new structures, he termed "epimorphosis"; the second, involving respecification of cells without growth, he labeled "morphallaxis." Both fragment duplication and regeneration involve epimorphosis. Duplication might, in principle, involve a cell-by-cell replication of each new cell from an analogue cell without respecification of preexisting cells. However, when marked clones were induced in discs and then scored in duplicating fragments, one half-element of the duplication was often found to contain substantially more marked cells than the other half (Postlethwait et al., 1971). Evidently, duplication involves growth and the readjustment of cell developmental capacities.

The finding that within each pair of complementary disc fragment classes, one group always regenerates missing structures and the other duplicates structures, was initially compatible with the notion of a simple gradient of developmental capacity within the disc. The explanation would be that pieces can always produce "lower" values of the gradient but not "higher" ones. Bisection of the gradient within a disc should therefore always produce a piece with the high point of the gradient (the regenerating piece) and a complementary piece (the duplicating piece) (Fig. 8.8*a*).

This model, however, is essentially a one-dimensional picture, whereas leg and wing discs are effectively two-dimensional structures. A two-dimensional sheet might harbor a monotonic gradient if the gradient ran smoothly from a high point along one edge down to a low point at the other but, in this circumstance, regeneration should always proceed in only one direction. Surgical experiments on leg discs show that the matter is not that simple:

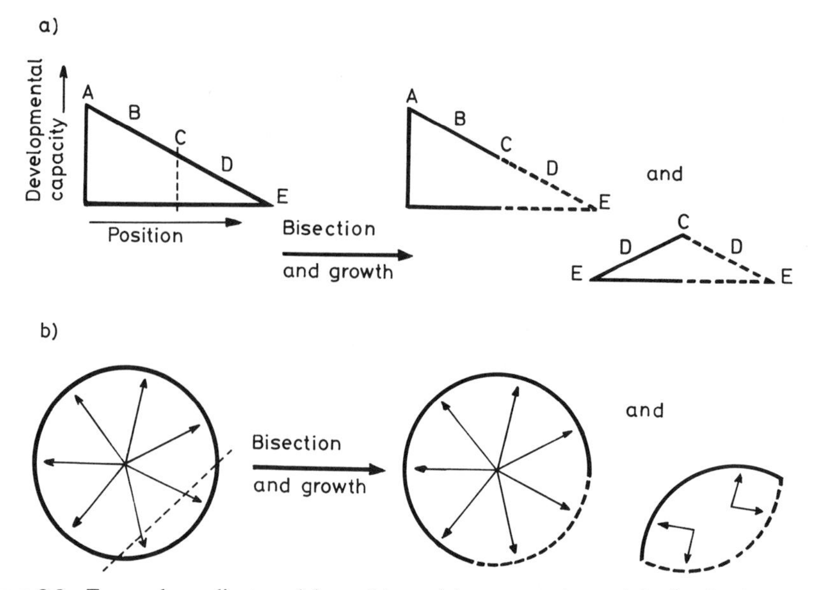

Figure 8.8 Two early gradient models used to explain regeneration and duplication in imaginal disc fragments. (*a*) The single monotonic gradient model; (*b*) the radial gradient model. (Adapted with permission: *a* from Bryant, 1971; *b* from Bryant, 1975.)

regeneration within leg disc halves can take place in either of two perpendicular directions (Schubiger, 1971).

Furthermore, when the regenerative capacity of wing disc fragments is tested, a complexity similar to that of the leg is seen. By slicing discs into various pairs of complementary fragments, Bryant (1975) was able to isolate a high point of regenerative capacity near the center of the disc. Cuts on any side of this high point produce peripheral fragments that duplicate rather than regenerate. However, possession of the high point is not necessary for fragment regeneration. Even in fragments that do not include the mapped high point, regeneration proceeds on the *sides* of the fragment that face away from the high point. In contrast, the sides that face toward it duplicate their elements. These results can be accommodated by the proposition that there are multiple gradients of developmental capacity, all running down from a high point at the center (Fig. 8.8*b*). The high point in the wing disc corresponds to the future distal tip of the wing.

However, this explanation has also proven inadequate. It predicts that central fragments, which contain the high point, should regenerate. In fact, they duplicate rather than regenerate, while the complementary peripheral fragments regenerate. Furthermore, all quarter-disc sectors are found to duplicate, while all three-quarter disc fragments regenerate. These findings indicate that a certain *amount* of material is required for regeneration. Finally, pieces that normally duplicate can be stimulated to regenerate by mixing in other, partially complementary pieces (Haynie and Bryant, 1976). This last finding indicates that regeneration involves an interaction between cells at different peripheral positions. Indeed, imaginal disc fragments, whether small or large, are observed to contract in upon themselves; such infolding would bring together regions that are never in contact in the unoperated disc (Reinhardt et al., 1977).

The polar coordinate model incorporates these various observations (Fig. 8.9). This model consists of one postulate concerning the positional specification of cells within a field and two rules of regenerative behavior when part of that field is removed. The postulate is that the position of each cell on a

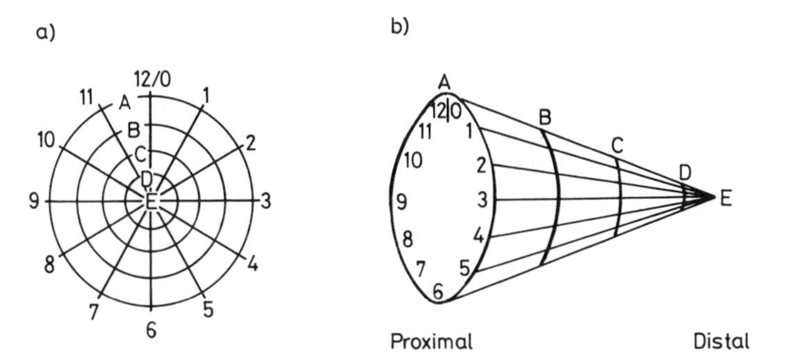

Figure 8.9 The polar coordinate model; see text for discussion. (Redrawn and reproduced with permission from Bryant et al., 1981, *Science* **212,** 993–1002. © 1981, AAAS.)

surface two-dimensional, circle-like field is specified uniquely by two coordinates with respect to a central point (E): the first coordinate is a radial distance from that central point, and the second is a circumferental position on the circle delimited by that radius. For the imaginal discs, the radial distance measures the distance along the p-d axis from E (the most distal point). The circumferential values are denoted by the arbitrarily given values of a clock face (0 and 12 being equivalent positions) and represent circumferential sites within the disc along the p-d axis. The lower part of the figure gives an expanded transformation of the field comparable to the physical transformation of the disc as it unfolds.

The first rule of regenerative behavior in the polar coordinate model is the one we have encountered in *Oncopeltus* segment regeneration, the "shortest intercalation rule": when cells with nonadjacent positional values are brought together, the process of growth restores all intermediate positional values by the shortest numerical route. The shortest intercalation applies to both circumferential and radial values.

The shortest intercalation rule explains the basic facts of duplication and regeneration in imaginal discs. Except for those bisecting cuts that produce pieces with equal numbers of positional values, every bisection will produce one piece with fewer and one with more than half of the positional values. The shortest intercalation rule explains why one-quarter sectors always duplicate and their complementary three-quarter pieces always regenerate. The smaller pieces always find their shortest route through duplication, the largest through regeneration. The rule also explains the stimulatory effects of separate pieces, with different positional values, on the regenerative capacity of those pieces that would normally duplicate; the two positionally different fragments interact, augmenting each other's set of positional values and inducing regeneration.

In thinking about circumferential positional values, however, it must be remembered that both the depicted number (12) and their spacing are completely arbitrary. It is convenient to draw these values as equally spaced, but assuming that the shortest intercalation rule describes physical reality, there may be uneven clustering of positional values in both the leg (Schubiger, 1971) and the wing (Karlsson, 1981a).

In the original formulation of the polar coordinate model, a second rule of regeneration was postulated: the "complete circle rule." The postulate was that restoration of radial values—proximodistal regeneration—can occur only when a complete set of circumferential values has first been restored. This rule was based on the initial observations but has subsequently been disproven. It is possible to obtain regeneration of more distal structures from poximal ones without a complete set of circumferential values. It has been possible, for instance, to construct synthetic legs in salamanders and cockroaches by surgical means from identical half-legs and to obtain distal regeneration when the legs are amputated. Such synthetic, symmetrical legs cannot have the complete set of positional values, by definition.

Bryant et al., have replaced the complete circle rule with the "distalization rule." This hypothesis explains the regeneration behavior of positionally "incomplete" appendages and is based on the premise, like that of the shortest intercalation rule, that wound closure brings normally nonadjacent positional values into contact. The rule states that when new cells are generated by a preexisting adjacent cell, the new cells adopt a radial positional value that is one degree more distal than that of the displacing cell. Each round of intercalary cell division thus produces progressively more distal cells, until the process of circumferential filling of missing values is complete (Fig. 8.10).

One significant feature of the distalization rule is that it eliminates the need to posit gradients as controlling devices in pattern restoration. Regeneration is viewed as proceeding step-by-step as a result of strictly local cell interactions. Each newly generated cell reads and responds to the positional values of its immediate neighbors; the complete sequence of these responses generates the complete sequence of regeneration.

This posited sequential, step-by-step regeneration of positional values explains the observed step-by-step acquisition of the capacity to form progressively more distant structures as growth proceeds outward from the surface of the cut (Schubiger, 1971; Kauffman and Ling, 1981; Karlsson, 1981b). By emphasizing the role of cell–cell contacts in the generation of positional information, the model focuses attention on local cell surface properties as the "signatures" of position.

The revised polar coordinate model can also explain the occurrence and structure of pattern *triplications,* which sometimes occur during recovery

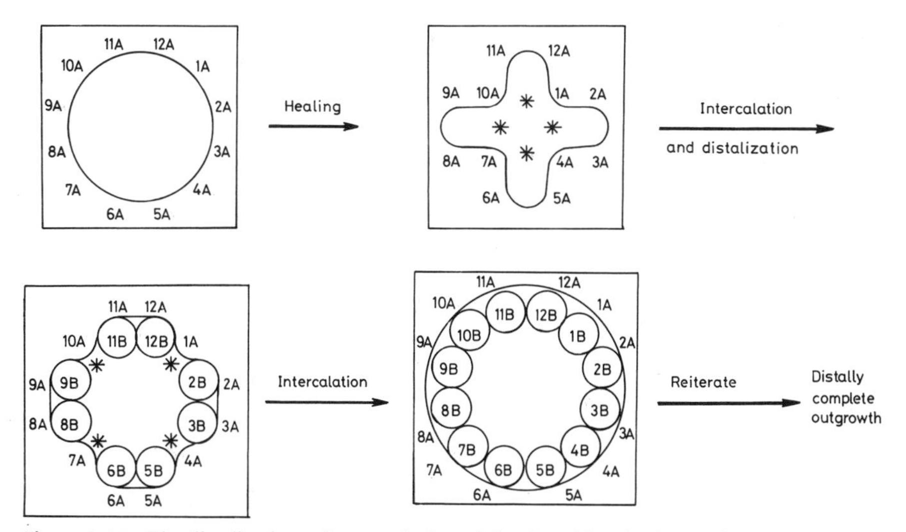

Figure 8.10 The distalization rule: completion of distal positional value set by outgrowth. See text for discussion. (Redrawn and reproduced with permission from Bryant et al., 1981, *Science* **212,** 993–1002. © 1981 AAAS.)

from injury. Triplications can form following certain surgical operations (see Bryant et al., 1981), localized UV-induced cell death (Girton and Berns, 1982), or heat pulses during development of ts cell lethals (Arking, 1975; Simpson and Schneiderman, 1975). Triplicated leg structures have been studied principally in the mutant *ts726,* where they can be induced with high efficiency by heat pulses during a narrow temporal interval surrounding the molt to third instar. Duplicated structures, in contrast, can be induced over a much longer period in this mutant, from first instar through early third instar (Girton, 1981). (The basis for this difference in the times of induction for triplications and duplications is not understood.)

Typically, two classes of leg triplication are formed, termed "diverging triplications" and "converging triplications" (Fig. 8.11). Both consist of one isolated normal leg appendage, termed the "orthodrome (O) element," and two fused elements, the "antidrome (A)" and the "paradrome (P)." In both classes, the O element is always of normal symmetry and orientation, the A element is in reverse symmetry to O, and the P element is of normal orientation; the fused A and P elements are thus always in mirror-image symmetry

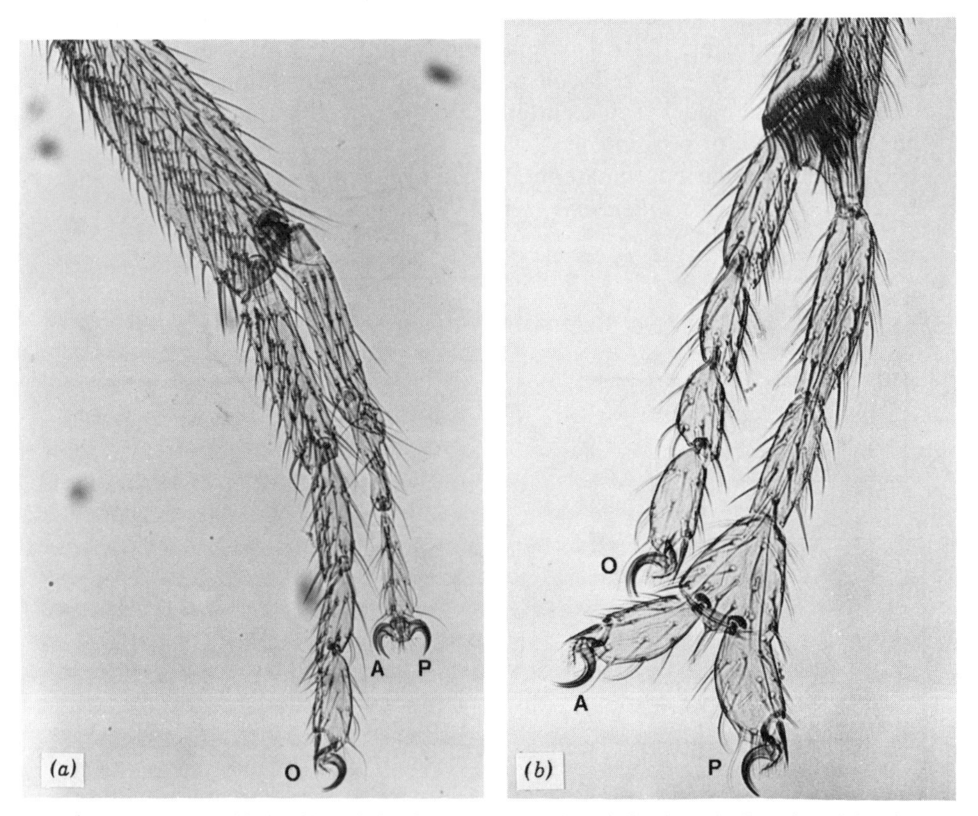

Figure 8.11 Two kinds of leg triplication. (*a*) Converging triplication; (*b*) diverging triplication. O, orthodrome; A, antidrome; P, paradrome. (Photographs kindly supplied by Dr. J. Girton; from Girton, 1981.)

with respect to one another. In diverging triplications, the A and P elements become more complete circumferentially in the distal direction, while in converging triplications, the A and P elements are fused distally.

The two patterns can be seen as the consequence of distal outgrowth following loss of positional values; diverging triplications may reflect a loss of a majority of these values, while converging triplications may follow from a loss of a minority of positional values following cell death (Girton, 1981). The two consequences are shown in Figure 8.12. In both diagrams, the

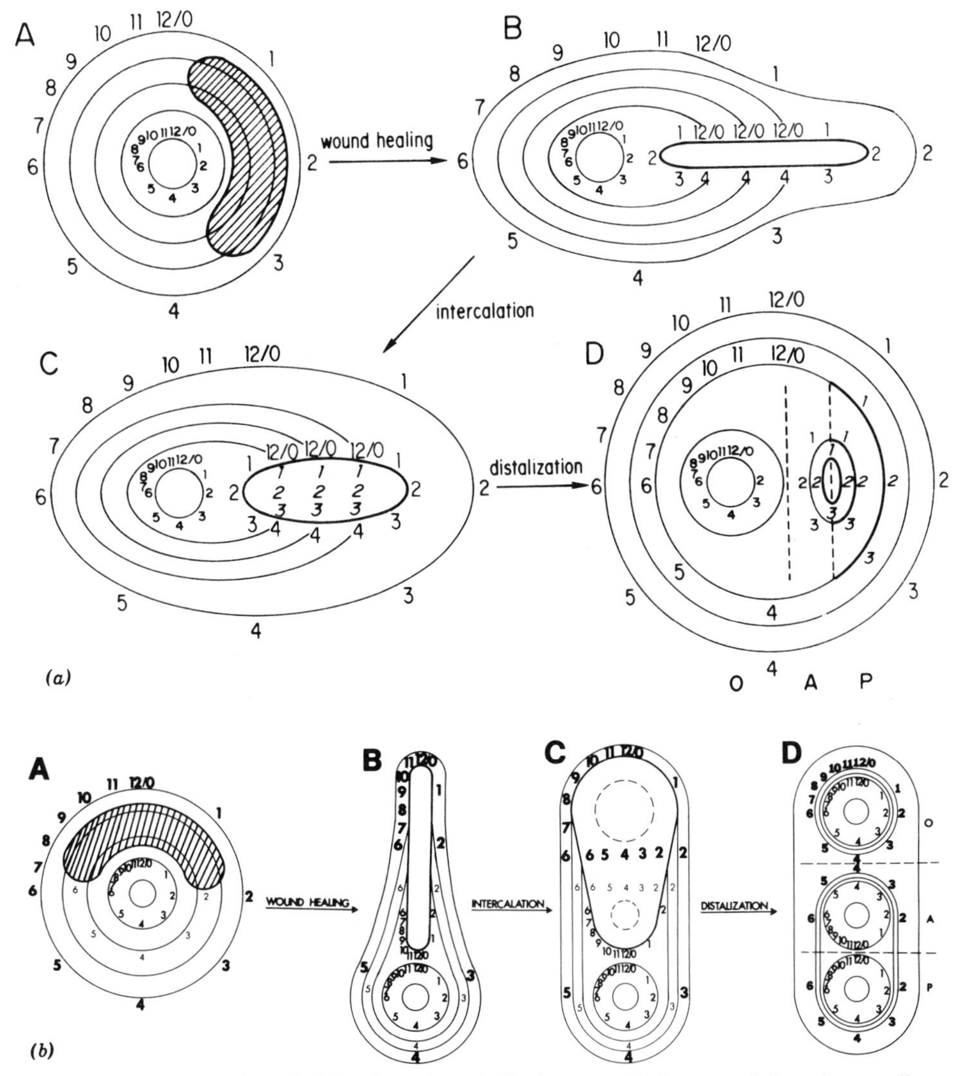

Figure 8.12 Formation of triplications through distal outgrowth, in terms of the polar coordinate model. (*a*) Converging triplication formation; (*b*) diverging triplication formation. (Figures courtesy of Dr. J. Girton; from Girton, 1981.)

majority of positional values are shown clustered in the upper medial quarter of the disc, since this quarter alone can regenerate the whole leg disc (Schubiger, 1971).

As can be seen in the figure, the elements that serve as the respective bases of the two kinds of triplication are different. In diverging triplications, the paradrome experiences the least regenerative growth and serves as the base; in converging triplications, the orthodrome acts as the base. If these predictions from the model are correct, then the P element in diverging and the O element in converging triplications should have the most cells of the three pattern elements shortly after triplication induction. A clonal analysis involving measurements of clone frequency—which gives an indication of cell number—bears out this prediction (Girton, 1983).

Some Implications

Perhaps the most significant aspect of the polar coordinate model is that it accounts for the construction of complex morphologies during normal growth: the process involves essentially a progressive intercalation of positional values along two axes during cell proliferation of the embryonic field. As new cells arise by division, they acquire either the same circumferential values as their neighbors or intermediate values and, at the same time, progressively more distal radial values. (The third dimension specification need not necessarily be along the p-d axis. Perhaps for some fields, situated medially in the developing animal, the third dimension is along the a-p axis).

Of course, each newly arisen cell has a particular lineage history as well as a unique position. However, the effects of variation in lineage history are smoothed out by the intercalation process; the indeterminancy of clone shape, in *Drosophila* and in the mouse, should therefore have no effect on the overall pattern. The early imposition of A-P compartment boundaries may be part of an enforcing mechanism for positional specification. The fact that these boundaries are breached by cells following wounding (Szabad et al., 1979b) indicates a built-in flexibility in the enforcement system when it is necessary to intercalate new values.

If regeneration utilizes the same positional cell markers that are employed during normal growth, then these cell markers should be present throughout growth. The existence of these cell properties might, in turn, be signified by similar regenerative behaviors at all stages. Indeed, Bownes (1975), using microcautery to induce localized damage in blastoderm embryos, found a few cases of thoracic structure duplications and one wing triplication in the adults that developed from surviving embryos. Postlethwait and Schneiderman (1973), employing x-ray treatment of embryos and first instar larvae, obtained an even higher frequency of duplicated imaginal structures. (The complementary events, successful regeneration of cells for deleted structures, would, of course, not be detected in such experiments.)

Furthermore, the sequence of developmental events in newly formed duplicated elements is highly similar to that in normal appendage development. Girton and Russell (1981) applied the standard techniques of clonal analysis to newly induced leg duplications, created by pulse exposure to high temperature in *ts726,* to ascertain the number of founder cells and the rate of cellular growth in the duplicated structures. Firstly, they found that each duplicate arises from approximately 7–20 founder cells, an estimated range similar to that for normal leg primordia. Secondly, the measured kinetics of growth, as estimated from the rate of clone size decrease, are virtually identical in the duplicate and its original (parental) leg structure. Thirdly, using the Minute technique, it was shown that the duplicated legs experience an early A-P compartmental restriction, just as normal leg primordia do. All of these findings suggest that the formation of a new leg primordium late in development employs much the same processes as those of normal development. Thus, pattern *re*-formation is highly similar to that used for positional specification throughout normal development.

The second major implication of the polar coordinate model is that of universality: all appendages, and perhaps all secondary fields in the embryo of a given species, use the same molecular markers to specify position. The supposition is the universality postulate of Wolpert transferred into the polar coordinate model. The evidence in support, however, is mixed. The different thoracic primordia may all have the same system, as indicated by homologous stimulation of regeneration by haltere and wing fragments of comparable position (Adler, 1979; Haynie, 1982). Furthermore, some fragments of the leg disc stimulate the regeneration of proximal wing disc fragments, while others do not (Haynie, 1982), indicating possible homologies between leg and wing. On the other hand, eye-antennal and genital discs do not seem to have this capability (Wilcox and Smith, 1977). The failure of antennal material to stimulate leg regeneration is surprising in light of the seemingly strong homologies between antenna and leg (Fig. 8.3).

The question of universality of positional values is intimately connected to a seemingly very different one, that of the basis of transdetermination. In early experiments on leg disc regeneration, Schubiger (1971) found that only fragments of the leg disc that are capable of regenerating show transdetermination. Strub (1977) subsequently reported that the presence of killed "feeder layer" cells can stimulate transdetermination of leg disc fragments. The feeder layer cells (prevented from further division by x irradiation) can be obtained from whole antennal, wing, or leg discs. Strub attributed the efficacy of killed feeder layer cells to the provision of complementary positional values; these values presumably stimulate growth, an essential precondition for transdetermination. A corollary is that all discs that provoke transdetermination in other discs have a common system of positional values. Hence, by this measure, antenna and leg *do* share such a system. Clearly, the question of universality of positional values is still to be resolved. Because the operational tests (based on the presence or absence of stimulated regeneration) may be too insensitive for what is being sought, an answer may be possible only when the molecular correlates of positional value have been identified.

Some Complications

Despite the success of the polar coordinate model in explaining the findings on pattern formation, some puzzling anomalies remain. The first of these has been mentioned: the presumptive unequal spacing of positional values within certain primordia. The heart of the model is the proposition that the pattern re-formation behavior of a fragment (of a disc, leg, etc.) is a function of the *number* of positional values it has. However, the only way to reconcile certain of the findings with the model is to posit that for certain secondary fields, the spacing of positional values is uneven, while being nearly equidistant for other fields.

The second difficulty stems from the supposedly fundamental dichotomous choice between duplication and regeneration. In reality, the behavior of imaginal disc fragments is more complex. Fragments that normally duplicate during short-term growth will often show some regeneration of pattern elements in addition to duplication when allowed prolonged growth (Kauffman and Ling, 1981; Karlsson, 1981b; Kirby et al., 1982). In contrast to normal regeneration behavior, in which the sequence of elements formed is a simple function of the distance from the cut edge, the long-term regenerative behavior of duplicating elements may "skip over" certain pattern elements (Kirby et al., 1982).

The significance of this finding has become connected to a second debate about the role of intercompartmental cooperation in regeneration. Schubiger and Schubiger (1978) suggested, on the basis of their experiments on leg discs, that distal regeneration requires the presence of cells from both the A and P compartments. Karlsson (1981b), on the basis of her observations of regeneration within duplicates, made a similar suggestion for circumferential regeneration. These conclusions have been questioned by others. Girton (1981) reported that some converging triplications are made of pattern elements belonging wholly to either anterior or posterior leg compartment material. Kirby et al. (1982) noted that a thin wedge of wholly anterior compartment material in the wing disc can undergo some regeneration of circumferential pattern elements within duplicates. These observations suggest that there is no strict requirement for material from both compartments to provoke regeneration.

A partial solution to the problem of regeneration within duplicates may follow from the recognition that this behavior is different from standard regeneration. In the first place, it takes place *after* duplicative pattern restoration has occurred. It therefore seems to be a secondary response. Furthermore, in the majority of duplicated implants showing some regeneration of pattern elements, the new structures arise near the line of mirror-image symmetry. Kirby et al. (1982) suggest that duplication induces growth instability

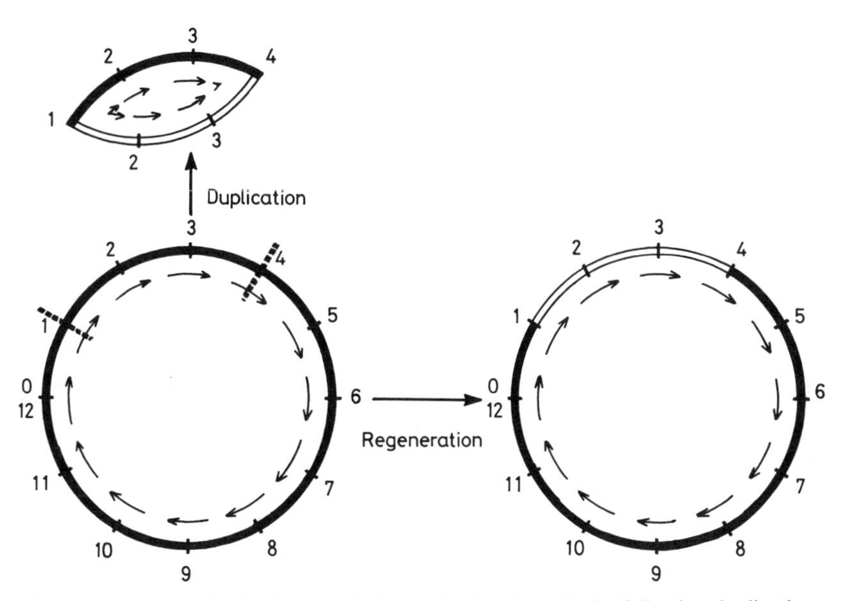

Figure 8.13 Hypothesized reversal of positional value polarity following duplication.

at the line of symmetry in duplicates, which in turn triggers regeneration. If, for instance, the sequence of circumferential values has a standard polarity characteristic of some kind, then duplication should produce localized regions of polarity reversal where the duplicates meet (Fig. 8.13). The clash of polarity reversal might prompt the regenerative response. (Note: this kind of polarity need have no connection with the polarity patterns of bristles and hairs described earlier.)

The suggestion of a vectorial quality associated with positional values adds an extra complexity to the polar coordinate model, but it too leaves some questions unanswered. Narrow strips of disc corresponding to the anterior or ventral region of the disc undergo secondary regeneration readily, yet equivalent-sized or larger pieces corresponding to the dorsal (notal) region do not. Evidently, there are regional properties that affect the regenerative capacity in duplicated fragments. Do these regional differences somehow reflect different numbers of positional values, or are some positional values unique in ways not predicted by the polar coordinate model?

PROGRAMMED CELL DEATH: A MODELING AGENT?

The polar coordinate model accounts for the general, aggregate properties of pattern formation but does not explain the origins of the detailed features of morphology. For that, other processes must be involved. One means of shape generation is differential growth in particular regions of an embryonic field, which almost certainly plays a major part in morphological structuring. However, modeling of two- or three-dimensional features might equally involve selective programmed cell death. There is a great deal of evidence, for instance, that localized cell death is a major contributory factor in the production of discrete digits in the vertebrate limb.

Programmed cell death may, in fact, be generally important in animal morphogenesis. Its importance in eliminating supernumerary cells in nematode development was discussed in Chapter 5. In *Drosophila,* many of the morphogenetic deficiencies associated with particular mutants may result from unleashed cell death. In some cases the precursor cells never arise, but in others the cells are born and then die. The wing mutant *vestigial (vg)* is an example. In an electron microscopic study, Fristrom (1968) showed that the wing discs of *vg* undergo massive cell death and degeneration. James and Bryant (1981) subsequently showed that cell death in the wing discs takes place specifically during the third larval instar. Significantly, they also observed some cell death, although much less, in wing discs of the wild-type. Other mutants, such as *scalloped* (*sc*) and *Beadex* (*Bx*), produce wings with ragged, incomplete edges. In these cases, it appears that the wing deficiencies are produced by localized cell death in the regions that give rise to the wing margins.

In these examples, and perhaps many more, the mutant defect may only be amplifying the extent of localized cell death that occurs during normal development. The augmentation may be by indirect and non-specific means (Girton and Bryant, 1980). From this perspective, it is the localization of cell death patches in the wild-type that is of primary interest. Presumably, the signal to die arises from the properties of the field itself (just as accentuated, localized cell proliferation must) and affects certain groups of cells that are for some reason "competent" to respond to it. A general solution to the problem of how regionalized cell death is built into the developmental program still awaits formulation.

A DEVELOPMENTAL GENETICS OF PATTERN FORMATION

Pattern formation is a subject at once central and elusive in developmental biology. The elusiveness of the genetic basis of pattern formation is particularly obvious; there are at present few reliable genetic techniques to deal with the phenomenon. However, genetic experiments have made some contributions. Genetic analysis, for instance, has eliminated the simplest prepattern-type explanations. Furthermore, clonal analysis has contributed a characterization of the growth dynamics and origins of supernumerary pattern elements.

Although the prospects for a genetic probing of pattern-forming mechanisms are slim at present, they are not nonexistent. The key postulate of positional information theory and of the polar coordinate hypothesis in particular is that there are certain common positional values that are used again and again within the various secondary embryonic fields of any animal system. If these positional values are solely concentrations of a universal morphogen gradient, then the matter will probably be forever beyond genetic analysis. However, it seems that however positional coordinates are initially established, they become fixed in or on cells, presumably as some form of cell surface label. All of the cutting-and-regeneration experiments make sense only if cells somehow "remember" their positional values. The simplest form of cell memory would be one embodying specific, qualitatively distinct molecules. If so, then genetic specification must play a part in the production of such molecules. The criterion for a positional value mutant in *Drosophila* is that it must show characteristic aberrations in all or most imaginal discs as a *function of position*. Have any such mutants been identified to date?

One candidate, of course, is *engrailed*. All of the mutations at this locus show marked or preferential aberration of imaginal disc posterior compartments (with the possible exception of the haltere). In the original suggestion of Morata and Lawrence (1975), one of *en*'s functions was specific cell labeling. This might be its only function, with all of the other aberrations reflecting indirect, regulative responses. Since the consequences of deletion of any particular positional value must be attempted intercalation of the missing value, it is to be expected that any positional value mutant would show a wide range of morphological aberrations in addition to the primary defect. Indeed, any such mutant would probably be an early lethal, and its position-specific effects could be detectable only in homozygous clones induced in heterozygotes, in a fashion similar to that of the *en* analysis. It is conceivable that some of the mutants identified first through their effects on larval segmental patterns (Table 6.2) might be positional value mutants. The possibility mentioned earlier that the wild-type alleles of these genes function in the establishment of the A-P compartment separation is not incompatible with a basic role in establishing positional values.

Ultimately, the identification of general positional cellular markers may come through the use of monoclonal antibodies raised against imaginal discs. Any monoclonals that bind to comparable geographical positions in many discs would be candidates for presumptive positional value-specific antisera. Searches of this kind have not yet proven successful—and indeed, if positional values are truly universal, it might be impossible to raise antibodies against them—but improvements in search methods may be the key. If genetic analysis pinpoints certain prospective mutants as possible positional value mutants, the candidate monoclonals could then be used as secondary screens to characterize the defects further.

PART FOUR

PRESENT AND FUTURE

NINE

CLONING DNA

THE ANALYSIS OF GENE STRUCTURE, EXPRESSION, AND FUNCTION

The development of techniques for cloning DNA sequences in the early 1970s constitutes the most important technical advance in modern genetics. Although the influence of the recombinant DNA revolution has been felt in every branch of genetics, it promises to be especially strong in the study of development. Instead of being forced to make indirect inferences about gene structure, expression, and action from the study of mutant phenotypes, investigators are increasingly able to ascertain these facts directly from analyses of the isolated genes. (These methods are also proving invaluable in characterizing the general properties of genome organization, as we have seen in Chapters 5 and 6.)

Before describing the application of recombinant DNA techniques to developmental questions, the techniques themselves require a brief introduction. Although the literature on this subject is already vast and still growing, the essence of cloning DNA sequences is fairly simple and consists of four fundamental steps: (1) a method for breaking long DNA molecules into smaller fragments; (2) inserting the collection of sequences by enzymological means into small host or "vector" DNA molecules that are capable of being replicated in bacteria; (3) introducing the resulting recombinant DNA molecules (each consisting of vector plus insert) into bacterial cells, which permits their replication and, hence, amplication; and (4) screening the different DNA clones for the particular sequence(s) of interest. If one is especially fortunate in being able to begin the process with a purified DNA of the desired sequence, then step 4 becomes unnecessary. In most instances, one begins with a heterogeneous mixture of DNA molecules and produces a correspondingly diverse set of DNA clones; in all of these cases, the screening procedure to eliminate the unwanted clones is of the utmost importance.

There are essentially two ways to fragment DNA molecules: physical shearing and enzymatic digestion. In general, only when large sequences, more than 30 kb, are needed is physical fragmentation employed. This results in DNA molecules with "ragged" ends that have to be enzymatically treated to make the fragments susceptible to insertion in the vector molecules. This is usually carried out by filling in the single-stranded ends, making them blunt-ended; the fragments can then be sealed into similar blunt-ended vector molecules.

The enzymatic digestion methods employ the so-called restriction enzymes, mentioned in Chapter 5, that cut at specific base sequences. Enzymatic means are normally the methods of choice because they permit precise cutting and precise excision of the desired sequences. The precision of enzymatic cutting stems from the fact that most of the enzymes cut only at specific short (6 or 4 bp) sequences and make staggered breaks in DNA leaving unpaired "sticky" ends of 2 or 4 bp. By also cutting the vector molecules with an enzyme of the same specificity, one can generate short complementary sequences in both the vector and the insert.

Cutting with the enzyme BamH1 will illustrate the process of recombinant DNA construction; the hypothetical situation is that of inserting *Drosophila* DNA segments into a small vector with a single BamH1 site. (The vectors are often constructed, by means of the appropriate genetic and enzymatic modification, to contain just a single site for the insertion of foreign DNA; the site chosen is one whose interruption does not impair the vector's replicative abilities.) BamH1 happens to cut DNA between two Gs at the specific recognition sequence CCTAGG, where the arrow marks the site of cutting. Note that this sequence, like most restriction enzyme sites, has dyad symmetry; that is, a 180° rotation produces the same double-stranded sequence, with consequent cutting on both strands. The effects of cutting both DNA samples (the DNA to be inserted and that of the vector) is illustrated in Figure 9.1. Complementary regions of four bases each are generated on both sets of molecules, permitting pairing and insertion of cut *Drosophila* DNA into the vector. Formation of the recombinant molecules by pairing of the short complementary ends is but the first step in the construction process. To make the annealed molecules functional, one then treats with the enzyme ligase, which seals the phosphodiester chain breaks at the sites of insertion.

Of course, other pairing reactions are also possible. The vector, for instance, might simply reseal. Hence, one requires stratagems to ensure the selection of true recombinant DNA molecules. One such technique employs a plasmid that possesses two genes for antibiotic resistance, with the insertion site placed in one of these resistance genes. (A plasmid is a small, DNA molecule that exists within the bacterial cell and that replicates independently of the bacterial chromosome without harm to the host cell.) With the insertion site in one of the resistance genes, recombinant plasmids made in vitro would be those that had lost that particular resistance marker but had retained the other.

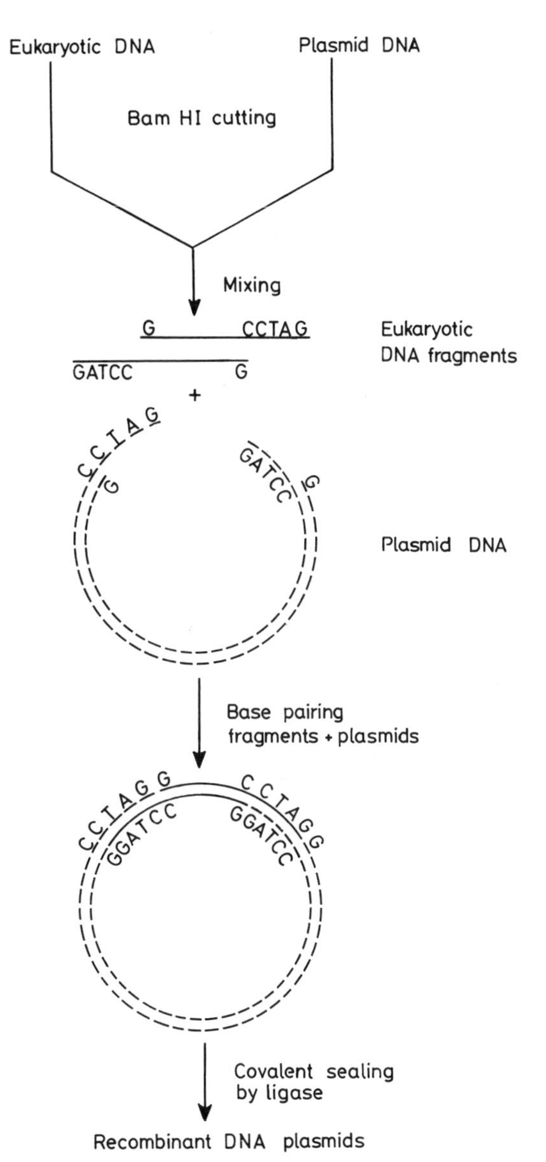

Figure 9.1 Cutting DNA with BamHl and making recombinant DNA molecules. See text for discussion.

If one does not know the sequence of the gene whose cloning is being attempted, there is the chance that the gene or sequence will contain a site sensitive to the restriction enzyme employed; in that case, cutting will inactivate the desired sequence. There are many ways to avoid this problem: one might, for instance, use only a partial digestion so that not every site is cut, or one might do the cutting with different enzymes on different samples and perform separate gene clonings for each kind of enzymatic digestion.

Completion of cutting and rejoining to make recombinant DNA molecules brings one to step 3, the placement of these molecules in bacterial host cells. The technique used depends upon the kind of vector employed, and there are two classes of vector. Plasmids have already been mentioned. The other kind of vector is a bacteriophage, namely, a bacterial virus. The typical bacteriophage or phage vector is derived from the bacterial virus lambda, one of the most thoroughly characterized organisms in molecular biology. If one has cloned sequences into bacteriophage, the recombinant DNA molecules must be "packaged" into infectious particles in vitro. This process is carried out by mixing the recombinant DNA phage in a lysate containing the protein components of the phage particle. When the conditions are right, the DNA molecules are assembled into functional virions, with an efficiency as high as 10^{-3}. (With tens of millions of DNA molecules in the mix, an efficiency of this order can produce many thousands of recombinant DNA phage particles per milliliter of solution.) Bacteria growing in culture can then be directly infected with these phage particles, and after numerous cycles of infection, the recombinant DNA molecules are correspondingly amplified. The collection of recombinant particles, if carrying many different DNA inserts, is termed a "library" or a "bank." (The factors that influence the choice of metaphor have not yet been analyzed, but that discussion takes one into the realm of psychology.) The recombinant phage particles can then be plated on a bacterial lawn, each focus of infection or plaque being pure for a particular recombinant phage, provided that the initial infection was at a low phage : bacteria ratio.

If, on the other hand, the recombinant DNA molecules consist of bacterial plasmids with inserted DNA, transfer into bacterial host cells is by means of bacterial transformation. *Escherichia coli* (the standard plasmid host) is treated with a calcium salt solution under suitable conditions, which make the cells permeable to exogenous DNA. These cells can then take up plasmid molecules without losing their viability and, upon plating, can give rise to single colonies directly. Each colony originating from a cell containing a recombinant DNA plasmid molecule gives rise to a recombinant plasmid clone. To prepare any such clone in bulk, one inoculates the colony in liquid culture and grows it to obtain a mass culture.

In any large collection of genetically heterogeneous recombinant phage or recombinant plasmid-carrying colonies, screening for the desired gene clone is essential. The particular procedure used is dictated in part by the source of the inserted DNA, whether the cloned sequences have been obtained directly from the chromosomal DNA ("genomic clones") or by copying an mRNA population with the enzyme reverse transcriptase to produce a set of DNA copies of transcripts ("cDNA clones"). In the majority of cases—all except those in which a relatively pure mRNA has been copied into cDNA—the population of starting DNA molecules for cloning is very heterogeneous, and it becomes necessary to screen large numbers of recombinant DNA molecules for the one of interest.

Imagine, for instance, that one is interested in obtaining a particular gene sequence of about 10 kb from the mouse genome, and one has begun by first digesting and then cloning the entire nuclear genome into 10- to 20-kb pieces into phage lambda. Because the mouse genome is about 3×10^6 kb of DNA, the probability of any one recombinant plasmid harboring a colony containing the desired 10-kb sequence is about $10/3 \times 10^6$ or about 0.33×10^{-5}. Clearly, one needs a winnowing process to find the gene of interest.

There are many different screening procedures, but only three types will be mentioned here. The most common procedure involves screening the recombinant plasmid colonies or recombinant phage plaques for hybridization to the sequence of an mRNA complementary to the gene (when this purified mRNA is available).

The Grunstein–Hogness (1975) technique was the first of such hybridization methods to be developed. In this procedure, groups of recombinant plasmid-carrying colonies, each grown on an agar plate, are replica plated to a master plate (for later retrieval) and to a nitrocellulose filter. On the latter filter, they are treated so as to make their DNA accessible for the subsequent hybridization and then baked. The filters are then exposed to a solution of heavily labelled cDNA copies of the RNA of choice, under conditions that favor annealing of complementary strands, and then washed to remove unbound label, dried, and placed under autoradiographic film. The colonies showing a high grain density are those that bear complementary DNA. The colonies are then picked from the master plate and grown. Similar procedures for isolating particular recombinant phage, using plaque hybridization, have been developed.

If the sequence of interest is known to be contained in an mRNA population, then there are distinct advantages in making a cDNA library. In particular, there is the elimination of a large amount of irrelevant DNA, perhaps more than 98–99%, not transcribed into mRNA. The labor of screening for the desired sequence is thus considerably reduced. An example of the use of cDNA banks to screen for genes with tissue-specific expression is shown in Figure 9.2.

However, having a cDNA clone is not equivalent to having the genomic clone. Firstly, the cDNA sequences are often short (100–800 bp), because of the nature of the synthesis reaction; hence, a given cDNA clone usually contains only a portion of the coding region of the gene of interest. Secondly, the cDNA clone necessarily contains only the nucleotide sequence present in the mRNA; it does not contain any flanking sequences either to

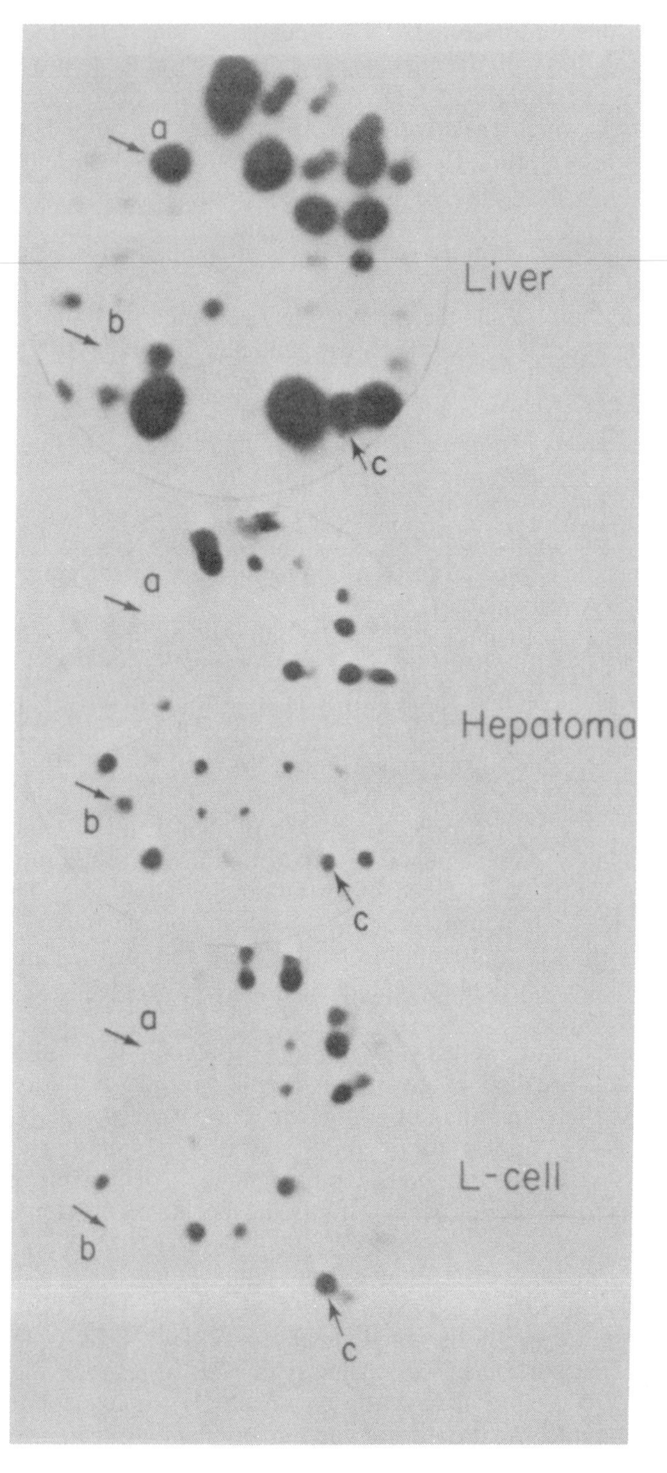
a
b
c
Liver
a
b
c
Hepatoma
a
b
c
L-cell

the left of the 5′ end of the message or to the right of the 3′ end. Such a DNA clone therefore lacks any nontranscribed controlling sequences. Furthermore, cDNA clones necessarily lack any introns present in the gene, since these are removed from the primary transcript. When starting with a purified cDNA clone, therefore, it is often necessary to retrieve the homologous (but more complete) genomic clone by screening for hybridization of these clones to the labeled cDNA. Although this procedure for finding the genomic clone may appear to be indirect, it often involves less work than a direct search.

Another general set of approaches involves screening cloned sequences in plasmids either for a particular biochemical function or for production of an immunologically recognizable protein. For instance, if one is attempting to clone a gene that encodes a particular enzyme, one can transform bacterial cells deficient for that activity with the library of recombinant plasmids and select for those that can carry out that enzymatic action; such procedures require special "expression vectors," designed to maximize expression of the cloned sequence. It is also possible to transform yeast cells, and these, being eukaryotic, are often better for obtaining expression of cloned eukaryotic genes.

A particularly elegant method for obtaining sequences, which is independent of activity, transcription, or even hybridization, is applicable to *Drosophila* only at this point: cutting out polytene chromosome bands of interest by microdissection and cloning the physically isolated sequences (Scalenghe et al., 1981). In this situation, one then needs some independent criterion to be certain that one has obtained the sequence of interest. Another set of approaches, using polytene chromosomes, involves chromosome "walking" and "jumping," described below.

In this chapter, three kinds of analysis employing cloned genes will be illustrated. They are (1) the exploration of the structure and expression of two developmentally significant clusters of genes, BX-C and ANT-C in *Drosophila*; (2) the analysis of the molecular basis of a genetic lesion of development, *oc*; and (3) the analysis of gene behavior in relationship to gene structure and position for cloned genes placed in the germ lines of fruit flies and mice by means of DNA transformation.

ANALYZING LARGE DNA REGIONS: PROBING THE BX-C AND ANT-C

The use of genetic stratagems to deduce the roles of the BX-C and ANT-C genes in setting segmental identities in *Drosophila* was discussed at length in

---

◀ **Figure 9.2** Screening of recombinant cDNA plasmids by colony hybridization to poly(A)$^+$ mRNA from three sources. The arrows indicate (*a*) a colony giving a signal with liver mRNA only, (*b*) a colony giving its strongest signal with hepatoma mRNA, and (*c*) a colony showing hybridization to mRNA from all three sources. (Redrawn and reproduced with permission from Derman et al., 1981; copyright MIT Press.)

Chapter 6. Molecular analysis of these two gene complexes has substantially confirmed the conclusions of the genetic studies and provided some surprising new insights.

The cloning of BX-C gene sequences was carried out by a combination of chromosome "walking" and "jumping" techniques. The principle of chromosome walking is simple: one takes a short, unique DNA sequence clone (or "probe"), labels it to high specific activity, and performs a hybridization to thousands of recombinant DNA clone–containing colonies or plaques. The colonies or plaques that hybridize to the probe are then identified, and the new clones are grown. The distal ends of these new DNA clones, not containing the initial probe sequence, are then used in a second round of colony or plaque hybridization to identify new neighboring cloned sequences. In principle, the process can be repeated ad infinitum. (Only the presence of repeated sequences can seriously sidetrack the walk.) By these means, one can isolate sequences that are progressively more distal from the initial sequence.

In *Drosophila,* one can then directly visualize the direction of the walk by doing in situ hybridizations with the successively isolated clones to suitably prepared polytene chromosome spreads. If the labeled cloned sequence in each hybridization is composed of unique sequence DNA, it will hybridize to one specific band or interband region; the site of hybridization is detectable autoradiographically. By examining the positions of the succession of clones, the direction of the walk becomes apparent.

Each cycle of a walk moves one down a chromosome just 10–20 kb; if one does not have a starting probe whose sequence is close to the one of interest, one may need an additional stratagem to move closer to the target sequence. One such stratagem is chromosome jumping: it involves the use of inversions that bring the target sequences closer to the sequences at hand. By cloning sequences in a walk from the noninversion-bearing strain, and then determining which clone shows in situ hybridization to two positions in the inversion strain, one identifies the DNA clone that spans the inversion breakpoint. This clone can then be used to identify clones from the noninversion strain that are close to or include the target sequence.

The combined chromosome walking/jumping procedure used to isolate the BX-C sequences is schematized in Figure 9.3; the details of the walk are described in Bender (1983a). To date, over 195 kb of the BX-C has been cloned, the region comprising the "left-hand" end (including the thoracic-segment essential genes) and part of the "abdominal" portion of the complex. With the wild-type sequences isolated, it then becomes possible to characterize many of the mutations in the complex, which were previously characterized only by their phenotypic effects. The surprising result is that most of the known mutants of the BX-C are mutant because of the insertion of pieces of DNA, either whole transposable elements or their fragments; a smaller number involve deletions (Bender et al., 1983b). Very few of the mutations are classical point mutations consisting of single base changes.

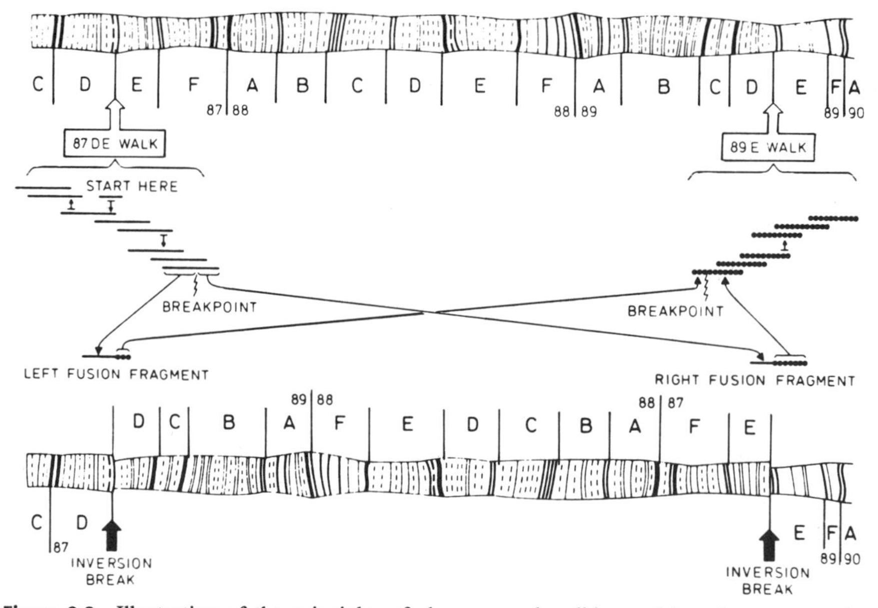

Figure 9.3 Illustration of the principles of chromosomal walking and jumping; see text for discussion. (Reproduced from Bender et al., 1983a. With permission from *Journal of Molecular Biology,* Vol. 168, copyright 1983 by: Academic Press Inc. (London) Limited.)

The prevalence of insertion mutations was discovered by means of "restriction enzyme mapping." This method involves digesting genomic DNA with different restriction enzymes in separate trials, electrophoresing the fragments on gels, and then hybridizing the fragments, transferred to a suitable matrix (usually nitrocellulose), to the individual DNA clones. Since each restriction enzyme used has a different base sequence specificity, each digestion yields a characteristic pattern of cuts and hence fragments. From a comparison of the different fragment patterns produced by the different enzymes, it is possible to determine where the specific enzyme cutting sites are located within the DNA sequence tested. The composite picture of restriction enzyme sites is termed the "restriction enzyme map." Each labeled DNA clone hybridizes only to those fragments with which it has homology, producing a characteristic band pattern on the autoradiograph. If any particular fragment from digested mutant DNA has altered its size—for instance, through insertion of a foreign sequence—this change is revealed on the autoradiograph by the altered position of the fragment. In general, point mutations will alter their position only if they alter particular restriction enzyme sites directly, preventing cutting at that site; however, such changes can be distinguished from insertions or deletions by the finding that one large "new" fragment is the size of two smaller fragments from the wild-type digestion.

The prevalence of insertion mutants among the BX-C mutants was deduced by this technique. Some of the sites of these insertions are shown in Figure 6.23. Equally surprising is the finding that many "revertants" of these mutants actually retain a piece of the inserted DNA. Because inserted DNA in coding regions should be fatal to the function of that coding region, the original mutants must carry the insertions not in coding regions but in introns or flanking regions. Evidently, DNA within introns can affect gene expression, perhaps by interfering with the normal splicing mechanisms.

By determining the outer limits of the DNA regions within which insertion mutations produce detectable mutant phenotypes, it has been possible to define the functional domains of two regions, *Ubx* and *bxd*. *Ubx* is 73 kb long, while *bxd* is 25 kb long (see Figure 6.23). Within the former region, *Ubx* mutants are generated by any form of chromosomal break (inversions, translocations, deletions, etc.) and by certain point or "pseudopoint" (very small deletions or insertions) mutations.

One of the surprising observations was the finding that both *bx* and *abx* mutations map within the *Ubx* region. What is one to make of a 73-kb "gene" that includes at least two other genes?

An important clue is provided by the transcripts produced by the *Ubx* region (as defined by their hybridization to clones within the region). All of the mature polyadenylated mRNAs that map to this region are considerably smaller than both the region itself and a very high molecular weight transcript from the region that has been detected (cited in North, 1983). It appears that the *Ubx* region is initially transcribed from beginning to end as a single transcript, presumably about 73 kb in length, and processed into a variety of smaller mRNAs; however, all of the mature mRNAs appear to have a common 5′ exon, and many share a 3′ exon. The results indicate that differential splicing plays an important part in the generation of mRNA diversity from this region. It is possible that particular mutant sites, such as those that give rise to *bx* or *abx* phenotypes, define exons that are included in some transcripts but not others; their mutational inactivation would give specific mutant phenotypes without affecting other functions from the same functional region. The variable utilization of different pieces of the genetic information of BX-C to generate different functions has implications for hypotheses of differential control of BX-C functions: a simple gradient of repressor cannot explain the full complexity of gene function expression in the region.

In addition, the possibility of differential splicing within the *Ubx* region has implications that extend beyond the BX-C. It may be that all "complex" loci are complex because they consist of sequences of exons that can be combined in different ways. Those mutations in complex loci that fail to complement with all others might be located in shared exons, while those that complement with some but not others may be in exons present in particular subsets of the total set of transcripts.

One genetic anomaly left unresolved by the findings of separate *Ubx* and *bxd* domains is the fact that inactivation of the former interferes with expression of the latter. The nature of this interaction remains unknown.

The discovery that the *Ubx* region is transcribed initially into a 73-kb transcript furnishes a reasonable explanation for the behavior of certain mutants of the RNA polymerase II gene. This gene encodes the RNA polymerase responsible for all mRNA synthesis. Certain viable mutants of this sex-linked gene (1–35.7) give a *Ubx*-like (*Ubl*) phenotype in heterozygotes (Mortin and Lefevre, 1981). Other mutations in this gene enhance a *Ubx* phenotype. Both phenotypes may reflect the inability of mutationally damaged polymerase to complete transcription of the *Ubx* region, a period that in eukaryotes would last for about 40 min. Not surprisingly, several of the *Ubl* alleles affect the expression of many other genes, either suppressing or enhancing the effects of various recessives, either in heterozygotes or in hemizygotes.

Chromosome walking and jumping strategies, similar to those employed for the BX-C, have revealed a comparable complexity for the ANT-C (Garber et al., 1983; Kaufman, 1983). The locus of *Antp*, like that of *Ubx*, is large, being about 100 kb in length, and gives rise to two or more much shorter, mature mRNAs, whose characteristics have been inferred from their corresponding cDNAs (Garber et al., 1983). One such mRNA, synthesized in embryos, contains four exons from four widely spaced locations within a genomic region of 100 kb. Together, the exons comprise 2.2 kb, or only 1/50th of the genomic *Antp* region. The other cDNA, synthesized from poly A^+ mRNA of 1-day-old pupae, consists of three exonic regions that span a region of 37 kb of genomic DNA; one of these exons is not present in the other transcript. As with the BX-C, differential splicing is implicated in the expression pattern of a complex genetic region.

The most exciting and potentially most important finding from the cloning of the two gene complexes has been the discovery of the "homeobox" (McGinnis et al., 1984a,b). This is a short region encoding a polypeptide domain 60 amino acids in length that is found in both the BX-C and the ANT-C and a small number of other genes including the pair-rule gene *ftz*. In *Antp, Ubx,* and *ftz*, it is located in the 3′ exon of the transcripts. The putative significance of the homeobox is at least twofold. Firstly, its sequence is very similar to that of DNA-binding domains in certain classical prokaryotic regulatory genes (Laughan and Scott, 1984); this may imply that genes containing a homeobox have a fundamental role in the regulation of transcription, supporting the idea that the genes of the BX-C and ANT-C are involved in the direct regulation of expression of other genes. Secondly, not only is the homeobox highly similar in the three characterized *Drosophila* genes in which it has been found—they are identical at 45 of the 60 amino acid positions—but the sequence is also found in other insects, annelids, and even vertebrates (frog, chicken, mouse, and human) (McGinnis et al., 1984b). In all of these animals, it is present in low copy number, as it is in the fruit fly. Sequences homologous to the homeobox are also found in yeast (in the yeast mating-type locus) but are not found in nematodes or sea urchins; they are thus widespread without being universal.

The full significance of the homeobox remains to be ascertained. McGinnis et al. (1984b) suggest that it may play a fundamental role in creating segmental patterns, since it is found in metameric animals but not in others, such as nematodes and sea urchins, which lack a segmental pattern. However, two qualifications must be kept in mind. The first is that most of the known homeobox sequences in *Drosophila* are found in genes that affect segmental character rather than segmentation capacity per se. The exception is the *ftz* homeobox; however, many of the other pair-rule genes seem to lack the homeobox. The second caution in ascribing a general *developmental* function to genes with homeoboxes is that homeobox-like sequences are found in yeast but not in echinoderms or nematodes, whose body plans have considerably more in common with the fruit fly than with *Saccharomyces*. Nevertheless, the elucidation of the significance of the homeobox promises to be an exceptionally interesting and informative story.

The existence of cloned gene segments from the two gene complexes has been significant in another way: it makes possible molecular tests of the genetic hypotheses proposed for the BX-C and ANT-C. One particularly useful technique is that of in situ hybridization for measuring the cellular concentrations and distributions for the transcripts of these genes. Results for *Ubx* transcripts have been reported by Akam (1983) and for *Antp* by Hafen et al. (1983) and Levine et al. (1983).

For both *Ubx* and *Antp*, expression begins in embryos at or shortly after blastoderm formation, during early gastrulation. Although labeling is detected in epithelial cells, the strongest labeling for both gene probes (the 5′ exon of *Ubx* and the entire 2.2-kb embryonic cDNA of *Antp*) is detected in the developing ventral nerve cord. As the segmental organization of the ventral nerve cord becomes manifest, the segmental specificity of expression becomes apparent. For the *Ubx* probe, there is no labeling observed either in the cephalic neural structures or in the pro- or mesothoracic ganglia; however, labeling is strong in the metathorax and even stronger in the ganglion of AB1, and then diminishes gradually until the ganglion of AB7 is reached (Fig. 9.4*a*). The result strongly supports the genetic inference that *Ubx* is not expressed in the mesothorax but in all posterior segments. (Whether it is expressed in the ganglion of AB8 is unclear; the signal may be below the level of resolution.)

In contrast to this pattern, *Antp* is most strongly expressed in the mesothoracic ganglion, with some labeling in the prothoracic and metathoracic ganglia (Fig. 9.4*a*). Again, the segmental specificity is consistent with the genetics: $Antp^+$ function is required and expressed in all three thoracic segments, with an especially important role in the mesothorax (in imaginal development, that of activating leg development). Furthermore, the discovery that there is abundant expression from the BX-C in the nervous

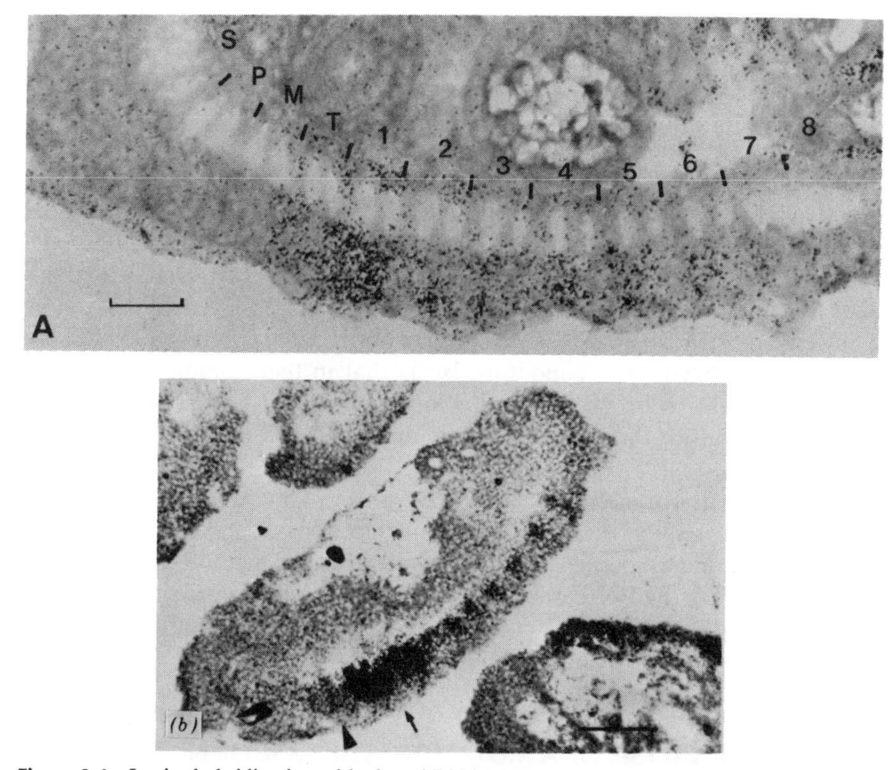

Figure 9.4 In situ hybridization with cloned DNA segments to measure transcript abundances in *Drosophila* embryos. (*a*) *Ubx* transcripts in a medial longitudinal section of a 14-hr embryo; the heaviest labeling is over the metathoracic (T) and first abdominal neuromeres. (Reproduced with permission from Akam, 1983.) (*b*) *Antp* transcripts in a sagittal section of a 10-hr embryo; labeling is heaviest over the meso- and metathoracic neuromeres (arrows). (Reproduced with permission from Levine et al., 1983.)

system is consistent with the finding that segmental transformation of the embryonic nervous system occurs in the absence of normal BX-C expression (Chapter 6).

In addition, the application of the technique to measuring transcripts in embryos has permitted an investigation of the relationship between BX-C and ANT-C activity. As discussed in Chapter 6, *Df(3)P9* embryos show a transformation of T3 through AB8 to a T2-segmental phenotype, and homozygous *Ubx* embryos have their T3 segment transformed to that of T2. If T2 development requires or is always accompanied by expression of *Antp,* the segmental labeling patterns with the *Antp* probe should mirror the phenotypic changes. Examination of the neural ganglia bears out this expectation: the *P9* embryos show strong in situ hybridization in all T2-type segments (with somewhat less in AB8), and *Ubx* embryos show hybridization in both T2-like segments (Hafen et al., 1984).

Similarly, the in situ patterns in imaginal discs reinforce the genetic observations. *Ubx* expression is strong in the third thoracic discs (haltere and third leg) and absent in the wing, genital disc, and head discs. There is only weak labeling of the second leg discs. The *Antp* probe detects labeling in all the discs of the three thoracic segments, again in accord with the genetics, and none in the cephalic discs or the genital disc of the wild-type.

The experiments have also produced some surprises. These include a heavy labeling by *Ubx* over the mesoderm, localized to the nuclei of the muscles, and a strong initial labeling by the *Antp* probe throughout the ventral nerve cord that becomes localized especially to the T2 ganglion. Does the initial early expression throughout the nerve cord reflect the origins of all segments from a T2-like segment, or is it functional in the biology of the *Drosophila* embryo? A second surprise is that in the thoracic discs, *Antp* transcripts seem to be preferentially localized in those regions that form proximal structures; this pattern was not anticipated from any of the genetic findings.

The present in situ methods can detect transcript levels as low as 100–150 copies per cell (Hafen et al., 1983), concentrations at the low end of the semiabundant range. Increases in the sensitivity of the method and the use of other probes from the BX-C and ANT-C regions should considerably expand our understanding of the molecular biology of these two gene clusters and their roles in setting segmental identity.

ANALYSIS OF A MUTATION: THE CASE OF *oc*

Understanding the nature of a mutational defect requires more than a thorough characterization of the phenotypic derangement; the genetic nature of the lesion must also be characterized. The molecular analysis of the mutation *oc* in *Drosophila* provides an excellent illustration of the power of recombinant DNA techniques to elucidate the genetic basis of complex mutant phenotypes.

The developmental effects of *oc,* a sex-linked mutation, were described in Chapter 3. Homozygous *oc* females lay defective eggs that are encased in fragile, defective chorions. In addition to female sterility, mutant flies of both sexes show a number of minor morphological abnormalities, including the elimination of the eye spot-like ocelli from which the mutant takes its name. The origins of the other abnormalities are unknown, but the aberration of *oc* eggs is a consequence of faulty chorion construction. The mutation exerts its effect through the follicle cells at a time when these cells are actively engaged in the elaboration of the chorionic layers. Analysis of chorion synthesis shows that the mutation causes a depression in synthesis of several of the major chorion proteins, out of the more than 20 that comprise the adult chorion.

The first clue to the basis of the *oc* defect in chorion synthesis was the discovery that two abundant poly A^+ mRNAs, produced by the follicle cells and coding for major chorionic proteins, are transcribed from genes on the X chromosome at or near the site of the *oc* mutation. The localization of the genes corresponding to these mRNAs involved several steps: labeled mRNAs from chorion-synthesizing follicle cells were first isolated; the more prominent individual mRNA bands were then exposed to spread *Drosophila* polytene chromosomes under conditions favoring their in situ hybridization; and the positions of the resulting individual grain clusters were ascertained from the standard cytogenetic map. With bulk-isolated poly A^+ mRNA, two sites of hybridization were revealed: one in the region of 7E11, the mapped site of *oc*, and the other at 12E (Fig. 9.5). When hybridization was performed with the most prominent mRNA species, isolated on polyacrylamide gels, both were found to label the 7E11 site; labeling of the 12E site may be caused by two of the smaller follicle cell–specific mRNAs (Spradling and Mahowald, 1979). The finding of a major site of hybridization near the *oc* site at 7E11 immediately suggested that the effect of the mutation on chorion protein synthesis was on the transcription of the mRNAs for these proteins. Analysis of the mRNAs and proteins of follicle cells from *oc* females confirmed this supposition: the synthesis of the two most prominent mRNA bands is decreased in these follicle cells, relative to wild-type, and there is a corresponding reduction in the synthesis of three major chorion proteins, those of MW 36,000, 38,000, and 19,000. Furthermore, it was shown, using naturally occurring electrophoretic variants of the two larger proteins, later designated s36 and s38, and a small deletion chromosome that covers the *oc* region, that this region directly encodes these two proteins; the reduction in synthesis affects only the alleles on the *oc*-bearing homologue (Spradling et al., 1979).

Evidently, *oc* produces a *cis*-acting inhibition of transcription of a small number of chorionic proteins. However, the results left unresolved the molecular basis of this effect. To answer this question, the DNA sequences surrounding the *oc* mutation had to be purified. To obtain these sequences, Spradling et al. (1980) first constructed a plasmid cDNA bank from bulk poly A^+ mRNA of ovarian follicles actively synthesizing the chorion proteins. The plasmid bank was then screened by colony hybridization to two in vitro ^{32}P-labeled mRNA preparations: one was from lysed follicle cells, the other from preblastoderm embryos. The latter source is enriched in maternal mRNAs and impoverished in follicle cell mRNAs. By screening for clones that hybridized preferentially to the follicle cell mRNA, these workers identified 34 colonies, or about 3% of the total, that were presumptive coding sequences for chorion sequences. Analysis of these clones revealed that they fell into four groups. Group I clones were found to hybridize to the mRNA encoding s38; group II clones hybridized to the mRNA encoding s36; and groups III and IV showed hybridization in situ to the third chromosome at 66D.

The cDNA clones were then used to probe the normal genome organization in wild-type follicle cells. Spradling and Mahowald (1980) found that in

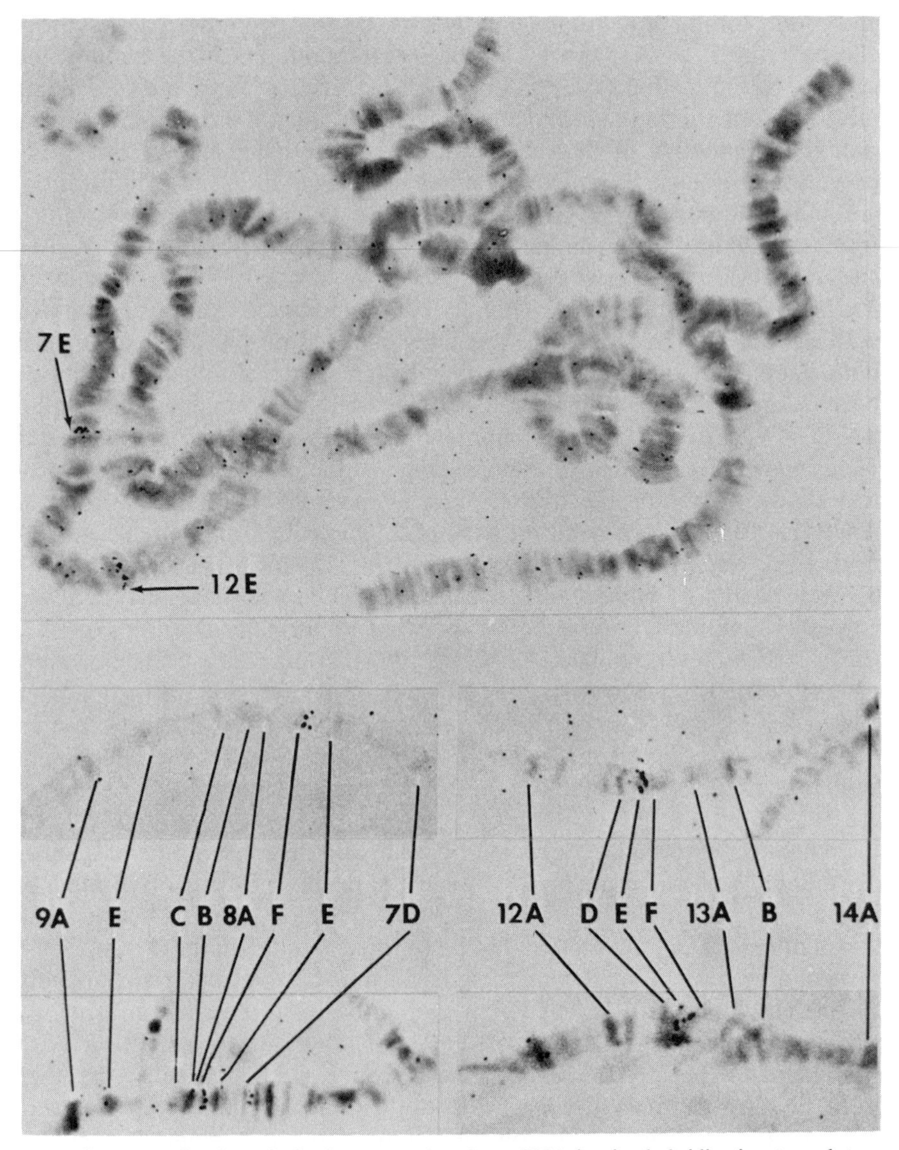

Figure 9.5 Localization of chorion gene sites by mRNA in situ hybridization to polytene chromosomes. (Photograph kindly provided by Dr. A.C. Spradling; reproduced with permission from Spradling and Mahowald, 1979; copyright MIT Press.)

these cells, during the period of chorion synthesis, gene sequences corresponding to cDNA clone groups I, II, and III are all present in higher copy number than is a control gene sequence uninvolved in chorion synthesis; the experiment is diagrammed in Figure 9.6. This gene amplification occurs only in follicle cells of late oocyte stages (10–14) and does not occur in other cell types.

Why should the chorion genes indulge in extra rounds of replication independently of other genes in the follicle cells? The answer is not known with certainty, but it probably reflects the need for extensive chorion protein synthesis during the construction of the chorion shell of the egg, the entire chorion being synthesized and assembled within the very short period of 5 hr. Although follicle cells in late ovarian follicle stages are highly polyploid, this level of dosage amplification may be insufficient to meet the biosynthetic demands for chorion synthesis; the extra gene dosage, provided by chorion gene amplification, would make this possible.

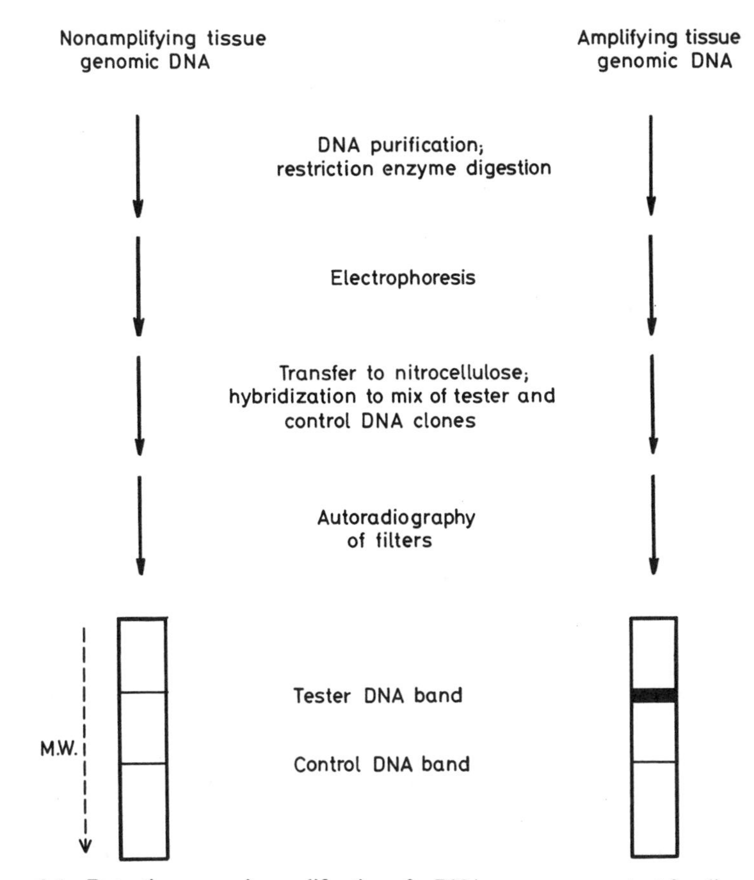

Figure 9.6 Detecting genomic amplification of a DNA sequence; see text for discussion.

When the same analysis is carried out on follicle cells from *oc* females, the results show that it is the amplification process that is defective in the mutant. Relative to the gene for the c18 protein, on the third chromosome, the amplification of the s36 and s38 genes is greatly diminished (Spradling and Mahowald, 1980). The *cis*-mediated reduction in transcription of the X-linked genes is now interpretable: this effect is indirect, reflecting the reduced levels of replication of the chorion gene cluster on the X chromosome.

The discovery that *oc* is a replication defect raises further problems that have been analyzed by recombinant DNA techniques. These questions concern the mechanism of the wild-type amplification and the manner of interference by *oc*. In standard chromosomes, replication starts at particular sites ("replication origins") and proceeds outward in both directions to produce growing "bubbles" or "eyes"—structural features that are apparent when the DNA is spread and examined by electron microscopy. Two fundamental questions about chorion gene amplification are whether it proceeds from a single origin and how far the region of amplification extends.

By performing a chromosomal walk, Spradling (1981) cloned a region of approximately 100 kb of the X chromosome, starting with the cDNA sequences for s36 and s38. Each cloned sequence was then prepared to high activity, along with a control DNA sequence, and filter hybridized to digested genomic DNA from stage 13 egg chambers. The control sequence provided an internal reference to measure the degree of amplification; hybridization was also performed with digested embryonic DNA to monitor the radioautographic pattern produced by cells without genomic amplification. The results showed that in the chorion gene region of the X chromosome in late follicle cells, the genomic sequence concentration declines over a length of 20–30 kb on either side of a single origin at the chorion gene position. A similar peak and gradual decline of amplification for the chorion genes on the third chromosome was also detected. Each chorion gene cluster marks the center of a single domain of replication that spreads bidirectionally and stops at variable distances from the center.

In addition, the position of each genomic clone isolated from the chromosomal walk was ascertained by in situ hybridization to polytene chromosomes. In the wild-type, each clone hybridizes to a single site. However, with hybridization to the chromosomes of the *oc* mutant, one clone was found to hybridize to two sites (Fig. 9.7). The result shows that an inversion interrupts this sequence in the mutant (Spradling and Mahowald, 1981). The *oc* lesion is therefore not a point mutation but an inversion, and the DNA clone derived from the wild-type that is split in the mutant covers one of the inversion breakpoints.

The inversion is the ultimate cause of the replication defect. When the DNA amplification test is carried out on egg chambers from *oc* females, the chorion gene sequences are found to be split into two groups; those left uninverted do not amplify at all, while those that are inverted (including the genes for s36 and s38) do so, but to a lesser extent. Evidently, only those

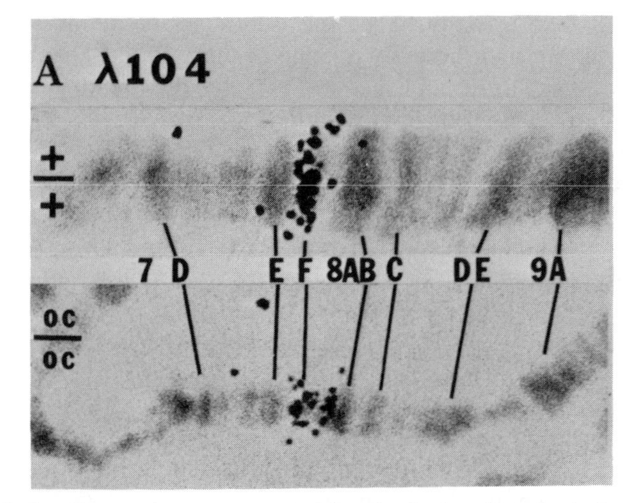

Figure 9.7 A cloned genomic sequence containing the distal *oc* "breakpoint"; note the single site of hybridization in the wild-type chromosome and the dual sites in the *oc* chromosome, indicating the splitting of the sequence. (Photograph kindly provided by Dr. A.P. Mahowald; reproduced from Spradling and Mahowald, 1981; copyright MIT Press.)

chorion genes shifted along with the origin of amplification retain the capacity to amplify, although to a reduced extent.

The analysis of the basis of the *oc* defect illustrates the resolving power of recombinant DNA techniques. Without these methods, the defect would have been classified as a partial transcriptional block, but the molecular nature of the block would have been incapable of resolution. The basis of the egg defect is now clear: it is the indirect result of an effect on replication. The pleiotropic effects associated with *oc* are not well understood but presumably reflect the disruption of other gene functions as a consequence of inversion.

DNA TRANSFORMATION OF FRUIT FLIES AND MICE

Genetic transformation of organisms was a serendipitous discovery of an English medical scientist, F. Griffiths, in the 1920s. Griffiths, who was investigating the epidemiology of bacterial pneumonia, found that mice inoculated with a mixture of live, nonvirulent pneumococcus and killed virulent pneumococcus developed lethal infections. When the infectious agent was recovered from the deceased test animals, it proved to have the coat polysaccharide typical of the dead strain. Evidently, the dead bacterial cells had somehow "transformed" the live, innocuous strain into one resembling itself. The phenomenon of transformation was soon taken up by Oswald Avery and his colleagues at the Rockefeller Institute in New York. The work of the Avery group showed that during transformation, a fraction of the recipient cells take up a substance released from the donor dead cells that permanently alters their heredity for one or more traits. The discovery that the substance was DNA, published in 1944, was the first demonstration that DNA is the molecular vehicle of heredity.

For more than three decades, genetic transformation was a phenomenon restricted to bacterial cells. Numerous attempts to transform eukaryotic cells with purified DNA for selected traits met with either unambiguous failure or irreproducible results. In recent years, the picture has altered. A variety of procedures have been developed that permit reliable transformation with yeast cells, human or mouse tissue culture cells, and *Drosophila* and mouse embryos. This list of transformable organisms will undoubtedly be extended in coming years.

Most transformation events in eukaryotes differ in one key respect from those that take place in bacteria. In bacterial cells, the segments of exogenous DNA that are incorporated into the host cell chromosomes physically replace the homologous chromosomal sequences by a process of recombination. For instance, if one transforms a *Bacillus* strain that cannot synthesize its own histidine (a his^-) strain with a solution of DNA from a his^+ strain (one that can make histidine), the his^+ recombinants contain only the + allele of the donor DNA; the his^- allele of the recipient has been eliminated by recombination. In eukaryotes, transformation involves addition of the exogenous DNA sequences without replacement of the homologous recipient alleles. Furthermore, multiple copies of the transforming DNA sequence are often present in the chromosomes of the successfully transformed cells or embryos. These copies may be clustered at one or a few chromosomal sites or may be dispersed and situated at many sites. Eukaryotic transformation is thus less precise than prokaryotic transformation, and each transformed line must be examined individually for the number and location of the added DNA sequences.

The advent of stable, heritable transformation of whole animals, in particular *Drosophila* and the mouse, is of potentially major significance for developmental genetics. For any gene that can be purified by recombinant DNA techniques, it will now be possible to determine precisely which features of its base sequence regulate its expression within the organism. Genes that can be purified can be altered in precise ways—shortened, lengthened, mutated at specific bases—and then reinserted into the organism. The altered expression characteristics of the gene can then reveal the important aspects of the gene structure/expression relationship. If methods of precise homologous replacement of genes become available, then the usefulness of transformation as an element of developmental analysis will be even greater; replacement of wild-type genes by "designed" mutant genes will permit definitive descriptions of the relationships between mutant gene structures and mutant developmental phenotypes.

In the transformation of whole animals, it is essential that the new genetic material be incorporated in the germ line of the animal. Failure to incorporate DNA in the germ line results, at best, in a somatic and germ line mosaic individual. However, transformation of the germ line produces a transformed line of organisms. Because the early embryogenesis and biology of germ line formation in fruit flies and mice differ substantially, two different strategies of germ line transformation have been employed for these organisms. These methods will therefore be presented separately. Some of the gene expression properties of transformed genes in the two animals will then be compared and contrasted.

Germ Line Transformation in *Drosophila*

The present technique for transforming *Drosophila* was developed by Spradling and Rubin (1982) and employs a naturally occurring, transposable element found in the fruit fly. Most of the transposable elements that have been identified in *Drosophila* are present in all strains and move at rates that are relatively low per generation, perhaps 10^{-3} to 10^{-4} per site per generation. However, a few classes of transposable elements are present only in certain strains. One of these, the family of so-called P elements, cause the syndrome known as "hybrid dysgenesis," which occurs in matings between P-bearing males and females that lack P elements (such P^- strains being termed "M strains"). The symptoms of hybrid dysgenesis include the induction of mutations, chromosomal rearrangement, and a failure of gonadal development. These symptoms are specific to the P male with M female cross; the reciprocal cross, M males with P females, produces normal progeny. The occurrence of hybrid dysgenesis is directly related to the release of P elements from their chromosomal sites when placed in an M cytoplasm (the M cytotype). In formal terms, the phenomenon is a form of zygotic induction: the transfer of a repressed episome into a cytoplasm free of repressor for that episome leads to a burst of episome replication. Excision of the P elements from their normal resident sites in the M cytotype is followed by their transposition to new sites in the M chromosomes; the dysgenesis-inducing cross is a form of directional transformation of M chromosomes by P-bearing chromosomes.

Spradling and Rubin's transformation system utilizes the P element's transposability as a means of carrying other DNA sequences into the germ line of recipient *Drosophila* eggs. When bacterial plasmids containing complete, 3-kb-long P elements are injected into the posterior polar region of late cleavage fruit fly embryos from M strains, the P elements leave their carrier plasmids and integrate into the chromosomes of the embryo's nuclei with measurable frequency. By injecting into the posterior region prior to pole cell formation, the procedure ensures that a certain fraction of the transpositions occur in the nuclei of future pole cells and hence in the germ line of the recipient embryos.

The surviving progeny of such injections that carry the P element can be detected in either of two ways. The first is by restriction enzyme digestion of their DNA and Southern blotting analysis, with hybridization to labeled P sequences. Only the transformed flies, coming from an M strain background, will show typical P sequences.

The other means is a genetic test. It is based on the fact that P elements are usually found in a range of sizes in the strains that contain them, from about 0.5 to 3.0 kb. The smaller fragments usually lack the P-encoded activity that facilitates transposition, so-called transposase, but retain the recombinogenic ends of the complete P element. Thus, a small P element inserted at a particular site, if exposed to the transposase activity of the complete P, can be induced to leave its chromosomal site. When such an incomplete P is situated within a gene, producing a mutant phenotype, the subsequent removal of the P can either restore wild-type activity or, if excision removes additional genetic material, can produce a more severe genetic deficiency. Spradling and Rubin used this mutability test to detect successful P element transposition into the target M chromosomes. The P-induced mutant employed was *singed weak* (sn^{w}), a mild P-caused mutation of the bristle locus *sn*. The progeny of the surviving M embryos were scored for the occurrence of either reversion (sn^{+}) or that of a more pronounced mutant phenotype, *singed extreme* (sn^{e}) (Fig. 9.8). All second-generation individuals showing mutability were subsequently found by the DNA hybridization test to harbor P elements.

To transform individuals with particular DNA sequences, the injection protocol was slightly modified. Target embryos were injected with a mixture of complete 3-kb P elements and a P-element "vehicle" containing the sequence of interest. The DNA vehicle is a defective, short (1.2-kb) element, itself incapable of synthesizing the transposase but containing the recom-

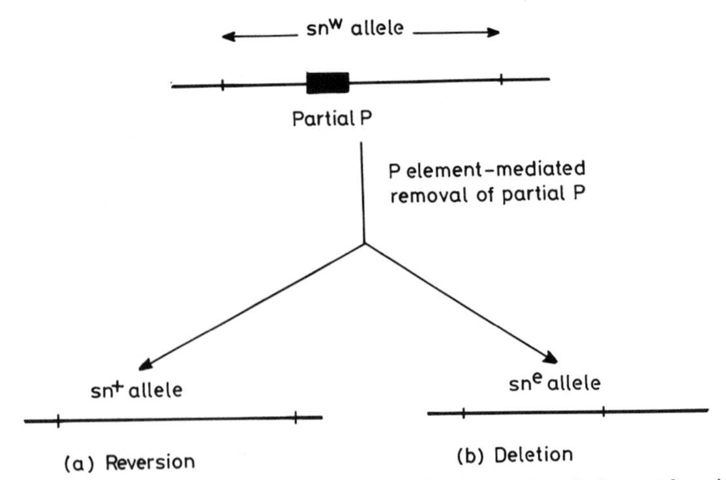

Figure 9.8 Consequences of induced transposition of an incomplete P element from its site in the *sn* gene: reversion to produce a wild-type phenotype or deletion to produce a more extreme mutant phenotype.

binogenic ends of the complete element; this element is inserted in a bacterial plasmid for cloning purposes, as is the complete P. The *Drosophila* gene to be carried into the germ line is then inserted within the defective P. Coinjection of the active P and the vehicle P permits the latter to transpose from its carrier plasmid into the germ line of the injected embryos. This system has also been termed "transduction," by analogy to the prokaryotic phenomenon in which bacterial genes are picked up by bacteriophage chromosomes in one host cell and moved to a second during a cycle of infection (Goldberg et al., 1983).

The first transduced *Drosophila* gene was the *rosy (ry)* gene, carried on an 8.1-kb inserted sequence (Rubin and Spradling, 1982). It was injected into M embryos of the *ry* genotype, and the surviving flies were mated to *ry* individuals. Production of ry^+ progeny in the experiment signifies successful transformation. The *ry* gene was chosen for two reasons: (1) only slight levels of gene activity, as little as 5% of wild-type, are enough to restore the wild-type phenotype, and (2) *ry* expression need not be autonomous to restore wild-type eye color because the *ry* metabolic product is diffusible, and small amounts taken up by eye imaginal discs can repair the mutant phenotype.

By the phenotype test, transformation is highly efficient; in the first experiments, 20–58% of the injected surviving flies had integrated one or more ry^+ genes into their germ line. Subsequent tests with other genes have given comparably high rates of transformation. From examination of polytene chromosomes of the transformed lines by means of in situ hybridization, in these experiments it appears to be a general rule that only one or a few copies are inserted per germ line. Furthermore, there are no strong preferences for insertion site; insertion is either random or nearly so.

The *ry* insertion experiment demands expression of the transduced sequence. If the marker DNA sequence does not give a visible phenotype, one can coinject the two DNAs into an M strain with a P-mutable site, for instance, sn^w. The mutable progeny lines obtained can then be screened by DNA hybridization tests for coinheritance of the sequence of interest (Rubin and Spradling, 1982). Between the methods of direct selection (as with *ry*) and secondary screening of mutable lines, it is now possible to transform the germ line of *Drosophila* with any gene sequence that can be isolated. In most of the experiments that have been reported, the gene expression properties of the transduced genes are similar to those of their normal analogues (see the section "Expression of Inserted Genes" below).

Germ Line Transformation in the Mouse

Transformation of the mouse germ line is simpler than that of the fruit fly but in some respects has been less satisfactory. Because germ line cells in the mammalian embryo are set aside comparatively late and in a poorly defined region of the embryo, no attempt is made to inject DNA into or near the germ line cells, as in *Drosophila*. Rather, the DNA to be injected is added to the mouse embryo early, immediately after fertilization, by injection of the recombinant plasmid or phage directly into the male pronucleus. A typical aliquot for injection carries several hundred copies of the DNA molecules (far fewer than in the *Drosophila* injections). The injected one-cell embryos are given a brief culture period in vitro and then implanted into foster mothers. Injection of DNA into the male pronucleus is possible in the mouse embryo because of the leisurely pace of events at the beginning of cleavage.

The use of selective methods to obtain transformants has not yet been attempted and may not prove feasible in the immediate future, given the small number of embryos available for each experiment. Tests for successful transformation have therefore been primarily by analysis of DNA, using restriction enzyme digestion and hybridization to marker DNA probes of the digested recipient DNA (or DNA of recipient progeny). By this measure, success has been variable, ranging from a few percent to 40% at best. The first-generation recipient mice are usually nonmosaic; apparently all tissues of a given transformant carry the integrated genes. Although there is usually a single site of integration, recipients often possess multiple tandem copies of the inserted sequence at this site. This pattern of integration may be generated by consecutive insertions of plasmids, with the first integrant providing a site for the insertion of the second by homologous recombination, and so forth. Typically, the number of inserted copies ranges from 1–3 to about 50. For the most part, this pattern is stable in successive generations, indicating that the integrated DNA is stable; in a few instances, rearrangements have been detected. As in the fruit fly, insertion seems to be at random, or at least there are no strongly preferred sites of integration.

The time of integration of the exogenous DNA into the chromosomes is unknown, but the general absence of mosaicism in the first-generation (recipient) mice suggests that it is very early, perhaps in the male pronucleus itself or during the first or second cleavage divisions. In contrast, the *Drosophila* transformation method almost invariably produces mosaic recipients, and the production of pure lines requires an additional generation of breeding. A second difference between the two systems is in the general patterns of expression: in the mouse, the inserted genes examined to date have subnormal expression capabilities. Some possible reasons for this difference are discussed in the next section.

Besides the direct interest of the transformation experiments for the study of expression of inserted genes, they have acquired a second significant aspect: insertion of the genes can be directly mutagenic with high frequency. Mutagenicity in this case appears to derive from insertion of the exogenous genes into normal functional genes, disrupting their structure. One recessive lethal, an insertion into a structural gene for collagen, was produced by a late insertion, into a postimplantation embryo, by a Moloney leukemia virus (Jaenisch, 1983). In another experiment, two out of six one-cell embryos injected with human growth hormone (HUGH)-containing plasmids were found to contain recessive lethals (Wagner et al., 1983). The latter frequency is substantially higher than anything approached by standard forms of mutagenesis. Whether or not the mutation frequency reflects some peculiarity of the HUGH gene or represents injection with foreign sequences in general at the one-cell stage remains to be determined. It may prove, however, that one of the oldest problems in mouse developmental genetics—the acquisition of new genetic source material—will be solved by the application of the most modern of techniques in genetic manipulation.

Expression of Inserted Genes

At the time of this writing, stable transformant *Drosophila* lines have been obtained for three genes: *ry* (Spradling and Rubin, 1983); *Alcohol dehydrogenase* (*Adh*) (Goldberg et al., 1983); and *dopa decarboxylase* (*Ddc*) (Scholnick et al., 1983). In all cases, the structural genes were placed in the middle of ample flanking sequences, the entire insert being placed within a defective P element, which in turn was placed within a bacterial plasmid; transformation was P-element mediated, as described above. In all cases, transformant lines showed normal expression of the integrated genes. With only a few exceptions, the genes are expressed in the appropriate tissues and at the developmental stages in which their normal chromosomal analogues are expressed.

These results have also partially answered the question of whether dosage compensation for X chromosome genes is an intrinsic property of genes or of the X chromosome. Earlier studies of rearrangements between chromosomal regions on the X and the autosomes indicated that compensation or its absence could not be transferred; the respective property appeared to be intrinsic to each gene (Lucchesi, 1977). However, these studies necessarily involved pieces of DNA hundreds of kilobases long. Using much smaller regions, each large enough to encompass a single autosomal gene, a different result is obtained for transduced genes: several *ry* integrant genes on the X and one *ddc* gene were found to acquire dosage compensation (Spradling and Rubin, 1983; Scholnick et al., 1983). Evidently, compensation involves special X chromosome sites that are liberally scattered throughout the chromosome and that can assume control over integrant genes.

Mouse transformation has also been carried out with several genes whose expression patterns have been monitored. These include the rabbit β-globins (Lacy et al., 1983), the chicken transferrin gene (McKnight et al., 1983), and the herpes virus thymidine kinase gene hooked to a mouse gene promoter (Brinster et al., 1981). In contrast to the fruit fly results, however, expression of the transformant genes is depressed relative to normal levels, and in some instances in which there is expression, it occurs in the "wrong" tissues.

The two sets of results present a puzzle when compared. There are at least four possible explanations:

1. In the fruit fly, genes have been selected for expression, but not in the mouse; the seeming difference therefore reflects only this difference of selection. Although this may be a contributory factor, searches for nonexpressed integrated genes in the fruit fly have failed to reveal them (Spradling and Rubin, 1983).
2. The difference reflects differential chemical processing of incoming genes, and specifically the occurrence of methylation during transformation of mouse embryos. Methylation has been implicated in higher eukaryotes as a means of inactivating gene expression, and *Drosophila* has few or no methylated genes. This explanation may have partial validity, like the one above, but fails to account for the fact that some transformant lines in the mouse *do* have activity; the transformed chicken transferrin genes are a strong case in point, although their activity is less than expected.
3. The fruit fly and mouse have different kinds of *cis*-acting elements, and in the mouse these elements may be at relatively greater distances. The evidence favors this interpretation: it is now known that in certain eukaryotes there are strong "enhancer" sequences that may lie several thousand base pairs away and that are required to stimulate transcription (Hogan, 1983). Some of these enhancer sequences may be tissue specific. If a cloned gene sequence does not contain such a sequence and integrates away from it, it has subnormal expression capability. The discovery of enhancer sequences marks another profound difference between prokaryotic and eukaryotic regulatory mechanisms.
4. Finally, the differences in expression between fruit fly and mouse transformed genes may disappear when a larger number of mouse genes are examined. The aberrantly expressed mouse genes are all in the abundant class of gene functions; transformation with gene sequences in the rare or intermediate abundant classes (as the tested *Drosophila* genes are) may yield a different picture.

PROSPECTS

The development of recombinant DNA technology has made the individual gene accessible to direct analysis and manipulation. Any gene that can be isolated can be sequenced, revealing its precise genetic content, and can be used as a probe to measure its own transcription in vivo. In addition, in those organisms whose germ line can be transformed with cloned DNA sequences, the way is open to analyze the control regions of each gene. By precise chemical deletion or in vitro mutagenesis, any region can be altered, and the consequences of in vivo expression can be ascertained.

Of the presently available transformation systems, that of *Drosophila* promises to be especially useful. For almost any gene, both the gene and its mutants can be obtained; and by careful alteration of the gene and transformation with the engineered copies of the mutant strain, precise correlations between structure, expression, and function can be obtained. Information on gene expression with transformed genes in the mouse is presently more difficult to obtain, but the system promises to reveal new facts about gene control in this organism.

Powerful as the new technology is, there is still one kind of information about genes that it does not provide—that of cellular role. One can sequence a gene, find the "open reading frame" (a stretch of potential protein-coding sequence), predict the amino acid sequence, and still be largely ignorant of the biochemical function of the protein. If the amino acid sequence resembles one of the known sequences of proteins of known function, one may have an important clue; computer searches are proving increasingly important in finding such relationships. The discovery of the homeobox illustrates the usefulness of such approaches. However, even should the homeobox prove to be a DNA-binding/transcription-regulating polypeptide domain, the target genes governed still remain to be identified.

However, if no relationships are found, one may still explore some aspects of the relevant cellular biology by localizing its likely cellular site. One can do this by synthesizing the protein in vitro, making antibodies to it, and then doing fluorescent labeling of slices of tissue that are known to express the gene. The site of labeling should reveal the cellular site of the protein, and this may provide a hint as to the protein's cellular role.

Ultimately, protein function may prove deducible from protein sequence and thus from DNA sequence. It is impossible to predict how rapidly such analytical developments may occur. However, their achievement would complete the recombinant DNA revolution: the secrets of the individual gene would at last be open to full inspection.

TEN

GENERALITY, GENETICS, AND FUTURE DIRECTIONS

According to present views, however, what is coded in the chromosome is the program to build that adult, that is, the instructions to manufacture its molecular structures and to put them into operation in time and space. However, the internal logic at work in the execution of the program remains completely unknown. It is generally admitted that a Laplacian demon able to examine the fertilized egg, its molecular structures and organization, would be able to describe the future adult. However, what kind of molecules besides DNA the demon would have to examine and what kind of algorithm it would have to use remain a complete mystery.

François Jacob, *The Possible and the Actual* (1982), pp. 43–44

Theories are nets cast to catch what we call "the world"; to rationalize, to explain, and to master it. We endeavor to make the mesh ever finer and finer.

Karl R. Popper, *The Logic of Scientific Discovery* (1959), p. 59

ARE THERE GENERAL PRINCIPLES OF GENETIC CONTROL IN DEVELOPMENT?

The aim of science is to reduce the disorder and complexity of the natural world to coherence in the form of testable concepts. By this measure, developmental biology is in an unhappy state. The plethora of observations of which the field consists are only loosely tied together, at best, by a handful of incomplete and disjointed hypotheses. Not only do we lack the necessary unifying concepts, but doubtless much of the information needed to formulate the ideas; even worse, we are largely ignorant of the criteria for recognizing the kinds of new facts that will permit productive generalization.

In the absence of illuminating general concepts, the question of whether there are common organizing principles in animal development is unanswerable at present. There are two reasons, however, for believing that such principles exist. The first is the fact of evolutionary relatedness of animals. Such relatedness guarantees that the closer the relationship between species, the less changed will be the fundamental biological properties of species. The whiteness of swans in the Northern Hemisphere is (as we know) no indicator of swan color in the Southern Hemisphere, but it would be very surprising if both white and black swans experienced different modes of development. Indeed, not only do all birds show obvious similarities to each other in embryogenesis, but these similarities are shared with mammalian embryos. Furthermore, the amphibian pattern, which appears so different from those of birds and mammals in the early stages, is probably little different in its fundamentals, even in those stages. Unlike those frog species that produce mesolecithal eggs with holoblastic cleavage, amphibian species that produce megalecithal eggs (like those of birds) produce blastodermal embryo discs similar to those of their (distant) avian cousins (del Pino and Elinson, 1983). The more distant the animal groups under comparison, of course, the greater the number of differences to be expected, but underneath the surface differences, there may be some fundamental commonalities in mechanism.

The second and stronger reason for suspecting the employment of general mechanisms in animal development is that many of the *elements* of development are observed in very different kinds of animals. A number of these elements have been noted in the course of this book. The involvement of gradients, or at least gradient-like morphogenetic systems, in the organization and fate-setting properties of many animal eggs is one example. A second is the fact that most developmental decisions, whether in nematodes, fruit flies, or mice, take place in groups of cells rather than in single cells. A third shared property involves the pattern restoration mechanism, which seems to obey the rules of the polar coordinate model in animals as distant as cockroaches and salamanders. Finally, the very different morphogenetic movements seen in different kinds of animal embryos may all be based on a simple, commonly shared mechanism of triggered, localized cell surface contraction (Odell et al., 1981). These shared elements of development may have been retained passively in evolution or because many represent optimal solutions to particular developmental problems; the essential point is the apparent fact of generality itself.

If many of the discernible elements of development are shared between distantly related animal groups, then the tremendous diversity of animal developmental patterns may reflect primarily differences in the relative *timing* of constituent events with respect to each other or to the end point of development. Thus, fixed (determined) cell states may arise early in so-called mosaic egg systems and late in regulative embryonic systems, but the process of restriction of cellular fate may have a highly similar cellular or chromosomal basis. The importance of differences in the relative timing of events in producing seemingly very disparate developmental outcomes has been emphasized by Johnson and Pratt (1983).

Developmental genetics is only one branch of developmental biology, but its concerns are increasingly central to the whole enterprise. The implicit goals of animal developmental genetics are to elucidate the principles of genetic control that operate in particular animal systems and to establish the elements of universality in genetic control between different systems. Much of the agenda of molecular biology is also devoted to these programmatic goals, but the approach is very different. Eukaryotic molecular biologists tend to focus on the molecular machinery that physically controls gene expression. Geneticists, on the other hand, are more concerned with the regulatory logic that governs the process.

These two emphases have given rise to different views of how the problem of genetic control will eventually be solved. Today, the most commonly held belief is that molecular approaches will suffice to reveal the answers. This belief follows from the assumption that because genetic findings can all be reduced *in principle* to molecular explanations, molecular biology provides the most direct route in solving the problems of genetic control. If this deduction is correct, the principles of genetic control will eventually "fall out" of the molecular data.

IS THERE A GENETIC "PROGRAM"?

This faith in the sufficiency of molecular biology is worth examining because it is intimately connected to the reigning metaphor of developmental genetics—that of the developmental program. The term "program" has been used throughout this book as a shorthand form for the sequential events of development. However, as commonly employed, the idea of the genetic program is a metaphor with much deeper implications. Indeed, it reflects the conjunction of two scientific revolutions in our time, those of molecular biology and cybernetics.

The word "program" means different things in different contexts, but in biology the reference is to computers: a program in this sense is a set of coded instructions for a digital computer. In this context, the genetic program is the genome itself, which may be regarded as containing the encoded instructions for development. In this view of things, development is simply the automatic computation and "readout" of these genetic instructions.

The idea of a genetic program of development evolved from the first successes of molecular biology in the 1950s and 1960s. It was the conscious application of the principles of information theory, borrowed from the then new science of cybernetics, that propelled the conceptual revolution in biology. The crucial insight, implicit in the Watson–Crick model of DNA structure, was that each gene is effectively a mini-program for the construction of a protein. For instance, if one takes a bacterial gene for the enzyme alkaline phosphatase (AP), copies it into mRNA, and then places the transcripts into the cytoplasmic mix from another organism (which contains the essential protein-synthesizing components), AP polypeptides are synthesized that are identical to that encoded by the bacterial genome. Because the DNA code and the decoding and processing instructions are universal, the gene program can be inserted into any cell computer and read to give a fixed result.

At first sight, the extrapolation from the gene-as-program (for polypeptide synthesis) to the genome-as-program (for development) seems reasonable because the genome contains all the necessary information for development. Furthermore, the genomes of different animals are ultimately responsible for the differences in development between these animals. However, despite the seductiveness of the extrapolation, it is probably misleading. In recent years, three of the men responsible for the molecular biological revolution have questioned the validity of the program metaphor in development (Stent, 1980; Brenner, 1981; Jacob, 1982).

The key differences between the putative developmental program and the gene program concern the *completeness* and *temporal sequence* of information utilization. When the gene program is decoded, *all* of that information is utilized (at least in prokaryotes) and processed in a strict linear sequence from a fixed initiation point. In contrast, each stage of development consists of the *selective* utilization of the information in the genome, with the information that is used at each step being scattered around the genome. Furthermore, the selected information at each point indirectly determines what information will be utilized at the next step. If the genome were a program in the sense that the individual gene is, it would contain instructions on where the information selection process is to begin and it would specify the rules for the successive selection steps. None of that information is in the genome, at least in any form that can be presently recognized.

In fact, genomes fail the test of universal readout in different cytoplasmic environments that individual genes so successfully pass. One can perform this test by creating interspecies hybrid genomes in certain animal groups, such as frogs. The behavior of these hybrid genomes is then examined in the two different maternal (egg) environments of the two constituent species by making the two reciprocal crosses. Often the embryos will undergo extensive development from one cross—which may be symbolized as species A ♂ × species B ♀—but not from the reciprocal cross—species B ♂ × species A ♀. In both crosses, the genetic material (the hybrid genome) is identical but the cytoplasmic context between them is slightly different; the mismatch between genome and regulatory components in one cross leads to developmental failure in that cross. Taken literally, the term "genetic program" places too much emphasis on the raw information content of the DNA and not enough on the intricacies of the reading and implementation of that information, processes that are not directly described by the DNA document.

If the answers to questions of development are not to be sought from DNA itself, might they nevertheless be obtained from the totality of studies on the control of gene expression? The answer, of course, may be "yes"; one can never eliminate in advance the possibility that more information will illuminate where less has failed to do so. However, the more likely answer is "no." Indeed, the growing body of studies on the control of gene expression in eukaryotes comprises what Karl Popper has referred to as a "metaphysical research program." A metaphysical research program is a body of research guided by an idea that is susceptible only to successive tests of verification but that, in some important sense, lacks explanatory value. The idea in this instance is that development is guided and accompanied by a changing pattern of gene expression. A quarter century of experimentation has amply confirmed this proposition, and another quarter century can be expected to do the same. What has singularly failed to emerge is a set of ideas that explain the underlying dynamics of the whole process. Nor is it clear how a preoccupation with the controls of gene expression will explain (1) what the biological functions of the gene products within cells are or (2) the interrelationships between gene products that determine individual cell-type properties. Finally, the phenomena of cell–cell interactions, which comprise such an important part of development, are so far removed from the processes of gene expression control that they are almost certain to be left unilluminated by the gene control studies.

THE ROLE OF GENETICS

Genetics provides the indispensable complement for molecular studies. Genetics has two different roles in developmental studies. The first is the use of cell markers for exploring cell behavior in development. In all those organisms in which mutant hunting is relatively difficult, such as frogs and mice, the principal function of genetics will continue to be in such cell-descriptive studies.

The second, and potentially more important, application is the elucidation of individual and collective gene product roles in development. Perforce, this will be possible on a large scale only in those animals, such as *Drosophila* and *Caenorhabditis,* in which large numbers of mutants can be isolated at will. In consequence, any generalizations about the genetic control of development must also necessarily be limited to these animals for the foreseeable future. Whether credible general principles of genetic control, extendable to vertebrates, will be derived from these studies remains to be seen.

At the very least, the principles of genetic control in one of these animal systems may begin to emerge in the next decade. However, even between the nematode and the fruit fly—which are two very different organisms, the latter being far more complex than the former—certain formal similarities of genetic control can already be detected, despite the incompleteness of the information. For instance, some of the underlying similarities in sex determination control have been discussed in Chapter 6. The finding of such commonalities in control between such distantly related animals suggests that there may be certain optimal genetic control strategies for "solving" certain developmental problems.

In addition, some of the molecular data on gene expression in various animals are consistent with the inferences about gene expression control in the fruit fly and the nematode that have been derived from strictly genetic approaches. A few tentative propositions can therefore be offered. Intriguingly, they differ from the suppositions of classical developmental genetics and the early post-Jacob–Monod formulations.

THEORIES AND CONUNDRUMS OF GENETIC CONTROL

The central hypothesis of classical developmental genetics was that each gene is expressed with a defined developmental phase specificity and a high degree of tissue specificity. The essential correlate is that different cellular phenotypes are produced by *qualitatively* different expressed gene sets with only partial overlap. As Jacob and Monod (1963) later expressed it: "We shall consider that two cells are differentiated with respect to each other if, while they harbor the same genome, the pattern of proteins which they synthesize is different." Nearly all of the thinking of the past quarter century has been predicated on and conditioned by this assumption.

However, the analysis of individual gene functions has brought this assumption into question. When the full range of phenotypic effects associated with different mutants of particular genes is determined, it is usually found that the amorphic phenotype is associated with a much lower degree of temporal and tissue specificity than that displayed by the first mutant. Thus, one may collect many "embryogenesis-specific" or "male-specific" mutants in *C. elegans,* only to find that most of the first set exhibit defects in postembryonic developmental stages and that most of the male-specific mutations produce abnormalities in hermaphrodites. Similarly, in *Drosophila,* most of the genes initially identified on the basis of their maternal effects prove, on closer inspection, to be required for stages of later development as well. To cite another example, most of the genes initially identified as performing imaginal disc-specific functions have proved to be required in larval tissues or for early embryogenesis.

The molecular findings from mRNA-DNA hybridization experiments further undermine the idea of phase and tissue specificity of gene action. In *Drosophila,* the mouse, and the chicken, it appears that most mRNAs are present in most cell types. Such mRNA ubiquity in cell types is not universal—the sea urchin is one exception—but it does appear to be a common pattern. For the animal species that exhibit highly similar cellular mRNA pools, there are some qualitative differences between cell types in the rare mRNA classes, but the most striking differences are in the representation of sequences in the abundant and semiabundant classes. The implication for development in these organisms is that different cellular phenotypes are produced either by a relatively small number of qualitative differences among the rare mRNAs, or primarily by quantitative differences in shared mRNA species, or by a combination of the two factors. In any event, the relationship between cell or tissue phenotype and the pattern of gene expression seems to be much subtler than previously imagined.

This subtlety of the gene set–phenotype relationship has serious implications for the way we view the regulation of gene expression in development. The first generation of regulatory models for eukaryotic cells (Monod and Jacob, 1962; Jacob and Monod, 1963) employed solely the elements of prokaryotic regulatory circuitry. The intent of these models was to demonstrate that the principles of regulatory circuitry derived from the study of prokaryotic cells could explain the basic dynamics of differentiation in eukaryotic cells. These models did not attempt to account for the particular structural or informational complexities of the eukaryotic cell. From the late 1960s to the mid-1970s, a second generation of regulatory theories, which specifically addressed these eukaryotic complexities, were fashioned (Britten and Davidson, 1969; Kauffman, 1971; Gierer, 1974; Wolpert and Lewis, 1975; Garcia-Bellido, 1975). These models differed in the relative attention they paid to the precise molecular details of gene regulation (and, in particular, transcription) versus the formal logic of regulation, but all shared certain assumptions or axioms.

The first of these axioms was the "binary choice" postulate—that each regulatory event consists of a choice between two mutually exclusive alternatives. This was stated explicitly in Kauffman's model but is implicit in all of the others. The second common feature of the theories was the combinatorial postulate, the idea that the phenotypic properties of cells/tissues/structures result from the summed total of the regulatory elements that are on and off. The third element was a focus on the mechanism for creating qualitatively new gene expression patterns, based on the on/off regulatory switch. The underlying assumption was that of classical developmental genetics—that phenotypic novelty necessarily reflects the expression of qualitatively new gene sets.

The gist of the molecular and genetic data seems to be that this assumption is untrue. If the different cells of an organism throughout development have qualitatively similar expressed gene sets, then much of the potential significance of the binary choice and the combinatorial postulates is lost. If most cells share a large set of commonly expressed genes, then developmental choices at the level of gene expression become less a matter of dramatic on/off switches than of progressive fine tuning (see below).

The fourth element in several of the theories was the assumption that there exist simple unidirectional hierarchies of control of gene expression (in particular, the models of Britten and Davidson and Garcia-Ballido). The simplest conceptions of control are those in which a given regulatory element switches on the expression of a second set of elements, which might trigger the expression of a third, and so on, culminating in the synthesis of the gene products of a battery of structural genes. However, where genetic data are available, this picture of a simple linear chain of command modules often does not fit the facts. The establishment of segment identity in *Drosophila* by zygotically acting genes provides a case in point. The classical hypothesis is that of a hierarchy of control that begins with an activator gene (*Pc*, for example), whose product is differentially distributed in the successive segment primordia such that differential activities of secondary control elements (the genes of the BX-C) are established. These secondary control elements then switch on the final sets of genes (the realizator genes), whose activity directly produces the segmental phenotype.

However, as we have seen, there is not just one gene, *Pc*, whose mutational inactivation leads to a repetitive (derepressed) AB8 phenotype, but many such genes. While a single repressor of the BX-C is an appealing notion (leaving aside the problems of diffusible repressor action described in Chapter 1), a half dozen or so such genes is too many for a simple hierarchical control hypothesis. Furthermore, the realizator genes, the ultimate "foot soldiers" in the hierarchy, whose differential activity directly establishes segment-specific differences, have remained singularly elusive.

Another possible example of the insufficiency of simple hierarchical schemes can be taken from the genetic biology of *Caenorhabditis*. There are several genes in this organism that mutate to a vulvaless (Vul) phenotype. If there were only one, it might make sense to speak of vulva development as "controlled" by that gene. With several, it is not at all clear which gene is controlling what.

Certainly, hierarchical *elements* of control must exist in development. The results noted above, however, suggest that the *global* properties may not always be simple hierarchies. There are at least two simple alternative explanations that can be listed. The first is that many gene products might combine to give one functional unit, analogous to a multienzyme complex; inactivation of one gene product could inactivate the unit as a whole. The other possible alternative to simple hierarchies is the idea of the network, in which several genes regulate each other in an interacting circuit: if the circuit is broken because one element is missing (e.g., through mutational inactivation), the circuit ceases to operate and a mutant phenotype is produced. In such a scheme, the *same* mutant phenotype is produced by deletion of any one of the activities in the circuit, just as the *Pc*-type phenotype is produced by mutation in many genes other than *Pc*. Of course, networks are much more difficult to analyze, and to comprehend, than unidirectional hierarchies. However, that is a problem for the investigator, not for the organism. One possible set of network structures in eukaryotic gene control has been discussed by Kauffman (1971).

The other regulatory feature whose existence the genetic data implies is a certain redundancy of genetic function: for many genetic functions, there are other genes that can partially substitute their roles. We have discussed the evidence for this conclusion in Chapter 5, in connection with the nematode postembryonic defective mutants. However, the same is likely to be true for *Drosophila,* many of whose severely hypomorphic or amorphic mutants exhibit variable penetrance and expressivity. Such variability is probably indicative of partial functional replacement. In the extreme version of this phenomenon, two functions may be identical, such that both must be mutationally inactivated in order to produce the mutant phenotype. The *lin-8; lin-9* double mutant possessing a Muv phenotype is probably an instance of this kind. The general existence of functional redundancy, whether partial or complete, is probably at the root of the earlier described phenomenon of "canalization," first described by C. H. Waddington (see Waddington, 1962). Canalization involves the stabilization of developmental pathways by multiple genetic factors within the genome, a form of genetic buffering.

The wider significance of functional redundancy may be in its implications for pathways of regulation. If much of the genetic information is expressed much of the time, then the regulation of gene expression probably entails progressive fine tuning. A large part of such fine tuning in any developmental pathway may consist of the differential biasing in expression of particular elements of redundancy. To put it in slightly different terms, much of genetic control may consist of "locking in" different pathways by building in the numbers and particular elements of functional redundancy. Correspondingly, understanding the hidden dynamics of a developmental pathway might therefore require tracking the time course of the expression of all these elements within the pathway. Brenner (1981) has described the regulatory process in these circumstances as one of progressive "optimization."

Eventually, information must be obtained about the fundamental units of regulation: the presumptive regulator genes and the batteries of genes whose expression they directly control. There is still remarkably little hard information about these matters. Several transcription-regulating genes have been found in various fungi, but few have been conclusively identified in higher eukaryotes. It is in *Drosophila* that one encounters the largest number of claims for regulatory genes. Although there is a heavy element of interpretation in many of these assignments, the bithorax and Antennapedia gene complexes probably include genuine examples. The discovery of the shared homeobox sequence in these gene clusters and its location in several other defined homeotic genes may provide the needed breakthrough in the analysis of gene regulation in development. In particular, the finding that part of the homeobox sequence corresponds to a well-defined DNA-binding domain of several prokaryotic regulatory proteins suggests that genes containing the homeobox are involved directly in the regulation of transcription of other genes. Whether this sequence is part of segmentation-regulating genes in higher animals, however, is unclear. Should the answer be affirmative, the case for certain universal principles in genetic control of animal development would be immeasurably strengthened. Nevertheless, the fact that a similar sequence is found in yeast cells suggests that the homeobox sequence may encode a protein domain only with a general DNA-binding function rather than being diagnostic of genes involved in segmentation.

Ultimately, the ease of genetic identification of regulatory genes in animals—apart from those containing the homeobox—may depend on whether there are simple one-to-one relationships between the individual batteries of expressed genes and recognizable cellular/regional phenotypes. If there are such relationships, as depicted in the top part of Figure 10.1, one may expect further genetic analysis to unearth them; molecular analysis could then be expected to confirm their identity. However, to the extent that phenotypes arise from the combined actions of several regulator genes (bottom part of Fig. 10.1), through the activation of new gene expression or the quantitative amplification of gene expression, the task of identification by genetic means will be that much more difficult. The greater the number of regulatory genes required to produce a phenotype, the harder it will be to isolate a particular phenotypic effect associated with a particular genetic element. Even in *Drosophila,* where many phenotypic transformations superficially seem to be associated with one gene, closer inspection reveals the probable involvement of multiple selector genes, whether in A-P compartmentation or in segmental specification. Although regulation of regional morphological phenotypes may be more subtle than binary on/off switches, the properties of each region do appear to reflect the combinatorial effects of selector genes.

FUTURE PATHS

It is apparent that neither the strict molecular reductionist approach by itself nor the use of pure genetic stratagems in isolation can solve the problem of the genetic control of development. If there are general principles of such control, and if the problem is to yield a solution, it will only come from a fusion of the two approaches. Inevitably, that will require a continued commitment to exploration with either or both of the two animal systems, *Drosophila* and *Caenorhabditis,* in which both genetic and molecular analyses are possible. Although it is always possible that the results from these systems will prove irrelevant to higher animal systems, the considerations outlined at the beginning of this chapter and the recent molecular findings of shared key gene sequences suggest otherwise.

Within the genetically tractable systems, what is most urgently needed is a systematic description of the molecular biology of cell divergence. As cells divide and diverge in cellular phenotype during development, what are the

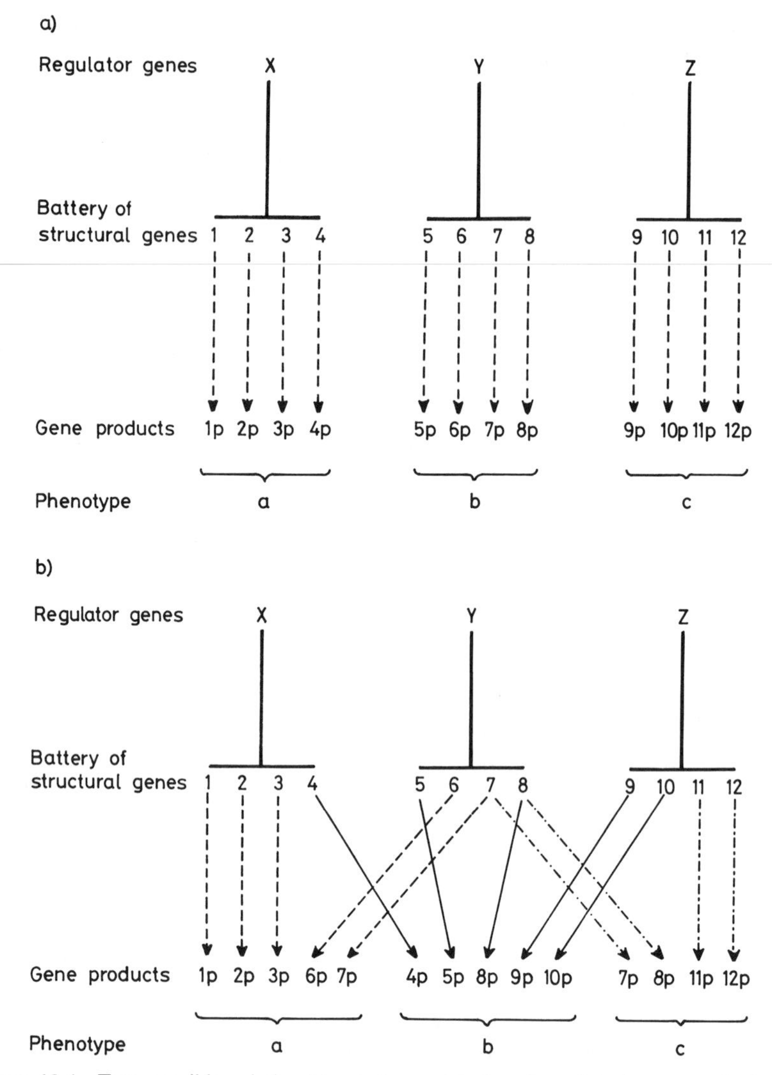

Figure 10.1 Two possible relationships between blocks of regulated genes and observable phenotypes. (*a*) A battery of structural genes activated by a single regulator gene produces a visibly recognizable phenotype; (*b*) each recognizable phenotype is produced by the combined effects of batteries of genes activated by two or more regulator genes.

molecular correlates of these changes? Traditionally, molecular biologists have tracked only selected aspects of molecular change within particular lineages. However, to assess the mechanisms by which cells come to differ from each other in development, it will be necessary to trace the molecular basis of divergence from its inception in the cell populations experiencing it.

Fortunately, it is becoming increasingly possible to map the changing molecular compositions of cells in situ by a variety of techniques. These include the use of monoclonal antibodies to map the changing distributions of surface antigens (McClay et al., 1983) and the technique of in situ hybridization (Chapter 9) to trace the changing internal cellular compositions of particular transcripts. Systematic measurements and surveys of these kinds will help to define whether there are general relationships between the changes in particular macromolecular classes and the divergence of cell populations or the appearance of new cell types. The in situ hybridization technique, in particular, can help define the relationship between quantitative changes in gene expression and the advent of new cellular phenotypes. For instance, it is important to know whether there are step function increases in transcript content for genes destined to be abundant or semiabundant in sequences in somatic lineages or only gradual increases. One might answer this question by making a cDNA bank from early embryos and from several fully differentiated tissues of the organism, then doing semiquantitative in situs and constructing profiles of the various increases and decreases in transcript content for all major gene transcript populations as a function of cellular developmental stage. Parallel studies on cell surface antigens could help to map the changes in external cell properties with reference to the changes in transcripts and visible cellular phenotypes. Of course, such surveys might only produce a mass of undigestible facts; yet, if there are general relationships between changes in composition for classes of macromolecules and the appearance of new cellular phenotypes, these approaches could reveal them.

To establish the functional significance of any such changes, genetic analysis is essential. Mutants that disrupt the pathway of interest can be studied in parallel with the wild-type; from the correlations in the presence or absence of specific molecular changes, inferences about the importance of those changes for the developmental phenotype can be inferred. For instance, a detailed description of macromolecular changes in the divergence of mesoderm from ectoderm in the early *Drosophila* embryo could provide the prelude to an analysis of the various dorsalizing mutants of the fruit fly (maternal or zygotic). The question to be answered is: Which changes are invariably abolished and which are uncoupled from mesodermal delimitation in the mutants?

In the second place, saturation mutant hunting can reveal the genes—in principle, *all* the genes—that are crucial for a particular developmental event. As noted above, the requirement for multiple genes to produce the same phenotype may reflect multifunctional complexes, regulatory networks, or the need for a large measure of functional redundancy. Purification of the genes, followed by their sequencing, and characterization of their temporal and spatial expression pattern can help us to decide between these alternatives. DNA sequencing can reveal whether they are part of a family of related genes; if they are so related, this structural similarity would be a strong indication of partial functional redundancy. If the genes are different, they might be part of an interactive regulatory network. In the latter event, characterization of the expression pattern of each of the genes might reveal something of the basis of the regulatory connections.

However, there is one large area in which genetics and molecular biology have yet to make a substantial contribution—that of understanding the organized cellular interactions that take place in development. Although many mutants have been observed to interfere with either morphogenesis or pattern formation, these observations have contributed little to an understanding of the underlying events. Although some possible genetic approaches to the analysis of pattern formation were discussed at the end of Chapter 8, it seems probable that a prerequisite for further understanding is a better set of observations and theories describing morphogenesis and pattern formation in cellular and biophysical/biochemical terms. The model of morphogenesis presented by Odell et al. (1981) is an example of this kind of approach. The evolution of a meaningful developmental genetics of morphogenesis and pattern formation may have to await further findings and concepts along these lines.

The maturation of a science usually shows two characteristics. In the first place, it tends to lose its distinctness and begins to merge with other disciplines. In the second place, it acquires a self-consistent body of theoretical ideas that explain its subject matter. By these criteria, developmental genetics in the mid-1980s is a half-matured science. On the one hand, it has lost the distinctiveness that characterized it until the early 1970s; today there are few boundaries between it and molecular and cellular biology. On the other hand, its theoretical underpinnings remain weak. Because the mechanisms of genetic control must in some way underlie developmental phenomena, developmental genetics has the potential to occupy as central a place in biology as quantum mechanics does in physics.

It will be the clarity and testability of the ideas proposed in the coming years that will determine how rapidly that place is occupied. In certain respects, the connections between genes and development seem clearer than ever. It is just possible that for some of the oldest questions in biology, one can begin to see the shapes of answers.

APPENDIX

Certain conventions in genetic nomenclature are general. Thus, gene names and allele designations are usually italicized, while the corresponding phenotypes are not, and wild-type alleles are designated by +, either directly (when the gene reference is unambiguous) or superscripted; for example, w^+ is the wild-type allele of the *Drosophila* white gene. Similarly, there are general designations for certain kinds of chromosome aberrations: In (inversion), T (translocation), Tp (transposition—a piece of chromosome moved into a site on another chromosome), Dp (duplication), and Df (deficiency). When used in reference to particular aberrations, the symbol becomes italicized; thus *Df(3)P9* refers to the P9 deletion on chromosome 3 (which deletes the BX-C). The heterozygous condition is indicated by a slash (e.g., w/w^+), while mutations or genes on different linkage groups are separated by semicolons (e.g., +/a; +/b symbolizes a dihybrid in which loci a and b are on different chromosomes). In general, recessive mutations are indicated by lowercase designations and dominants by uppercase designations of one to four letters.

The nomenclatural rules for *C. elegans* have been briefly described (pp. 47–48). Alleles are indicated by italicized one or two letters (representing the laboratory of origin) and an allele number. When several mutations are present on a chromosome, they are listed from left to right, as on the standard genetic map. The full set of nomenclatural rules are described by Horvitz et al. (1979). A crude genetic map is shown in Figure App. 1; only some of the genes mentioned in this book are listed, and positions are relative, not to scale. The map is based on recombination measurements. However, measuring recombination in a hermaphroditic organism is less simple than in a standard outcrossing system. The basic principle is illustrated in Table App. 1 and the principles are discussed at greater length in Brenner (1974).

The nomenclatural rules for *Drosophila* are somewhat more complex than those for the nematode, reflecting the longer history of the fruit fly as an experimental genetic organism. Wild-type alleles are given by a + superscript and mutant alleles also by a superscript. Lethals are denoted by l,

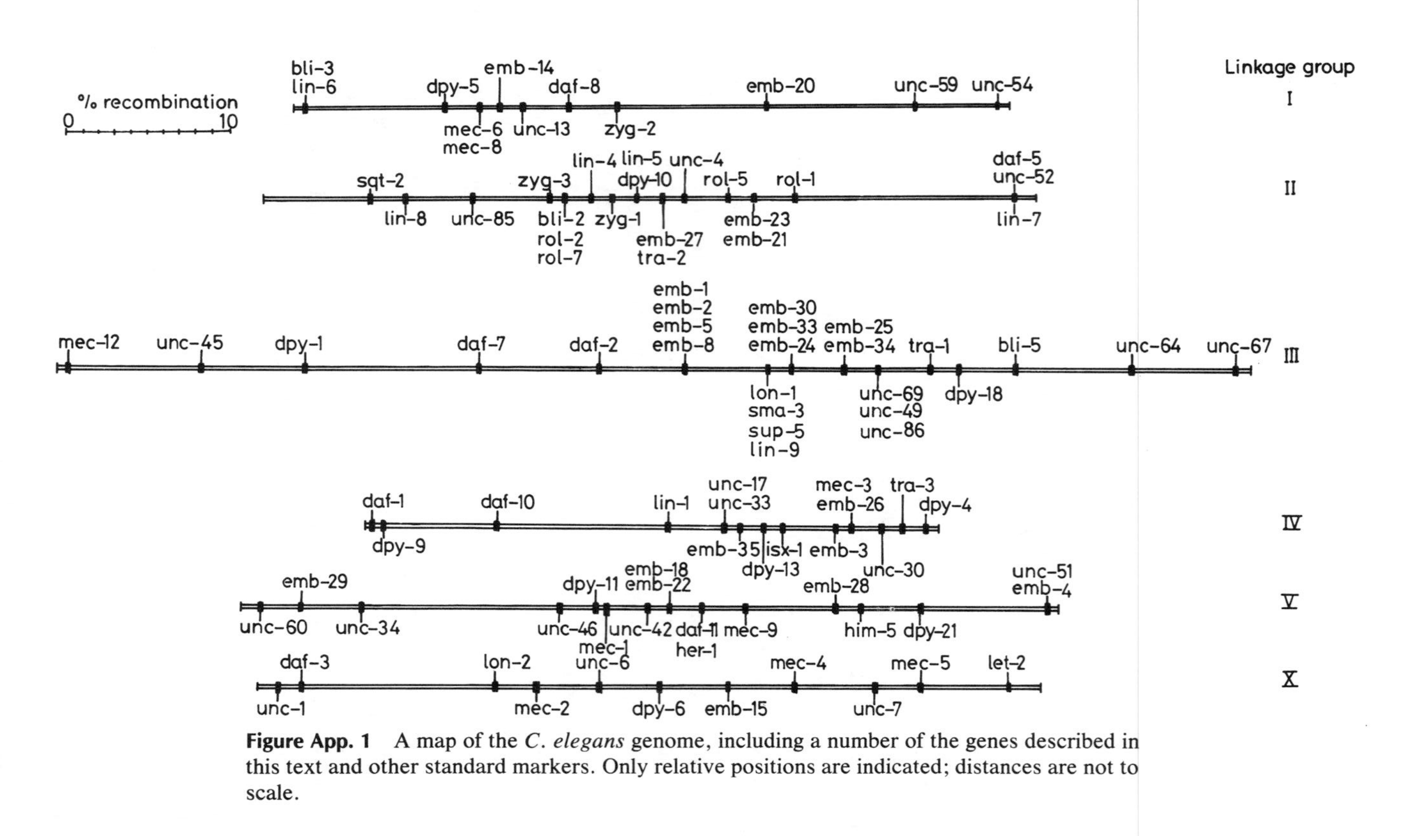

Figure App. 1 A map of the *C. elegans* genome, including a number of the genes described in this text and other standard markers. Only relative positions are indicated; distances are not to scale.

Table App. 1 Measuring the Map Distance Between Two Genes in *Caenorhabditis elegans*[a]

Sperm Chromosome Frequencies		Egg Chromosome Frequencies			
		Parental ($1 - p$, total)		Recombinant (p, total)	
		$++$	ab	$+a$	$+b$
		$\frac{1}{2}(1 - p)$	$\frac{1}{2}(1 - p)$	$\frac{1}{2}p$	$\frac{1}{2}p$
Parental ($1 - p$, total)	$++$	W	W	W	W
	$\frac{1}{2}(1 - p)$	$\frac{1}{4}(1 - p)^2$	$\frac{1}{4}(1 - p)^2$	$\frac{1}{4}(1 - p)p$	$\frac{1}{4}(1 - p)p$
	ab	W	AB	A	B
	$\frac{1}{2}(1 - p)$	$\frac{1}{4}(1 - p)^2$	$\frac{1}{4}(1 - p)^2$	$\frac{1}{4}(1 - p)p$	$\frac{1}{4}(1 - p)p$
Recombinant (p, total)	$a+$	W	A	A	W
	$\frac{1}{2}p$	$\frac{1}{4}(1 - p)p$	$\frac{1}{4}(1 - p)p$	$\frac{1}{4}p^2$	$\frac{1}{4}p^2$
	$+b$	W	B	W	B
	$\frac{1}{2}p$	$\frac{1}{4}(1 - p)p$	$\frac{1}{4}(1 - p)p$	$\frac{1}{4}p^2$	$\frac{1}{4}p^2$

[a] The parental genotype: the *cis* heterozygote $++/ab$.

Notes: a and b are the recessive mutant alleles; A and B are their corresponding mutant phenotypes, and W is the wild-type phenotype; p is the map distance (percentage of recombinant chromosomes).

Let R stand for the total recombinants, A and B. Adding the A and B frequencies, one obtains $R = p(1 - p) + \frac{1}{2}p^2$. Expanding and solving the quadratic for p yields $p = 1 - \sqrt{1 - 2R}$.

One can arrive at the same conclusion in a slightly different way. It can be seen from the table that half of the recombinant chromosomes end up in zygotes that give a wild-type phenotype. It follows that the frequency of zygotes produced by matings involving only *nonrecombinant* chromosomes is $1 - 2R$, and since this frequency is the square of the nonrecombinant chromosome frequency itself, it follows that the recombinant chromosome frequency, which equals 1 minus the frequency of the nonrecombinant chromosomes, is $p = 1 - \sqrt{1 - 2R}$. For semidominant mutations, in which every zygote containing a recombinant is distinguishable, the relationship becomes $p = 1 - \sqrt{1 - R}$.

An example will illustrate the procedure: The X-linked recessive mutants *dpy-6* and *unc-3* were put into a *cis* double heterozygote: *dpy-6 unc-3*/++ (wild-type in phenotype) and the segregants scored. The results were: 826 wild-type; unc, dpy, 213; unc, 110; dpy, 97. The last two classes are the recombinant classes, and comprise $R = 207/1246 = 15.8\%$. Solving for p, $p = 18.3\%$ (example cited from Brenner, 1974).

When scoring both recombinant types is difficult, one can calculate R simply as the scorable recombinant class divided by either parental class (when both parental classes are equal).

followed by the number of the chromosome bearing the lethal, and an individual allele designation. For instance, $l(3)c^{43}$ is a particular ts lethal on the third chromosome. Deficiencies are indicated as noted above, while duplications (Dps) are designated by the chromosomal origin and present site of the duplicated material, followed by the particular name of the duplication; for example $Dp\ (1;2)rb^{71g}$ denotes a particular piece of the X chromosome (chromosome 1) carrying rb^+ that has been moved to the second chromosome. Inversions in either of the large autosomes are pericentric if they involve material on both chromosome arms [e.g., In (2LR)] or paracentric if

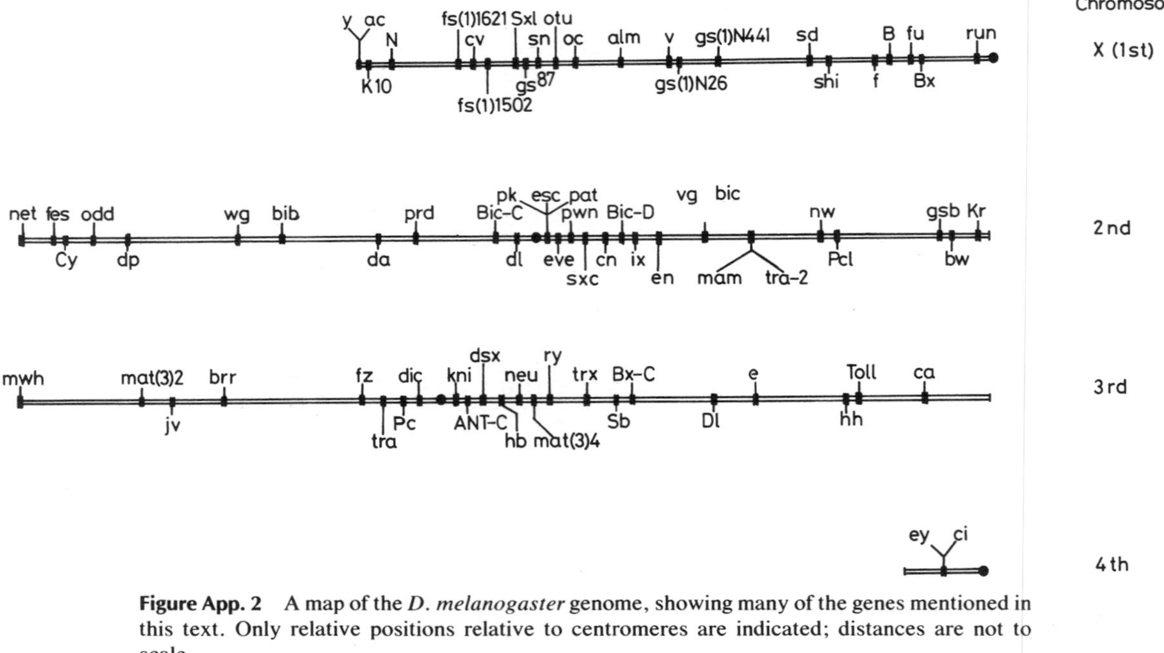

Figure App. 2 A map of the *D. melanogaster* genome, showing many of the genes mentioned in this text. Only relative positions relative to centromeres are indicated; distances are not to scale.

they involve just one arm [e.g., In (2L) inverts material solely on the left arm]. The complete set of nomenclatural rules for *Drosophila* is given in Lindsley and Grell (1968), and a map of the *melanogaster* genome is shown in Figure App. 2.

The nomenclatural rules for mouse genetics were briefly described in Chapter 4 and are given in full in Lyon (1981). A detailed genetic map is shown in Figure App. 3.

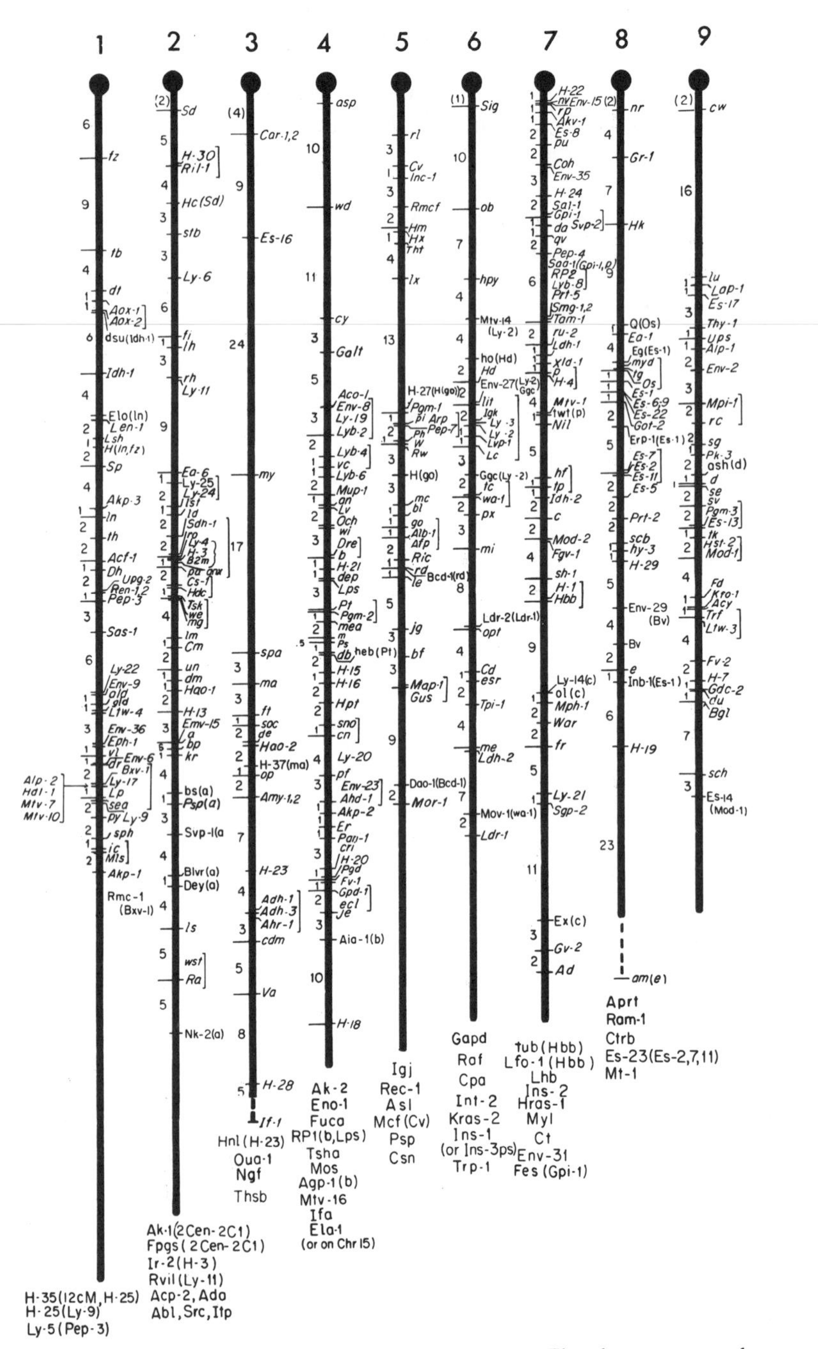

Figure App. 3 A map of the genome of the laboratory mouse. The chromosomes, shown as solid vertical bars, are drawn to their proportionate lengths based on an estimated total haploid genome distance of 1600 map units (centimorgans; centromeres are represented by knobs. The extensions in hatched lines indicate that lengths exceed predicted lengths. Gene symbols shown to the right of chromosomes and numbers to the left indicate recombination distances. Genes

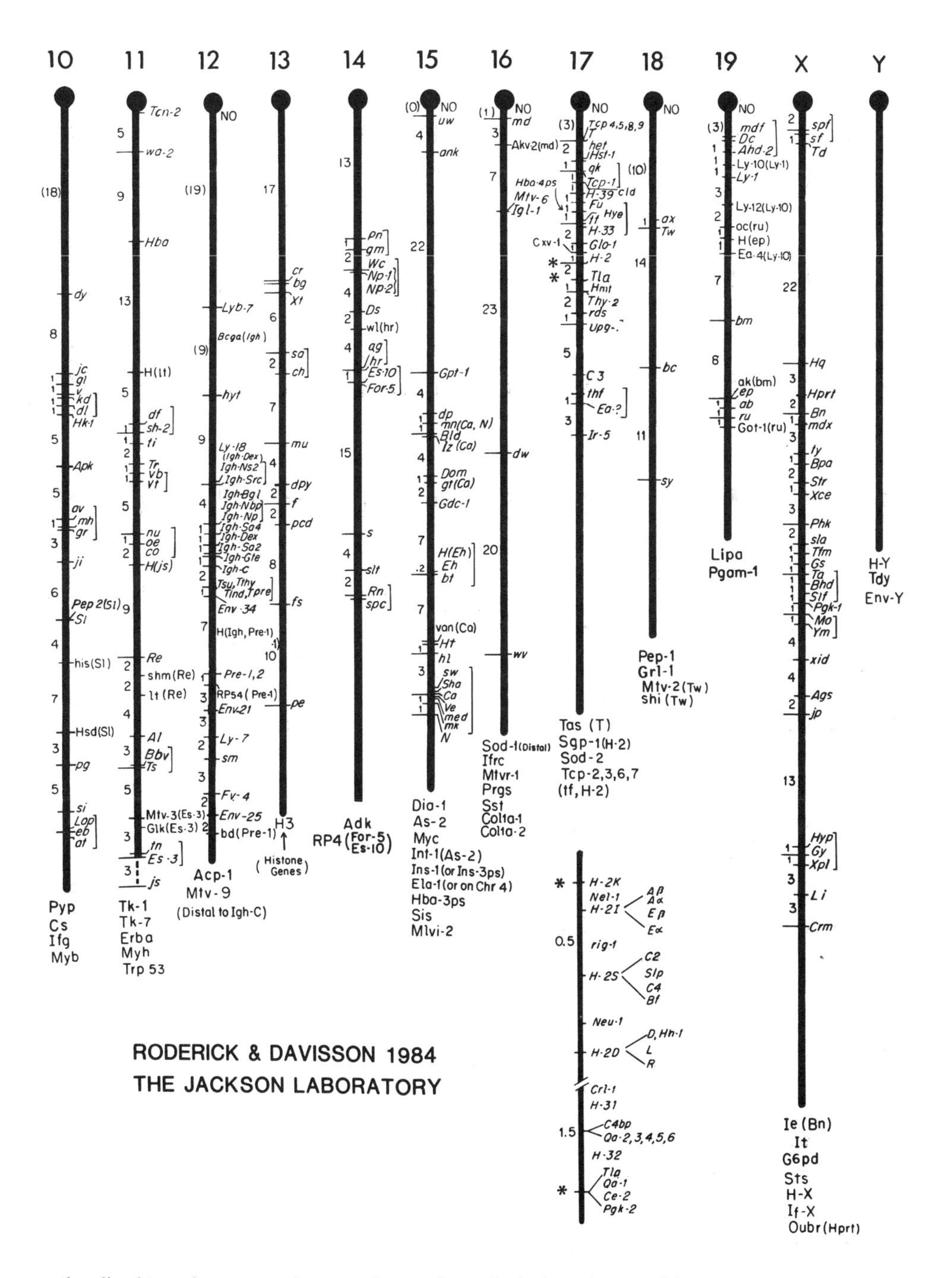

localized to a chromosome by somatic genetic methods, but whose positions are not known, are shown at the bottom of the chromosome. Genes whose positions are known with great certainty are shown by lines extending through the chromosomal bars. (Figure and information courtesy of Drs. T.H. Roderick and M.T. Davisson, The Jackson Laboratory, Bar Harbor, Maine.)

GLOSSARY

allele One of several distinguishable forms of a gene.

alloenzyme A distinguishable allelic form of an enzyme.

allophenic mouse A mouse developed from the cells of two or more embryos; a mouse chimera.

amorph A mutant allele of a gene having no phenotypically detectable activity.

aneuploidy The state of having one, two, or a small number of chromosomes more or less than the standard number of chromosomes for the species.

antimorph An allele of a gene whose phenotypic effect is opposite or antagonistic to the wild-type gene activity.

autonomy (also *cell autonomy*) The confinement of a recessive mutant gene's effect to the cell or clone homozygous for that mutation.

autosome A chromosome that is not one of the sex chromosomes.

backcross A cross between a strain, usually bearing recessive alleles, and its offspring by a previous cross.

balanced lethal system The condition of having two nonallelic lethal genes on opposite members of a pair of nonrecombining homologous chromosomes.

chimera An individual containing cells derived from two or more different zygotes, usually of different genotype.

clone A set of genetically identical members, either cells, organisms, or genes, all descended directly from a single group progenitor.

compartment A group of somatic cells whose progeny are restricted to a particular region of the developing animal.

complementary DNA (cDNA) The complementary sequences synthesized from mRNA, usually by the enzyme reverse transcriptase.

complementation The restoration, in whole or in part, of a wild-type phenotype in cells or organisms containing two mutations on opposite homologues (in *trans*).

complexity The summed information in a set of DNA sequences, given in base pairs, in which each sequence, regardless of repetition frequency, is counted once.

deficiency (or deletion) The loss of a DNA sequence from a chromosome; the portion that is missing.

diploid The condition of having two complete, homologous sets of chromosomes.

dominant An allele of a gene that produces the same phenotypic effect in heterozygotes as in homozygotes.

electromorph An electrophoretically distinguishable allelic form of an enzyme.

episome A small genetic element capable of replicating independently in the cytoplasm or as an integrated part of a chromosome.

epistasis The condition in which one gene mutation masks the expression of a nonallelic gene mutation.

exon A portion of a eukaryotic gene that codes for a polypeptide sequence, located between noncoding portions (introns).

expressivity The strength of mutant gene expression compared to that of the extreme mutant phenotype.

F_1 The first-generation offspring of a cross between two strains.

F_2 The second-generation offspring following a cross between two strains; produced by intercrossing the F_1.

flanking sequence The sequences on either side (both 5′ and 3′) of a polypeptide coding sequence.

gamete A mature germ line cell; in animals, either a spermatozoan or an ovum.

gene A stretch of DNA (or of RNA in RNA viruses) possessing a definable function; usually but not always a DNA sequence specifying a polypeptide chain.

genome The complete haploid set of chromosomes of an organism.

genotype The genetic constitution of an organism or a cell within that organism.

germ line The reproductive cells within a complex organism that give rise to the sex cells or gametes involved directly in reproduction.

haploid The condition of having a single complete set of chromosomes, the chromosomal constitution of the gametes.

haplo-insufficiency The condition in which a single dose of a wild-type gene in a diploid organism is insufficient to give a wild-type character for that gene.

hemizygote A diploid individual that carries only one copy of a gene, usually a deficiency heterozygote for that gene or the heterogametic sex (e.g., XY individuals in an organism with X-Y sex determination).

heterochromatin The highly compacted, densely staining regions of eukaryotic chromosomes that are transcriptionally inactive.

heterogametic sex The sex that produces two different kinds of gametes, distinguished by their sex chromosome constitution, for example, X- and Y-bearing sperm.

heterozygote A diploid individual carrying two different allelic forms of a gene.

homeotic mutation A mutation causing the replacement of one part of an individual by a recognizable structure or region that normally develops in a different location.

homozygote A diploid individual carrying two identical alleles of a gene.

housekeeping gene A gene encoding a protein that carries out a general cellular metabolic function.

hypomorph A mutant gene producing reduced, but not nil, gene activity.

intron A noncoding portion of a eukaryotic gene inserted between two coding regions (exons).

inversion A chromosome rearrangement in which a section has been turned through 180° to give a reversal of gene order in that section.

linkage An association in gene inheritance such that parental allele combinations tend to be inherited.

locus The position of a gene on a chromosome; the gene itself.

maternal effect gene A gene expressed in oogenesis, capable of mutating to give a maternal effect in the progeny.

maternal effect mutation A mutation whose expression in the female gamete–producing parent creates an effect, usually deleterious, in the progeny regardless of progeny genotype.

meiosis The two successive nuclear divisions that accompany gamete formation in diploid organisms and reduce the diploid state in gamete precursor cells to the haploid state in gametes.

messenger RNA (mRNA) The complementary RNA copy of a protein-coding gene.

midrepetitive sequence A sequence present in multiple copies, often slightly different from one another, in the genome of a eukaryote. A typical midrepetitive sequence is present in several hundred copies.

mosaic An individual composed of cells of different genotypes, although all were derived initially from a single-genotype zygote.

mutant An individual manifesting the expression of an altered gene. Also, adjectival: the state of bearing a mutation, referring either to a gene or to an individual.

mutation (1) The process by which a gene undergoes a transmissible alteration; (2) the hereditary alteration in the gene itself.

neomorph A mutant gene producing a qualitatively novel phenotypic effect.

nonautonomy (cellular nonautonomy) The expression of either the mutant or wild-type character in a mosaic or chimera in cells of the other genotype.

null allele A mutant allele producing no wild-type gene product or activity. Sometimes used synonymously for "amorph," but if classification involves a biochemical test, categorization as a null allele is a more definitive test of nil activity.

operator A repressor-binding site adjacent to one or more structural genes.

penetrance The percentage of mutant individuals that exhibit the mutant phenotype.

phene A single, distinguishable characteristic of an organism, usually one associated with a mutant condition.

phenocopy An individual whose phenotype has been altered by external factors during development, mimicking a known mutant condition.

phenotype The appearance of an organism with respect to one or more specific gene-affected characters; the phenotype is produced by the interaction of genotype and environment.

plasmid A self-replicating circular DNA molecule found in bacteria; used in cloning exogenous DNA sequences.

pleiotropy The multiple phenotypic effects produced by the expression of a single mutant gene.

polyploidy The condition of possessing an integral number of gene sets greater than two.

polytene The condition of multiple, aligned chromatids in certain "giant" chromosomes; found in certain insects and plants.

posttranscriptional control The regulated, delayed translation of preformed mRNAs.

posttranslational modification The addition or removal of certain metabolite groups (such as phosphates) to preformed polypeptides.

recessive A mutant allele that produces a phenotypic effect only in homozygotes.

reciprocal crosses A pair of crosses of the form X♂ × Y♀ and Y♂ × X♀, where X and Y are two strains of differing genotype.

recombination The process of genetic exchange, usually of reciprocal, homologous chromosome sections between paired chromosomes.

regulator gene A gene whose product directly controls the expression of other genes.

regulator site A DNA sequence that binds the product of a regulator gene and controls the expression in *cis* of neighboring genes.

replicon A unit of DNA replication possessing an independent site of initiation, or origin, of replication.

repressor The gene product of a regulator gene whose primary and direct action is the inhibition of transcription or the translation of one or more structural genes.

restriction enzyme An endonuclease that cleaves at a particular short DNA sequence; each such enzyme is found in and isolated from particular bacterial species.

reverse transcriptase An enzyme encoded by certain RNA tumor viruses that copies RNA into single-stranded complementary DNA.

satellite sequence A simple DNA sequence reiterated thousands to millions of times in the genome and usually located in or around the centromere.

selector gene A positive-control eukaryotic regulator gene that activates the expression of a battery of general cell function realizator genes; originally defined as acting within delimited cell compartments (Garcia-Bellido, 1975).

sex chromosomes The two chromosomes that share some homology that are dissimilar in the heterogametic sex.

sex linkage The coinheritance of a gene and a sex chromosome, usually reflecting gene location on the X chromosome in animals with X-Y sex determination.

single copy (or unique) sequence A DNA sequence present only once, or a very few times, per haploid genome; defined operationally by renaturation kinetics.

structural gene A gene that codes for a polypeptide chain involved in some aspect of metabolism or cell structure; a gene whose expression is controlled by a regulator gene.

suppressor mutation A mutation that abolishes the phenotypic expression of a mutation in one or more other genes.

temperature-sensitive mutant A mutant whose phenotypic manifestation is temperature dependent; usually refers to a heat-sensitive mutant.

translocation A chromosome rearrangement involving the interchange of segments between nonhomologous chromosomes.

wild-type The character or strain taken as the representative genetic standard or nonmutant type.

zygotic genome The genome of the embryo or the fertilized cell from which the embryo derives; as distinguished from the genome of the mother from which the embryo derives (the maternal genome).

zygotic induction The freeing from repression of a chromosomally-integrated episome upon transfer during bacterial mating to a repressor-free cytoplasm.

REFERENCES

Adler, P. N. (1979). Position-specific interactions between cells of the imaginal wing and haltere discs of *Drosophila melanogaster*. *Dev. Biol.* **70,** 262–267.

Affara, N., Jacquet, M., Jakob, H., Jacob, F., and Gros, F. (1977). Comparison of polysomal polyadenylated RNA from embryonal carcinoma and committed myogenic and erythropoietic lines. *Cell* **12,** 509–520.

Affara, N., and Dauras, P. (1979). Regulation of a group of abundant mRNA sequences during Friend cell differentiation. *Dev. Biol.* **72,** 110–125.

Agrell, I. (1964). Natural division synchrony and mitotic gradients in metazoan tissues. In E. Zeuthen (Ed.), *Synchrony in Cell Division and Growth*. New York: Wiley-Interscience.

Akam, M. E. (1983). The location of *Ultrabithorax* transcripts in *Drosophila* tissue sections. *EMBO J.* **2,** 2075–2084.

Alberts, B., Bray, D., Lewis, J., Raff, M., Roberts, K., and Watson, J. D. (1983). *The Molecular Biology of the Cell*. New York: Garland.

Albertson, D. G., Sulston, J. E., and White, J. G. (1978). Cell cycling and DNA replication in a mutant blocked in cell division in the nematode *Caenorhabditis elegans*. *Dev. Biol.* **63,** 165–178.

Alderson, T. (1965). Chemically induced delayed germinal mutation in *Drosophila*. *Nature* **207,** 164–167.

Ambros, V., and Horvitz, H. R. (1974). Heterochronic mutants of the nematode *Caenorhabditis elegans*. *Science* **226,** 409–416.

Anderson, K. V., and Lengyel, J. A. (1979). Rates of synthesis of major classes of RNA in *Drosophila* embryos. *Dev. Biol.* **70,** 217–231.

Anderson, K. V., and Nusslein-Volhard, C. (1984). Genetic analysis of dorsal-ventral embryonic pattern in *Drosophila*. In G. M. Malacinski and S. Bryant (Eds.), *Pattern Formation*. New York: Macmillan.

Aristotle (1963). *Generation of Animals*. Cambridge, Mass.: Harvard University Press.

Arking, R. (1975). Temperature-sensitive cell-lethal mutants of *Drosophila*: isolation and characterization. *Genetics* **80,** 519–537.

Arthur, C. G., Weide, C. M., Vincent, W. S., and Goldstein, E. S. (1979). mRNA sequence diversity during early embryogenesis in *Drosophila melanogaster*. *Exp. Cell Res.* **121,** 87–94.

Ashburner, M., and Bonner, J. J. (1979). The induction of gene activity in *Drosophila* by heat shock. *Cell* **17,** 241–254.

Babu, P. (1974). Biochemical genetics of *Caenorhabditis elegans*. *Mol. Gen. Genet.* **135,** 39–44.

Bachvarova, R. (1974). Incorporation of tritiated adenosine into mouse ovum RNA. *Dev. Biol.* **40,** 52–58.

Bachvarova, R., and De Leon, V. (1980). Polyadenylated RNA of mouse ova and loss of maternal RNA in early development. *Dev. Biol.* **74,** 1–8.

Baker, B. S., and Ridge, K. A. (1980). Sex and the single cell. I. On the action of major loci affecting sex determination in *Drosophila melanogaster*. *Genetics* **94,** 383–423.

Baker, B. S., and Belote, J. M. (1983). Sex determination and dosage compensation in *Drosophila melanogaster*. *Ann. Rev. Genet.* **17,** 345–393.

Baker, B. S., Smith, D. A., and Gatti, M. (1982). Region-specific effects on chromosome integrity of mutations at essential loci in *Drosophila melanogaster*. *Proc. Natl. Acad. Sci. USA* **79,** 1205–1209.

Baker, T. G. (1982). Oogenesis and ovulation. In C. R. Austin and R. V. Short (Eds.), *Germ Cells and Fertilization*. Cambridge: Cambridge University Press.

Baker, W. K. (1978). A clonal analysis reveals early developmental restrictions in the *Drosophila* head. *Dev. Biol.* **62,** 447–463.

Bakken, A. (1973). A cytological and genetic study of oogenesis in *Drosophila melanogaster*. *Dev. Biol.* **33,** 100–122.

Bard, J. B. L. (1981). A model for generating aspects of zebra and other mammalian coat patterns. *J. Theor. Biol.* **93,** 363–385.

Beadle, G. W., and Tatum, E. L. (1941). Genetic control of biochemical reactions in *Neurospera*. *Proc. Natl. Acad. Sci. USA* **27,** 499–506.

Beddington, R. S. P. (1982). An autoradiographic analysis of tissue potency in different regions of the embryonic ectoderm during gastrulation in the mouse. *J. Embryol. Exp. Morphol.* **69,** 265–285.

Belote, J. M., and Lucchesi, J. C. (1980a). Male-specific lethal mutations of *Drosophila melanogaster*. *Genetics* **96,** 165–186.

Belote, J. M., and Lucchesi, J. C. (1980b). Control of X-chromosome transcription by the maleless gene in *Drosophila*. *Nature* **285,** 573–575.

Belote, J. M., and Baker, B. S. (1982). Sex determination in *Drosophila melanogaster*: analysis of *transformer-2,* a sex-transforming locus. *Proc. Natl. Acad. Sci. USA* **79,** 1568–1572.

Bender, W., Spierer, P., and Hogness, D. S. (1983a). Chromosomal walking and jumping to isolate DNA from the *Ace* and *rosy* loci and the bithorax complex in *Drosophila melanogaster*. *J. Mol. Biol.* **168,** 17–33.

Bender, W., Akam, M., Karch, E., Beachy, P. A., Peifer, M., Spierer, P., Lewis, E. B., and Hogness, D. S. (1983b). Molecular genetics of the bithorax complex in *Drosophila melanogaster*. *Science (Wash.)* **221,** 23–29.

Bennett, D. (1975). The T-locus of the mouse. *Cell* **6,** 441–454.

Biessman, H. (1981). Changes in the abundance of polyadenylated RNAs in development of *Drosophila melanogaster* analyzed with cloned DNA fragments. *Dev. Biol.* **83,** 62–68.

Bischoff, W. L., and Lucchesi, J. C. (1971). Genetic organization in *Drosophila melanogaster:* Complementation and fine structure analysis of the deep orange locus. *Genetics* **69,** 453–466.

Bishop, D. L., and King, R. C. (1984). An ultrastructural study of ovarian development in the otu^7 mutant of *Drosophila melanogaster*. *J. Cell Sci.* **67,** 87–119.

Blumenthal, A. B., Kriegstein, H. R., and Hogness, D. S. (1974). The units of DNA replication in *Drosophila melanogaster* chromosomes. *Cold Spring Harbor Symp. Quant. Biol.* **38,** 205–233.

Bodenstein, D. (1950). The postembryonic development of *Drosophila*. In M. Demerec (Ed.), *The Biology of Drosophila*. New York: Wiley-Interscience.

Boveri, T. (1899). Die entwicklung von *Ascaris megalocephala* mit besonderer rucksicht auf die kernverhaltnisse. In *Festschrift Kupfer*. Jena: G. Fischer.

Boveri, T. (1910). Über die Teilung centrifugierter eier von *Ascaris megalocephala*. *Arch. Entwicklungsmechanik* **30,** 101–125.

Bownes, M. (1975). Adult deficiencies and duplications of head and thoracic structures resulting from microcautery of blastoderm stage *Drosophila* embryos. *J. Embryol. Exp. Morphol.* **34,** 33–54.

Bownes, M., and Sang, J. H. (1974a). Experimental manipulations of early *Drosophila* embryos. I. Adult and embryonic defects resulting from microcautery at nuclear multiplication and blastoderm stages. *J. Embryol. Exp. Morphol.* **32,** 253–272.

Bownes, M., and Sang, J. H. (1974b). Experimental manipulations of early *Drosophila* embryos. II. Adult and embryonic defects resulting from the removal of blastoderm cells by pricking. *J. Embryol. Exp. Morphol.* **32,** 273–285.

Boycott, R. E., and Diver, C. (1923). On the inheritance of sinistrality in *Limnaea peregra. Proc. R. Soc. Lond.* **95B,** 207–213.

Bradley, A., Evans, M., Kaufman, M. H., and Robertson, E. (1984). Formation of germ-line chimaeras from embryo-derived teratocarcinoma cell lines. *Nature* **309,** 255–256.

Braude, P. R. (1979). Time-dependent effects of α-amanitin on blastocyst formation in the mouse. *J. Embryol. Exp. Morphol.* **52,** 193–202.

Braude, P., Pelham, H., Flach, G., and Lobatto, R. (1979). Post-transcriptional control in the early mouse embryo. *Nature* **282,** 102–105.

Brennan, M. D., Weiner, A. J., Goralski, T. J., and Mahowald, A. P. (1982). The follicle cells are a major site of vitellogenin synthesis in *Drosophila melanogaster. Dev. Biol.* **89,** 225–236.

Brenner, S. (1974). The genetics of *Caenorhabditis elegans. Genetics* **77,** 71–94.

Brenner, S. (1981). Genes and development. In C. W. Lloyd and D. A. Rees (Eds.), *Cellular Controls in Differentiation.* New York: Academic Press.

Bridges, C. B. (1925). Sex in relation to chromosomes and genes. *Am. Nat.* **59,** 127–137.

Bridges, C. B. (1935). Salivary chromosome maps. *J. Hered.* **26,** 60–64.

Brinster, R. L. (1974). The effect of cells transferred into the mouse blastocyst on subsequent development. *J. Exp. Med.* **140,** 1049–1056.

Brinster, R. L., Chen, H. Y., Trumbauer, M., Senear, A. W., Warren, R., and Palmiter, R. D. (1981). Somatic expression of herpes thymidine kinase in mice following injection of a fusion gene into eggs. *Cell* **27,** 223–231.

Britten, R. J., and Kohne, D. E. (1968). Repeated sequences in DNA. *Science* **161,** 529–540.

Britten, R. J., and Davidson, E. H. (1969). Gene regulation for higher cells: a theory. *Science* **165,** 349–357.

Brower, D. L., Lawrence, P. A., and Wilcox, M. (1981). Clonal analysis of the undifferentiated wing disc of *Drosophila. Dev. Biol.* **86,** 448–455.

Brown, D. D. (1981). Gene expression eukaryotes. *Science* **211,** 667–674.

Brown, D. D. (1984). The role of stable complexes that repress and activate eucaryotic genes. *Cell* **37,** 359–365.

Brown, E. H., and King, R. C. (1972). Studies on the events resulting in the formation of an egg chamber in *Drosophila melanogaster.* In *Invertebrate Oogenesis,* vol. 1. New York: MSS Information Corp.

Bryant, P. J. (1970). Cell lineage relationships in the imaginal wing disc of *Drosophila melanogaster. Dev. Biol.* **22,** 389–411.

Bryant, P. J. (1971). Regeneration and duplication following operations *in situ* on the imaginal discs of *Drosophila melanogaster. Dev. Biol.* **26,** 637–651.

Bryant, P. J. (1975). Regeneration and duplication in imaginal discs. In R. Porter and J. Rivers (Eds.), *Cell Patterning*. Amsterdam: Elsevier/North-Holland.

Bryant, P. J. (1978). Pattern formation in imaginal discs. In M. Ashburner and T. R. F. Wright (Eds.), *The Genetics and Biology of Drosophila*. London: Academic Press.

Bryant, P. J., and Schneiderman, H. A. (1969). Cell lineage, growth, and determination in the imaginal leg discs of *Drosophila melanogaster*. *Dev. Biol.* **20,** 263–290.

Bryant, S. V., French, V., and Bryant, P. J. (1981). Distal regeneration and symmetry. *Science* **212,** 993–1002.

Bulfield, G. (1981). Inborn errors of metabolism in the mouse. In R. J. Berry (Ed.), *Biology of the House Mouse*. Symposia of the Zoological Society of London, no. 47. London: Academic Press.

Bull, A. L. (1966). *Bicaudal,* a genetic factor which affects the polarity of the embryo in *Drosophila melanogaster*. *J. Exp. Zool.* **161,** 221–242.

Burgoyne, P. S. (1982). Genetic homology and crossing over in the X and Y chromosomes of mammals. *Hum. Genet.* **61,** 85–90.

Byerly, L., Cassada, R. C., and Russell, R. L. (1976). The life cycle of the nematode *Caenorhabditis elegans* I. Wild-type growth and reproduction. *Dev. Biol.* **51,** 23–33.

Calarco, P. G., and Epstein, C. J. (1973). Cell surface changes during pre-implantation development in the mouse. *Dev. Biol.* **32,** 208–213.

Callan, H. G. (1974). DNA replication in the chromosomes of eukaryotes. *Cold Spring Harbor Symp. Quant. Biol.* **38,** 195–203.

Capdevila, M. P., and Garcia-Bellido, A. (1974). Development and genetic analysis of *bithorax* phenocopies in *Drosophila*. *Nature* **250,** 500–502.

Capdevila, M. P., and Garcia-Bellido, A. (1981). Genes involved in the activation of the bithorax complex of *Drosophila*. *Wilh. Roux's Arch.* **190,** 339–350.

Carpenter, A. T. C. (1975). Electron microscopy of meiosis in *Drosophila melanogaster* females. I. Structure, arrangement, and temporal change of the synaptinemal complex in wild-type. *Chromosoma (Berl.)* **51,** 157–182.

Cassada, R. C., and Russell, R. L. (1975). The dauer larva, or post-embryonic developmental variant of the nematode *Caenorhabditis elegans*. *Dev. Biol.* **46,** 326–342.

Cassada, R., Issenghi, E., Culotti, M., and von Ehrenstein, G. (1981). Genetic analysis of temperature-sensitive embryogenesis mutants in *Caenorhabditis elegans*. *Dev. Biol.* **84,** 193–205.

Castle, W. E., and Little, C. C. (1910). On a modified Mendelian ratio among yellow mice. *Science* **32,** 868–870.

Catalano, G., Eilbeck, C., Monroy, A., and Parisi, E. (1979). A model for early segregation of territories in the ascidian egg. In N. LeDouarin (Ed.), *Cell Lineage, Stem Cells, and Cell Determination*. Amsterdam: Elsevier/North-Holland.

Cattanach, B. M., Pollard, C. E., and Hawkes, S. G. (1971). Sex-reversed mice: XX and XO males. *Cytogenetics* **10,** 318–337.

Chalfie, M., and Sulston, J. E. (1981). Developmental genetics of the mechanosensory neurons of *Caenorhabditis elegans*. *Dev. Biol.* **82,** 358–370.

Chalfie, M., Horvitz, H. R., and Sulston, J. E. (1981). Mutations that lead to reiterations in the cell lineages of *C. elegans*. *Cell* **24,** 59–69.

Chan, L-N., and Gehring, W. (1971). Determination of blastoderm cells in *Drosophila melanogaster*. *Proc. Natl. Acad. Sci. USA* **68,** 2217–2221.

Chandra, H. S. (1984). A model for mammalian male determination based on a passive Y chromosome. *Mol. Gen. Genet.* **193,** 384–388.

Chapman, V. M., Ansell, J. D., and McLaren, A. (1972). Trophoblast cell differentiation in the mouse: expression of glucose phosphate isomerase (GPI-1) electrophoretic variants in transferred and chimaeric embryos. *Dev. Biol.* **29,** 48–54.

Chapman, V. M., Adler, D., Labarca, C., and Wudl, L. (1976). Genetic variation of β-glucuronidase during early embryogenesis. In M. H. Johnson (Ed.), *Early Development of Mammals,* Ciba Foundation Symposium. Amsterdam: Elsevier/North-Holland.

Chesley, P. (1935). Development of the short-tailed mutant in the house mouse. *J. Exp. Zool.* **70,** 429–459.

Church, R. B., and Schultz, G. A. (1974). Differential gene activity in the pre- and postimplantation mammalian embryo. *Curr. Top. Dev. Biol.* **8,** 179–202.

Cleffman, G. (1963). Agouti pigment cells *in situ* and *in vitro*. *Ann. N.Y. Acad. Sci.* **100,** 749–761.

Clegg, K. B., and Piko, L. (1982). RNA synthesis and cytoplasmic polyadenylation in the one-cell mouse embryo. *Nature* **295,** 342–345.

Cline, T. W. (1978). Two closely linked mutations in *Drosophila melanogaster* that are lethal in opposite sexes and interact with *daughterless*. *Genetics* **90,** 683–698.

Cline, T. W. (1979). A male-specific lethal mutation in *Drosophila melanogaster* that transforms sex. *Dev. Biol.* **72,** 266–275.

Cole, E. S., and Palka, J. (1982). The patterns of campaniform sensillae on the wing and haltere of *Drosophila melanogaster* and several of its homoeotic mutants. *J. Embryol. Exp. Morphol.* **71,** 41–61.

Coleman, D. L. (1962). Effect of genic substitution in the incorporation of tyrosine into melanin of mouse skin. *Arch. Biochem. Biophys.* **96,** 562–568.

Conway, K., Feiock, K., and Hunt, R. K. (1980). Polyclones and patterns in growing *Xenopus* eye. *Curr. Top. Dev. Biol.* **15,** 217–317.

Cori, C. F., Gluecksohn-Waelsch, S., Shaw, P. A., and Robinson, C. (1983). Correction of a genetically caused enzyme defect by somatic cell hybridization. *Proc. Natl. Acad. Sci. USA* **80,** 6611–6614.

Counce, S. J. (1956). Studies on female-sterility genes in *Drosophila melanogaster*. I. The effects of the gene *deep orange* on embryonic development. *Z. Induktive Abstammungs-Vererbungslehre* **87,** 443–461.

Counce, S. J., and Ede, D. A. (1957). The effect on embryogenesis of a sex-linked female-sterility factor in *Drosophila melanogaster*. *J. Embryol. Exp. Morphol.* **5,** 404–421.

Cox, G. N., Laufer, J. S., Kusch, M., and Edgar, R. S. (1980). Genetic and phenotypic characterization of roller mutants of *Caenorhabditis elegans*. *Genetics* **95,** 317–339.

Cox, G. N., Kusch, M., DeNevi, K., and Edgar, R. S. (1981a). Temporal regulation of cuticle synthesis during development of *Caenorhabditis elegans*. *Dev. Biol.* **84,** 277–285.

Cox, G. N., Staprons, S., and Edgar, R. S. (1981b). The cuticle of *Caenorhabditis elegans*. II. Stage-specific changes in ultrastructure and protein composition during postembryonic development. *Dev. Biol.* **86,** 456–470.

Crain, W. R., Eden, F. C., Pearson, W. R., Davidson, E. H., and Britten, R. J. (1976). Absence of short period interspersion of repetitive and non-repetitive sequences in the DNA of *Drosophila melanogaster*. *Chromosoma* (*Berl.*) **56,** 309–326.

Crick, F. H. C., and Lawrence, P. A. (1975). Compartments and polyclones in insect development. *Science* **189,** 340–347.

Cuénot, L. (1908). Sur quelques anomalies apparentes des proportions Mendeliennes. *Notes Renne,* **Ib, 9,** 7–15.

Dalcq, A. M. (1957). *Introduction to General Embryology*. London: Oxford University Press.

Damjanov, I., and Solter, D. (1982). Maternally transmitted factors modify development and malignancy of teratomas in mice. *Nature* **296,** 95–96.

Davidson, E. H. (1976). *Gene Activity in Early Development.* New York: Academic Press.

Davidson, E. H., and Britten, R. J. (1979). Regulation of gene expression: possible role of repetitive sequences. *Science* **204,** 1052–1059.

Davidson, E. H., Galau, G. A., Angerer, R. C., and Bullen, R. J. (1975). Comparative aspects of DNA organization in metazoa. *Chromosoma* (*Berl.*) **75,** 253–259.

Davidson, E. H., Jacobs, H. T., and Britten, R. J. (1983a). Very short repeats and coordinate induction of genes. *Nature* **301,** 468–470.

Davidson, E. H., Jacobs, H. T., Thomas, T. L., Hough-Evans, B. R., and Britten, R. J. (1983b). Poly (A) RNA of the egg cytoplasm: Structual resemblance to the nuclear RNA of somatic cells. In R. Porter and J. Whelan (Eds.), *Molecular Biology of Egg Maturation.* London: Pitman Press.

Deak, I. (1980). Embryogenesis in *Drosophila:* can a single mechanism explain the developmental segregation of germ-line and somatic cells and their subsequent determination? *Wilh. Roux Arch.* **188,** 179–185.

del Pino, E. M., and Elinson, R. P. (1983). A novel developmental pattern for frogs: gastrulation produces an embryonic disk. *Nature* **306,** 589–591.

Denell, R., Hummels, R. K., Wakimoto, B. T., and Kauffman, T. C. (1981). Developmental studies of lethality associated with the Antennapedia gene complex in *Drosophila melanogaster. Dev. Biol.* **81,** 43–50.

Deppe, U., Schierenberg, E., Cole, T., Krieg, C., Schmitt, D, Yoder, B., and von Ehrenstein, G. (1978). Cell lineages of the embryo of the nematode *Caenorhabditis elegans. Proc. Natl. Acad. Sci. USA* **75,** 376–380.

Derman, E., Krauter, K., Waller, L. Weinberger, C., Ray, M., and Darnll, J. E., Jr. (1981). Transcriptional control in the production of liver-specific mRNAs. *Cell* **23,** 731–739.

Di Berardino, M. A. (1980). Genetic stability and modulation of metazoan nuclei transplanted into eggs and oocytes. *Differentiation* **17,** 17–30.

Doane, W. W. (1973). Role of hormones in insect development. In S. J. Counce and C. H. Waddington (Eds.), *Developmental Systems: Insects,* vol. 2. London: Academic Press.

Dobrovolskaia-Zavovskaia, N. (1927). Sur la mortification spontanée de la queue chez la souris nouveau-née et sur l'existence d'un caractère (facteur) héreditaire "non-viable." *Comp. Rend. Soc. Biol.* **97,** 114–116.

Dowsett, A. P. (1983). Closely related species of *Drosophila* can contain different libraries of middle repetitive DNA sequences. *Chromosoma* (*Berl.*) **88,** 104–108.

Dubendorfer, K., and Nothiger, R. (1982). A clonal analysis of cell lineage and growth in the male and female genital disc of *Drosophila melanogaster. Wilh. Roux Arch.* **191,** 42–55.

Ducibella, T. (1977). Surface changes of the developing trophoblast cell. In M. H. Johnson (Ed.), *Development in Mammals,* vol. 1 Amsterdam: Elsevier/North-Holland.

Duncan, I. M. (1982). *Polycomblike:* a gene that appears to be required for the normal expression of the bithorax and Antennapedia complexes of *Drosophila melanogaster. Genetics* **102,** 49–70.

Duncan, I. M., and Lewis, E. B. (1982). Genetic control of body segment differentiation in *Drosophila.* In S. Subtelny and P. B. Green (Eds.), *Developmental Order: Its Origin and Regulation.* New York: Alan R. Liss.

Dunn, L. C., and Bennett, D. (1971). Lethal alleles near locus T in house mouse populations on the Jutland peninsula, Denmark. *Evolution* **25,** 451–453.

Eberlein, S., and Russell, M. A. (1983). Effects of deficiencies in the *engrailed* region of *Drosophila melanogaster. Dev. Biol.* **100,** 227–237.

Eicher, E. (1983). Primary sex determining genes in mice. In R. P. Ammaru and G. E. Seidel (Eds.), *Prospects for Sexing Mammalian Sperm.* Boulder: Colorado University Press.

Emmons, S. W., Klass, M. R., and Hirsh, D. (1979). Analysis of the constancy of DNA sequences during development and evolution of the nematode *Caenorhabditis elegans*. *Proc. Natl. Acad. Sci. USA* **76,** 1333–1337.

Emmons, S., Rosensweig, B., and Hirsh, D. (1980). Arrangement of repeated sequences in the DNA of the nematode *Caenorhabditis elegans*. *J. Mol. Biol.* **144,** 481–500.

Englesberg, E., and Wilcox, G. (1974). Regulation: positive control. *Ann. Rev. Genet.* **8,** 219–242.

Esworthy, S., and Chapman, V. M. (1981). The expression of β-galactosidase during preimplantation mouse embryogenesis. *Dev. Genet.* **2,** 1–12.

Evans, M. H., and Kaufman, M. H. (1981). Establishment in culture of pluripotential cells from mouse embryos. *Nature* **292,** 154–156.

Eyal-Giladi, H., Ginsberg, M., and Farberov, A. (1981). Avian primordial germ cells are of epiblastic origin. *J. Embryol. Exp. Morphol.* **65,** 139–147.

Fausto-Sterling, A., and Smith-Schiess, H. (1982). Interactions between *fused* and *engrailed,* two mutations affecting pattern formation in *Drosophila melanogaster*. *Genetics* **101,** 71–80.

Ferrus, A., and Kankel, D. R. (1981). Cell lineage relationships in *Drosophila melanogaster:* the relationships of cuticular to internal tissues. *Dev. Biol.* **85,** 485–504.

Finnerty, V. (1976). Genetic units of *Drosophila*—simple cistrons. In M. Ashburner and E. Novitski (Eds.), *The Genetics and Biology of Drosophila*. London: Academic Press.

Flach, G., Johnson, M. H., Braude, P. R., Taylor, R. A. S., and Bolton, V. N. (1982). The transition from maternal to embryonic control in the two-cell mouse embryo. *EMBO J.* **1,** 681–686.

Flanagan, J. R. (1976). A computer program automating construction of fate maps of *Drosophila*. *Dev. Biol.* **53,** 142–146.

Foe, V. E., Forrest, H., Wilkinson, L., and Laird, C. (1982). Morphological analysis of transcription in insect embryos. In R. C. King and H. Akai (Eds.), *Insect Ultrastructure,* vol. 1. New York: Plenum Press.

Foe, V. E., and Alberts, B. M. (1983). Studies of nuclear and cytoplasmic behavior during the five mitotic cycles that precede gastrulation in *Drosophila* embryogenesis. *J. Cell Sci.* **61,** 31–70.

Ford, P. J., and Southern, E. M. (1973). Different sequences for 5S RNA in kidney cells and ovaries of *X. laevis*. *Nature New Biol.* **241,** 7–12.

Freeman, G. (1979). The multiple roles which cell division can play in the localization of developmental potential. In S. Subtelny and I. R. Konigsberg (Eds)., *Determinants of Spatial Organization*. New York: Academic Press.

French, V., Bryant, P. J., and Bryant, S. (1976). Pattern regulation in epimorphic fields. *Science* **193,** 969–981.

Fristrom, D. (1968) Cellular degeneration in wing development of the mutant *vestigial* of *Drosophila melanogaster*. *J. Cell Biol.* **39,** 488–491.

Fullilove, S. L., and Jacobson, A. G. (1971). Nuclear elongation and cytokinesis in *Drosophila montana*. *Dev. Biol.* **26,** 560–577.

Fullilove, S. L., Jacobson, A. G., and Turner, F. R. (1978). Embryonic development: descriptive. In M. Ashburner and T. R. F. Wright (Eds.), *The Genetics and Biology of Drosophila*. London: Academic Press.

Galau, G., Klein, W. H., Davis, M. M., Wold, B. J., Britten, B. J., and Davidson, E. H. (1976). Structural gene sets active in embryos and adult tissues of the sea urchin. *Cell* **7,** 487–506.

Gans, M., Audit, C., and Masson, M. (1975). Isolation and characterization of sex-linked female-sterile mutants in *Drosophila melanogaster*. *Genetics* **81,** 863–704.

Gans, M., Forquignon, F., and Masson, M. (1980). The role of dosage in the region 7D1–7D5–6 of the X chromosome in the production of homoeotic transformations in *Drosophila melanogaster*. *Genetics* **96,** 887–902.

Garber, R. J., Kuroiwa, A., and Gehring, W. J. (1983). Genomic and cDNA clones of the homoeotic locus *Antennapedia* in *Drosophila*. *EMBO J.* **2,** 2027–2036.

Garcia-Bellido, A. (1975). Genetic control of wing disc development in *Drosophila*. In R. Porter and K. Elliott (Eds.), *Cell Patterning,* CIBA Symposium 29. Amsterdam: Elsevier/North-Holland.

Garcia-Bellido, A. (1977). Homoeotic and atavic mutations in insects. *Amer. Zool.* **17,** 613–629.

Garcia-Bellido, A., and Merriam, J. R. (1969). Cell lineage of the imaginal discs in *Drosophila* gynandromorphs. *J. Exp. Zool.* **170,** 61–76.

Garcia-Bellido, A., and Merriam, J. R. (1971a). Parameters of the wing imaginal disc development of *Drosophila melanogaster*. *Dev. Biol.* **24,** 61–87.

Garcia-Bellido, A., and Merriam, J. R. (1971b). Genetic analysis of cell heredity in imaginal discs of *Drosophila melanogaster*. *Proc. Natl. Acad. Sci. USA,* **68,** 2222–2226.

Garcia-Bellido, A., and Santamaria, P. (1972). Developmental analysis of the wing disc in the mutant *engrailed* of *Drosophila melanogaster*. *Genetics* **72,** 87–104.

Garcia-Bellido, A., and Santamaria, P. (1978). Developmental analysis of the achaete-scute system of *Drosophila melanogaster*. *Genetics* **88,** 469–486.

Garcia-Bellido, A., and Moscoso del Prado, J. (1979). Genetic analysis of maternal information in *Drosophila*. *Nature* **278,** 346–348.

Garcia-Bellido, A., and Robbins, L. G. (1983). Viability of female germ-line cells homozygous for zygotic lethals in *Drosophila melanogaster*. *Genetics* **103,** 235–247.

Garcia-Bellido, A., Ripoll, P., and Morata, G. (1973). Developmental compartmentalization of the wing disc of *Drosophila*. *Nature New Biol.* **245,** 251–253.

Garcia-Bellido, A., Ripoll, P., and Morata, G. (1976). Developmental compartmentalization in the dorsal mesothoracic disc of *Drosophila*. *Dev. Biol.* **48,** 132–147.

Gardner, R. L. (1968). Mouse chimaeras obtained by the injection of cells into the blastocyst. *Nature* (*Lond.*) **220,** 596–597.

Gardner, R. L. (1977). Developmental potency of normal and neoplastic cells of the early mouse embryo. In L. de Grouchy (Ed.), *Birth Defects*. Amsterdam: Excerpta Medica International Congress Service, no. 432.

Gardner, R. L. (1978). The relationship between cell lineage and differentiation in the early mouse embryo. In W. J. Gehring (Ed.), *Genetic Mosaics and Differentiation*. Heidelberg: Springer-Verlag.

Gardner, R. L., and Johnson, M. H. (1975). Investigation of cellular interaction and deployment in the early mammalian embryo using interspecific chimaeras between the rat and the mouse. In R. Porter and K. Elliott (Eds.), *Cell Patterning,* Ciba Foundation Symposium 29. Amsterdam: Elsevier/North-Holland.

Gardner, R. L., and Papaiannou, V. E. (1975). Differentiation in the trophectoderm and inner cell mass. In M. Balls and A. E. Wild (Eds.), *The Early Development of Mammals*. Cambridge: Cambridge University Press.

Gardner, R. L., and Rossant, J. (1976). Determination during embryogenesis. In K. Elliott and M. O'Connor (Eds.), *Embryogenesis in Mammals*. Amsterdam: Elsevier/North-Holland.

Gardner, R. L., and Rossant, J. (1979). Investigation of the fate of 4.5 day post-coitum mouse inner cell mass cells by blastocyst injection. *J. Embryol. Exp. Morphol.* **52,** 141–152.

Garen, A., and Gehring, W. (1972). Repair of the lethal developmental defect in deep orange (*dor*) embryos of *Drosophila* by injection of normal egg cytoplasm. *Proc. Natl. Acad. Sci. USA* **69,** 2982–2985.

Garner, W., and McLaren, A. (1974). Cell distribution in chimaeric mouse embryos before implantation. *J. Embryol. Exp. Morphol.* **32,** 495–503.

Gatti, M., and Pimpinelli, S. (1983). Cytological and genetic analysis of the Y chromosome of *Drosophila melanogaster*. I. Organization of the fertility factors. *Chromosoma (Berl.)* **88,** 349–373.

Gearhart, J. D., and Mintz, B. (1972). Clonal origins of somites and their muscle derivatives: evidence from allophenic mice. *Dev. Biol.* **29,** 27–37.

Gehring, W. J. (1967). Clonal analysis of determination dynamics in cultures of imaginal discs in *Drosophila melanogaster. Dev. Biol.* **16,** 438–456.

Gehring, W. J. (1972). The stability of the determined state in cultures of imaginal disks in *Drosophila*. In H. Ursprung and R. Nothiger (Eds.), *Results and Problems in Cell Differentiation,* vol. 5. Berlin: Springer-Verlag.

Gehring, W. J. (1976). Determination of primordial disc cells and the hypothesis of stepwise determination. In P. A. Lawrence (Ed.), *Insect Development*. Oxford: Blackwell.

Gehring, W. J. (1978). Imaginal discs: determination. In M. Ashburner and T. R. F. Wright (Eds.), *The Genetics and Biology of Drosophila,* vol. 2c. London: Academic Press.

Gerasimova, T. I., and Smirnova, S. G. (1979). Maternal effect for genes encoding 6-phosphogluconate dehydrogenase and glucose-6-phosphate dehydrogenase in *Drosophila melanogaster. Dev. Genet.* **1,** 97–107.

Ghysen, A., and Richelle, J. (1979). Determination of sensory bristles and pattern formation in *Drosophila*. II. The achaete-scute locus. *Dev. Biol.* **70,** 438–452.

Gierer, A. (1974). Molecular models and combinatorial principles in cell differentiation and morphogenesis. *Cold Spring Harbor Symp. Quant. Biol.* **38,** 951–961.

Ginelli, E., Di Lennia, R., Corneo, G. (1977). The organization of DNA sequences in the mouse genome. *Chromosoma (Berl.)* **61,** 215–226.

Girton, J. R. (1981). Pattern triplications produced by a cell-lethal mutation in *Drosophila. Dev. Biol.* **84,** 164–172.

Girton, J. R. (1983). Morphological and somatic clonal analyses of pattern triplication. *Dev. Biol.* **99,** 202–209.

Girton, J. R., and Bryant, P. J. (1980). The use of cell lethal mutations in the study of *Drosophila* development. *Dev. Biol.* **77,** 233–243.

Girton, J. R., and Russell, M. A. (1980). A clonal analysis of pattern duplication in a temperature-sensitive cell-lethal mutant of *Drosophila melanogaster. Dev. Biol.* **77,** 1–21.

Girton, J. R., and Russell, M. A. (1981). An analysis of compartmentalization in pattern duplications induced by a cell-lethal mutation in *Drosophila Dev. Biol.* **85,** 55–64.

Girton, J. R., and Berns, M. W. (1982). Pattern abnormalities induced in *Drosophila* imaginal discs by an ultraviolet laser microbeam. *Dev. Biol.* **91,** 73–77.

Golden, J. W., and Riddle, D. L. (1982). A pheromone influences larval development in the nematode *Caenorhabditis elegans. Science* **218,** 578–580.

Golden, J. W., and Riddle, D. L. (1984). The *Caenorhabditis elegans* dauer larva: development effects of pheromone, food and temperature. *Dev. Biol.* **102,** 368–378.

Gluecksohn-Waelsch, S. (1979). Genetic control of morphogenetic and biochemical differentiation: lethal albino deletions in the mouse. *Cell* **16,** 225–237.

Gluecksohn-Waelsch, S. (1983). Genetic control of differentiation. *Teratocarc. Stem Cells* **10,** 3–13.

Gluecksohn-Waelsch, S., and Erickson, R. P. (1970). The T-locus of the mouse: implications for mechanisms of development. In A. A. Moscona and A. Monroy (Eds.), *Current Topics in Developmental Biology,* vol. 5. New York: Academic Press.

Goldberg, D. A., Poskony, J. W., and Maniatis, T. (1983). Correct developmental expression of a cloned alcohol dehydrogenase gene transduced into the *Drosophila* germ line. *Cell* **34,** 59–73.

Goldstein, E. S. (1978). Translated and sequestered untranslated messages in *Drosophila* oocytes and embryos. *Dev. Biol.* **63,** 59–66.

Gollin, S. M., and King, R. C. (1981). Studies of *fs(1)1621,* a mutation producing ovarian tumors in *Drosophila melanogaster. Dev. Genet.* **2,** 203–218.

Gossett, L. A., and Hecht, R. M. (1980). A squash technique demonstrating embryonic nuclear cleavage of the nematode *Caenorhabditis elegans. J. Histochem. Cytochem.* **28,** 507–510.

Gossett, L. A., Hecht, R. M., and Epstein, H. F. (1982). Muscle differentiation in normal and cleavage-arrested mutant embryos of *Caenorhabditis elegans. Cell* **30,** 193–204.

Graham, C. F. (1973). The necessary conditions for gene expression during early mammalian development. In F. H. Ruddle (Ed.), *Genetic Mechanisms of Development.* New York: Academic Press.

Graham, C. F., and Lehtonen, E. (1979). Analysis of cell behaviour in the pre-implantation mouse lineage. In N. Le Douarin (Ed.), *Cell Lineage, Stem Cells, and Cell Determination.* Amsterdam: Elsevier/North-Holland.

Graziosi, G., and Roberts, D. B. (1977). Protein synthesis in the early *Drosophila* embryo: analysis of the protein species synthesized. *J. Embryol. Exp. Morphol.* **41,** 101–110.

Green, S. H. (1981). Segment-specific organization of leg motoneurones is transformed in *bithorax* mutants of *Drosophila. Nature* **292,** 152–154.

Greenwald, I. S., and Horvitz, H. R. (1980). unc93(1500): a behavioral mutant of *Caenorhabditis elegans* that defines a gene with a wild-type null phenotype. *Genetics* **96,** 147–164.

Greenwald, I. S., Sternberg, P. W., and Horvitz, H. R. (1983). The *lin-12* locus specifies cell fates in *Caenorhabditis elegans. Cell* **34,** 435–444.

Gregg, B. C., and Snow, M. H. L. (1983). Axial abnormalities following disturbed growth in mitomycin C–treated embryos. *J. Embryol. Exp. Morphol.* **73,** 135–149.

Grobstein, C. (1963). Cytodifferentiation and macromolecular synthesis. In M. Locke (Ed.), *Cytodifferentiation and Macromolecular Synthesis.* New York: Academic Press.

Grunstein, M., and Hogness, D. (1975). Colony hybridization: a method for the isolation of cloned DNAs that contain a specific gene. *Proc. Natl. Acad. Sci. USA* **72,** 3961–3965.

Gubb, D. (1985). Pattern formation during development and the design of the adult cuticle in *Drosophila.* In M. Bownes (Ed.), *Metamorphosis.* Oxford: Oxford University Press.

Gubb, D., and Garcia-Bellido, A. (1982). A genetic analysis of the determination of cuticular polarity during development in *Drosophila melanogaster. J. Embryol. Exp. Morphol.* **68,** 37–57.

Gurdon, J. B., Laskey, R. A., and Reeves, O. R. (1975). The developmental capacity of nuclei transplanted from keratinized cells of adult frogs. *J. Embryol. Exp. Morphol.* **34,** 93–102.

Gutzheit, H. O. (1979). Expression of the zygotic genome in blastoderm stage embryos of *Drosophila*: analysis of a specific protein. *Wilh. Roux Arch.* **188,** 153–156.

Gutzheit, H. O., and Gehring, W. J. (1979). Localized protein synthesis of specific proteins during oogenesis and early embryogenesis in *Drosophila melanogaster. Wilh. Roux Arch.* **187,** 151–165.

Hadorn, E. (1937). Transplantation of gonads from lethal to normal larvae in *Drosophila melanogaster. Proc. Natl. Acad. Sci. USA* **23,** 478–484.

Hadorn, E. (1948). Gene action in growth and differentiation of lethal mutants. *Society of Experimental Biology Symposium,* vol. 2. Cambridge: Cambridge University Press.

Hadorn, E. (1961). *Developmental Genetics and Lethal Factors.* New York: Wiley-Interscience.

Hadorn, E. (1965). Problems of determination and transdetermination. *Brookhaven Symp.* **18,** 148–161.

Hadorn, E. (1978). Transdetermination. In M. Ashburner and T. R. F. Wright (Eds.), *The Genetics and Biology of Drosophila,* vol. 2c. London: Academic Press.

Hadorn, E., Gsell, R., and Schultz, J. (1970). Stability of a position-effect variegation in normal and transdetermined larval blastemas from *Drosophila melanogaster. Proc. Natl. Acad. Sci. USA* **65,** 633–637.

Hafen, E., Levine, M., and Gehring, W. J. (1984a). Regulation of *Antennapedia* transcript distribution by the bithorax complex in *Drosophila. Nature* **307,** 287–289.

Hafen, E., Kuroiwa, A., and Gehring, W. J. (1984b). Spatial distribution of transcripts from the segmentation gene *fushi tarazu.* during *Drosophila* embryonic development. *Cell* **37,** 833–841.

Hafen, E., Levine, M., Garber, R. L., and Gehring, W. J. (1983). An improved *in situ* hybridization method for the detection of cellular RNAs in *Drosophila* tissue sections and its application for localizing transcripts of the homoeotic *Antennapedia* gene complex. *EMBO J.* **2,** 617–623.

Hall, J. C., Gelbart, W. M., and Kankel, D. A. (1976). Mosaic systems. In M. Ashburner and E. Novitski (Eds.), *Genetics and Biology of Drosophila,* vol. 1a. London: Academic Press.

Hall, L. M. C., Mason, P. J., and Spierer, P. (1983). Transcripts, genes and bands in 315,000 base-pairs of *Drosophila* DNA. *J. Mol. Biol.* **169,** 83–96.

Hand, R. (1978). Eucaryotic DNA: organization of the genome for replication. *Cell* **15,** 317–325.

Handyside, A. H. (1978). Time of commitment of inside cells isolated from pre-implantation mouse embryos. *J. Embryol. Exp. Morphol.* **45,** 37–53.

Handyside, A. H., and Johnson, M. H. (1978). Temporal and spatial patterns of the synthesis of tissue-specific polypeptides in the pre-implantation mouse embryo. *J. Embryol. Exp. Morphol.* **44,** 191–199.

Hannah, Alava, A., and Stern, C. (1957). The sex combs in males and intersexes of *Drosophila melanogaster. J. Exp. Zool.* **134,** 533–556.

Hartwell, L. H., Culotti, J., and Reid, B. (1970). Genetic control of the cell-division cycle in yeast. I. Detection of mutants. *Proc. Natl. Acad. Sci. USA* **66,** 352–359.

Hastie, N. D., and Bishop, J. O. (1976). The expression of three abundance classes of messenger RNA in mouse tissues. *Cell* **9,** 761–774.

Hayes, P. (1982). Mutant analysis of determination in *Drosophila.* Ph.D. Thesis, Department of Genetics, University of Alberta, Edmonton, Canada.

Haynie, J. L. (1982). Homologies of positional information in thoracic imaginal discs of *Drosophila melanogaster. Wilh. Roux Arch.* **191,** 293–300.

Haynie, J. L., and Bryant, P. J. (1976). Intercalary regeneration in imaginal wing disk of *Drosophila melanogaster. Nature (Lond.)* **259,** 659–662.

Hecht, R. M., Gossett, L. A., and Jeffery, W. R. (1981). Ontogeny of maternal and newly transcribed mRNA analyzed by *in situ* hybridization during development of *Caenorhabditis elegans. Dev. Biol.* **83,** 374–379.

Held, L. I., Jr. (1979). Pattern as a function of cell number and cell size on the second leg basitarsus of *Drosophila. Wilh. Roux Arch.* **187,** 105–127.

Heller, D. T., Cahill, D. M., and Schultz, R. M. (1981). Behavioural studies of mammalian oogenesis: metabolic cooperativity between granulosa cells and growing mouse oocytes. *Dev. Biol.* **84,** 455–464.

Herbert, M. C., and Graham, C. F. (1974). Cell determination and biochemical differentiation of the early mammalian embryo. In A. A. Moscona and A. Monroy (Eds.), *Current Topics in Developmental Biology,* vol. 8. New York: Academic Press.

Herman, R. K. (1984). Analysis of genetic mosaics of the nematode *Caenorhabditis elegans*. *Genetics* **108,** 165–180.

Herman, R. K., and Horvitz, H. R. (1980). Genetic analysis of *Caenorhabditis elegans*. In B. M. Zuckerman (Ed.), *Nematodes as Model Biological Systems,* vol. 1 of *Behavioral and Developmental Models*. New York: Academic Press.

Herman, R. K., Albertson, D. G., and Brenner, S. (1976). Chromosome rearrangements in *Caenorhabditis elegans*. *Genetics* **83,** 91–105.

Herskowitz, I. (1982). Cellular differentiation, cell lineages, and transposable genetic cassettes in yeast. In A. A. Moscona and A. Monroy (Eds.), *Current Topics in Developmental Biology,* vol. 18. New York: Academic Press.

Higgins, B. J., and Hirsh, D. (1977). Roller mutants of the nematode *Caenorhabditis elegans*. *Mol. Gen. Genet.* **150,** 63–72.

Hildreth, P. E. (1965). *Doublesex,* a recessive gene that transforms both males and females of *Drosophila* into intersexes. *Genetics* **51,** 659–678.

Hillman, N., Sherman, M. I., and Graham, C. F. (1972). The effect of spatial arrangement on cell determination during mouse development. *J. Embryol. Exp. Morphol.* **28,** 263–278.

Hirsh, D. (1979). Temperature-sensitive maternal effect mutants of early development in *Caenorhabditis elegans*. In S. Subtelny and I. R. Konigsberg (Eds.), *Determinants of Spatial Organization*. New York: Academic Press.

Hirsh, D., and Vanderslice, R. (1976). Temperature-sensitive developmental mutants of *Caenorhabditis elegans*. *Dev. Biol.* **49,** 220–235.

Hirsh, D., Oppenheim, D., and Klass, M. (1976). Development of the reproductive system of *Caenorhabditis elegans*. *Dev. Biol.* **49,** 200–219.

Hodgkin, J. (1980). More sex determination mutants of *Caenorhabditis elegans*. *Genetics* **96,** 649–664.

Hodgkin, J. A. (1983a). Male phenotypes and mating efficiency in *Caenorhabditis elegans*. *Genetics* **103,** 43–64.

Hodgkin, J. A. (1983b). Two types of sex determination in a nematode. *Nature* **304,** 267–268.

Hodgkin, J. A., and Brenner, S. (1977). Mutations causing transformation of sexual phenotype in the nematode *Caenorhabditis elegans*. *Genetics* **86,** 275–281.

Hodgkin, J. A., Horvitz, H. R., and Brenner S. (1979). Nondisjunction mutants of the nematode *Caenorhabditis elegans*. *Genetics* **91,** 67–94.

Hogan, B. (1983). Enhancers, chromosome position effects, and transgenic mice. *Nature* **306,** 313–314.

Hogan, B., and Tilly, R. (1978). *In vitro* development of inner cell masses isolated immunosurgically from mouse blastocysts. I. Inner cell masses from 3.5 day p.c. blastocysts incubated for 24 hr before immunosurgery. *J. Embryol. Exp. Morphol.* **45,** 93–105.

Horstadius, S. (1973). *Experimental Embryology of Echinoderms*. Oxford: Clarendon Press.

Horvitz, H. R., and Sulston, J. E. (1980). Isolation and genetic characterization of cell-lineage mutants of the nematode *Caenorhabditis elegans*. *Genetics* **96,** 435–454.

Horvitz, H. R., Brenner, S., Hodgkin, J., and Herman, R. K. (1979). A uniform genetic nomenclature for the nematode *Caenorhabditis elegans*. *Mol. Gen. Genet.* **175,** 129–133.

Hotta, Y., and Benzer, S. (1973). Mapping of behavior in *Drosophila* mosaics. In F. Ruddle (Ed.), *Genetic Mechanisms of Development*. New York: Academic Press.

Hough-Evans, B. R., Wold, B. J., Ernst, S. G., Britten, R. J., and Davidson, E. H. (1977). Appearance and persistence of maternal RNA sequences in sea urchin development. *Dev. Biol.* **60,** 258–277.

Hough-Evans, B. R., Jacobs-Lorena, M., Cummings, M. R., Britten, R. J., and Davidson, E. H. (1980). Complexity of RNA in eggs of *Drosophila melanogaster* and *Musca domestica*. *Genetics* **95,** 81–94.

Hull, D. L. (1974). *Philosophy of Biological Science*. Englewood Cliffs, N.J.: Prentice-Hall.

Hyafil, F., Babinet, C., and Jacob, F. (1981). Cell–cell interactions in early embryogenesis: a molecular approach to the role of calcium. *Cell* **26,** 447–454.

Illmensee, K. (1972). Developmental potencies of nuclei from cleavage, preblastoderm, and syncytial blastoderm transplanted into unfertilized eggs of *Drosophila melanogaster*. *Wilh. Roux Arch.* **170,** 267–298.

Illmensee, K. (1976). Nuclear and cytoplasmic transplantation in *Drosophila*. In P. A. Lawrence (Ed.), *Insect Development*. Oxford: Blackwell.

Illmensee, K. (1978). *Drosophila* chimaeras and the problem of determination. In W. J. Gehring (Ed.), *Genetic Mosaics and Cell Differentiation,* vol. 9 of *Results and Problems in Differentiation*. Berlin: Springer-Verlag.

Illmensee, K., and Mahowald, A. P. (1974). Transplantation of posterior polar plasm in *Drosophila*. Induction of germ cells at the anterior pole of the egg. *Proc. Natl. Acad. Sci. USA* **71,** 1016–1020.

Illmensee, K., and Mahowald, A. P. (1976). The autonomous function of germ plasm in a somatic region of the *Drosophila* egg. *Exp. Cell Res.* **97,** 127–140.

Illmensee, K., and Hoppe, P. (1981). Nuclear transplantation in *Mus musculus:* developmental potential of nuclei from preimplantation embryos. *Cell* **23,** 9–18.

Illmensee, K., Mahowald, A. P., and Loomis, M. R. (1976). The ontogeny of germ plasm during oogenesis in *Drosophila*. *Dev. Biol.* **49,** 40–65.

Ingham, P. W. (1981). *Trithorax:* a new homoeotic mutation of *Drosophila melanogaster*. II. The role of trx^{+} after embryogenesis. *Wilh. Roux Arch.* **190,** 365–369.

Ingham, P. W. (1983). Differential expression of bithorax complex genes in the absence of the *extra sex combs* and *trithorax* genes. *Nature* **306,** 591–593.

Ingham, P. W. (1984). A gene which regulates the bithorax complex differentially in larval and adult cells of *Drosophila*. *Cell* **37,** 815–823.

Ingham, P. W., and Whittle, R. (1980). Trithorax: a new homoeotic mutation of *Drosophila melanogaster* causing transformations of abdominal and thoracic imaginal segments. I. Putative role during embryogenesis. *Mol. Gen. Genet.* **179,** 607–614.

Izquierdo, M., and Bishop, J. O. (1979). An analysis of cytoplasmic RNA populations in *Drosophila melanogaster,* Oregon R. *Biochem. Genet.* **17,** 473–497.

Jackle, H., and Kalthoff, K. (1980). Synthesis of a posterior indicator protein in normal embryos and double abdomens of *Smittia* sp. (Chironomidae, Diptera). *Proc. Natl. Acad. Sci. USA* **77,** 6700–6704.

Jackle, H., and Kalthoff, K. (1981). Proteins foretelling head or abdomen development in the embryo of *Smittia* spec. (Chironomidae, Diptera). *Dev. Biol.* **85,** 287–298.

Jacob, F. (1982). *The Possible and the Actual.* New York: Pantheon Books.

Jacob, F., and Monod, J. (1961). Genetic regulatory mechanisms in the synthesis of proteins. *J. Mol. Biol.* **3,** 318–356.

Jacob, F., and Monod, J. (1963). Genetic repression, allosteric inhibition, and cellular differentiation. In M. Locke (Ed.), *Cytodifferentiation and Macromolecular Synthesis*. New York: Academic Press.

Jaenisch, R. (1983). Retroviruses and mouse embryos: a model system in which to study gene expression in development and differentiation. In R. Porter and J. Whelan (Eds.), *Molecular Biology of Egg Maturation*. Ciba Foundation Symposium 98. London: Pitman Books.

James, A. A., and Bryant, P. J. (1981). Mutations causing pattern deficiencies and duplications in the imaginal wing disc of *Drosophila melanogaster*. *Dev. Biol.* **85,** 39–54.

Janning, W. (1978). Gynandromorph fate maps in *Drosophila*. In W. J. Gehring, (Ed.), *Genetic Mosaics and Cell Differentiation*. Berlin: Springer-Verlag.

Janning, W., Pfreudt, J., and Tiemann, R. (1979). The distribution of anlagen in the early embryo of *Drosophila*. In N. Le Douarin (Ed.), *Cell Lineages, Stem Cells, and Cell Determination*. Amsterdam: Elsevier/North-Holland.

Jiminez, F., and Campos-Ortega, J. A. (1981). A cell arrangement specific to the thoracic ganglia in the central nervous system of the *Drosophila* embryo: its behavior in homoeotic mutants. *Wilh. Roux Arch.* **190,** 370–373.

Jiminez, F., and Campos-Ortega, J. A. (1982). Maternal effects of zygotic mutants affecting early neurogenesis in *Drosophila. Wilh. Roux Arch.* **191,** 191–201.

Johnson, C. C., and King, R. C. (1974). Oogenesis in the *ocelliless* mutant of *Drosophila melanogaster* Meigen (Diptera: Drosophilidae). *Int. J. Insect. Morphol. Embryol.* **3,** 385–395.

Johnson, J. H., and King, R. C. (1972). Studies on *fes,* a mutation affecting cytokinesis in *Drosophila melanogaster. Biol. Bull.* **143,** 525–547.

Johnson, K., and Hirsh, D. (1979). Patterns of proteins synthesized during development of *Caenorhabditis elegans. Dev. Biol.* **70,** 241–248.

Johnson, M. H. (1979). Intrinsic and extrinsic factors in pre-implantation development. *J. Reprod. Fertil.* **55,** 255–265.

Johnson, M. H., and Rossant, J. (1981). Molecular studies on cells of the trophectodermal lineage of the postimplantation mouse embryo. *J. Embryol. Exp. Morphol.* **61,** 103–116.

Johnson, M. H., and Ziomek, C. A. (1981). The foundation of two distinct cell lineages within the mouse morula. *Cell* **24,** 71–80.

Johnson, M. H., and Pratt, H. P. M. (1983). Cytoplasmic localizations and cell interactions in the formation of the mouse blastocyst. In W. R. Jeffery and R. A. Raff (Eds.), *Time, Space, and Pattern in Development*. New York: Alan R. Liss.

Johnson, M. H., and Ziomek, C. A. (1983). Cell interactions influence the fate of mouse blastomeres undergoing the transition from the 16- to the 32-cell stage. *Dev. Biol.* **95,** 211–218.

Johnson, M. H., Pratt, H. P. M., and Handyside, A. H. (1981). The generation and recognition of positional information in the preimplantation mouse embryo. In S. R. Glasser and D. W. Bullock (Eds.), *Cellular and Molecular Aspects of Implantation*. New York: Plenum Press.

Jones, K. W., and Singh, L. (1982). Conserved sex-associated repeated DNA sequences in vertebrates. In G. A. Dover and R. B. Flavell (Eds.), *Genome Evolution*. London: Academic Press.

Judd, B. H. (1976). Genetic units of *Drosophila*. In M. Ashburner and E. Novitski (Eds.), *The Genetics and Biology of Drosophila,* vol. 1b. London: Academic Press.

Judd, B. H. (1977). The nature of the module of genetic function in *Drosophila*. In E. M. Bradbury and K. Javaherian (Eds.), *The Organization and Expression of the Eukaryotic Genome*. London: Academic Press.

Judd, B. H., Shen, M. W., and Kaufman, T. C. (1972). The anatomy and function of a segment of the X chromosome of *D. melanogaster. Genetics* **71,** 139–156.

Judson, H. F. (1979). *The Eighth Day of Creation: Makers of the Revolution in Biology*. London: Jonathan Cape.

Kalthoff, K. (1979). Analysis of a morphogenetic determinant in an insect embryo (*Smittia* spec., Chironomidae, Diptera. In S. Subtelny and I. Konigsberg (Eds.), *Determinants of Spatial Organization*. New York: Academic Press.

Kalthoff, K. (1983). Cytoplasmic determinants in dipteran eggs. In W. R. Jeffery and R. A. Raff (Eds.), *Time, Space, and Pattern in Embryonic Development*. New York: Alan R. Liss.

Karlsson, J. (1981a). The distribution of regenerative potential in the wing disc of *Drosophila. J. Embryol. Exp. Morphol.* **61,** 303–316.

Karlsson, J. (1981b). Sequence of regeneration in the *Drosophila* wing disc. *J. Embryol. Exp. Morphol.* **65,** 37–47.

Karn, J., Brenner, S., Barnett, L., and Cesareni, G. (1980). Novel bacteriophage lambda cloning vector. *Proc. Natl. Acad. Sci. USA* **77,** 5172–5176.

Kauffman, S. (1971). Gene regulation networks: a theory for their global structure and behaviors. *Curr. Topics Dev. Biol.* **7,** 145–182.

Kauffman, S. A. (1973). Control circuits for determination and transdetermination. *Science* **181,** 310–318.

Kauffman, S. A. (1980). Heterotopic transplantation in the syncytial blastoderm of *Drosophila:* evidence for anterior and posterior nuclear commitments. *Wilh. Roux Arch.* **189,** 135–145.

Kauffman, S. A., and Ling, E. (1981). Regeneration by complementary wing disc fragments of *Drosophila melanogaster. Dev. Biol.* **82,** 238–257.

Kauffman, S. A., Shymko, R. M., and Trabert, K. (1978). Control of segmental compartment formation in *Drosophila. Science,* **199,** 259–270.

Kaufman, T. C. (1983). The genetic regulation of segmentation in *Drosophila melanogaster.* In W. R. Jeffery and R. A. Raff (Eds.), *Time, Space, and Pattern in Development.* New York: Alan R. Liss.

Kaufman, T. C., Lewis, R., and Wakimoto, B. (1980). Cytogenetic analysis of chromosome 3 in *Drosophila melanogaster:* the homoeotic gene complex in polytene chromosome interval 84A-B. *Genetics* **94,** 115–133.

Kelly, S. J. (1975). Studies of the potency of the early cleavage blastomeres of the mouse. In M. Balls and A. E. Wild (Eds.), *The Early Development of Mammals.* Cambridge: Cambridge University Press.

Kelly, S. J., Mulnard, J. G., and Graham, C. F. (1978). Cell division and cell allocation in early mouse development. *J. Embryol. Exp. Morphol.* **48,** 37–51.

Kemler, R., Babinet, C., Condamine, H., Gachelin, G., Guenet, J. L., and Jacob, F. (1976). Embryonal carcinoma antigen and the *T/t* locus of the mouse. *Proc. Natl. Acad. Sci. USA* **73,** 4080–4084.

Kimble, J. E. (1981a). Alterations in cell lineage following laser ablation of cells in the somatic gonad of *Caenorhabditis elegans. Dev. Biol.* **87,** 286–300.

Kimble, J. E. (1981b). Strategies for the control of pattern formation in *Caenorhabditis elegans. Phil. Trans. R. Soc. Lond. B,* **295,** 539–551.

Kimble, J. E., and Hirsh, D. (1979). The postembryonic cell lineages of the hermaphrodite and male gonads in *Caenorhabditis elegans. Dev. Biol.* **70,** 396–417.

Kimble, J. E., and White, J. G. (1981). On the control of germ cell development in *Caenorhabditis elegans. Dev. Biol.* **81,** 208–219.

Kimble, J. E., and Sharrock, W. J. (1983). Tissue-specific synthesis of yolk proteins in *Caenorhabditis elegans. Dev. Biol.* **96,** 189–196.

Kimble, J., Sulston, J., and White, J. (1979). Regulative development in the postembryonic lineages of *Caenorhabditis elegans.* In N. Le Doumarin (Ed.), *Cell Lineage, Stem Cells, and Determination.* Amsterdam: Elsevier/North-Holland.

Kimble, J. E., Hodgkin, J., Smith, T., and Smith, J. (1982). Suppression of an amber mutation by microinjection of suppressor tRNA in *Caenorhabditis elegans. Nature* **299,** 456–458.

King, R. C. (1970). *Oogenesis in Drosophila.* New York: Academic Press.

King, R. C. (1979). Aberrant fusomes in the ovarian cystocytes of the fs(1)231 mutant of *Drosophila melanogaster* Meigen (Diptera, Drosophilidae). *Int. J. Insect Morphol. Embryol.* **8,** 297–309.

King, R. C., and Bodenstein, D. (1965). The transplantation of ovaries between genetically sterile and wild type *Drosophila melanogaster. Z. Naturforsch.* **20b,** 292–297.

King, R. C., and Mohler, J. D. (1975). The genetic analysis of oogenesis in *Drosophila melanogaster.* In R. C. King (Ed.), *Handbook of Genetics,* vol. 3. New York: Plenum Press.

King, R. C., and Riley, S. F. (1982). Ovarian pathologies generated by various alleles of the *otu* locus in *Drosophila melanogaster. Dev. Genet.* **3,** 69–89.

King, R. C., Koch, E. A., and Cassens, G. A. (1961). The effect of temperature upon the hereditary ovarian tumors of the *fes* mutant of *Drosophila melanogaster. Growth* **25,** 45–65.

King, R. C., Cassidy, J. D., and Rousset, A. (1982). The formation of clones of interconnected cells during gametogenesis in insects. In R. C. King and H. Akai (Eds.), *Insect Ultrastructure*. New York: Plenum Press.

King, T. C., and Briggs, R. (1956). Serial transplantation of embryonic nuclei. *Cold Spring Harbor Symp. Quant. Biol.* **21,** 271–290.

Kirby, B. S., Bryant, P. J., and Schneiderman, H. A. (1982). Regeneration following duplication in imaginal wing disc fragments of *Drosophila melanogaster. Dev. Biol.* **90,** 259–271.

Klass, M., Wolf, N., and Hirsh, D. (1976). Development of the male reproductive system and sexual transformation in the nematode *Caenorhabditis elegans. Dev. Biol.* **52,** 1–18.

Knowland, J. S., and Graham, C. F. (1972). RNA synthesis at the two-cell stage of mouse development. *J. Embryol. Exp. Morphol.* **27,** 167–176.

Kornberg, T. (1981a). *engrailed:* a gene controlling compartment and segment formation in *Drosophila. Proc. Natl. Acad. Sci. USA* **78,** 1095–1099.

Kornberg, T. (1981b). Compartments in the abdomen of *Drosophila* and the role of the *engrailed* locus. *Dev. Biol.* **86,** 363–372.

Krieg, C., Cole, T., Deppe, U., Schierenberg, E., Schmitt, D., Yoder, B., and von Ehrenstein, G. (1978). The cellular anatomy of embryos of the nematode *Caenorhabditis elegans:* analysis and reconstruction of serial section electron micrographs. *Dev. Biol.* **65,** 193–215.

Kuhn, D. T., Fogerty, S. C., Eskens, A. A. C., and Sprey, Th. E. (1983). Developmental compartments in the *Drosophila melanogaster* wing disc. *Dev. Biol.* **95,** 399–413.

Lacy, E., Roberts, S., Evans, E. P., Burtenshaw, M. D., and Costantini, F. D. (1983). A foreign β-globin gene in transgenic mice: integration at abnormal positions and expression in inappropriate tissues. *Cell* **34,** 343–358.

Lamb, M., and Laird, C. D. (1976). Increase in nuclear poly(A)-containing RNA at syncytial blastoderm in *Drosophila melanogaster* embryos. *Dev. Biol.* **52,** 31–42.

Laufer, J. S., and von Ehrenstein, G. (1981). Nematode development after removal of egg cytoplasm: absence of localized unbound determinants. *Science* **211,** 402–405.

Laufer, J. S., Bazzicalupo, P., and Wood, W. B. (1980). Segregation of developmental potential in early embryos of *Caenorhabditis elegans. Cell* **19,** 569–577.

Laugé, E. (1980). Sex determination. In M. Ashburner and T. R. F. Wright (Eds.), *The Genetics and Biology of Drosophila,* vol. 2d. London: Academic Press.

Laughan, A., and Scott, M. P. (1984). Sequence of a *Drosophila* segmentation gene: protein structure homology with DNA-binding proteins. *Nature* **310,** 25–31.

Lawrence, P. A. (1973a). The development of spatial patterns in the integument of insects. In S. J. Counce and C. H. Waddington (Eds.), *Developmental Systems: Insects,* vol. 2. London: Academic Press.

Lawrence, P. A. (1973b). A clonal analysis of segment development in *Oncopeltus* (Hemiptera). *J. Embryol. Exp. Morphol.* **30,** 681–699.

Lawrence, P. A. (1981). The cellular basis of segmentation in insects. *Cell* **26,** 3–10.

Lawrence, P. A. (1982). Cell lineage of the thoracic muscles of *Drosophila. Cell* **29,** 493–503.

Lawrence, P. A., and Morata, G. (1976a). Compartments in the wing of *Drosophila:* a study of the *engrailed* gene. *Dev. Biol.* **50,** 321–337.

Lawrence, P. A., and Morata, G. (1976b). The compartment hypothesis. In P. A. Lawrence (Ed.), *Insect Development*. Oxford: Blackwell Scientific.

Lawrence, P. A., and Morata, G. (1977). The early development of mesothoracic compartments in *Drosophila*. An analysis of cell lineage and fate mapping and an assortment of methods. *Dev. Biol.* **56,** 40–51.

Lawrence, P. A., and Morata, G. (1979a). Pattern formation and compartments in the tarsus of *Drosophila*. In S. Subtelny and I. R. Konigsberg (Eds.), *Determinants of Spatial Organization*. New York: Academic Press.

Lawrence, P. A., and Morata, G. (1979b). Early development of the thoracic discs of *Drosophila*. *Wilh. Roux Arch.* **187,** 375–379.

Lawrence, P. A., and Brower, D. L. (1982). Myoblasts from *Drosophila* wing discs can contribute to developing muscles throughout the fly. *Nature* **295,** 55–57.

Lawrence, P. A., and Struhl, G. (1982). Further studies of the *engrailed* phenotype in *Drosophila*. *EMBO J.* **1,** 827–833.

Lawrence, P. A., and Morata, G. (1983). The elements of the bithorax complex. *Cell* **35,** 595–601.

Lee, L-W., and Gerhart, J. C. (1973). Dependence of transdetermination frequency on the developmental stage of cultured imaginal discs of *Drosophila melanogaster*. *Dev. Biol.* **35,** 62–82.

Lefevre, G., Jr. (1974). The relationship between genes and polytene chromosome bands. *Ann. Rev. Genet.* **8,** 51–62.

Lehmann, R., Dietrich, U., Jiminez, F., and Campos-Ortega, J. A. (1981). Mutations of early neurogenesis in *Drosophila*. *Wilh. Roux Arch.* **190,** 226–229.

Lehmann, R., Jiminez, F., Dietrich, U., and Campos-Ortega, J. A. (1983). On the phenotype and early development of mutants of early neurogenesis in *Drosophila melanogaster*. *Wilh. Roux Arch.* **192,** 62–74.

Levey, I. L., Stull, G. B., and Brinster, R. L. (1978). Poly (A) and synthesis of polyadenylated RNA in the preimplantation mouse embryo. *Dev. Biol.* **64,** 140–148.

Levine, M., Hafen, E., Garber, R. L., and Gehring, W. J. (1983). Spatial distribution of *Antennapedia* transcripts during *Drosophila* development. *EMBO J.* **2,** 2037–2046.

Levy, L. S., and Manning, J. E. (1981). Messenger RNA sequence complexity and homology in developmental stages of *Drosophila*. *Dev. Biol.* **85,** 141–149.

Levy, W. B., and McCarthy, B. J. (1975). Messenger RNA complexity in *Drosophila melanogaster*. *Biochemistry* **14,** 2440–2446.

Lewis, E. B. (1964). Genetic control and regulation of developmental pathways. In M. Locke (Ed.), *The Chromosomes in Development*. New York: Academic Press.

Lewis, E. B. (1978). A gene complex controlling segmentation in *Drosophila*. *Nature* **276,** 565–570.

Lewis, E. B. (1981). Developmental genetics of the bithorax complex in *Drosophila*. In D. D. Brown and C. F. Fox (Eds.), *Developmental Biology Using Purified Genes*. ICN-UCLA Symposia in Molecular and Cellular Biology. New York: Academic Press.

Lewis, J. H. (1975). Fate maps and the patterns of cell division: a calculation for the chick-wing bud. *J. Embryol. Exp. Morphol.* **33,** 419–434.

Lewis, R. A., Kaufman, T. C., Denell, R. E., and Tallerico, P. (1980a). Genetic analysis of the Antennapedia gene complex (ANT-C) and adjacent chromosomal regions of *Drosophila melanogaster*. I. Polytene chromosome segments 84B-D. *Genetics* **95,** 367–381.

Lewis, R. A., Wakimoto, B. T., Denell, R. E., and Kaufman, T. C. (1980b). Genetic analysis of the Antennapedia gene complex (ANT-C) and adjacent chromosomal regions of *Drosophila melanogaster*. II. Polytene chromosome segments 84A-84B1,2. *Genetics* **95,** 383–397.

Lifschytz, E., and Lindsley, D. L. (1972). The role of X chromosome inactivation during spermatogenesis. *Proc. Natl. Acad. Sci. USA* **69,** 182–186.

Lindsley, D. L., and Grell, E. H. (1968). Genetic variations of *Drosophila melanogaster.* Washington, D.C.: Carnegie Institute Publ. 627.

Lohs-Schardin, M. (1982). *Dicephalic*—a *Drosophila* mutant affecting polarity in follicle organization and embryonic patterning. *Wilh. Roux Arch.* **191,** 28–36.

Lohs-Schardin, M., Sander, K., Cremer, C., Cremer, T., and Zorn, C. (1979a). Localized ultraviolet laser microbeam irradiation of early *Drosophila* embryos: fate maps based on location and frequency of adult defects. *Dev. Biol.* **68,** 533–545.

Lohs-Schardin, M., Cremer, C., and Nusslein-Volhard, C. (1979b). A fate map for the larval epidermis of *Drosophila melanogaster.* Localized cuticle defects following irradiation of the blastoderm with an ultraviolet laser microbeam. *Dev. Biol.* **73,** 239–255.

Lovett, J., and Goldstein, E. S. (1977). The cytoplasmic distribution and characterization of poly A^+ RNA in oocytes and embryos of *Drosophila. Dev. Biol.* **61,** 70–78.

Loyd, J. E., Raff, E. C., and Raff, R. A. (1981). Site and timing of synthesis of tubulin and other proteins during oogenesis in *Drosophila melanogaster. Dev. Biol.* **86,** 272–284.

Lucchesi, J. C. (1977). Dosage compensation: transcription-level regulation of X-linked genes in *Drosophila. Am. Zool.* **17,** 685–693.

Lucchesi, J. C. (1983). The relationship between gene dosage, gene expression, and sex in *Drosophila. Dev. Genet.* **3,** 275–282.

Lucchesi, J. C., and Skripsky, T. (1981). The link between dosage compensation and sex differentiation in *Drosophila melanogaster. Chromosoma (Lond.)* **82,** 217–227.

Lyon, M. F. (1961). Gene action in the X-chromosome of the mouse (*Mus musculus* L.). *Nature* **190,** 372–373.

Lyon, M. F. (1981). Nomenclature. In H. L. Foster, J. D. Small, and J. G. Fox (Eds.) *The Mouse in Biomedical Research,* vol. 1. History, Genetics, and Wild Mice. New York: Academic Press.

Lyon, M. F., and Hawkes, S. G. (1970). X-linked gene for testicular feminization in the mouse. *Nature* **227,** 1217–1219.

Lyon, M. F., Glenister, P. H., and Lamoureux, M. L. (1975). Normal spermatozoa from androgen-resistant germ cells of chimaeric mice and the role of androgen in spermatogenesis. *Nature* **258,** 620–622.

Lyon, M. F., Evans, E. P., Jarvis, S. E., and Sayers, I. (1979). t-haplotypes of the mouse may involve a change in intercalary DNA. *Nature* **279,** 38–42.

MacLeod, A. R., Karn, J., and Brenner, S. (1981). Molecular analysis of the *unc-54* myosin heavy-chain gene of *Caenorhabditis elegans. Nature* **291,** 386–390.

McCarrey, J. R., and Abbott, U. K. (1979). Mechanisms of genetic sex determination, gonadal sex differentiation, and germ-cell development in animals. *Adv. Genet.* **20,** 217–290.

McClay, D. R., Cannon, G. W., Wessel, G. M., Fink, R. D., and Marchase, R. B. (1983). Patterns of antigenic expression in early sea urchin development. In W. R. Jeffery and R. A. Raff (Eds.), *Time, Space, and Pattern in Development.* New York: Alan R. Liss.

McClintock, B. (1951). Chromosome organization and genic expression. *Cold Spring Harbor Symp. Quant. Biol.* **16,** 13–47.

McClintock, B. (1956). Controlling elements and the gene. *Cold Spring Harbor Symp. Quant. Biol.* **21,** 197–216.

McGinnis, W., Levine, M. S., Hafen, E., Kuroiwa, A., and Gehring, W. J. (1984a). A conserved DNA sequence in homoeotic genes of the *Drosophila* Antennapedia and bithorax complexes. *Nature* **308,** 428–433.

McGinnis, W., Garber, R. L., Wirz, J., Kuroiwa, A., and Gehring, W. J. (1984b). A homologous protein-coding sequence in *Drosophila* homoeotic genes and its conservation in other metazoans. *Cell* **37,** 403–408.

McKnight, G. S., Hammer, R. E., Kuegel, E. A., and Brinster, R. L. (1983). Expression of the chicken transferrin gene in transgenic mice. *Cell* **34,** 335–341.

McKnight, S. L., and Miller, O. L., Jr. (1976). Ultrastructural patterns of RNA synthesis during early embryogenesis of *Drosophila melanogaster*. *Cell* **8,** 305–319.

McLaren, A. (1972). The numerology of development. *Nature* **239,** 274–276.

McLaren, A. (1976a). *Mammalian Chimaeras*. Cambridge: Cambridge University Press.

McLaren, A. (1976b). Genetics of the early mouse embryo. *Ann. Rev. Genet.* **10,** 361–388.

McLaren, A. (1979). The impact of pre-fertilization events on post-fertilization development in mammals. In D. R. Newth and M. Balls (Eds.), *Maternal Effects in Development*. Cambridge: Cambridge University Press.

McLaren, A. (1981a). *Germ Cells and Soma: A New Look at an Old Problem*. New Haven, Conn.: Yale University Press.

McLaren, A. (1981b). Analysis of maternal effects on development in mammals. *J. Repro. Fert.* **62,** 591–596.

McLaren, A. (1983a). Primordial germ cells in mice. *Bibliotheca Anat.* **24,** 59–66.

McLaren, A. (1983b). Sex reversal in the mouse. *Differentiation* **23,** S93–S98.

McLaren, A., and Monk, M. (1982). Fertile females produced by inactivation of an X-chromosome of "sex-reversed" mice. *Nature* **300,** 446–448.

McMahon, A., Fosten, M., and Monk, M. (1981). Random X-inactivation in female primordial germ cells in the mouse. *J. Embryol. Exp. Morphol.* **64,** 251–258.

McMahon, A., Fosten, M., and Monk, M. (1983). X-chromosome inactivation mosaicism in the three germ layers and the germ line of the mouse embryo. *J. Embryol. Exp. Morphol.* **74,** 207–220.

Madhavan, M., and Schneiderman, H. A. (1977). Histological analysis of the dynamics of growth and imaginal discs and histoblast nests during the larval development of *Drosophila melanogaster*. *Wilh. Roux Arch.* **183,** 269–305.

Madl, J., and Herman, R. K. (1979). Polyploids and sex determination in *Caenorhabditis elegans*. *Genetics* **93,** 393–402.

Mahowald, A. P. (1962). Fine structure of pole cells and polar granules in *Drosophila melanogaster*. *J. Exp. Zool.* **151,** 201–215.

Mahowald, A. P. (1983). Genetic analysis of oogenesis and determination. In W. R. Jeffery and R. A. Raff (Eds.), *Time, Space, and Pattern in Embryonic Development*. New York: Alan R. Liss.

Mahowald, A. P., and Turner, F. R. (1978). Scanning electron microscopy of *Drosophila* embryos. *Scanning Electron Microsc.* **11,** 11–19.

Mahowald, A. P., and Kambysellis, M. P. (1980). Oogenesis. In M. Ashburner and T. R. F. Wright (Eds.), *The Genetics and Biology of Drosophila,* vol. 2c. London: Academic Press.

Mahowald, A. P., Allis, C. D., Karrer, K. M., Underwood, E. M., and Wareing, G. L. (1979a). Germ plasm and pole cells of *Drosophila*. In S. Subtelny and I. R. Konigsberg, (Eds.), *Determinants of Spatial Organization*. New York: Academic Press.

Mahowald, A. P., Caulton, J. H., and Gehring, W. J. (1979b). Ultrastructural studies of oocytes and embryos derived from female flies carrying the *grandchildless* mutation in *Drosophila subobscura*. *Dev. Biol.* **69,** 118–132.

Manning, J. E., Schmid, C. W., and Davidson, N. (1975). Interspersion of repetitive and nonrepetitive DNA sequences in the *Drosophila* melanogaster genome. *Cell* **4,** 141–155.

Mariol, M-C. (1981). Genetic and developmental studies of a new grandchildless mutant of *Drosophila melanogaster*. *Mol. Gen. Genet.* **181,** 505–511.

Marsh, J. L., and Wieschaus, E. (1978). Is sex determination in germ line and soma controlled by separate genetic mechanisms? *Nature* **272,** 249–251.

Marsh, J. L., van Deusen, E. B., Wieschaus, E., and Gehring, W. J. (1977). Germ line dependence of the *deep orange* maternal effect in *Drosophila*. *Dev. Biol.* **56,** 195–199.

Martin, G. R. (1980). Teratocarcinomas and mammalian embryogenesis. *Science* **209,** 768–776.

Martin, G. R., and Evans, M. J. (1975). Differentiation of clonal lines of teratocarcinoma cells: formation of embryoid bodies *in vitro*. *Proc. Natl. Acad. Sci. USA* **78,** 7634–7638.

Meinhardt, H. (1977). A model of pattern formation in insect embryogenesis. *J. Cell Sci.* **23,** 117–139.

Meinhardt, H. (1982). *Models of Biological Pattern Formation.* London: Academic Press.

Meneely, P. M., and Herman, R. K. (1979). Lethals, steriles, and deficiencies in a region of the X chromosome of *Caenorhabditis elegans*. *Genetics* **92,** 99–115.

Meneely, P. M., and Wood, W. B. (1984). An autosomal gene that affects X chromosome expression and sex determination in *Caenorhabditis elegans*. *Genetics* **106,** 29–44.

Mermod, J. J., Schatz, G., and Crippa, M. (1980). Specific control of messenger translation in *Drosophila* oocytes and embryos. *Dev. Biol.* **75,** 177–186.

Merriam, J. R. (1978). Estimating primordial cell numbers in *Drosophila* imaginal discs and histoblasts. In W. J. Gehring (Ed.), *Genetic Mosaics and Cell Differentiation.* Berlin: Springer-Verlag.

Miller, O. J., and Miller, D. (1975). Cytogenetics of the mouse. *Ann. Rev. Genet.* **9,** 285–303.

Mintz, B. (1962). Experimental recombination of genotypically mosaic mouse embryos. *Am. Zool.* **2,** 432.

Mintz, B. (1964). Synthetic processes and early development in the mammalian egg. *J. Exp. Zool.* **157,** 85–100.

Mintz, B. (1967). Gene control of mammalian pigmentary differentiation. I. Clonal origin of melanocytes. *Proc. Natl. Acad. Sci. USA* **58,** 344–351.

Mintz, B. (1974). Gene control of mammalian differentiation. *Ann. Rev. Genet.* **8,** 411–470.

Mintz, B., and Baker, W. W. (1967). Normal mammalian muscle differentiation and gene control of isocitrate dehydrogenase synthesis. *Proc. Natl. Acad. Sci. USA* **58,** 592–598.

Mintz, B., and Illmensee, K. (1975). Normal genetically mosaic mice produced from malignant teratocarcinoma cells. *Proc. Natl. Acad. Sci. USA* **72,** 3585–3589.

Mintz, B., and Stewart, T. A. (1981). Successive generations of mice produced from an established culture line of euploid teratocarcinoma cells. *Proc. Natl. Acad. Sci. USA* **78,** 6314–6318.

Mirsky, A. E., and Ris, H. (1951). The desoxyribonucleic acid content of animal cells and its evolutionary significance. *J. Gen. Physiol.* **34,** 451–462.

Mitchell, H. K., and Lipps, L. S. (1978). Heat shock and phenocopy induction in *Drosophila*. *Cell* **15,** 907–919.

Mitchell, H. K., and Petersen, N. S. (1981). Rapid changes in gene expression in differentiating tissues of *Drosophila*. *Dev. Biol.* **85,** 233–242.

Miwa, J., Schierenberg, E., Miwa, S., and von Ehrenstein, G. (1980). Genetics and mode of expression of temperature-sensitive mutations arresting embryonic development in *Caenorhabditis elegans*. *Dev. Biol.* **76,** 160–174.

Modlinski, J. A. (1981). The fate of inner cell mass and trophectoderm nuclei transplanted to fertilized mouse eggs. *Nature* **292,** 342–343.

Mohler, J. D. (1977). Developmental genetics of the *Drosophila* egg. I. Identification of 59 sex-linked cistrons with maternal effects on embryonic development. *Genetics* **85,** 259–272.

Monod, J., and Jacob, F. (1962). General conclusions: teleonomic mechanisms in cellular metabolism, growth, and differentiation. *Cold Spring Harbor Symp. Quant. Biol.* **26,** 389–401.

Moore, W. J., and Mintz, B. (1972). Clonal model of vertebral column and skull development derived from genetically mosaic skeletons in allophenic mice. *Dev. Biol.* **27,** 55–70.

Morata, G., and Lawrence, P. A. (1975). Control of compartment development by the *engrailed* gene in *Drosophila. Nature* **255,** 614–617.

Morata, G., and Ripoll, P. (1975). Minutes: mutants of *Drosophila* autonomously affecting cell division rate. *Dev. Biol.* **42,** 211–221.

Morata, G., and Lawrence, P. A. (1977). Homoeotic genes, compartments and cell determination in *Drosophila. Nature,* **265,** 211–216.

Morata, G., and Lawrence, P. A. (1978). Anterior and posterior compartments in the head of *Drosophila. Nature* **274,** 473–474.

Morata, G., and Lawrence, P. A. (1979). Development of the eye-antenna imaginal discs of *Drosophila. Dev. Biol.* **70,** 355–371.

Morata, G., and Kerridge, S. (1981). Sequential functions of the bithorax complex of *Drosophila. Nature* **290,** 778–781.

Morgan, T. H. (1934). *Embryology and Genetics.* New York: Columbia University Press.

Morgan, T. H., and Bridges, C. B. (1919). The origin of gynandromorphs. Washington, D.C.: Carnegie Institution. publ. **278.**

Morse, H. C., III (1981). The laboratory mouse: a historical perspective. In H. L. Foster, J. D. Small, and J. G. Fox (Eds.), *The Mouse in Biomedical Research,* vol. 1, History, Genetics, and Wild Mice. New York: Academic Press.

Mortin, M. A., and Lefevre, G., Jr. (1981). An RNA polymerase II mutation in *Drosophila melanogaster* that mimics *Ultrabithorax. Chromosoma* (*Lond.*) **82,** 237–247.

Moyzis, R. K., Bonnet, J., Li, D. W., and Ts'o, P. O. P. (1981a). An alternative view of mammalian DNA sequence organization. I. Repetitive sequence interspersion in Syrian hamster DNA: a model system. *J. Mol. Biol.* **153,** 841–870.

Moyzis, R. K., Bonnet, J., Li, D. W., and Ts'o, P. O. P. (1981b). An alternative view of mammalian DNA sequence organization. II. Short repetitive sequences are organized into scrambled tandem clusters in Syrian hamster DNA. *J. Mol. Biol.* **153,** 871–896.

Muller, H. J. (1932). Further studies on the nature and causes of gene mutations. *Proc. 6th Int. Cong. Genet.* **1,** 213–255.

Murphy, C. (1974). Cell death and autonomous gene action in lethals affecting imaginal discs in *Drosophila melanogaster. Dev. Biol.* **39,** 23–36.

Nelson, G. A., Lew, K. K., and Ward, S. (1978). Intersex, a temperature-sensitive mutant of the nematode *Caenorhabditis elegans. Dev. Biol.* **66,** 386–409.

Nesbitt, M. N. (1971). X chromosome inactivation mosaicism in the mouse. *Dev. Biol.* **26,** 252–263.

Nesbitt, M. N. (1974). Chimeras vs X inactivation mosaics: significance of differences in pigment distribution. *Dev. Biol.* **38,** 202–207.

Nesbitt, M. N., and Gartler, S. M. (1971). The applications of genetic mosaicism to developmental problems. *Ann. Rev. Genet.* **5,** 143–162.

Newport, J., and Kirschner, M. (1982a). A major developmental transition in early *Xenopus* embryos. I. Characterization and timing of cellular changes at the midblastula stage. *Cell* **30,** 675–686.

Newport, J., and Kirschner, M. (1982b). A major developmental transition in early *Xenopus* embryos. II. Control of the onset of transcription. *Cell* **30,** 687–697.

Newrock, K. M., Alfageme, C. A., Nardi, R. V., and Cohen, L. H. (1977). Histone changes during chromatin remodelling in embryogenesis. *Cold Spring Harbor Symp. Quant. Biol.* **42,** 421–431.

Nigon, V. (1965). Developpment et reproduction des nematodes. In P-P. Grasse (Ed.), *Traité de Zoologie,* vol. 4, part 2. Paris: Masson.

Niki, Y., and Okada, M. (1981). Isolation and characterization of *grandchildless*-like mutants in *Drosophila melanogaster. Wilh. Roux Arch.* **190,** 1–10.

Nissani, M. (1977). Cell lineage analysis of germ cells of *Drosophila melanogaster. Nature* **265,** 729–731.

North, G. (1983). Cloning the genes that specify fruit flies. *Nature* **303,** 134–136.

Nothiger, R. (1972). The larval development of imaginal discs. In H. Ursprung and R. Nothiger (Eds.), *The Biology of Imaginal Discs: Results and Problems in Cell Differentiation,* vol. 5, Berlin: Springer-Verlag.

Nothiger, R. (1976). Clonal analysis in imaginal discs. In P. A. Lawrence (Ed.), *Insect Development.* Oxford: Blackwell.

Nothiger, R., Dubendorfer, A., and Epper, F. (1977). Gynandromorphs reveal two separate primordia for male and female genitalia in *Drosophila melanogaster. Wilh. Roux Arch.* **181,** 367–373.

Nusslein-Volhard, C. (1977). Genetic analysis of pattern-formation in the embryo of *Drosophila melanogaster.* Characterization of the maternal-effect mutant *Bicaudal. Wilh. Roux Arch.* **183,** 249–268.

Nusslein-Volhard, C. (1979a). Maternal effect mutations that alter the spatial coordinates of the embryo of *Drosophila melanogaster.* In S. Subtelny and I. R. Konigsberg (Eds.), *Determinants of Spatial Organization.* New York: Academic Press.

Nusslein-Volhard, C. (1979b). Pattern mutants in *Drosophila* embryogenesis. In N. Le Douarin (Ed.), *Cell Lineage, Stem Cells, and Cell Determination.* Amsterdam: Elsevier/North-Holland.

Nusslein-Volhard, C., and Wieschaus, E. (1980). Mutations affecting segment number and polarity in *Drosophila. Nature* **287,** 795–801.

Nusslein-Volhard, C., Wieschaus, E., and Jurgens, G. (1982). Segmentation in *Drosophila,* a genetic analysis. *Verh. Dtsch. Zool. Ges.,* 91–104.

Nusslein-Volhard, C., Lohs-Schardin, M., Sander, K., and Cremer, C. (1980). A dorso-ventral shift of embryonic primordia in a new maternal effect mutant of *Drosophila. Nature* **283,** 474–476.

O'Brien, S. J. (1973). On estimating functional gene number in eucaryotes. *Nature New Biol.* **242,** 52–54.

Odell, G. M., Oster, G., Albrech, P., and Burnside, B. (1981). The mechanical basis of morphogenesis. I. Epithelial folding and invagination. *Dev. Biol.* **85,** 446–462.

Ohno, S. (1976). Major regulatory genes for mammalian sexual development. *Cell* **7,** 315–321.

Okada, M., Kleinman, I. A., and Schneiderman, H. A. (1974a). Repair of a genetically-caused defect in oogenesis in *Drosophila melanogaster* by transplantation of cytoplasm from wild-type eggs and by injection of pyrimidine nucleosides. *Dev. Biol.* **37,** 55–62.

Okada, M., Kleinman, I. A., and Schneiderman, H. A. (1974b). Chimaeric *Drosophila* adults produced by transplantation of nuclei into specific regions of fertilized eggs. *Dev. Biol.* **39,** 286–294.

Oppenheimer, J. M. (1967). *Essays in the History of Embryology and Biology.* Cambridge, Mass.: MIT Press.

Ouweneel, W. J. (1976). Developmental genetics of homoeosis. *Adv. Genet.* **25,** 179–248.

Paigen, K. (1980). Temporal genes and other developmental regulators in mammals. In T. Leighton and W. F. Loomis (Eds.), *The Molecular Genetics of Development.* New York: Academic Press.

Papaionnou, V., and Gardner, R. L. (1979). Investigation of the lethal yellow A^y/A^y embryo using mouse chimaeras. *J. Embryol. Exp. Morphol.* **52,** 153–163.

Parks, H. B. (1936). Cleavage, patterns in *Drosophila* and mosaic formation. *Ann. Entomol. Soc. Am.* **29,** 350–392.

Pedersen, R. A., and Spindle, A. I. (1980). Role of the blastocoele microenvironment in early mouse embryo differentiation. *Nature* **284,** 550–552.

Pedersen, R. A., and Spindle, A. I. (1981). Cellular and genetic analysis of mouse blastocyst development. In S. R. Glasser and D. W. Bullock (Eds.), *Cellular and Molecular Aspects of Implantation.* New York: Plenum Press.

Petersen, N. S., and Mitchell, H. K. (1982). Effects of heat shock on gene expression during development: induction and prevention of the multihair phenocopy in *Drosophila.* In M. J. Schlesinger, M. Ashburner, and A. Tissieres (Eds.), *Heat Shock: From Bacteria to Man.* Cold Spring Harbor, N. Y.: Cold Spring Harbor Laboratory.

Piko, L. (1975). Expression of mitochondrial and nuclear genes during early development. In M. Balls and A. E. Wild (Eds.), *The Early Development of Mammals.* Cambridge: Cambridge University Press.

Poodry, C. A., and Schneiderman, H. A. (1976). Pattern formation in *Drosophila melanogaster:* the effects of mutations on polarity in the developing leg. *Wilh. Roux Arch.* **180,** 175–188.

Poole, T. W. (1980). Dermal–epidermal interactions and the action of alleles at the agouti locus in the mouse. II. The viable yellow (A^{vy}) and mottled agouti (a^{m}) alleles. *Dev. Biol.* **80,** 495–560.

Ponder, B. A. J., Wilkinson, M. M., and Wood, M. (1983). H-2 antigens as markers of cellular genotype in chimaeric mice. *J. Embryol. Exp. Morphol.* **76,** 83–93.

Popper, K. R. (1959). *The Logic of Scientific Discovery.* New York: Basic Books.

Postlethwait, J. H. (1974). Development of the temperature-sensitive homoeotic mutant *Opthalmoptera* in *Drosophila melanogaster. Dev. Biol.* **36,** 212–216.

Postlethwait, J. H. (1978). Clonal analysis of *Drosophila* cuticular patterns. In M. Ashburner and T. R. F. Wright (Eds.), *The Genetics and Biology of Drosophila,* vol 2c. London: Academic Press.

Postlethwait, J. H., and Schneiderman, H. A. (1969). A clonal analysis of determination in *Antennapedia,* a homoeotic mutant of *Drosophila melanogaster. Proc. Natl. Acad. Sci. USA* **64,** 176–183.

Postlethwait, J. H., and Schneiderman, H. A. (1971). Pattern formation and determination in the antenna of the homoeotic mutant *Antennapedia* of *Drosophila melanogaster. Dev. Biol.* **25,** 606–640.

Postlethwait, J. H., and Schneiderman, H. A. (1973). Pattern formation in imaginal discs of *Drosophila melanogaster* after irradiation of embryos and young larvae. *Dev. Biol.* **32,** 345–360.

Postlethwait, J. H., Poodry, C. A., and Schneiderman, H. A. (1971). Cellular dynamics of pattern duplication in imaginal discs of *Drosophila melanogaster. Dev. Biol.* **26,** 125–132.

Poulson, D. F. (1940). The effects of certain X-chromosome deficiency on the embryonic development of *Drosophila melanogaster. J. Exp. Zool.* **83,** 271–325.

Poulson, D. F. (1950). Histogenesis, organogenesis, and differentiation in the embryo of *Drosophila* Meigen. In M. Demerec (Ed.), *The Biology of Drosophila.* New York: Wiley-Interscience.

Pratt, H. P. M., Bolton, V. N., and Gudgeon, K. A. (1983). The legacy from the oocyte and its role in controlling early development of the mouse embryo. In R. Porter and J. Whelan (Eds.), *Molecular Biology of Egg Maturation.* London: Pitman Books.

Rabinowitz, M. (1941). Studies on the cytology and early embryology of the egg of *Drosophila melanogaster. J. Morphol.* **69,** 1–49.

Raff, R. A. (1983). Localization and temporal control of expression of maternal histone mRNA in sea urchin embryos. In W. R Jeffery and R. A. Raff (Eds.), *Time, Space, and Pattern in Embryonic Development.* New York: Alan R. Liss.

Reed, C. T., Murphy, C., and Fristrom, D. (1975). The ultrastructure of the differentiating pupal leg of *Drosophila melanogaster*. *Wilh. Roux Arch.* **178,** 285–302.

Reinhardt, C. A., Hodgkin, N. A., and Bryant, P. J. (1977). Wound healing in the imaginal discs of *Drosophila*. I. Scanning electron microscopy of normal and healing wing discs. *Dev. Biol.* **60,** 238–257.

Rice, R. B., and Garen, A. (1975). Localized defects of blastoderm formation in maternal effect mutants in *Drosophila*. *Dev. Biol.* **43,** 277–286.

Rice, R. B., Rice, F. A., and Garen, A. (1979). Adult abnormalities resulting from a localized blastoderm defect in a maternal-effect mutant of *Drosophila*. *Dev. Biol.* **69,** 194–201.

Richelle, J., and Ghysen, A. (1979). Determination of sensory bristles and pattern formation in *Drosophila*. I. A model. *Dev. Biol.* **70,** 418–437.

Riddle, D. L. (1977). A genetic pathway for dauer larva formation in *Caenorhabditis elegans*. *Stadler Symp.* **9,** 101–120.

Riddle, D. L., Swanson, M. M., and Albert, P. (1981). Interacting genes in nematode dauer larva formation. *Nature* **290,** 668–671.

Ripoll, P., and Garcia-Bellido, A. (1979). Viability of homozygous deficiencies in somatic cells of *Drosophila melanogaster*. *Genetics* **91,** 443–453.

Robbins, L. G. (1980). Maternal–zygotic lethal interactions in *Drosophila melanogaster*: the effects of deficiencies in the zeste-white region of the X chromosome. *Genetics* **96,** 187–200.

Roberts, D. B., and Graziosi, G. (1977). Protein synthesis in the early *Drosophila* embryo: analysis of the protein species synthesized. *J. Embryol. Exp. Morpol.* **41,** 101–110.

Roberts, P. (1961). Bristle formation controlled by the achaete locus in genetic mosaics of *Drosophila melanogaster*. *Genetics* **46,** 1241–1243.

Rodgers, M. E., and Shearn, A. (1977). Patterns of protein synthesis in imaginal discs of *Drosophila melanogaster*. *Cell* **12,** 915–921.

Rossant, J. (1975a). Investigation of the determinative state of the mouse inner cell mass. I. Aggregation of isolated inner cell masses with morulae. *J. Embryol. Exp. Morph.* **33,** 979–990.

Rossant, J. (1975b). Investigation of the determinative state. II. The fate of isolated inner cell masses transferred to the oviduct. *J. Embryol. Exp. Morphol.* **33,** 991–1001.

Rubin, G. M., and Spradling, A. C. (1982). Genetic transformation of *Drosophila* with transposable element vectors. *Science* **218,** 348–353.

Rugh, R. (1968). *The Mouse: Its Reproduction and Development*. Minneapolis: Burgess.

Sadler, J. R., and Novick, A. (1965). The properties of repressor and the kinetics of its action. *J. Mol. Biol.* **12,** 305–327.

Sakoyama, Y., and Okubo, S. (1981). Two dimensional gel patterns of protein species during development of *Drosophila* embryos. *Dev. Biol.* **81,** 361–365.

Sanchez, L., and Nothiger, R. (1982). Clonal analysis of *Sex-lethal,* a gene needed for female sexual development in *Drosophila melanogaster*. *Wilh. Roux Arch.* **191,** 211–214.

Sanchez, L., and Nothiger, R. (1983). Sex determination and dosage compensation in *Drosophila melanogaster*: production of male clones in XX females. *EMBO J.* **2,** 485–491.

Sander, K. (1975). Pattern specification in the insect embryo. In R. Porter and J. Rivers (Eds.), *Cell Patterning*. Amsterdam: Elsevier/North-Holland.

Sander, K. (1976). Specification of the basic body pattern in insect embryogenesis. *Adv. Insect Physiol.* **12,** 125–238.

Sander, K., Lohs-Schardin, M., and Bowmann, M. (1980). Embryogenesis in a *Drosophila* mutant expressing half the normal segment number. *Nature* **287,** 841–843.

Sandler, L. (1977). Evidence for a set of closely linked autosomal genes that interact with sex-chromosome heterochromatin in *Drosophila melanogaster*. *Genetics* **86,** 567–582.

Santamaria, P., and Nusslein-Volhard, C. (1983). Partial rescue of *dorsal,* a maternal effect mutation affecting the dorso-ventral pattern of the *Drosophila* embryo, by the injection of wild-type cytoplasm. *EMBO J.* **2,** 1695–1699.

Sanyal, S., and Zeilmaker, G. H. (1977). Cell lineage in the retinal development of mice studied in experimental chimaeras. *Nature* **265,** 731–733.

Satoh, N. (1979). On the "clock" mechanism determining the time of tissue-specific enzyme development during ascidian embryogenesis. I. Acetylcholinesterase development in cleavage-arrested embryos. *J. Embryol. Exp. Morphol.* **54,** 131–139.

Satoh, N., and Ikegami, S. (1981). A definite number of aphidicolin-sensitive cell-cyclic events are required for acetylcholinesterase development in the presumptive muscle cells of the ascidian embryo. *J. Embryol. Exp. Morphol.* **61,** 1–13.

Scalenghe, F., Turco, E., Edstrom, E., Pirrota, V., and Melli, M. (1981). Microdissection and cloning of DNA from a specific region of *Drosophila melanogaster* polytene chromosomes. *Chromosoma (Berl.)* **82,** 205–216.

Scherer, G., Telford, J., Baldari, C., and Pirrota, V. (1981). Isolation of cloned genes differentially expressed at early and late stages of *Drosophila* embryonic development. *Dev. Biol.* **86,** 438–447.

Schierenberg, E., Miwa, J., and von Ehrenstein, G. (1980). Cell lineages and developmental defects of temperature-sensitive embryonic arrest mutants in *Caenorhabditis elegans. Dev. Biol.* **76,** 141–159.

Schneiderman, H., and Bryant, P. J. (1971). Genetic analysis of developmental mechanism in *Drosophila. Nature* **234,** 187–194.

Scholnick, S. B., Morgan, B. A., and Hirsh, J. (1983). The cloned dopa decarboxylase gene is developmentally regulated when reintegrated into the *Drosophila* genome. *Cell* **34,** 37–45.

Schubiger, G. (1971). Regeneration, duplication, and transdetermination in fragments of the leg disc of *Drosophila melanogaster. Dev. Biol.* **26,** 277–295.

Schubiger, G. (1976). Adult differentiation from partial *Drosophila* embryos after egg ligation during stages of nuclear multiplication and cellular blastoderm. *Dev. Biol.* **50,** 476–488.

Schubiger, G., and Schubiger, M. (1978). Distal transformation in *Drosophila* leg imaginal disc fragments. *Dev. Biol.* **67,** 286–295.

Schubiger, G., and Newman, S. M. (1981). Determination in *Drosophila* embryos. *Am. Zool.* **22,** 47–55.

Schultz, R. M., LeTourneau, G. E., and Wassarman, P. M. (1979). Program of early development in the mammal: changes in the patterns and absolute rates of tubulin and total protein synthesis during oocyte growth in the mouse. *Dev. Biol.* **73,** 120–133.

Schupbach, T. (1982). Autosomal mutations that interfere with sex determination in somatic cells of *Drosophila* have no direct effect on the germline. *Dev. Biol.* **89,** 117–127.

Searle, A. G. (1968). *Comparative Genetics of Coat Color in Mammals.* London: Logos Press.

Seybold, W. D., and Sullivan, D. T. (1978). Protein synthetic patterns during differentiation of imaginal discs *in vitro. Dev. Biol.* **65,** 69–80.

Sharrock, W. J. (1983). Yolk proteins of *Caenorhabditis elegans. Dev. Biol.* **96,** 182–188.

Shearn, A. (1977). Mutational dissection of imaginal disc development in *Drosophila melanogaster. Am. Zool.* **17,** 585–594.

Shearn, A., and Garen, A. (1974). Genetic control of imaginal disc development in *Drosophila. Proc. Natl. Acad. Sci. USA* **71,** 1393–1397.

Shearn, A., Rice, R., Garen, A., and Gehring, W. (1971). Imaginal disc abnormalities in lethal mutants of *Drosophila. Proc. Natl. Acad. Sci. USA* **68,** 2594–2598.

Shearn, A., Hersperger, G., and Hersperger, E. (1978a). Genetic analysis of two allelic temperature-sensitive mutants of *Drosophila melanogaster,* both of which are zygotic and maternal effect lethals. *Genetics* **89,** 341–353.

Shearn, A., Hersperger, G., Hersperger, E., Pentz, E. S., and Denker, P. (1978b). Multiple allele approach to the study of genes in *Drosophila melanogaster* that are involved in imaginal disc development. *Genetics* **89,** 355–370.

Siddiqui, S. S., and Babu, P. (1980). Genetic mosaics of *Caenorhabditis elegans*: a tissue-specific fluorescent mutant. *Science* **210,** 330–332.

Silver, L. M. (1981). A structural gene (*Tcp-1*) within the mouse *t*-complex is separable from effects on tail length and lethality but may be associated with effects on spermatogenesis. *Genet. Res. (Camb.)* **38,** 115–123.

Silver, L. M., and Artzt, K. (1981). Recombination suppression of mouse *t*-haplotypes due to chromatin mismatching. *Nature* **290,** 68–70.

Silver, L. M., Artzt, K., and Bennett, D. (1979). A major testicular cell protein specified by a mouse T/t complex gene. *Cell* **17,** 275–284.

Silvers, W. K. (1979). *The Coat Colors of Mice*. New York: Springer-Verlag.

Silvers, W. K., and Russell, E. S. (1955). An experimental approach to action of genes at the agouti locus in the mouse. *J. Exp. Zool.* **130,** 199–220.

Silvers, W. K., Gasser, D. L., and Eicher, E. M. (1982). H-Y antigen, serologically detectable male antigen and sex determination. *Cell* **28,** 439–440.

Simpson, P. (1979). Parameters of cell competition in the compartments of the wing disc of *Drosophila*. *Dev. Biol.* **69,** 182–193.

Simpson, P. (1983). Maternal–zygotic gene interactions during formation of the dorso-ventral pattern in *Drosophila* embryos. *Genetics* **105,** 615–632.

Simpson, P., and Schneiderman, H. A. (1975). Isolation of temperature-sensitive mutations blocking clone development in *Drosophila melanogaster,* and effects of a temperature-sensitive cell lethal mutation on pattern formation in imaginal discs. *Wilh. Roux Arch.* **178,** 247–275.

Sina, B. J., and Pellegrini, M. (1982). Genomic clones coding for some of the initial genes expressed during *Drosophila* development. *Proc. Natl. Acad. Sci. USA* **79,** 7351–7355.

Singh, L., and Jones, K. W. (1982). Sex reversal in the mouse (*Mus musculus*) is caused by a recurrent nonreciprocal crossover involving the X and an aberrant Y chromosome. *Cell* **28,** 205–216.

Slack, J. M. W. (1983). *From Egg to Embryo*. Cambridge: Cambridge University Press.

Smith, R., and McLaren, A. (1977). Factors affecting the time of formation of the mouse blastocoele. *J. Embryol. Exp. Morphol.* **41,** 79–92.

Snell, G. D., and Stevens, L. C. (1966). Early embryology. In E. L. Green (Ed.), *Biology of the Laboratory Mouse*. New York: McGraw-Hill.

Snow, M. H. L. (1981). Autonomous development of parts isolated from primitive-streak-stage mouse embryos. Is development clonal? *J. Embryol. Exp. Morphol.* **65,** 269–287.

Snow, M. H. L., and Tam, P. P. L. (1979). Is compensatory growth a complicating factor in mouse teratology? *Nature* **279,** 555–557.

Snow, M. H. L., and Monk, M. (1983). Emergence and migration of mouse primordial germ cells. In A. McLaren and C. C. Wylie (Eds.), *Current Problems in Germ Cell Differentiation*. Cambridge: Cambridge University Press.

Snow, M. H. L., Tam, P. P. L., and McLaren, A. (1981). On the control and regulation of size and morphogenesis in mammalian embryos. In S. Subtelny (Ed.), *Levels of Genetic Control in Development*. Cambridge: Cambridge University Press.

Sonnenblick, B. P. (1950). The early embryology of *Drosophila melanogaster*. In M. Demerec (Ed.), *The Biology of Drosophila*. New York: Wiley-Interscience.

Spiegelman, M. (1976). Electron microscopy of cell associations in T-locus mutants. In K. Elliott and M. O'Connor (Eds.), *Embryogenesis in Mammals*. Amsterdam: Elsevier/North Holland.

Spiegelman, S. (1948). Differentiation as the controlled production of unique enzymatic patterns. In *Growth in Relation to Differentiation and Morphogenesis,* vol 2, Symposium of the Society for Experimental Biology. Cambridge: Cambridge University Press.

Spradling, A. C. (1981). The organization and amplification of two chromosomal domains containing *Drosophila* chorion genes. *Cell* **27,** 193–201.

Spradling, A. C., and Mahowald, A. P. (1979). Identification and genetic localization of mRNAs from ovarian-follicle cells of *Drosophila melanogaster. Cell* **16,** 589–598.

Spradling, A. C., and Mahowald, A. P. (1980). Amplification of genes for chorion proteins during oogenesis in *Drosophila melanogaster. Proc. Natl. Acad. Sci. USA* **77,** 1096–1100.

Spradling, A. C., and Mahowald, A. P. (1981). A chromosome inversion alters the pattern of specific DNA replication in *Drosophila* follicle cells. *Cell* **27,** 203–209.

Spradling, A. C., and Rubin, G. M. (1981). *Drosophila* genome organization: conserved and dynamic aspects. *Ann. Rev. Genet.* **15,** 219–264.

Spradling, A. C., and Rubin, G. M. (1982). Transposition of cloned P elements into *Drosophila* germ line chromosomes. *Science* **218,** 341–347.

Spradling, A. C., and Rubin, G. M. (1983). The effect of chromosomal position on the expression of the *Drosophila* xanthine dehydrogenase gene. *Cell* **34,** 47–57.

Spradling, A. C., Wareing, G. L., and Mahowald, A. P. (1979). *Drosophila* bearing the *ocelliless* mutation underproduces two major chorion proteins, both of which map near this gene. *Cell* **16,** 609–616.

Spradling, A. C., Digan, M. E., Mahowald, A. P., Scott, M., and Craig, E. A. (1980). Two clusters of genes for major chorion proteins of *Drosophila melanogaster. Cell* **19,** 905–914.

Spurway, H. (1948). Genetics and cytology of *Drosophila pseudoobscura.* IV. An extreme example of delay in gene action, causing sterility. *J. Genet.* **49,** 126–140.

Steiner, E. (1976). Establishment of compartments in the developing leg imaginal discs of *Drosophila melanogaster. Wilh. Roux Arch.* **180,** 9–30.

Stent, G. S. (1980). Genetic approach to developmental neurobiology. *Trends Neurosci.* **3,** 49–51.

Stern, C. (1954). Two or three bristles. *Am. Scientist* **42,** 213–247.

Stewart, M., Murphy, C., and Fristrom, J. W. (1972). The recovery and preliminary characterization of X chromosome mutants affecting imaginal discs of *Drosophila melanogaster. Dev. Biol.* **27,** 71–83.

Strobel, E., Dunsmuir, P., and Rubin, G. M. (1979). Polymorphisms in the chromosomal locations of elements of the *412, copia,* and *297* dispersed repeated gene families in *Drosophila. Cell* **17,** 429–439.

Strome, S., and Wood, W. B. (1982). Immunofluorescence visualization of germ-line–specific cytoplasmic granules in embryos, larvae, and adults of *Caenorhabditis elegans. Proc. Natl. Acad. Sci. USA* **79,** 1558–1562.

Strome, S., and Wood, W. B. (1983). Generation of asymmetry and segregation of germ-line granules in early *C. elegans* embryos. *Cell* **35,** 15–25.

Strub, S. (1977). Pattern regulation and transdetermination in *Drosophila* imaginal leg disc reaggregates. *Nature* **269,** 688–691.

Struhl, G. (1977). Developmental compartments in the proboscis of *Drosophila. Nature* **270,** 723–725.

Struhl, G. (1981a). Anterior and posterior compartments in the proboscis of *Drosophila. Dev. Biol.* **84,** 372–385.

Struhl, G. (1981b). A blastoderm fate map of compartments and segments of the *Drosophila* head. *Dev. Biol.* **84,** 386–396.

Struhl, G. (1981c). A gene product required for correct initiation of segmental determination in *Drosophila. Nature* **293,** 36–41.

Struhl, G. (1981d). A homoeotic mutation transforming leg to antenna in *Drosophila. Nature* **292,** 635–638.

Struhl, G. (1982a). *Spineless-aristapedia:* a homoeotic gene that does not control the development of specific compartments in *Drosophila. Genetics* **102,** 737–749.

Struhl, G. (1982b). Genes controlling segmental specification in the *Drosophila* thorax. *Proc. Natl. Acad. Sci. USA* **79,** 7380–7384.

Struhl, G. (1983). Role of the esc^+ product in ensuring the selective repression of segment-specific homoeotic genes in *Drosophila. J. Emb. Exp. Morphol.* **76,** 297–330.

Struhl, G., and Brower, D. (1982). Early role of the esc^+ gene product in the determination of segments in *Drosophila. Cell* **31,** 285–292.

Sturtevant, A. H. (1923). Inheritance of direction of shell coiling in *Limnaea. Science* **58,** 269–270.

Sturtevant, A. H. (1929). The claret mutant type of *Drosophila simulans:* a study of chromosome elimination and of cell lineage. *Z. Wiss Zool.* **135,** 323–356.

Sulston, J. E. (1976). Post-embryonic development in the ventral cord of *Caenorhabditis elegans. Proc. R. Soc. London, B* **275,** 287–297.

Sulston, J. E., and Brenner, S. (1974). The DNA of *Caenorhabditis elegans. Genetics* **77,** 95–104.

Sulston, J. E., and Horvitz, H. R. (1977). Post-embryonic cell lineages of the nematode, *Caenorhabditis elegans. Dev. Biol.* **56,** 110–156.

Sulston, J. E., and White, J. G. (1980). Regulation and cell autonomy during postembryonic development of *Caenorhabditis elegans. Dev. Biol.* **78,** 577–597.

Sulston, J. E., and Horvitz, H. R. (1981). Abnormal cell lineages in mutants of the nematode *Caenorhabditis elegans. Dev. Biol.* **82,** 41–55.

Sulston, J. E., Albertson, D. G., and Thomson, J. N. (1980). The *Caenorhabditis elegans* male: postembryonic development of nongonadal structures. *Dev. Biol.* **78,** 542–576.

Sulston, J. E., Schierenberg, E., White J. G., Thomson, J. N., and von Ehrenstein, G. (1983). The embryonic cell lineage of the nematode *Caenorhabditis elegans. Dev. Biol.* **100,** 64–119.

Suzuki, D. T. (1970). Temperature-sensitive mutations in *Drosophila melanogaster. Science* **170,** 695–706.

Suzuki, D. T., and Griffiths, A. J. F. (1981). *An Introduction to Genetic Analysis,* 2nd ed. San Francisco: W. H. Freeman and Co.

Swanson, M. M., and Riddle, D. L. (1981). Critical periods in the development of the *Caenorhabditis elegans* dauer larva. *Dev. Biol.* **84,** 27–40.

Szabad, J., and Bryant, P. J. (1982). The mode of action of "discless" mutations in *Drosophila melanogaster. Dev. Biol.* **93,** 240–256.

Szabad, J., Schupbach, T., and Wieschaus, E. (1979a). Cell lineage and development in the larval epidermis of *Drosophila melanogaster.* **73,** 256–271.

Szabad, J., Simpson, P., and Nothiger, R. (1979b). Regeneration and compartments in *Drosophila. J. Embryol. Exp. Morphol.* **49,** 229–241.

Tam, P. P. L. (1981). The control of somitogenesis in mouse embryos. *J. Embryol. Exp. Morphol.* **65,** 103–128.

Tam, P. P. L., and Snow, M. H. L. (1981). Proliferation and migration of primordial germ cells during compensatory growth in mouse embryos. *J. Embryol. Exp. Morphol.* **64,** 133–147.

Tanaka, Y. (1953). Genetics of the silkworm *Bombyx mori.* In M. Demerce (Ed.), *Advances in Genetics,* vol 5. New York: Academic Press.

Tarkowski, A. K. (1961). Mouse chimaeras developed from fused eggs. *Nature* **190,** 857–860.

Tarkowski, A. K., and Wroblewska, J. (1967). Development of blastomeres of mouse eggs isolated at the 4- and 8-cell stage. *J. Embryol. Exp. Morphol.* **18,** 155–180.

Teugels, E., and Ghysen, A. (1983). Independence of the numbers of legs and leg ganglia in *Drosophila* bithorax mutants. *Nature* **304,** 440–442.

Thorig, G. E. W., Heinstra, P. W. H., and Scharloo, W. (1981). The action of the Notch locus in *Drosophila melanogaster.* II. Biochemical effects of recessive lethals on mitochondrial enzymes. *Genetics* **99,** 65–74.

Tonegawa, S. (1983). Somatic generation of antibody diversity. *Nature* **302,** 575–581.

Trumbly, R. J., and Jarry, B. (1983). Stage-specific protein synthesis during early embryogenesis in *Drosophila melanogaster. EMBO J.* **2,** 1281–1290.

Turner, F. R., and Mahowald, A. P. (1976). Scanning E. M. of *Drosophila* embryogenesis. I. The structure of the egg envelopes and the formation of the cellular blastoderm. *Dev. Biol.* **50,** 95–108.

Underwood, E. M., and Mahowald, A. P. (1980). The chorion defect of the ocelliless mutation is caused by abnormal follicle cell function. *Dev. Genet.* **1,** 247–256.

Underwood, E. M., Caulton, J. H., Allis, C. D., and Mahowald, A. P. (1980a). Developmental fate of pole cells in *Drosophila melanogaster. Dev. Biol.* **77,** 303–314.

Underwood, E. M., Turner, F. R., and Mahowald, A. P. (1980b). Analysis of cell movements and fate mapping during early embryogenesis in *Drosophila melanogaster. Dev. Biol.* **74,** 286–301.

Ursprung, H. (1972). The fine structure of imaginal discs. In H. Ursprung and R. Nothiger (Eds.), *The Biology of Imaginal Discs. Results and Problems in Cell Differentiation,* vol. 5. Berlin: Springer-Verlag.

van Blerkom, J. (1981). Structural relationships and post-translational modification of stage-specific proteins synthesized during early pre-implantation development in the mouse. *Proc. Natl. Acad. Sci. USA* **78,** 7629–7633.

van Blerkom, J., Barton, S. C., and Johnson, M. H. (1976). Molecular differentiation in the pre-implantation embryo. *Nature* **259,** 319–321.

van Blerkom, J., Janzen, R., and Runner, M. N. (1982). The patterns of protein synthesis during foetal and neonatal organ development in the mouse are remarkably similar. *J. Embryol. Exp. Morphol.* **72,** 97–116.

van der Meer, J. M., Kemner, W., and Miyamoto, D. M. (1982). Mitotic waves and embryonic pattern formation: no correlation in *Callosobruchus (Coleoptera). Wilh. Roux Arch.* **191,** 355–365.

van Deusen, E. B. (1976). Sex determination in germ line chimaeras of *Drosophila melanogaster. J. Embryol. Exp. Morphol.* **37,** 173–185.

Vanderslice, R., and Hirsh, D. (1976). Temperature-sensitive zygote defective mutants of *Caenorhabditis elegans. Dev. Biol.* **49,** 236–249.

Vogel, O. (1977). Anomalies in the expression of abdominal denticle belts of partial larvae produced by fragmentation of the *Drosophila melanogaster* egg. *Wilh. Roux Arch.* **182,** 33–38.

von Ehrenstein, G., Schierenberg, E., and Miwa, J. (1979). Cell lineages of the wild-type and of temperature-sensitive embryonic arrest mutants of *Caenorhabditis elegans.* In N. Le Douarin (Ed.), *Cell Lineage, Stem Cells, and Cell Determination.* Amsterdam: Elsevier/North-Holland.

Waddington, C. H. (1962). *The Strategy of the Genes.* London: George Allen and Unwin.

Waddington, C. H. (1973). The morphogenesis of patterns in *Drosophila.* In S. Counce and C. H. Waddington (Eds.), *Developmental Systems: Insects.* London: Academic Press.

Wagner, E. F., Covarrubian, L., Stewart, T. A., and Mintz, B. (1983). Prenatal lethalities in mice homozygous for human growth hormone gene sequences integrated in the germ line. *Cell* **35,** 647–655.

Wakimoto, B. T., and Kaufman, T. C. (1981). Analysis of larval segmentation in lethal genotypes associated with the Antennapedia gene complex in *Drosophila melanogaster. Dev. Biol.* **81,** 51–64.

Ward, S., and Carrel, J. S. (1979). Fertilization and sperm competition in the nematode *Caenorhabditis elegans. Dev. Biol.* **73,** 304–321.

Waring, G. L., and Mahowald, A. P. (1979). Identification and time of synthesis of chorion proteins in *Drosophila melanogaster. Cell* **16,** 599–607.

Washburn, L. L., and Eicher, M. (1983). Sex reversal in XY mice caused by dominant mutation on chromosome 17. *Nature* **303,** 338–339.

Wassarman, P. M., and Josefowicz, W. J. (1978). Oocyte development in the mouse: an ultrastructural comparison of oocytes isolated at various stages of growth and meiotic competence. *J. Morphol.* **156,** 209–236.

Wassarman, P. M., and Mrozak, S. C. (1981). Program of early development in the mammal: synthesis and intracellular migration of histone H4 during oogenesis in the mouse. *Dev. Biol.* **84,** 364–371.

Wegman, T. G., and Gilman, J. G. (1970). Chimaerism for three genetic systems in tetraparental mice. *Dev. Biol.* **21,** 281–291.

Weintraub, H., Fling, S. J., Leffak, I. M., Groudine, M., and Grainger, R. M. (1977). The generation and propagation of variegated chromosome structures. *Cold Spring Harbor Symp. Quant. Biol.* **42,** 401–407.

Weisblat, D. A., Sawyer, R. T., and Stent, G. S. (1978). Cell lineage analysis by intracellular injection of a tracer enzyme. *Science* **202,** 1295–1297.

Wells, D. E., Showman, R. M., Klein, W. H., and Raff, R. A. (1981). Delayed recruitment of maternal histone H3 in sea urchin embryos. *Nature* **292,** 477–478.

Wensink, P. C., Tabata, S., and Pachl, C. (1979). The clustered and scrambled arrangement of moderately repetitive elements in *Drosophila* DNA. *Cell* **18,** 1231–1246.

West, J. D. (1976). Clonal development of the retinal epithelium in mouse chimaeras and X-inactivation mosaics. *J. Embryol. Exp. Morphol.* **35,** 445–461.

West, J. D. (1978). Analysis of clonal growth using chimaeras and mosaics. In M. H. Johnson (Ed.), *Development in Mammals,* vol. 3. Amsterdam: Elsevier/North-Holland.

White, J., Albertson, D., and Anness, M. (1978). Connectivity changes in a class of motor neurones during the development of a nematode. *Nature* **271,** 764–766.

Whittaker, J. R. (1973). Segregation during ascidian embryogenesis of egg cytoplasmic information for tissue-specific enzyme development. *Proc. Natl. Acad. Sci. USA* **70,** 2096–2100.

Whittaker, J. R. (1979). Cytoplasmic determinants of tissue differentiation in the ascidian egg. In S. Subtelny and I. R. Konigsberg (Eds.), *Determinants of Spatial Organization.* New York: Academic Press.

Wieschaus, E. (1978a). The use of mosaics to study oogenesis in *Drosophila melanogaster.* In S. Subtelny and I. Sussex (Eds.), *The Clonal Basis of Development.* New York: Academic Press.

Wieschaus, E. (1978b). Cell lineage relationships in the *Drosophila* embryo. In W. J. Gehring (Ed.), *Genetic Mosaics and Cellular Differentiation.* Berlin: Springer-Verlag.

Wieschaus, E. (1979). *fs(l)K10,* a female-sterile mutation altering the pattern of both the egg coverings and the resultant embryos in *Drosophila.* In N. LeDouarin (Ed.), *Cell Lineage, Stem Cells, and Differentiation.* Amsterdam: Elsevier/North-Holland.

Wieschaus, E., and Gehring, W. J. (1976a). Gynandromorph analysis of the thoracic disc primordia in *Drosophila melanogaster. Wilh. Roux Arch.* **180,** 31–46.

Wieschaus, E., and Gehring, W. J. (1976b). Clonal analysis of primordial disc cells in the early embryo of *Drosophila melanogaster*. *Dev. Biol.* **50,** 249–263.

Wieschaus, E., and Szabad, J. (1979). The development and function of the female germ line in *Drosophila melanogaster:* a cell lineage study. *Dev. Biol.* **68,** 29–46.

Wieschaus, E., and Nothiger, R. (1982). The role of the transformer genes in the development of genitalia and analia of *Drosophila melanogaster*. *Dev. Biol.* **90,** 320–324.

Wieschaus, E., Marsh, J. L., and Gehring, W. J. (1978). *fs(1)K10,* a germline-dependent female sterile mutation causing abnormal chorion morphology in *D. melanogaster*. *Wilh. Roux Arch.* **184,** 75–82.

Wieschaus, E., Audit, C., and Masson, M. (1981). A clonal analysis of the roles of somatic cells and germ line during oogenesis in *Drosophila*. *Dev. Biol.* **88,** 92–103.

Wigglesworth, V. B. (1940). Local and general factors in the development of "pattern" in *Rodnius prolixus* (Hemiptera). *J. Exp. Biol.* **17,** 180–200.

Wilcox, M., and Smith, R. J. (1977). Regenerative interactions between *Drosophila* imaginal discs of different types. *Dev. Biol.* **60,** 287–297.

Wilcox, M., Brower, D. L., and Smith, R. J. (1981). A position-specific cell surface antigen in the *Drosophila* wing imaginal disc. *Cell* **25,** 159–164.

Wilkins, A. S. (1976). Replicative patterning and determination. *Differentiation* **5,** 15–19.

Williams, K. L., and Newell, P. C. (1976). A genetic study of aggregation in the cellular slime mould *Dictyostelium discoideum* using complementation analysis. *Genetics* **82,** 287–307.

Wills, N., Gestleland, R. F., Karn, J., Barnett, L., Bolten, S., and Waterston, R. H. (1983). The genes *sup-7X* and *sup-5III* of *C. elegans* suppress amber nonsense mutations via altered transfer RNA. *Cell* **33,** 575–583.

Wilson, A. C., Maxson, L. R., and Sarich, V. M. (1974). Two types of molecular evolution. Evidence from studies of interspecific hybridization. *Proc. Natl. Acad. Sci. USA* **71,** 2843–2847.

Wilson, E. B. (1925). *The Cell in Development and Heredity*. New York: Macmillan.

Wilson, I. B., and Stern, M. S. (1975). Organization in the preimplantation embryo. In M. Balls and A. E. Wild (Eds.), *The Early Development of Mammals*. Cambridge: Cambridge University Press.

Wolpert, L. (1969). Positional information and the spatial pattern of cellular differentiation. *J. Theor. Biol.* **25,** 1–47.

Wolpert, L. (1971). Positional information and pattern formation. In A. A. Moscona and A. Monroy (Eds.), *Current Topics in Developmental Biology,* vol. 7. New York: Academic Press.

Wolpert, L., and Lewis, J. H. (1975). Towards a theory of development. *Fed. Proc.* **34,** 14–20.

Wood, W. B., Hecht, R., Carr, S., Vanderslice, R., Wolf, N., and Hirsh, D. (1980). Parental effects and phenotypic characterization of mutations that affect early development in *Caenorhabditis elegans*. *Dev. Biol.* **74,** 446–469.

Wood, W. B., Strome S., and Laufer, J. S. (1983). Localization and determination in embryos of *Caenorhabditis elegans*. In W. R. Jeffery and R. A. Raff (Eds.), *Time, Space, and Pattern in Embryonic Development*. New York: Alan R. Liss.

Woodruff, R. C., and Ashburner, M. (1979). The genetics of a small autosomal region of *Drosophila melanogaster* containing the structural gene for alcohol dehydrogenase. II. Lethal mutations in the region. *Genetics* **92,** 133–149.

Wright, D. A., and Lwrence, P. A. (1981a). Regeneration of the segment boundary in *Oncopeltus*. *Dev. Biol.* **85,** 317–327.

Wright, D. A., and Lawrence, P. A. (1981b). Regeneration of segment boundaries in *Oncopeltus:* cell lineage. *Dev. Biol.* **85,** 328–333.

Wright, T. R. F. (1970). The genetics of embryogenesis in *Drosophila*. *Adv. Genet.* **15,** 262–395.

Young, B. D., Birnie, G. D., and Paul, J. (1976). Complexity and specificity of polysomal poly $(A)^+$ RNA in mouse tissues. *Biochemistry* **15,** 2823–2830.

Young, M. W. (1979). Middle repetitive DNA: a fluid component of the *Drosophila* genome. *Proc. Natl. Acad. Sci. USA* **76,** 6274–6278.

Young, M. W., and Judd, B. H (1978). Nonessential sequences, genes and the polytene chromosome bands of *D. melanogaster*. *Genetics* **88,** 723–742.

Zalokar, M. (1976). Autoradiographic study of protein and RNA formation during early development of *Drosophila* eggs. *Dev. Biol.* **49,** 425–437.

Zalokar, M., and Erk, I. (1976). Division and migration of nuclei duing embryogenesis of *Drosophila melanogaster*. *J. Microsc. Biol. Cell.* **25,** 97–106.

Zalokar, M., Audit, C., and Erk, I. (1975). Developmental defects of female-sterile mutants of *Drosophila*. *Dev. Biol.* **47,** 419–432.

Zalokar, M., Erk, I., and Santamaria, P. (1980). Distribution of ring-X chromosomes in the blastoderm of gynandromorphic *D. melanogaster*. *Cell* **19,** 133–141.

Zamboni, L. (1972). Comparative studies on the ultrastructure of mammalian oocytes. In J. D. Biggers and A. W. Schultz (Eds.), *Oogenesis*. Baltimore: University Park Press.

Zimmerman, J. L., Fouts, D. L., and Manning, J. E. (1980). Evidence for a complex class of nonadenylated mRNA in *Drosophila*. *Genetics* **95,** 673–691.

INDEX

Numbers in *italics* indicate pages on which illustrations appear.